CREATIVE
PROBLEM SOLVING

THINKING SKILLS FOR A
CHANGING WORLD

CREATIVE
PROBLEM SOLVING

THINKING SKILLS FOR A CHANGING WORLD

EDWARD LUMSDAINE

Dean, College of Engineering
Michigan Technological University

MONIKA LUMSDAINE

Visiting Scholar
Michigan Technological University

McGRAW-HILL, INC.

New York St. Louis San Francisco Auckland Bogotá Caracas Lisbon London Madrid
Mexico City Milan Montreal New Delhi San Juan Singapore Sydney Tokyo Toronto

CREATIVE PROBLEM SOLVING
Thinking Skills for a Changing World

This book is printed on acid-free paper.

3 4 5 6 7 8 9 0   DOC/DOC   9 9 8 7 6 5

ISBN 0-07-039091-6

The editors were B. J. Clark and Margery Luhrs.
The production supervisor was Denise L. Puryear.
R. R. Donnelley & Sons Company was printer and binder.

Library of Congress Catalog Card Number: 94-77733

Permissions and Copyrights

Paul MacCready's two lengthy quotes in the preface are used by permission of the author.

Table 1-1 was adapted with the permission of Macmillan College Publishing Company from *Careers in Engineering and Technology,* fourth edition, by George C. Beakley, Donovan L. Evans, and Deloss H. Bowers. Copyright © 1987 by George C. Beakley.

The Ned Herrmann materials presented in Chapter 3, including the HBDI profiles, are used by permission of the inventor, Ned Herrmann. Copyright © 1986 by Ned Herrmann. The color graphics on the cover show the Lumsdaine creative problem-solving model and associated mindsets (with ranges indicated by color) superimposed on the Ned Herrmann four-quadrant model of thinking preferences. These mindset metaphors are "ideals" and are independent of the occupational norms published by Ned Herrmann.

The "what if" story in Chapter 4 by Roger Von Oech is used by permission of the author; it is condensed from the original told in *A Whack on the Side of the Head,* Warner Books, New York, 1983.

Table 4-1 was adapted from *De Bono's Thinking Course* by Edward de Bono (pp. 4-10). Copyright © 1986 by Pentacor B.V. Reprinted with permission by Facts on File, Inc., New York.

The old legend retold by Iron Eyes Cody in Chapter 7 of this book is reprinted with permission from *Guideposts* Magazine. Copyright © 1988 by Guideposts Associates, Inc., Carmel, New York 10512. The story originally appeared in the July 1988 issue under the title "Words to Grow On."

The engineering ethics case studies in Chapter 7 are used by permission of *Engineering Times.* They are from the February 1989, April 1989, and June 1989 issues.

Material on the Pugh method (Chapter 10), FMEA (Appendix B), and FTA (Appendix C) were obtained from the American Supplier Institute, Dearborn, Michigan, and are used by permission.

The two BLONDIE comic strips in Chapter 11 are reprinted by special permission of King Features Syndicate; we found them originally in the *Toledo Blade* on September 14 and 15, 1991.

To our children Andrew, Anne, Alfred, and Arnold
for inspiring us to keep growing.

To Wendy, Jim, and Sarah, as well as Kevin and Nancy,
who have all become a special part of our family.

To Benjamin and Emily, our first grandchildren,
who are teaching us to see the world anew.

ABOUT THE AUTHORS

Edward: I grew up in Shanghai and made my way to the United States by working for nearly two years on a Danish freighter. I joined the U.S. Air Force, and on the G.I. bill four years later I entered Ventura Junior College, where I met Monika.

Monika: I grew up in Switzerland and was sponsored by a Rotary family in California to attend my first year of college. Some time after Ed and I got married, we transferred to New Mexico State University where I earned a B.S. in mathematics and had three children while Ed earned his B.S., M.S., and a doctorate in mechanical engineering.

Edward: After graduation, I first worked as a research engineer for the Boeing Company and then taught at South Dakota State University, where our youngest son was born. Then I taught at the University of Tennessee and served at New Mexico State University as professor, research engineer, and director of the New Mexico Solar Energy Institute.

Monika: In New Mexico, Ed got me interested in solar energy design, and one of my first plans won a national award. I had my own research project, designed the visitors and operations center of the photovoltaic facility in Lovington as well as several private homes, and taught a number of workshops on solar home design. We designed and built our own unique passive solar home, and Ed encouraged me to start my own company, E & M Lumsdaine Solar Consultants, Inc.—I am still its president.

Edward: Back at the University of Tennessee, I became director of the Energy, Environment, and Resources Center, and Monika designed and built another solar house for us. During these years, I did quite a bit of work overseas, lecturing for UNESCO and the U.S. Information Service and as visiting professor in Egypt, Qatar, and Taiwan. In 1982 I was named dean of engineering at the University of Michigan-Dearborn. The Detroit area provided many opportunities for becoming closely involved with manufacturing companies and product quality, and I began developing special continuing education courses. I had been concerned for years about the way we were educating the next generation of engineers, and my growing interest in continuous improvement led me into teaching creative thinking and problem solving. I continued these activities at the University of Toledo when I became dean of engineering there in 1988.

Monika: At Toledo, I helped in team-teaching the creative problem-solving classes and developing the written materials (including a manual for the popular math/science Saturday academy program for secondary school students and their parents). We are continuing our team-teaching of creative problem solving for freshmen at Michigan Technological University, where Ed was appointed dean in 1993; we are also teaching faculty workshops all over the country.

Edward: We are occasionally asked who did what as we coauthored our book. This question is rather difficult to answer, because the book is truly a synergistic team effort. We have different thinking preferences—since Monika is the more experienced designer and word processor, she is the one who did the final layout and desk-top execution of the manuscript. She also did the mindmapping.

Monika: But Ed wrote the chapter on MATHEMATICA. The stories and ideas he comes up with in classes and workshops have naturally found their way into the book, which mirrors the innovative approach that he takes in his teaching. We have both read most of the reference books listed at the end of each chapter. When one of us drafts up some notes, the other supplements with further insight, a process we usually continue for several rounds. When we teach together, we are able to give each other immediate positive feedback. This conversational biography is but one example of the "teaming" we enjoy doing. Our book is truly the result of creative thinking, creative problem solving, and teamwork between the two of us spanning many years.

CONTENTS

PART THREE — APPLICATIONS

APPENDICES

LIST OF ACTIVITIES IN THE TEXT

Note: The first number indicates the chapter, the second number the order within the chapter for all activities, figures, and tables.

LIST OF FIGURES

LIST OF TABLES

Three people were working on a construction project. When asked what they were doing, the first person said: "I'm just doing a job." The second worker replied: "I'm nailing plywood to the joists," while the third answered with a smile: "We are working together to build a home." Or picture this scene: A mass of young people are milling around on a large field. Piles of various building components are scattered all over the place: bricks, boards, door jambs, rolls of wire, an odd assortment of pipes—it is difficult to make sense of all these disorganized heaps. Periodically, a truck drives up and haphazardly dumps another load. What is going on here?

The piles of building materials are a metaphor for the basic way we deliver education to our students in this country. We package knowledge and learning into separate, self-contained units. We do not teach students how each brick can be joined to others; we do not supply the mortar, glue, and nails; we do not show them how to make connections and build a whole through working together and using the most appropriate tools. Some will discover these relationships on their own and build a beautiful edifice while still young and idealistic; some may find the connections only later in life; and some may never understand the whole when given all the parts.

With this book, we want to help learners of all ages understand their work, their studies, and their thinking abilities in a wider context. Our world needs thinkers who can see how problems and solutions are connected with people and cultural values, with ecology, with global economic concerns, with quality and ethics, with the future and the past. Our world needs doers who can integrate creativity with analytical thinking, who can see the whole picture, who can communicate and work in teams, who can use technological tools, who can think critically, and who can innovate. These are precisely the goals of broad-based education. And this is the approach taken in this book—with these thinking skills you can become a more effective problem solver.

We live in a rapidly changing world, and creative thinking helps us cope with change. We feel it is our responsibility as teachers to educate people who will be able to succeed in the high-tech environment of the twenty-first century. We believe that everyone is creative and can learn to be even more creative—creativity can be unlocked and nurtured in a supportive environment. Everyone can learn to express and apply their creative ideas to the benefit of their community. In essence, we can learn to make better use of the capabilities of the whole brain. Paul Mac-Cready, the inventor of such low-energy aircraft as the Gossamer Condor, the Gossamer Albatross, the Gossamer Penguin, and the Solar Challenger, said in the Fall 1987 issue of *Engineering and Science* (California Institute of Technology): "No single technological advance will be the key to a safe and comfortable long-term future for civilization. Rather, the key, if any exists, will lie in getting large numbers of human minds to operate creatively and from a broad, open-minded perspective, to cope with the new challenges."

Interestingly enough, the idea for this book did not originate in the university environment—it came from industry. Several years ago we were asked to develop a workshop and manual on brainstorming and problem solving because many engineers in industry lacked creative problem-solving skills. As we studied the subject, we came to perceive the root of the problem—many students in our schools are not taught an essential thinking skill needed in the workplace to function well in today's rapidly changing world. In addition, we were looking for a creative solution to the challenge of attracting and keeping a broader diversity of students in engineering. We found that an approach that integrated left-brain and right-brain thinking and learning processes would meet our needs.

Thus we began teaching different versions of the early workshop to engineers and managers in industry, to manufacturing engineering professors, to public school teachers, to a group of 14-year-olds from inner-city Detroit, and to Ohio high school students through the Governor's Institute. The workshop format was expanded into a first-year introductory course for engineering students, and we also taught the concepts as part of a ten-week Saturday morning math/science academy to sixth-, seventh-, and eighth-grade gifted students in Toledo (and their parents). The first edition of the textbook published by McGraw-Hill in its College Custom Series was named *Creative Problem Solving: An Introductory Course for Engineering Students.* It was used at several universities all over the country. It was also adopted for a senior course in teachers' education at the University of Toledo. Students from different fields are taking creative problem solving as an elective because it is useful—and fun. Our goal is to help students and instructors grow and learn together to find new and better ways to make the education experience responsive to the problems, conditions, and challenges of our world.

Because the first edition of the book received such wide-ranging interest, we designed the contents of the second edition in the College Custom Series to benefit a broader audience of curious students of all ages in the arts and sciences, in education, in business management and related professions, and in law and medicine, in addition to those in engineering and technology. The focus truly became centered on "thinking skills for a changing world." For the present revised edition of *Creative Problem Solving: Thinking Skills for a Changing World,* we completely changed the format and improved the quality of the text to make learning easier. The redesigned mindmaps provide a better overview of the material covered in the respective chapters. Also, we have included an update of our research results on student thinking preferences, and we have substituted a discussion of E-mail for the now obsolete instructions on how to do a card-catalog library search.

The book is divided into three sections, each with a different emphasis. Part One concentrates on developing creative thinking skills, including visualization for enhanced memory and learning. It introduces the four-quadrant model of thinking preferences created by Ned Herrmann. An understanding of brain dominance enables students to appreciate the strengths that "different" thinkers can bring to a problem-solving team. Many ideas and exercises are included on how to break down the mental barriers that prevent people from synergistically using the capabilities of their whole brain for optimum thinking and problem solving. Throughout the book, key exercises are identified that need to be done to develop different thinking skills and gain practice with the creative problem solving process. The book is a self-study guide, yet contains many team activities and exercises.

Part Two covers the creative problem-solving process and the related mindset metaphors in detail. Instructions are provided so that teams of students can take a problem through the entire process. For problem definition, detectives search for clues and causes, while explorers investigate the broader context and trends to come up with a definition of the *real* problem. Artists working in teams brainstorm many solutions to the defined problem; engineers through the process of creative idea evaluation develop these original ideas into better, more practical concepts, and judges use carefully thought-out criteria to rank these ideas and make decisions based on the merits of the best solutions. The producers then develop work plans; they sell and implement the optimum solution; they also monitor the implementation to check that the problem is actually solved. Case studies and examples illustrate each step of the process.

Part Three presents four areas of application for creative thinking and problem solving—in using computers, in finding a "best" solution or design concept through team consensus and optimization (known as the Pugh method of creative evaluation), in improved communications, and in coming up with an invention. Although the Pugh method was developed to help design teams find a superior solution through an iterative process aimed at eliminating all weaknesses and flaws from proposed designs, this approach can be applied to any situation where the best of many potential solutions has to be found. The last chapter builds bridges between creativity, culture, and technology. The Appendices include summaries of more technical methods used for total quality management; they give a glimpse of the education and training done by industry to improve the competence of its work force and its capabilities for competing in the global marketplace.

Changing technology affects us profoundly by shaping our society and how we think about our world. Consider what a powerful metaphor and model computers have become for the human mind—this technology has certainly caused an irreversible change in our culture in the way knowledge is created, stored, and transmitted. But we must keep in mind—continuing in Dr. MacCready's words—that "In the end, technology does not exist by itself. Rather, it fits into a global, ethical framework, where serious, complex questions and concerns arise related to the survival of humankind, nature, and civilization. It is appropriate that those of us involved in the development and use of technology devote attention to consequences and solutions (whether or not the solutions involve technology). We must not succeed in our various short-term goals and find that we thereby lose the grander game and forfeit a sustainable future."

In creative problem solving, you have a thinking tool that will help you succeed in a changing world. You will be able to develop optimal solutions to problems if you practice and apply these skills.

ACKNOWLEDGMENTS

We are indebted to so many people—known and unknown—for what we have learned about creativity and for many ideas we have incorporated into this book on creative problem solving. It is the ninth version of material we have prepared and taught on the subject—and we had much help along the way.

We began with *Creative Problem Solving/Brainstorming*—a workshop for engineers and managers in industry—commissioned by the American Supplier Institute (ASI) of Dearborn, Michigan, and published in February 1987. Bruce Simpson of ASI furnished much of the source material. Willie Moore of Ford Motor Company made suggestions on how the content could be coordinated with methods already used at Ford. Bill Spurgeon, director of manufacturing engineering at the University of Michigan-Dearborn, supplied additional useful information.

In the early classes and workshops we taught, we were given many interesting examples and thoughtful insights by college faculty, working engineers, and high school students. We used these contributions in the next three versions—a training manual for workshop instructors, a customized edition for Lucas County, Ohio, teachers, and a manual for managers and engineers in industry. The fifth version was a manual designed for two pilot classes of the creative problem solving/engineering orientation course taught at the University of Toledo (UT) during the fall quarter of 1989. Many of the quotes on creativity still in our book came from these two classes of lively students. Sylvia Pinkerman, pre-engineering counselor at UT, supplied materials on study skills and time management. The sixth version was a customized workshop manual for Dana Corporation managers, engineers, and trainers. It underwent major reorganization and practical additions to emerge as the first edition of our textbook published in the College Custom Series by McGraw-Hill. We would like to thank these enthusiastic participants—who were much more than ordinary students—for sharing their thoughts, for questioning us, and for giving us creative ideas and stepping-stones for further growth.

Creative Problem Solving: An Introductory Course for Engineering Students grew out of our workshop and teaching experience. Additionally, we have read scores of books and listened to many tapes and lectures to learn more about the wide range of topics we are teaching. We owe a debt of gratitude to many authors for the ideas and interesting quotes that have found their way into our own lecture material. When we were ready to publish our book, we wanted to give credit to all these contributors. Alas, we found that we were unable to identify the sources of many of these items. For this we apologize. We have learned the material—appreciatively—but we do not remember who has taught it to us originally. But these ideas are so enriching that we want to pass them on to students and readers. We have added brief comments to the references listed at the end of each chapter, identifying these whenever possible as the source of concepts, ideas, and special vocabulary that we have incorporated into our text. At the stimulating Creativity Institute at the University of Wisconsin/Whitewater in the summer of 1987, we received much encouragement and many ideas on how to practice creative thinking from a broad spectrum of instructors and participants. Again, it is impossible to acknowledge individual contributions—they have become a part of our subconscious database. However, one dynamic person who stands out is Dr. Roger Von Oech; he really did give us "a whack on the side of the head." Two special people have made an impact on our thinking. We appreciate Paul MacCready's interest in our work; his designs of low-energy vehicles are wonderful examples of his creative spirit and concern for a sustainable future. Ned Herrmann, the creator of brain dominance technology, is a tremendous inspiration to us for his enthusiasm and work in all aspects of creativity and the brain. We want to thank Laura Herrmann, a member of his staff, for carefully reviewing Chapter 3 and making many helpful suggestions.

Many students in our creative problem-solving classes gave valuable feedback which enabled us to take a quantum leap for the eighth version of our book, now strongly focused on thinking skills for a changing world. At the University of Toledo, Jim Machen contributed ideas for improving the negative thinking exercise and the section on patents; Jennifer Voitle assisted with the MATHEMATICA material; and Carol Reed wrote the material on communicating in cyberspace. We thank Ellen Stanton, one of our students from industry, for permission to include her "thinking report" case study. Pat McNichols from the Lucas County Office of Education contributed some interesting math problems. Ken Hardy, a Toledo teacher with a wonderful sense of humor, sketched many of the illustrations for the second edition, and Geoffrey Ahlers, an artist from Copper City, Michigan, made the drawings for the present edition, based on Ken's ideas. Brigit Koenig, Monika's high-school friend from Switzerland, arrived in the nick of time to give advice on desktop publishing page layouts in the final stages of manuscript preparation.

As authors, we want to acknowledge the contributions made by the reviewers of our book. We have taken their comments very seriously—even when we had to disagree, we were able to appreciate and learn from these different viewpoints. The quality of the present edition owes much to their efforts. The reviewers were: Bill Cawley, a consultant in Richland, Washington; Gerald Jakubowski and Michael Mulvihill at Loyola Marymount University; Fred Janzow at Southeast Missouri State University; Carolyne Kincy at the University of Arkansas; Richard Lewis and James Nelson at Louisiana Tech University; and James Ross at NASA-Ames.

It was a pleasure to work with Margaret Hollander, our editor for the College Custom Series. For the present edition we are indebted to B. J. Clark, executive editor for engineering, for his patience while we changed jobs, location, and word processing equipment. We are grateful to Margery Luhrs, editing supervisor, for keeping us organized, and to Ann Craig for final proofreading and giving us further ideas for improvement—creative writers can only thrive when they have such dedicated people watching over the details.

Above all, the more we learn about the brain, creativity, and the process of invention, the more we stand in awe before the Mind of God, the Great Designer and Source of all Creativity.

Ed and Monika Lumsdaine

Part One

Creative Thinking

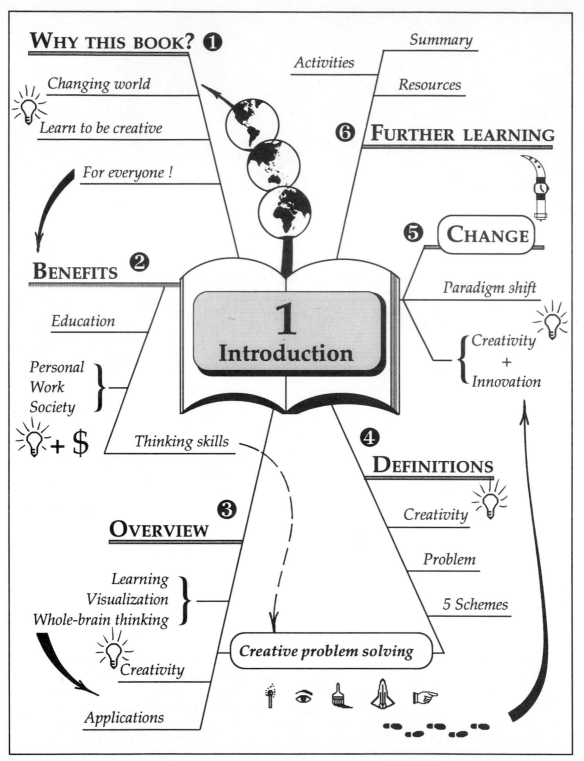

Mindmap of Chapter 1. Refer to Chapter 2 (pp. 55-56) to learn more about mindmapping.

1

INTRODUCTION

WHAT YOU CAN LEARN FROM THIS CHAPTER:
- Why you should study this book.
- Motivation for learning—the benefits of having creative thinking skills.
- Topical overview, chapter organization, and how to learn.
- Definitions and comparisons of problem-solving techniques.
- Understanding the connections between change, paradigm shift, creative thinking, and innovation.
- Resources for further learning: references, diagnostic tools and exercises, and summary.

WHY SHOULD YOU STUDY THIS BOOK?

We live in a world that is changing rapidly, and in times of change, creative thinking is the key that lets us cope, adapt, and succeed. During such times of change, the usual approaches and routine methods are no longer adequate for optimum problem solving and innovation. We need a framework that will encourage exploration, investigation, flexibility, and play with ideas as well as promote idea synthesis and constructive judgment. Creative problem solving is such a framework which employs many different thinking skills and thinking tools. In creative problem solving, we use the whole brain.

This book will give you an introduction to creative problem solving. You will become more aware and supportive of creativity. By participating in the activities and exercises (alone or better yet in a group), you will build your creative problem-solving skills. This is a practical book; you will find a presentation of basic principles, but the focus is on learning through examples and hands-on activities. You need to bring something to this book—curiosity and a willingness to become actively involved in learning, because these thinking skills cannot become your own through passive observation. Think of yourself as participating in a workshop, which means a seminar for intensive study and applications requiring a real mental effort. You will have to interact with other people as you explore questions, discuss ideas, do exercises, and brainstorm projects. Be prepared to enjoy yourself, and bring along an open mind!

Help yourself—have an open mind!
Michelle Dominique, engineering student.

There is no weapon—no tool—as powerful as the human mind. Sharpen and use it!
Angela Wallington, engineering student.

This book is based on the premise that creative thinking skills can be taught. Some people assume that creativity is a talent—you either have it or you don't. Together with many others working in the field of creativity and education, we believe that the creativity in each person can be unlocked, nurtured, and developed. People can learn to expand their creative thinking and integrate this ability with other higher-level thinking skills to become capable problem solvers. This can be likened to playing a piano. Some people have a natural talent. But with motivation, encouragement, effort, and steady practice, most people can learn to be very competent piano players with a joyful appreciation for music far exceeding those with an untrained talent.

Although earlier versions of this book were written for first-year engineering students, the examples and illustrations used here are not highly technical. Students and professionals in business, medicine, law, and many other careers can benefit equally from learning these thinking skills. In addition, the material will help develop a better appreciation for the role of engineering in our society. Such an understanding is necessary if we are to be a technologically literate population that can make informed decisions on the many serious issues that are connected with science and technology in a democratic society.

BENEFITS OF CREATIVE THINKING AND PROBLEM-SOLVING SKILLS

Why should you develop creative problem-solving skills? Will it be worth your investment in time and effort? You are the one to answer the second question after we present the advantages of having and using these skills.

Creative problem solving is needed to make up for shortcomings in your education. Our school systems, all the way from first grade through college, have tended to emphasize the use of our minds for storing information instead of developing its marvelous power for producing new ideas and turning these into reality. We have been taught the mechanics or "cookbook" methods of problem solving, also known as the plug and chug approach. But today we live in a world that is changing, a world that is very complex, a world that faces many difficult problems—thus everyone needs flexible, critical, and creative thinking skills to cope with these problems and find solutions that can improve the physical and social environment. With creative thinking, we can explore the fundamental changes our schools must make to improve education for all students.

Also, because calculators and computers are now widely available, many mathematical problems can be solved routinely without much thinking. People mistakenly believe that computers allow us to solve complex problems with the same problem-solving skills that we are accustomed to using. But to properly take advantage of the computer's capabilities we must substantially expand our creative problem-solving abilities. Our productivity will be enhanced when we use our brain to question, explore, invent, discover, and create—in other words, when we use the brain for creative thinking, a task that computers cannot do.

Creative problem solving is needed to help manufacturing companies become more competitive in the global marketplace. Manufacturing creates the wealth in a country; when this base declines, so does the economy and people's standard of living. We need to invent better products as well as better ways of doing things. With creative thinking, businesses will be able to develop and implement new manufacturing, quality-control, and management techniques that will enable them to satisfy customers all over the world. We do not mean that we need more products that encourage consumption and waste of resources; instead, we need to research, invent, and produce products that have "value added"—that are technologically appropriate to culture and the environment while making life better for individuals and communities.

ⓘ *THREE-MINUTE ACTIVITY 1-1: PROBLEMS*

With two other people, brainstorm and jot down problems that are in the news right now which could use a creative approach because the old ways of dealing with these problems are just not working. Also think about some personal problems that would make good "targets" for creative problem solving. Be sure to observe the three key rules of brainstorming: 1. Get as many ideas as possible. 2. Encourage wild and crazy ideas. 3. No criticism of ideas—defer judgment until later in the problem-solving process.

What are the direct, personal benefits of having creative problem solving skills? Creative problem solving can enrich our life because these thinking skills can be used at home, in our recreational pursuits, and especially in our relationships with other people. We can become more successful in our jobs, and we can make our careers more challenging and exciting. We will be energized with a sense of adventure, surprise, enjoyment, and fun as we learn to explore dreams and possibilities. Through creative problem solving, we generate new ideas and innovative solutions for a given need or problem. These ideas will be different and often of much higher quality than those we could get if we used the way it has always been done. Creative problem solving gives balance to our thinking since it integrates <u>analytical</u> and imaginative thinking. Intuitive and interpersonal thinking are as important as critical and structured thinking for achieving the best results.

Creative minds result in unique lives.
Dawn Rinehart, high school student.

To be competitive, industries need creative problem solvers as team leaders, managers, and engineers; in sales and product development; in service and customer relations. When people work together to find the best solutions to problems, goals are set, the needs of the customer are identified, much trial and error is eliminated, and the working environment is invigorated. We are no longer in a position where we can afford to wait until we have a crisis that forces us to react. Through creative thinking, "If it ain't broke don't fix it" attitudes are replaced by goals that say: "If it ain't broke make it even better." With creative problem solving we have the power to direct change! We can create organizational climates that are supportive of inventive thinking and innovation.

As an engineering manager in a Fortune 500 company, I value this approach of teaching teamwork and creative problem solving to complement the 3 R's. I would love to be able to hire engineers with these "whole-brain approach" skills. We as a nation would be well served if all of our educated youths came to industy packing the skills taught in this book.
John Faust, parent of an eighth-grade daughter who learned creative problem solving in Saturday Academy, a precollege program developed by the College of Engineering at the University of Toledo.

Society needs creative thinkers if we are to find solutions to the many technical, ecological, economic, educational, social, and political problems that we face in the industrialized and in the developing world. With creative, flexible thinking, leaders and many groups of people working together can find and implement new ideas to solve these problems. Creative thinking skills are needed to help people accept and cope with change. Table 1-1 shows the results of a survey that illustrates how industrial companies rank creative thinking skills. Are you surprised at the subjects that are listed in the top ranks? You have to go down to the twelfth item on the list before you encounter a subject dealing specifically with engineering topics.

🕐 THREE-MINUTE ACTIVITY 1-2: SKILLS COMPARISON

The top eleven subjects from Table1-1 involve thinking and communication skills in various proportions. Discuss with one or two people where each of these subjects would fit on a continuous line where communication is on the right and thinking skills are on the left. A subject that has equal components or aspects of communication and thinking would be at the midpoint of the line.

THINKING _____|_____ COMMUNICATION
SKILLS   SKILLS

The future of the world rests not on today's youth, but on today's creativity.
Chris DelSignore, engineering student.

We are entering an economic environment that will reward those who can adapt to change and punish those who can't—or won't!
Cordell Reed, senior vice president, Commonwealth Edison Company, Chicago.

Table 1-1
Subjects Most Needed for Careers in Engineering and Technology

Rank	Subject	Rank	Subject
1	Management practices	20	Applications programming
2	Technical writing	21	Psychology
3	Probability and statistics	22	Reliability
4	Public speaking	23	Vector analysis
5	Creative and innovative thinking	24	Electronics and circuit design
6	Working with people	25	Laplace transforms
7	Visualization	26	Solid-state physics
8	Reading (speed, idea discernment)	27	Electromechanical energy
9	Communicating with people		transformation
10	Finance, marketing, accounting	28	Matrix algebra
11	Use of computers	29	Computer systems engineering
12	Heat transfer	30	Operations research
13	Instrumentation and measurements	31	Law principles: patents, contracts
14	Data processing	32	Information and control systems
15	Systems programming	33	Numerical analysis
16	Economics	34	Physics of fluids
17	Ordinary differential equations	35	Thermodynamics
18	Logic	36	Electromagnetics
19	Engineering economics	37	Human engineering

Reprinted with the permission of Macmillan College Publishing Company from *Careers in Engineering and Technology*, 4th edition, by George C. Beakley, Donovan L. Evans, and Deloss H. Bowers. Copyright © 1987 by George C. Beakley.

The American Society for Training and Development in April 1990 reported the results of a survey on needs in human resource development. Executives from 93 percent of the nation's Fortune 500 companies and major private firms indicated that they provided or were planning to provide training in basic workplace skills in addition to reading, writing, and computation. Key items mentioned were knowing how to learn, communication, adaptability, personal management, and group effectiveness. The specific skills receiving the most attention in these companies are: problem solving (58 percent), teamwork (51 percent), interpersonal skills (48 percent), oral communication (45 percent), listening (43 percent), writing (41 percent), and goal setting (33 percent). Why do these companies spend so much money to provide this training? Among the reasons listed are: a need to upgrade skill levels of current employees, changes in the work processes, changes in technology, and lack of education and skills in new employees. Better educated and trained employees are needed because of increasing competitive pressure from all around the world. Companies (and their employees) are no longer just competing against other companies in the United States; the competition is now worldwide and includes people who receive a much more thorough basic education than we provide to most children in this country.

Examples of Creative Problem Solving

To find examples of problem solving of all kinds, we have only to look in magazines or the daily newspaper. What is especially interesting is to observe the reaction that creative ideas frequently encounter. For example, when the police department in Bowling Green used an undercover narcotics agent in the disguise of a Santa Claus to nab a drug dealer, this creative (and successful) approach caused quite an uproar in this Ohio community. Training in creative thinking is thus needed not only for problem solving but also for building understanding and broader support for innovative solutions. Three examples are briefly summarized: the first one from industry points out the importance of teamwork; the second shows what happens when support is lacking; and the third demonstrates the global extent and potential for creativity and cooperation—problems cannot be attacked in isolation but need to be solved within the social context. These solutions can thus build bridges, not barriers, between different cultures.

Industry: In May 1992, the Boeing Company announced that it is developing a new passenger airplane, the Boeing-777. This is the first new airplane for the company since 1978; it will be the world's largest twin jet with very powerful but quiet engines. What is remarkable in this design is that the customers (United Airlines, All Nippon Airways, and other carriers) participated in intensive group sessions to define the new airplane's configuration. Such input to the design process should result in a final product that is responsive to the needs of the world's airlines and the flying public. More than 200 design/build teams are responsible for major systems and are staffed by experts from many diverse disciplines (including suppliers and customers). According to James O'Sullivan, British Airways' chief project engineer for the B-777, "The benefits of teamwork include speed of response, education, synergy of shared goals—better planning—discipline, new perspectives on old problems, and an improved final product."

Education: David Broder, a newspaper columnist writing out of Washington, D.C., reported in March 1992 on an education bill before the U.S. Senate where $500 million were to be spent on improving the quality of the nation's schools. The debate was not about the funding but on how to encourage change and initiate innovation in what is said to be one of the most change-averse institutions in the country—the public school system. The Bush administration proposed to finance a nationwide network of "break-the-mold" schools, one for each congressional district. Educators and other talented people could submit proposals for radical redesign of school buildings, curricula, schedules, and teaching methods. The winners of the competition could then test their approaches, with parents and students judging the results. The responses to this creative idea were very interesting: Democrats feared that these schools would become political footballs, and county school superintendents saw the idea as a threat against "the old order." Thus a compromise between Senator Edward Kennedy (the floor leader on the measure) and Education Secretary Lamar Alexander was negotiated (and supported by the education lobby): one-quarter of these funds could be used by states to finance innovative schools—but instead of ideas from fresh minds, the requests would have to come from each state's chief school official who would then direct the project. But giving the innovation money to the same people who may be part of the problem in educational achievement is futile. Broder concludes: "Offering that money to innovators inside the system—or to outsiders ready to challenge the status quo—might have worked no better; there are no guarantees for such experiments. But there is at least a chance that models might be created that excite and motivate and educate youngsters. And that, in turn, might have unleashed a demand for similar changes in other schools."

Economic development in the Third World: This example very briefly summarizes an article by Audrey Liounis, "Silk for Cocaine," published in *OMNI* magazine in July 1992. Try to read the article with its many interesting details to gain a more complete understanding of the problem's historical and cultural setting and its connections to U.S. politics and finance. The setting of the story is the lush but primitive Cauca Valley high in the Andes of Colombia where, to improve their economic conditions, the campesinos started to sell coca leaves to drug producers, giving up raising coffee and food crops—until prices dropped and they could no longer support their families. The starving farmers abandoned their land and moved to overcrowded Bogotá slums or joined the guerrillas—after experiencing the ruthless destruction by government forces or seeing their children addicted to drugs. Patricia Conway, a farmer's advocate and educator who was working as a volunteer nurse in the slums of Bogotá, met one of the Cauca families. After visiting the valley and buying a piece of land there, she got the idea for the *Silk for Life* project and formed a small economic development community. Without any help (and with a lot of ridicule from officials in Colombia and Washington, D.C., including USAID as well as the Colombian Coffee Federation who had pioneered the raising of silkworms as an economically viable alternative to coca) about forty families painstakingly hauled in mulberry canes and in three years harvested their first crop of cocoons which they sold through established middlemen.

When Chinese visitors advised them to buy back unused cocoons, they learned the degumming process, and the women began to hand-spin the silk into yarn. When this beautiful silk piled up in the farmer's huts, Patricia Conway decided to find a market in the United States. She met Kerry Evans, a Milwaukee weaver who was teaching Hmong refugees and other disadvantaged women to weave. Using this Andean silk alone or in combination with Wisconsin wool, they developed a flourishing industry, and the women appreciate the connection between helping the Andean farmers raise a cash crop and thus fighting the drugs that are hurting the young in their own communities. The Cauca valley farmers now have plans to become self-sufficient; they have built a dam for a small generator (for a silk-reeling machine), and they are obtaining silkworm eggs directly from a Korean producer. Their goal is to become economically successful so that they can attract teachers and health-care workers to their community (where schools so far only cover the first three grades). A silk company in the United States has offered to buy one million pounds of Andean silk. Patricia Conway believes that "The link between industrial consumers and small family farmers will revolutionize the way the world does business in the next decade, where farmers and factory owners understand how the web of life works and how each delicate strand respects the others and protects the entire web." Mulberry tree culture is ecologically gentle, small-scale, and easy; with initial help and a favorable climate, it can yield a cash crop within a year, while paying at least three times more than coca leaf. Requests for assistance to start similar silk-producing projects are coming from Africa and many places in South America.

This story has a postscript—the Colombian government is now spraying the Cauca Valley with the herbicide glyposate in its attempts to eradicate coca as well as the more recently planted poppy fields, thereby threatening the mulberry trees and the silkworm crop (since the worms refuse to eat leaves that have been treated with chemicals). The United States spends hundreds of millions of dollars in narcotics interdiction in Colombia, in addition to sending covert troops. As the article concludes, "The Bush administration's Andean Initiative is a narrowly focused policy on drug eradication to the exclusion of the fundamental issue: agricultural alternatives that are successful and provide farmers with a viable income." It will be interesting to watch how this policy will change in the Clinton administration.

HOW TO LEARN FROM THIS BOOK:
OVERVIEW, ORGANIZATION, AND HINTS

This book is divided into three parts, as summarized in Table 1-2. In Part 1, we want to give you an understanding of creativity and its relationship to the brain and the environment. You may study Chapter 4 first, if you wish. In Part 2, we will focus on the distinct steps in the creative problem-solving process; thus these chapters need to be studied sequentially. In Part 3, some areas of application for creative problem solving will be presented. These topics can be investigated at different stages while you are learning the creative problem-solving process; you may choose the application topics in whatever order is most appealing to you.

Part 1—Creative thinking. You will see the advantages of creative problem solving as a comprehensive thinking process, especially as related to change and innovation. You will investigate the connections between visualization, learning, and memory. You will learn about the Herrmann four-quadrant model of brain dominances or ways of knowing and how they relate to creative problem solving. And you will look at mental blocks to creative thinking and how to overcome them.

Part 2—The creative problem-solving process. The stages of the creative problem-solving process are presented in terms of associated mindsets or thinking skills. You will learn how to define a problem by adopting the explorer's and detective's mindset. You will generate many ideas through brainstorming using the artist's mindset. You will then make these ideas better and more practical in the engineer's mindset, and you will determine the best ideas in the judge's mindset. Finally, you will adopt the producer's mindset to put the best solution into action.

Table 1-2
Overview of Topics

PART 1: CREATIVE THINKING

1. **INTRODUCTION**
   Diagnostic tools and exercises.

2. **VISUALIZATION AND MEMORY**
   Sketching and memory exercises. Thinking languages activities.

3. **THE 4-QUADRANT BRAIN MODEL OF THINKING PREFERENCES**
   Identifying your team's thinking preferences. Claiming your space.

4. **OVERCOMING MENTAL BARRIERS TO CREATIVE THINKING**
   Writing exercises for creative thinking. Negativism score sheet.

PART 2: CREATIVE PROBLEM SOLVING

5. **PROBLEM DEFINITION**
   The detective's mindset: customer survey and Pareto analysis.
   The explorer's mindset: contextual thinking exercises; study of trends.

6. **IDEA GENERATION, BRAINSTORMING, AND TEAMWORK**
   The artist's mindset: brainstorming and force-fitting exercises.

7. **CREATIVE EVALUATION, JUDGMENT, AND CRITICAL THINKING**
   The engineer's mindset: making ideas better and more practical.
   The judge's mindset: choosing the best solution. Critical thinking.

8. **SOLUTION IMPLEMENTATION**
   The producer's mindset: action plans and working for acceptance.

PART 3: APPLICATIONS

9. **COMPUTERS AND CREATIVE THINKING**
   Exploring MATHEMATICA.

10. **THE PUGH METHOD**
    Evaluating design concepts (or other ideas) to find the optimal solution.

11. **COMMUNICATIONS AND CREATIVE PROBLEM SOLVING**
    Presenting an effective 30-second message. Exploring cyberspace: E-mail.

12. **CULTURE, TECHNOLOGY, INVENTIONS, AND PATENTS**
    How to do a patent search. Using creative problem solving for inventing.

Part 3—Applications. You will see how creative thinking and symbolic manipulation let us use computers to enhance thinking, learning, communication, and many other aspects of the workplace. You will learn how the Pugh method of creative design concept evaluation converges to superior solutions. You will find that communication is enhanced when creative problem solving is applied in negotiation and in preparing concise messages. And you will be able to draw connections between creative problem solving, inventing, culture, and technology.

Overview: Table 1-2 shows a sequential outline in boldface print of the topics that are covered in this book. The main focus of the associated activities and exercises is indicated in regular-face print. Each topic represents roughly a three- to five-hour study unit consisting of the introduction of the topic, a selection of exercises or applications, and a review.

Chapter organization: Each chapter starts with a list of the learning objectives. A different kind of overview is provided with the mindmaps located at the beginning of the first eight chapters and at the start of Part 3 of the book. You will learn more about the purposes of the mindmap (and how to construct one) in Chapter 2. The body of the main text is subdivided into major subtopics; these are indicated by a header in capital letters. At the end of the chapter you will find a list of books and other materials that can provide additional information on the topics covered in the chapter. As part of the learning resources, most chapters have a collection of activities and exercises for practicing and applying different thinking skills. If you are studying this book on your own, assign yourself as a minimum the exercises and activities marked with a **key**. Advanced exercises are also included; these are indicated with a pair of **asterisks** (* ... *) and will require extra time, but they offer an opportunity for more in-depth thinking and extension of learning. The first eight chapters have a summary which can be used as a tool for review.

> *The learning challenges for teenagers and executives are the same:*
> *learning to do a job well; facing and enduring hardships; learning from role models;*
> *learning from mistakes; and learning in the classroom.*
> Michael M. Lombardo, Center for Creative Leadership, Greensboro, North Carolina.

Hints for more effective learning: Three techniques are used to encourage your mind to switch from routine reading to an active thinking mode. (1) We will pose questions in the text—pause a moment to think how you might answer these questions before continuing on with the material! (2) The illustrations as well as symbols and quotes extending into the margins invite you to make connections between these materials and the text for better learning and recall—use your imagination! (3) As you read through each chapter, watch out for brief assignments given in a rounded box—take the time for these activities and team discussions to make your learning more effective! Working in a team enhances thinking and creative problem solving in a number of ways, from increasing the quantity and quality of ideas to easier acceptance of the best solution for implementation and change. Many assignments in this book integrate academic and social aspects because learning should not be compartmentalized apart from the rest of your life. If possible, study with one or two friends or colleagues. *Whenever italicized font is used in the text, it is a reminder for you to stop reading and do something.*

The two-questions quiz: What else can you do to make your learning as interactive as possible? One very useful technique is to get into the habit of asking questions. As you read, think up questions about the topics you are studying. Jot them down in a notebook or in the margin of the text. If you cannot find answers to your questions further into the book, ask an instructor, discuss the questions with someone else, or find additional material on the topic in the library—starting with the references given for each chapter. At the end of reading a chapter (or of attending a lecture or a class), ask yourself these two questions: "What is the most important thing I have just learned? What is an important question I still have?" Write your answers to this two-questions quiz down! You will be amazed how this technique will sharpen your attention and thereby increase learning and retention, not just of the material in this book, but in any subject.

WARNING!
If you want your life to go on as usual,
don't read and learn from this book—
because learning to think creatively will change your life!

 FOUR-MINUTE ACTIVITY 1-3: LEARNING AND PROBLEM SOLVING
Alone or with another person, try to come up with several different answers to the question:
What is the relationship between learning and problem solving?

DEFINITIONS AND COMPARISON OF PROBLEM-SOLVING TECHNIQUES

Before we go any further, we need to make sure that we use the same vocabulary and understanding about key terms and concepts involved in creative problem solving. In particular, we need to understand what creativity is, what we mean by the word "problem," and how creative problem solving differs from the methods you commonly use to solve problems.

What Is Creativity?

We can think about creativity in many different ways. Is it something in the environment that encourages creativity? Is it the sum total of internal mental processing that makes up creativity? Or, to be valid, does creativity require a tangible output—a product or application? We believe that all three of these aspects are involved in creativity. Thus in the first part of the book, the spotlight will be on the influence of environment and its interaction with creative thinking and the brain. In the second part, we will focus on the thinking skills needed in creative problem solving, and in the third part, we will concentrate on some special applications for creative thinking. It is fun to describe aspects of creativity as "slogans," and you will find such quotes extending into the margins throughout the book. But let us begin with a working definition of creativity; as you become more familiar with the subject, you can come up with a definition of your own.

> ### CREATIVITY
> *is playing with imagination and possibilities,*
> *leading to new and meaningful connections and outcomes*
> *while interacting with ideas, people, and the environment.*

Creativity is a dynamic activity that involves conscious and subconscious mental processing. Creativity involves the whole brain! Ned Herrmann, author of *The Creative Brain*, defines creativity this way:

> My own thinking is that creativity in its fullest sense involves both generating an idea and manifesting it—making something happen as a result. To strengthen creative ability, you need to apply the idea in some form that enables both the experience itself and your own reaction and others' to reinforce your performance. As you and others applaud your creative endeavors, you are likely to become more creative.

When a creative idea has been widely implemented in such a way that it has led to a permanent change, we can say that innovation has occurred. When we study the development of innovation in technology over the last 1000 years, it is fascinating to note in many instances that just hearing about an invention or advance in a faraway place can lead to a blossoming of creativity and innovation in another culture. Creativity rarely happens in isolation—it needs other people's minds, ideas, and inventions. Thus in the broadest perspective, creativity is expressed in the quality of the solutions we come up with when we solve problems.

Mater artium necessitas.
Necessity is the mother of invention.
Ancient saying.

Wei ji.
The Chinese symbol for crisis is made up of two words:
danger + opportunity.

What Is a Problem?

Now what do we mean when we use the word "problem" in this book? A problem is not only something that is not working right or an assignment that teachers give students to work out—a problem is anything that could be made different or better through some change. A problem is finding the best birthday gift ever for the most important person in your life; a problem could be building or inventing something that fills a specific need; or a problem could be finding a better way of managing an organization or providing a service. A problem as defined here has two aspects: it can involve a difficulty (or danger), or it can represent an opportunity (or challenge). Both aspects are usually present in a problem, although one may be more apparent on the surface. It is easy to overlook the opportunity aspects when dealing with an emergency, yet once the crisis has been dealt with, do not stop there—this may just be a perfect chance to introduce a policy of continuous improvement or make a fundamental change leading to true innovation.

Unfortunately, creative thinking is not taught or emphasized in our schools and universities, where the curriculum concentrates almost exclusively on feeding the students information. Mathematics and science courses do provide some training in analytical thinking. Yet it has been estimated that about 80 percent of all problems in life need to be approached with creative thinking. Also, students are not trained in developing critical thinking skills. As we shall see, the process of creative problem solving involves all three types of thinking: analytical, creative, and critical. The key is to employ these in the most appropriate sequence to solve problems well.

> *One of the ingredients of survival will be flexibility, tolerance of ambiguity, and creativity in facing issues*
> *that will unfold, gain in complexity, and mutate as we grapple with them.*
> *Hunter Lovins, president and executive director, Rocky Mountain Institute, Snowmass, Colorado.*

> *To find solutions in school you need knowledge; to find solutions in life you need creativity.*
> *Edward Lumsdaine.*

⏱ *THREE-MINUTE ACTIVITY 1-4: PROBLEM SOLVING*

In a brief paragraph, describe the method that you use most often to solve problems. If you are in a class or group, share your answer with one or two people sitting next to you. Compare your approach with the schemes outlined in Table 1-3.

Comparison of Different Schemes for Solving Problems

Now we want to do a comparison of different problem-solving methods. Creative problem solving is somewhat different from the problem-solving schemes that you may already be familiar with from mathematics, science, or psychology. It is also different from the problem-solving process taught in traditional engineering courses or used in industry. This comparison is summarized in Table 1-3. Where does your problem-solving approach fit in? In working with many groups of adults as well as students, we very rarely find a person who is using the scientific method to solve everyday problems. Many people use a variation of Polya's method, while others rely on the intuitive "aha" technique—they wait until an idea or solution "pops" into their mind. Engineers combine analytical tools, sketches, and modeling when solving problems. Creative problem solving employs aspects of all these problem-solving methods. When solving problems, some people may even use unguided experimentation, trial and error, or guessing—these approaches to problem solving are not listed in Table 1-3 since they are not techniques that are taught in some discipline.

Table 1-3 **Problem-Solving Schemes of Various Disciplines**					
SCIENTIFIC METHOD SCIENCE	CREATIVE THINKING PSYCHOLOGY	POLYA'S METHOD MATH	ANALYTICAL THINKING ENGINEERING	8-D METHOD INDUSTRY	CREATIVE PROBLEM SOLVING MANY PROBLEMS
Inductive data analysis and hypothesis.	Exploration of resources.	What is the problem?	Define and sketch system. Identify unknowns.	1. Use a team approach. 2. Define the problem.	Problem definition: data collection and analysis/exploration of trends and context.
Deduction of possible solutions.	Incubation— possibilities.	Plan the solution.	Model the problem.	3. Deal with the emergency. 4. Find root causes.	Idea generation —> many ideas.
					Creative idea evaluation —> better ideas.
Test alternate solutions.	Illumination— definite decision on solution.	Look at alternatives.	Conduct analysis and experiments.	5. Test corrective action and devise best action plan.	Idea judgment and decision making —> best solution.
Implement best solution.	Verification and modifications.	Carry out the plan. Check the results.	Evaluate the final results.	6. Implement plan. 7. Prevent problem recurrence. 8. Congratulate team.	Solution implementation and follow-up. What was learned?

The **scientific method** uses inductive data analysis to arrive at a hypothesis. For example, let us say that we are in the business of manufacturing brakes for trucks. We are having a problem: some brakes fail after a relatively short time. We examine the data and hypothesize that heat build-up in the brake rotor disk causes the problem. We design a number of different disk brake configurations with fins and holes to allow the brake to cool faster. We run a series of tests on these prototypes and pick the brake design that seems to solve the problem best. An interesting feature of the scientific method is that it must be on the lookout for data that will *disprove* the hypothesis. The results of problem solving with the scientific method are then presented sequentially, whereas the actual process is much less straightforward and consists of many detours.

Based on the scheme described by researcher Graham Wallas, psychologists regard **creative thinking** as a process where the available resources and information are explored first. The mind then subconsciously incubates ideas and possibilities until—quite suddenly—a definite decision on the solution emerges. This is the "aha" phenomenon. The conscious mind verifies this solution and makes minor modifications as required to make it practical. But since the first idea that comes to mind may not necessarily be a superior idea, a method that invites many different ideas before making a judgment may result in a higher-quality solution.

George Polya has devised a set of steps for **solving mathematical problems**. First, we need to ask: What is the problem? Then we plan the solution and look for alternate ways on how we may be able to get there. Finally, we carry out the plan and check the results. We have all been taught this method in some way in school; our problem is that we use it for other types of problems where such an analytical approach does not work well because it discourages contextual, holistic thinking.

In **engineering problem solving**, we first define the system and identify the key elements before the appropriate physical laws can be applied. The system is sketched. In electrical engineering, this is called a circuit diagram; in mechanics, it is a free-body diagram; in thermodynamics, the system is defined in terms of a control volume. Next, the known and unknown quantities are listed separately, and the problem is modeled. Engineering students learn how to do this mathematically in junior- and senior-level courses. In computer-aided engineering (CAE), this process is done graphically. Then the model is analyzed. Tests may be needed to determine the accuracy of the model and assumptions made in modeling. Other items that need checking are the units and the "reasonableness" of the answer. It helps to learn to make quick estimates on the order of magnitude the answer is expected to have. To solve the problem of the overheating disk brake rotor with engineering analysis, the heating, cooling, and internal stresses in the disk rotor are modeled with mathematical equations and then confirmed with tests. Industry now uses tools such as the Taguchi method of design of experiments to optimize product design and testing.

The **8-D method (eight disciplines of problem solving)** is a procedure used by Ford Motor Company. It is focused on finding the root causes of problems and devising corrective actions through teamwork and analytical thinking. The 8-D method and engineering problem solving are useful approaches for problems such as finding and fixing a "clunk" in an engine. They do not generate innovative concepts or contextual solutions to problems in the field, because these require creative thinking. For example, analytical methods can come up with an alternator design that can dump heat more efficiently to prevent damage to the rectifier— creative thinking comes up with a solution that places the heat-sensitive rectifier outside the alternator, where it is cooled by ambient air flow. Creative problem solving might even consider a redesign of the car's entire power system and engine.

Creative problem solving has five distinct steps that correspond with different, distinct mindsets or thinking modes. Creative problem solving is a sequence of successive phases of divergent thinking followed by convergent thinking. As detectives, we collect as much information about a problem area as possible, then analyze the data and condense it to its major causes or factors. As explorers, we brainstorm the context of the problem and related issues, then converge our thinking to a problem definition statement expressed as a positive goal. As divergent-thinking artists, we use brainstorming to get many "wild and crazy" ideas. Divergent thinking is at first continued in the engineer's mindset, as ideas are elaborated. But soon the focus shifts to idea synthesis and <u>convergence</u> to better and more practical solutions. As judges, we use <u>divergent</u> thinking to explore criteria and make constructive improvements to the final ideas in order to overcome flaws. This is followed by convergent thinking which results in decision making and selection of the best idea for implementation. Implementation in itself is a new problem that requires creative problem solving. Thus, as producers, we are again involved in alternate periods of divergent and convergent thinking. Creative problem solving is thorough and takes time. The quality of ideas improves if the mind is given enough time to incubate and think through the problem.

> *Divergent thinking is an effort to search, to stretch our thinking, and to consider many possibilities and directions. Convergent thinking is an effort to screen, select, or choose the most important or promising possibilities, closing in on one or a few items.*
> Scott G. Isaksen and Donald J. Treffinger, Creative Problem Solving: The Basic Course, *p. 17.*

At each step, creative problem solving may employ aspects of all five methods in Table 1-3. However, the main difference lies in its emphasis on generating creative ideas and improving these ideas into better solutions. Each step of the creative problem-solving process can be likened to a tool box—many different techniques are available during each phase to enhance the process and achieve an optimum result, depending on the type of problem and its context, the time and resources available, the experiences and training of the team members, the organizational culture, and the objectives of the problem solving.

 THREE-MINUTE ACTIVITY 1-5: DEFINITION
Write a brief definition of creative problem solving from what you have learned so far in this chapter. Then get together with two other people who have worked out this assignment and try to combine your definition. If you are in a large group, continue this process once or twice more, perhaps with the aid of an overhead projector.

THE PROBLEM OF CHANGE AND PARADIGM SHIFT

We opened this book by saying that creative thinking is needed to cope with change. If change were only an occasional, insignificant happening in your life, you may not see the need for having creative thinking skills. So let's pause for a moment to think about change. Change is a natural part of living—consider the cycles in nature and the change of seasons; think of infancy, childhood, adolescence, young adulthood with career choices and marriage, a growing family, an empty nest, grandparenthood, retirement, and the death of loved ones—all tremendous changes. But in recent years, the rate of social and technological changes has greatly accelerated. Table 1-4 is a list of changes that have affected our world and life in the United States during the last twenty years or so—you can probably think of others.

Table 1-4
Important Changes in the World Since 1970

- We truly are "Spaceship Earth." Ecological concerns are: acid rain, the ozone level, the greenhouse effect, rain forest preservation, water quality (including the oceans), endangered species, recycling of materials, and waste disposal.

- Energy choices are difficult; oil use has environmental, economic, and political costs, nuclear energy has problems with safe operation of aging plants, radioactive waste disposal, and high costs due to regulation, research, and decommissioning of plants. Coal use involves environmental and health problems.

- U.S. manufacturing declines; Japan, Europe, and Korea have become leading manufacturers—employees (and their education) now compete in a global marketplace.

- Total quality management represents a revolution in manufacturing; zero defects is now the standard; continuous improvement is an attitude leading to innovation.

- Powerful personal computers are widely available at reasonable cost. Satellites allow real-time communication worldwide; information has become a key resource.

- Women have become important in the work force and in public life. Union power is declining. Minority rights are becoming widely guaranteed.

- Nationalism is rising; communism is disintegrating; the developing world has global influence.

- A loss of values is seen in the decline of marriages and values in TV programming, a decreasing respect for authority (police, courts, government), and increasing drug use, child abuse, and violent crime.

- High-tech health care is available, but access to health care and insurance coverage is not. AIDS, homosexuality, and abortion rights are difficult social issues. Exercise, low-fat foods, and a smoke-free environment have become important.

- Video games, videotapes, TV, and convenience foods have changed family life and recreation.

- In the U.S,. large federal deficits and deregulation have become acceptable.

- A high school (or even a college) education is no guarantee of a job.

With so many major changes, it is interesting to think about the opposite angle. What are some things that have had little or no change? As a society we have decided that some things are so valuable in their original form that they should not be changed—such as the U.S. Constitution. But even here changes have been made through the amendments. Some things change very little, such as the way baseball is played—although in the early days, it took "seven balls for a walk." But what about our educational institutions? School systems and universities are examples of institutions that are very resistant to change, even when it is evident that they are no longer working well in a changing world. U.S. industry as a whole has been quite slow during the last decade in recognizing the need for change toward higher-quality products and meeting the customer's needs. Also, because creative ideas demand change, many people face quite a battle to get these ideas accepted and implemented. Why is there often such a resistance to change—unless a major crisis makes change imperative?

A classic illustration of this resistance is told by Joel Barker in his videotape on *Discovering the Future: The Business of Paradigms.* Here is a condensed version:

> The Swiss watch manufacturing industry in the 1970's had about 65 percent of the world market in watches and over 80 percent of the profits. Yet between 1979 and 1982, they had to cut employment from 65,000 to 15,000, and their market share fell to less than 10 percent. What caused such a rapid decline? The answer of course is the invention of the quartz watch. Do you know who invented the quartz watch? A team of researchers at the Federal Watch Research Center in Neuchâtel, Switzerland, created the first prototype in 1967, but when they presented the model to the Swiss manufacturers, the idea was rejected, instead of patented. When the model was exhibited the following year, Seiko people saw its possibilities. Which nation has the largest share of the watch market now?

Why did the Swiss watch manufacturers not recognize the potential of this invention? In its January 14, 1980, issue (p. 68), *Fortune* magazine wrote:

> They simply refused to adjust to one of the biggest technological changes in the history of time-keeping, the development of an electronic watch. ... Swiss companies were so tied to traditional technology that they couldn't—or wouldn't—see the opportunities offered by the electronic revolution. It was a classic case of vested interests blocking innovation.

This story has a sequel. When Joel Barker returned to Switzerland years later and asked one of these manufacturers how he felt about this loss of jobs, he received the astonishing reply that losing all those jobs was nothing. The man then went on to explain the bigger picture: because this early opportunity to get workers trained in small electronics was missed, Switzerland lost out on the next development, which was much more important, namely the manufacture of small electronics components for computers and instruments—a much larger employment market. Joel Barker explains what happened to the Swiss in terms of paradigms and paradigm shift. Their old, successful watchmaking paradigm simply blocked them from being able to recognize a different way of keeping time. Paradigms tell you what the game is and how to play it successfully according to the "rules" (even though the rules are not usually spelled out). When a paradigm shifts, past success can be a barrier to future success because it can blind you to visions of the future and possible alternatives. When a paradigm shifts, everyone goes "back to zero."

Paradigms as tools for problem solving have a life cycle in the shape of a typical S-curve as shown in Figure 1-1. In the early phase (Segment A), problem solving is slow because of the learning curve and because only a few pioneers are beginning to use the paradigm. During the main phase (Segment B), problem solving with the paradigm is quite successful and is getting well established, although some "impossible" problems are set aside in the hopes that further development with increased experience, refinement, and precision will help in solving these cases. In the last phase (Segment C), problem solving becomes more costly, more time-consuming, and less satisfactory, not only because the problems solved in this stage are the more difficult problems but also because the solutions no longer fit the larger context because of changes that have occurred elsewhere.

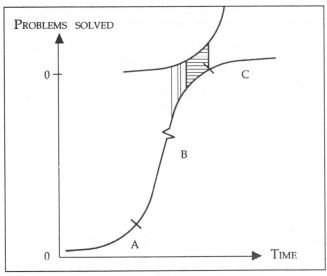

Figure 1-1 The paradigm life-cycle curve.

*A paradigm is a set of rules and regulations that defines boundaries
and helps us be successful within those boundaries,
where success is measured by the problems solved using these rules and regulations.*
Joel Barker, futurist.

Unresolved problems create a feeling of uneasiness and uncertainty—a climate that encourages outsiders to look for a new paradigm, even though the current paradigm is still very useful and doing well in solving most problems in its field. This stage of creative thinking by the outsider is shown by the thin vertical lines in Figure 1-1. Once these so-called paradigm shifters are beginning to be successful in solving problems the new way, they are joined by the paradigm pioneers, the people who are adopting the new paradigm and helping to bring about change. This shift may happen over a period of time, as indicated by the hatched area. Note that problem solving now has shifted to a new S-curve. Paradigm pioneers take the risks and reap the benefits. When an organization or a person delays making a decision to adopt a new paradigm, it will take a much larger effort and higher costs to catch up and recapture a competitive position later.

Here is an example relevant to students. When students turn in homework and reports, some use the old paradigms of writing by hand or typewriter, hopefully (from the instructor's point of view) turning in a second or third version that is reasonably neat, well-organized, and has most spelling mistakes eliminated. But today, these students are competing with students who are using the new report-writing paradigm—a word processor. These students can go through ten or more rewrites, editing, and corrections; they can use "spell-checker" and other tools to put out a higher-quality product. Instructors cannot help but give such a report a higher grade, everything being equal, just because it is much easier to read and thus takes less time to grade. But no matter how good a hand-written report is, it could be improved with the new paradigm because it will be easier to proofread for grammar, spelling, and logic for the writer as well. Now at the very beginning, when switching from hand-writing to word processing, efficiency is lost because learning requires an initial investment in time and effort. Even later the gain may not necessarily be an advantage in time, but definitely an advantage in quality. As we shall see, quality is no longer a fixed standard but has expanding boundaries. Writing with the computer now requires desktop publishing skills.

 Five-Minute Activity 1-6: Paradigm Shift

Depending on your interest, select one of the following:

a. With a teacher or student, discuss where along the paradigm curve you would place the educational systems in your community. Have some paradigms already been discovered through tinkering or breakthrough thinking by people who did not follow the "rules"? Are these new paradigms being adopted by paradigm pioneers?

b. With a colleague in your line of work, identify a paradigm shift that has occurred in your organization or in your industry within the last ten years. Discuss what happened. Describe how easy or difficult it was to make the change.

c. With another person, select an area of technology and innovation that is personally impacting you. Sketch a paradigm shift curve and mark the position of your chosen subject. If it is located in Segment A on the curve, discuss the specific advantages and disadvantages of having made the jump. If it is located in Segment B, discuss the possibilities of new paradigms and how they may be discovered. If it is located in Segment C, describe what changes must be made to shift to a new paradigm already being pioneered. Ideas: Are you thinking about getting a home fax machine? Are you using roller blades to get around town? Does the no-smoking rule on airplanes annoy or please you?

Some examples of paradigm shift are very striking, such as going from donkey cart or horse-drawn carriage to the automobile, or traveling long distances by airplane instead of bus, rail, or ocean liner. In real-time communications, paradigm shifts have made it possible to send increasingly detailed and accurate messages over greater and greater distances as people have progressed from shouting, smoke, fire, drum, and flag signals to electrically transmitted impulses such as the telegraph, telephone, fax, and live video by wire, optical fiber, and communications satellite. Note that paradigm shift is different in scope from continuous improvement. Two hundred horses hitched together, no matter how powerful and fast, cannot get a carriage to go from zero to 50 miles per hour in 5 seconds, although they represent the same horsepower as a modern automobile engine. Continuous improvement certainly results in improved problem solving, but innovation comes with a paradigm shift or creative leap. However, what is often overlooked is the benefit of an attitude of continuous improvement: it prepares the mind to recognize good ideas and to become a paradigm shifter or paradigm pioneer.

Creative thinking is the key in all phases of paradigm shift. Creativity is exhibited by paradigm shifters and their innovative ideas. However, to be a paradigm pioneer able to adopt, adapt, and take advantage of the new rules also requires creative thinking, because taking risks, changing directions, and developing and following a vision take flexibility and feeling comfortable with change. Creative thinking will let us recognize good ideas in others, so we can support them and take early advantage of the opportunities they represent. Joel Barker, in his most recent book *Future Edge*, gives the example of creative thinking and paradigm shift in a hand-grenade company. Its president noticed that air bags in cars "go off on impact" and "blow up." He perceived a potential application for his company's expertise. His engineers in less than a year developed a trigger for an automobile air bag that would cost less than $50. The trigger is by far the most expensive component in the air bag system, which at present costs around $600. Can you guess the reaction the hand-grenade people received when they presented their innovative trigger idea to engineers at one of the Detroit automakers? They were sent away to look for interest elsewhere. Toyota in Japan and Jaguar in Europe are now testing this invention.

To prevent us from having an inflexible mind that is incapable of recognizing a coming paradigm shift, we must develop a habit of frequently asking ourselves the paradigm shift question posed by Joel Barker.

THE PARADIGM SHIFT QUESTION

"What is impossible to do in my field or organization, but if it could be done, would fundamentally change what I do?"

If you look at your education or your career in these terms, trends and developments will not surprise you or pass you by. You will be prepared to become a paradigm pioneer and take advantage of opportunities to innovate. Indeed, we live in a world that is changing rapidly, and in times of change and paradigm shift, we need creative thinking to cope, to succeed, to lead. In the next chapter, you will be introduced to a paradigm for learning and memorizing that may be very unfamiliar to you—visualization. Chapter 4 presents practical ideas on how to overcome mental blocks and paradigms that keep you from thinking creatively.

Creativity—like sugar—makes life sweeter.
Wendy Gallant, engineering student.

Learning is the new form of labor.
Shoshana Zuboff, professor of organizational behavior
and human resource management, Harvard Business School.

Jeder Tag muss das Denken verändern. Each day must change your thinking.
Hans Erni, Swiss artist, in an interview on his 85th birthday.

RESOURCES FOR FURTHER LEARNING

Learning takes time and effort—it is not a passive activity. And there is no end to learning. In high school, students still have the perception that teachers are responsible for making sure the students learn the subject. In college, instructors and professors are responsible for presenting the subject in a way that makes it interesting and challenging, for knowing the subject well, for discovering new applications and new knowledge. It is primarily the students' responsibility to learn, to find their way around, to buy the textbooks and use them, to seek out additional resources like computer labs, software, seminars, and special speakers.

To keep up with the rapid changes in technology and the growth of knowledge, everyone has to develop a habit of lifelong learning. Because technical knowledge becomes outdated quickly, continuous learning is crucial for people in many fields if they want to retain their professional skills. Also, when we continue to learn new things as we get older, we remain good thinkers; our brains do not atrophy. Mental activity and learning are as important to mental fitness as regular exercise is to physical fitness; both are needed for a healthy, productive life. For learning more about creative problem solving and related subjects, we recommend the following resources; ideas from each of these sources have been incorporated into this book. Additional books and materials are listed at the end of each chapter.

Once you have acquired a habit of exploration, creative ideas can be found almost anywhere—not just in all types of books, but in professional journals, news-magazines, and the daily paper; in conversations with children, your friends, and relatives; and especially as they surface in your own mind. If you are not a good reader, you can still explore new subjects. Check out audio- or videotapes, or even films, from libraries and bookstores. KEEP LEARNING!

References

1-1   James L. Adams, *The Care and Feeding of Ideas—A Guide to Encouraging Creativity*, Addison-Wesley, Reading, Massachusetts, 1986. This book gives many interesting facts about the workings of the human brain and a realistic view of its capabilities and limitations.

1-2   James L. Adams, *Conceptual Blockbusting—A Guide to Better Ideas*, third edition, Addison-Wesley, Reading, Massachusetts, 1986. This book includes thought-provoking sections on mental "languages" and includes good examples and illustrations.

1-3   *American Heritage of Invention & Technology*, published quarterly by Forbes (with General Motors as its sole sponsor and advertiser). Each issue of this magazine has articles about the history of technological innovation and stories about inventors and their creative achievements.

1-4   Joel A. Barker, *Future Edge: Discovering the New Paradigms of Success*, Morrow, New York, 1992. The book includes powerful messages about overcoming the resistance to change and creating an innovative environment from his two *Discovering the Future* videotapes: *The Business of Paradigms* and *The Power of Vision*, Charthouse International Learning Corporation, 221 River Ridge Circle, Burnsville, Minnesota 55337. You may be able to borrow these very expensive tapes for educational purposes from a University Extension Service or from a large corporation.

1-5   Edward de Bono, *De Bono's Thinking Course*, Facts on File, New York, 1982. The tools and techniques of good thinking are outlined in a practical, easy-to-follow manner.

1-6   John Fabian, *Creative Thinking & Problem Solving*, Lewis Publishers, Chelsea, Michigan, 1990. This book targets scientists, engineers, and project leaders. His breakthrough discovery process has four phases: determine the target, search for options, check for fit, and take action, thus putting creative problem solving into a somewhat different framework and vocabulary.

1-7   Ned Herrmann, *The Creative Brain*, revised edition, Brain Books, Lake Lure, North Carolina, 1990. This beautiful, creative, large-sized book presents the four-quadrant model of thinking preferences. It contains a multitude of useful applications in education, management and productivity, communications, and personal relationships. It is especially recommended for managers who want to build a creative climate in their organizations and to anyone who wants to gain insight into his or her own thinking patterns. A prepaid diagnostic tool, the Herrmann Brain Dominance Instrument (HBDI), is included in the price of the hardcover edition.

1-8   Scott G. Isaksen and Donald J. Treffinger, *Creative Problem Solving: The Basic Course*, Bearly Limited, Buffalo, New York, 1985. This softcover workbook emphasizes the cycles of divergent and convergent thinking in the five steps of creative problem solving: mess and data (fact)-finding, problem-finding, idea-finding, solution-finding, and acceptance-finding. We did not run across this work until after the first edition of our book was published. It includes a brief history of the development and research in creativity and problem solving.

1-9   Don Koberg and Jim Bagnall, *All New Universal Traveler*, Kaufmann, Los Altos, California, 1981. This softcover book is a veritable "horn of plenty" for creative ideas, approaches to solving problems, and processes of reaching goals, all presented within a travel metaphor.

1-10   Michael LeBoeuf, *Imagineering—How to Profit from Your Creative Powers*, Berkley Books, New York, 1980. This practical paperback shows how to motivate yourself to be more creative.

1-11   Robert H. McKim, *Experiences in Visual Thinking*, second edition, PWS Publishers, Boston, Massachusetts, 1980. This softcover book contains many exercises for flexible thinking; it is used as text in a first-year engineering course at Stanford University and is highly recommended.

1-12   Ron Meiss, *Total Quality in the Real World: From Ideas to Action*, Ron Meiss & Associates, Kansas City, Missouri, 1991. This step-by-step interactive TQM manual lets you examine the parts of total quality and determine how they apply to your organization. The book is easy to read and contains many practical and motivational ideas that can be applied immediately.

1-13   Alex F. Osborn, *Applied Imagination—The Principles and Problems of Creative Problem-Solving*, third revised edition, Scribner's, New York, 1963. This classic work by the "inventor" of brainstorming is well worth reading.

1-14   Arnold Pacey, *Technology in World Civilization—A Thousand-Year History*, MIT Press, Cambridge, Massachusetts, 1990. This is an excellent presentation of the development of technology in many cultures, as well as the cross-cultural flow of ideas that can lead to creative thinking and innovations, in both large-scale industrial and appropriate "survival" technologies.

1-15   Arthur B. Van Gundy, Jr., *Managing Group Creativity—A Modular Approach to Problem Solving*, American Management Associations, New York, 1984. This contains rather technical discussions of various brainstorming methods.

1-16   Roger Von Oech, *A Kick in the Seat of the Pants*, Harper and Row, New York, 1986. This softcover book plays around with four mindsets used in creative problem solving. It takes you through the entire process from thinking up creative ideas to implementing them.

1-17   Roger Von Oech, *A Whack on the Side of the Head*, Warner Books, New York, 1983. This fun book on creative thinking and mental blocks comes with the *Creative Whack Pack*, a deck of topical cards to encourage people to think and play with ideas in new ways as they solve problems.

DIAGNOSTIC TOOLS AND EXERCISES

1-1 DIAGNOSTIC QUIZ: HOW CREATIVE ARE YOU?

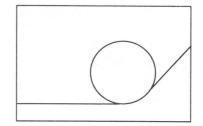

a. *Briefly describe the processes or approaches you use most frequently to solve (1) math problems, and (2) "life" problems.*

b. *Briefly summarize your previous training (if any) in creative thinking and brainstorming.*

c. *In a few sentences, describe what you expect to get out of studying creative problem solving.*

d. *What do you see in the figure on the right? You can give more than one answer.*

1-2 TWO-QUESTIONS QUIZ: WHAT HAVE YOU LEARNED?
(a) What are the most important ideas you learned from this chapter? (b) What is an important question you still have?

1-3 ANALYZE YOUR STUDY HABITS
Think over your past experience with school. Which subjects or learning experiences were the most enjoyable? Are you creating a learning environment for your homework and studying that will make it easier for you to learn any subject? For example, if you are people-oriented, do you get together with like-minded friends and do some serious group study? If you have difficulty with study habits or study skills, review the material in these references:

- •David B. Ellis, *Becoming a Master Student*, College Survival, Inc., Rapid City, SD 57709, 1985 (344 pages).
- •Marcia K. Johnson et al., *How to Succeed in College*, William Kaufmann, Inc., Los Altos, CA 94022, 1982.
- •Uelaine Lengefeld, *Study Skills Strategies: Guide to Critical Thinking*, Crisp Publications, Los Altos, CA, 1986.
- •Claude W. Olney, *Where There's a Will There's an "A"* (library videotape or cassette plus manual).

1-4 SETTING GOALS
Why are you studying this book—what do you expect to learn? After reading the introduction, make a list of some personal benefits that you would like to gain by learning creative thinking and problem-solving skills. Also make a list of some short-term goals (one year or less) and long-term goals (five to ten years) in your life. Be sure to include these aspects of a balanced life: spiritual, family, career, social, self, health, leisure, money. Do you think that creative thinking and problem-solving skills will help you achieve your goals? Now consider your weekly schedule. Have you set aside sufficient time to study this book and do practice problems? Are you incorporating into your weekly schedule activities that are related to your accomplishing your short-term as well as some of your long-term goals? If you need help with time management and goal-setting, go to Section 5 in Chapter 8.

1-5 READING LIST
Make a list of books that especially interest you on the subject of creativity and problem solving— books that you may want to read when you have time. Check the library for additional titles. If you are not yet familiar with the computer system at your library for doing a search, put this at the top of your list of things to do. Make an implementation plan—schedule the steps and time needed to get an interesting book as well as sufficient free time for reading at least a portion of the book. Enjoy!

1-6 OUTSIDE MATERIALS RELATED TO CREATIVITY AND LEARNING
During an entire week, pay attention to anything that relates to the topics presented in this chapter. Do you hear a TV news report mention some government action regarding the quality of education in this country? Do you notice a newspaper or magazine article about a creative learning project being done in an inner-city school? Do you participate in a discussion on the influence of computers on learning? Make a folder with clippings and notes.

> *If people viewed your mind as an open book, would they see blank pages? Improve your mind! Read!*
> Charles F. Nopper, Jr., engineering student.

1-7   LEGO™ "MACHINE"
As a creative thinking exercise, build a vehicle or machine from Lego tiles.

1-8   NAME TAG AND LOGO
Design a creative name tag or logo for yourself, incorporating some design or symbol that expresses something meaningful about yourself and your interests.

1-9   WORD PROCESSOR
If you do not yet know how to use a word processor, decide that it is high time for you to make this paradigm shift. Identify a short course or class that you can take. If you are in high school or college, investigate the computer labs available to students (type of equipment and hours). Try out different computers, then "stake out" a model that seems comfortable to you. Experiment with different word processing software, such as MacWrite, WriteNow, or Microsoft Word. Begin to use the computer immediately for your writing assignments. Write a one-paragraph description on how you got started in word processing (or why it took you so long).

1-10   *CREATIVITY BULLETIN BOARD*
Over the course of a few weeks or months, make up a bulletin board with comics, cartoons, jokes, and puns that illustrate creative thinking and give a positive message about learning.

1-11   MERCURY FEVER THERMOMETER SHAKEDOWN
Imagine that you have a fever and that your hands are bandaged. Your doctor wants you to chart your temperature every two hours, but you are unable to get a firm grip on the mercury thermometer to shake it down (the only instrument at hand). You are home alone during the day. How would you solve this problem? Hint: You can brainstorm ideas with friends—try to come up with at least five different and unconventional ideas.

1-12   *SKILL STICKS® STRUCTURE (GROUP PROJECT)*
Skill sticks are available in hobby shops in packs of 150 for approximately $2; they are thin, wooden, prenotched sticks that can be assembled whole or easily broken into shorter pieces. Make bundles of 20 sticks each (with a rubber band); each person participating in this activity receives one bundle. Have a tape or ruler to measure results. The assignment is to build the tallest self-supporting structure. Students can discover two concepts: Sometimes we have to break a "whole" into parts in order to be able to use it in a new way, and the assignment can be solved much better when people work together. Set a time limit (say 10 minutes) to determine the winning team. Research has shown that students who experiment and "fail" a lot in the early stages of such a project come up with the best results in the end. Be prepared for some unusual solutions—we recently had a creative person who built a self-supporting structure downward. This project can be extended by asking teams of five people to build the tallest and strongest tower with 200 sticks and 10 thin rubber bands. The towers can be fastened to plywood bases and tested on a shaker table. The tallest structure that withstands 4 minutes of shaking (at the lowest setting) without damage wins. Before the experiment begins, ask what the builders expect will happen to their structures during the test. Some principles of earthquake-proof design can be investigated and taught with this experiment.

1-13   *TINKERTOY INVENTION (GROUP PROJECT)*
Divide a class or group into teams of three people each. Hand out a handful of tinkertoy pieces selected at random for each team. The teams are to invent a model of a new and useful product with two moving parts, come up with a name for the product, and do a "sales" presentation. Give them 20 minutes to complete the assignment.

---

*Asterisks denote advanced exercises requiring extra time but providing more in-depth learning.

1-14 *DIAGNOSTIC TOOL: GROUP CREATIVE THINKING*

Hand out a piece of plain notepaper to each person in a group. The assignment is to enclose the largest area possible with the paper. Do not give any other explanations. Observe what will happen. People will look around and perhaps timidly begin talking with others. Some may start folding and tearing their sheet of paper. Some may ask questions, "Can we team up; can we use scissors and tape?" The answer to both questions is, "yes." One good solution to this problem is for the group to cut or tear each paper into a spiral and then to fasten all these strips together into a long loop. The group—to complete the assignment—will then have to go outside the room and "enclose" the building, parking lot, or a field. Or they can do a theoretical calculation of the largest circle that could be enclosed with the combined strips. What makes this assignment difficult is that very few people will solve the problem by working together. Write a brief summary of your experience with the assignment and how the group solved the problem. What have you learned from this activity?

1-15 *THE CAMEL AND THE BANANAS*

A man has 2000 kg of bananas to take to market in a town 100 kilometers away. His only means of transport is an ornery camel who can carry 100 kilograms of bananas at a time; unfortunately, this peculiar beast eats one kilogram of bananas per kilometer when carrying the banana bundles. Is there a way that the man can get some bananas to market? If you can figure out how this could be done, about how many kilometers total will the camel have to travel, and about how many kilograms of bananas will end up in town?

1-16 *ICE-BREAKER ACTIVITIES*

The following problems are fun to do when you have a group of strangers you want to get talking to each other. Within a specified amount of time, say 10 minutes, assign two or three different (nonconsecutive) problems to each team of three or four people, taking into account that the last two problems are more difficult than the others. When the time is up, a spokesperson from each team can introduce the team members and explain the team's solutions.

a. *The sum of two whole numbers is 24. One of the numbers is three times as large as the other. What are the numbers? (Find at least one solution that does not use algebra.)*

b. *The sum of two coins is $1.10. One of the coins is not a dime. What are the two coins?*

c. *You have a cold and have suddenly and completely lost your voice. How would you let your family members know that they should answer the phone and take a message for you? (Find at least three very different solutions.)*

d. *A year has 12 months, and some months have 30 days and some have 31 days. How many months have 28 days?*

e. *History is strange—they have a 4th of July in the Republic of Tajikistan. Explain.*

f. *What happens once in a minute, twice in a moment, and never in a year?*

g. *The butcher at Tony's Meat Market is 41 years old, 6 feet 2 inches tall, wears a size 17 shirt and a size 12 shoe. What does he weigh?*

h. *Can a man marry his widow's sister? Why or why not?*

i. *How many letters are in Mississippi?*

j. *How many people are in your entire group or class? If each person shook everyone else's hand just once, how many handshakes would it take? How would you calculate this, and can you think of a shortcut?*

k. *Imagine that you have eight blocks—seven weigh exactly the same; one is just a little bit lighter. Your company wants to have zero defects to maintain its reputation for quality. You want to be efficient and not waste time weighing each block to find the defective one. You have heard that it is possible to find the odd block with just two weighings. Prove that this really can be done.*

CHAPTER 1 — SUMMARY

The Purpose of This Book: People need creative thinking skills to cope with and succeed in a rapidly changing world. This is a practical book with many exercises and applications. The only prerequisite is an open mind. Creative thinking can be taught; everyone can benefit from learning these skills.

Benefits of Creative Problem-Solving Skills: Creative problem solving is needed to make up for shortcomings in our educational systems since we are mostly taught plug-and-chug approaches to problem solving. Also, to properly use the capabilities of computers, we must expand our creative problem-solving abilities. Creative problem solving is needed to enable manufacturers and businesses to provide quality products and services of value in the global marketplace. *Personal benefits*—Enriched life, successful career, and balanced thinking. *Benefits for business and industry*—Through teamwork, meeting customer needs, and continuous improvement, companies innovate and compete worldwide. *Benefits for society*—Solutions to many serious economic, social, and environmental problems are found when people work together creatively.

Definitions: *Creativity* is playing with imagination and possibilities, then making new and meaningful connections while interacting with ideas, people, and the environment. This process results in a product or application that in turn will encourage more creativity. It can happen within an individual, as interaction between people, or as idea transfer between widely separated cultures. A *problem* is anything that can be made better through some change—it usually involves aspects of danger and of opportunity. To solve a problem well, we need to use analytical, creative, and critical thinking in the most appropriate sequence.

Problem-Solving Schemes: Math, science, industry, and engineering methods are analytical; creative thinking (from psychology) is a subconscious process. *Creative problem solving* combines aspects of all the other approaches. Divergent and convergent thinking are used in steps with different mindsets: the explorer and detective for problem definition; the artist for idea generation; the engineer for creative idea evaluation and synthesis; the judge for idea judgment and decision making; the producer for putting the best solutions into action.

Change and Paradigm Shift: Despite major changes that have happened in the last twenty years, people who are comfortable or successful often resist change or are unable to recognize a paradigm shift. "A *paradigm* is a set of rules and regulations that defines boundaries and helps us be successful within those boundaries, where success is measured by the problems solved using these rules and regulations" (Joel Barker). The best time to seek new paradigms is while the current paradigm is still useful in solving problems. Through creative thinking, paradigm shifters come up with new ideas to solve "impossible" problems. Through creative thinking, paradigm pioneers recognize the value of these new ideas and take the risk to adopt them—often on faith since little proof exists that this will really work.

Keep Learning! Learning is a lifelong occupation, duty, joy, and adventure. Take the time to discover, practice, and apply new knowledge and skills.

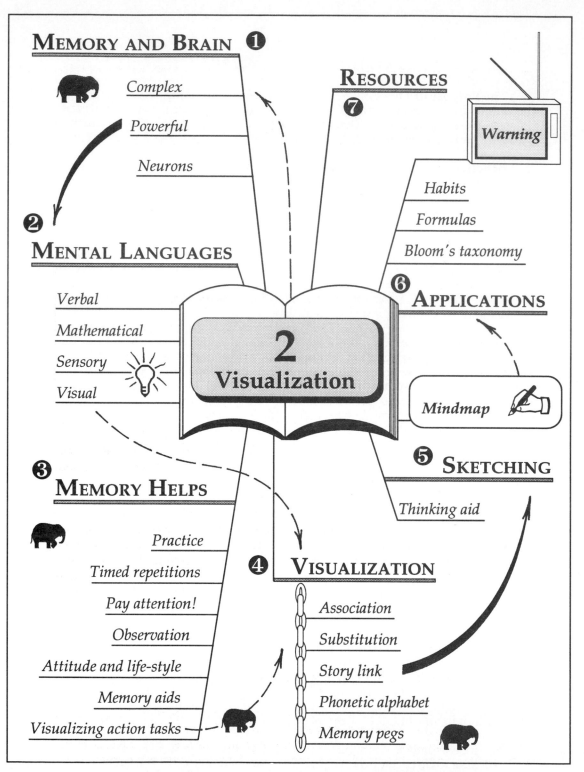

MEMORY AND BRAIN ❶

Complex

Powerful

Neurons

❷

MENTAL LANGUAGES

Verbal

Mathematical

Sensory

Visual

RESOURCES

❼

Warning

Habits

Formulas

Bloom's taxonomy

❻ **APPLICATIONS**

Mindmap

2
Visualization

❸

MEMORY HELPS

Practice

Timed repetitions

Pay attention!

Observation

Attitude and life-style

Memory aids

Visualizing action tasks

❹ **VISUALIZATION**

Association

Substitution

Story link

Phonetic alphabet

Memory pegs

❺ **SKETCHING**

Thinking aid

Mindmap of Chapter 2. Refer to Chapter 2 (pp. 55-56) to learn more about mindmapping.

2

VISUALIZATION AND MEMORY

WHAT YOU CAN LEARN FROM THIS CHAPTER:
- *Memory as a complex function of the brain.*
- *Mental "languages": verbal, mathematical, visual, and sensory thinking.*
- *First steps for improving your memory.*
- *Visualization and memory techniques: association, substitute word, story link, phonetic alphabet, and peg system.*
- *Sketching, a tool for visualizing and thinking, mindmaps and diagrams.*
- *Applications: Bloom's taxonomy, formulas, changing a habit, and watching television.*
- *Further learning: books and other resources; exercises and activities; summary.*

Our educational system concentrates on teaching students information and facts about many different subjects. Many tests (often multiple choice) then ask students to recall this information. The first step in learning involves memorizing information—we have to make "deposits" into our memory banks to enable us to do higher levels of thinking. The main technique used in schools to get students to memorize is through rote learning and repetition. The mental activity of visualization is ignored, yet this is a very powerful tool for remembering because of the way the brain functions. In this chapter, we will explore the connections between memory, visualization, thinking, and learning. Powerful memory tools based on visualization will be demonstrated. And you will see how sketching and other graphing techniques can be used to improve your thinking processes.

MEMORY AND THE BRAIN

Scientists used to think that people commonly used only about 10 percent of the brain's capacity. However, researchers at the University of California-Los Angeles now estimate that we use less than 2 percent. The memory capacity of the latest, most powerful computers is only about one four-millionth of a human brain, and to build a computer with the capacity of a human brain (at 1988 prices)—if that were possible—would cost roughly twenty trillion dollars. Simply put, all the computers existing in 1987 had a combined capacity of about one-quarter of one average human brain. Although computation is one specialized task that computers can do much faster and more accurately than the human brain, we have an amazing array of unbelievably complex thinking abilities.

The major actors in the brain and nervous system are the neurons. Neuropsychologists see the neuron as an independent and unique cell which is not physically connected to other neurons. Think of each neuron as being an information processing system. It is estimated that a human brain consists of as many as 180 billion neurons. A neuron is not an average-looking, smooth, round-as-a-ball cell. Instead, it has a very large number of tentaclelike protrusions called dendrites, which make it possible for each neuron to receive synapses (signals) from as many as 1,000 to 15,000 neighboring neurons. Some of the synapse connections between neurons are determined by genes, but a large majority are made by experience. In the first five years of life, these connections occur most easily. However, there is no evidence (short of some diseases like Alzheimer's) that we lose our ability to make new connections as we get older. Scientists think that the number of possible connections between neurons in a human brain exceeds the number of atoms in the known universe. Neurons differ from other cells in the body in that they do not replicate themselves.

When does a physical brain—which in a human adult is about the size of a grapefruit weighing three pounds—become a thinking mind? This happens through the complex interactions between the neurons in chemical and microelectrical processes that are not yet understood. However, it has been determined that much of our knowledge, attitudes, and abilities are wired in and controlled by neural networks that have been determined by experience. When people say that they are too old to learn a new skill or subject, it is because they are no longer willing to spend the time and effort. Think of how long it takes to learn math in school. If we are to learn a similar amount of new material at a later age, it would take just as long. This is why learning new habits and problem-solving skills takes a considerable effort, since we have to establish new connections in our neural networks that will override our old habitual patterns.

In his book *The Care and Feeding of Ideas*, James L. Adams uses these words (on pp. 53-54) to describe the connection between our beliefs about creative thinking and the work required:

> Conscious effort is necessary to pursue new directions. Perspiration is, in fact, an excellent investment. … Perhaps the most common inhibition to creativity is our usual reliance upon traditional problem-solving routines and the fantasy that creative problem solving should be easier, rather than more difficult, than producing answers to routine problems.

The brain is the last and greatest biological frontier;
it is the most complex thing we have yet discovered in our universe.
James Watson, codiscoverer of the double helix in DNA.

There are profound and important differences between information and
knowledge, but the nature of those differences is still obscure.
Jeremy Campbell, The Improbable Machine.

Memory is the vital source of all aspects of human intelligence, imagination, and accomplishment.
Jack Maguire, Care and Feeding of the Brain.

As reported in the April 20, 1992, issue of *Newsweek,* powerful new devices are being developed that peer through the skull to "see" the brain at work. Computer-assisted tomography (CAT) and magnetic resonance imaging (MRI) can distinguish minute brain structures; positron emission tomography (PET) and single-photon emission computerized tomography (SPECT) track blood flow, a sign of brain activity, whereas superconducting quantum interference devices (SQUID) pick up magnetic fields, another mark of brain function. An "ecoplanar" MRI machine being developed at Massachusetts General Hospital will be capable of snapping pictures of the brain at 45-millisecond intervals. Brain researchers are preparing to make visible the thoughts, feelings, and memories as they blink on and leap from one tiny clump of neurons to others elsewhere in the brain, unfolding or exploding into ideas, a creative leap, or a unique insight.

These research tools are finding that memory is much more complex than previously thought. Different types of memories are located in different parts of the brain, and the way something is learned seems to determine the location of the memory. A face or other information "encoded" in a visual image is remembered in the region of the brain concerned with spacial information processing. If you have learned about hammers by touching and using one, this knowledge and memory is preserved in the part of the brain that was involved in the original experience—those that control movement and touch. Skill memory (how to do something) is quite different from semantic memory (learning isolated facts and ideas) which again is quite different from what is called episodic memory (experiences involving complex circumstances and settings). Recognition is different from recall: you may understand many more words in a foreign language than you are able to use for speaking.

It is tempting to draw analogies between computer memories and the human brain. When talking about the short-term or "scratch-pad" memory, we can liken it to the work on a computer screen. Unless it is "saved" into the hard-disk or long-term memory, it can disappear at the touch of a button. When you look up a telephone number in the book and then dial the number, you are using short-term memory. If the number is busy and you have to redial, you will most likely find that you have forgotten the number and will have to look it up again. The short-term memory is subject to the primacy and recency effects: when presented with strings of information, it will remember those items at the beginning and at the end best. However, human minds have many properties that are very much different and far more complex than computers. Human memory may be less reliable than computer memory, but it is much vaster. But most importantly, the human mind is not just an information processor; it is able to associate things in many different ways, and these connections can be very unexpected and creative. Basically, the brain is experience-based, not a logic machine. When our brain learns "formulas" and "scripts," it not only can use these to help us navigate life efficiently, it is also able to adapt, change, and move these scripts around, either in response to changing circumstances or in response to imagination. Unfortunately, many schools primarily teach passing tests and plugging into the formulas, not exploring different ways of thinking, changing the scripts, and being flexible in the creative use of knowledge. In the next section, we will explore some of these different ways of thinking.

MENTAL "LANGUAGES"

Imagine the following scenario: It is your sister's birthday, and you are giving a party at your house. The doorbell rings, and when you open the door, a friend enters carrying a small gym bag. This friend—being something of a joker—announces that your sister must guess what the gift is by drawing it from your description. But you are to be blindfolded, and then you are invited to touch the gift by inserting one hand into the bag. As you explore the mystery object (which seems to have a somewhat irregular geometric shape), you are asked to give a running commentary about its attributes, without naming the object. Although this sounds like an easy exercise, you will find it in most cases surprisingly difficult. It is not a simple task to identify shape by feel, and it is even more difficult to describe this type of an object verbally. And making a sketch of the mystery object will be a very baffling task for most people, because we are not taught good sketching skills.

In our culture, great emphasis is placed on verbal and written language skills. We have only to think of the typical aptitude or I.Q. test. But verbal thinking, which is linear and sequential, is not the only thinking mode suitable for problem solving. When we are dealing with quantities, a verbal approach becomes very complicated, whereas a mathematical approach can easily solve the problem. Another thinking mode—visualization—is now receiving increased attention. We need to learn to really "see" things around us; we need to practice imagination (or making "mental pictures"), and we need to develop our skills in graphically representing our ideas. When we sketch images and make diagrams about data and idea relationships, we help our thinking processes—seeing, imagining, sketching, and thinking are all connected. Sketching also helps in communicating ideas and information to others. Which can you remember better, directions to a certain location conveyed verbally or directions given with a roughly sketched map?

A well-armed problem solver is fluent in many mental languages (verbal, mathematical, visual, and sensory) and is able to use them interchangeably.
James L. Adams, Conceptual Blockbusting, *third edition.*

When I started teaching and did demonstrations,
I found that I could either talk or draw, but I couldn't do both at once.
Betty Edwards, Drawing on the Right Side of the Brain.

Mathematical thinking can be enhanced through visualization. We need to develop a "feel" for size and quantities and how they are related to each other. For example, what is the difference between one million and one billion, or one billion and one trillion? If we use time as a measure of comparison, one million seconds is approximately eleven and a half days; one billion seconds is almost 32 years, and one trillion seconds is over 30,000 years, long before the rise of human civilizations. How do viruses, atoms, and protons compare to each other in size? A virus compares to a person's size in the same proportion as a person relates to the size of the Earth. An atom is to a person as a person is to the Earth's orbit around the sun. And a proton is to a person as a person is to the distance from Earth to its nearest star, Alpha Centauri. Which is the smallest: virus, atom, or proton, and are you able to get a "feel" or idea as to the scale of comparative "smallness"?

It is useful to develop a mental image of some common, large quantities. Get two rolls of pennies or dimes and experience what "100 of something" looks and feels like. Examine a brick wall in your neighborhood—how large a wall area contains 1000 bricks? What is easier to compare: a quarter-inch segment on a line that is nine yards long, or the area of a small, dark fingerprint on a standard-size white sheet of paper, or one cubic centimeter (about one teaspoon) in one liter (or one quart) of milk? Each of these represents roughly one part of one thousand.

⌚ *THREE-MINUTE ACTIVITY 2-1: VISUALIZING QUANTITIES*
With two people, brainstorm different ideas on how to visualize 1,000, 10,000 , 100,000, and one million. Then select the best ideas and write them down here:

1,000 can be visualized as _____

10,000 can be visualized as _____

100,000 can be visualized as _____

1,000,000 can be visualized as _____

An extension of mathematical thinking combined with visualization is *scientific visualization*. This is the visual display of multidimensional, time-dependent, or complex information. Scientific visualization applies computer software to interpreting scientific and engineering data from physical and chemical interactions of materials and processes. Visualization through computing is used to increase productivity as well as prepare the way for major scientific breakthroughs. Research in this area began in disciplines such as medical imaging and then expanded into fields such as mechanical computer-aided engineering where algorithms have been developed and are now commercially available to display three-dimensional behavior of solids, allowing engineers to better understand what is happening inside the modeled solid under different simulated conditions. Some of these techniques can be combined with animation for dynamic analysis. Chapter 9 demonstrates how computers can visualize mathematical relationships.

For creative thinking, other sensory modes besides visual thinking are important also. We need to be aware of input from our senses of smell, touch, taste, and hearing, as well as from muscular sensations in our body. This information may have a direct bearing on the problem that needs to be solved, whether you are inventing a prize-winning recipe, devising a marketing strategy for a new toothpaste, designing a piece of furniture, or investigating why a baby is crying. And a physical environment that includes pleasant sights, sounds, and smells, as well as different touch and kinesthetic sensations indirectly stimulates and enhances brain function for creative thinking because these stimuli are processed primarily in the right hemisphere of the brain. Attaching sensory information to things that need to be remembered is a powerful memory tool. Watch a few television commercials and note how much sensory information is conveyed visually, verbally, and with sound effects and rhythm to help you remember the product being advertised.

> *The next time you're stumped, build a tree in your mind.*
> *Brandon Slotterbeck, engineering student.*

FIVE-MINUTE ACTIVITY 2-2: DUCKS AND LAMBS
Use different mental languages to solve this problem in at least three different ways:
A farmer's child got a present of 8 animals (ducks and lambs) with a total of 22 legs.
Determine the number of ducks received.

MEMORY—FIRST THINGS FIRST

Most adults when polled will say that they have a bad memory, and many will blame this on aging. However, most people's memory (or brain) can function much better than they think. They simply have never learned to use the best techniques that work *with* the brain and its design. Not using these tested methods is like ignoring the operating manual to a complicated piece of machinery.

Practice new techniques: The techniques that have been developed to enhance memory are all very effective. But they are rarely used because most of us prefer to rely on the habits we already have rather than learn new ones. Learning something new takes effort because we have to make new neural networks in our brains. Initially, thinking in a new way takes much more energy than after these paths have become ingrained, as shown on MRI scans. Recent research found that the human brain is much more plastic—changeable—than was believed earlier. Thus you can use this dynamic characteristic of the brain to intentionally alter the way it is organized and functions. When you decide to develop and practice certain mental or physical abilities, you are changing the physical structure of the brain. Once you have built in a structure, it will help you think or remember in this particular new way. Repetition reinforces learning because it strengthens the new structures and connections in the brain.

Whatever you practice, you will perform.
Jerry Lucas, memory expert and gold-medal winning basketball star.

Practice is the best of all instructors.
Publilius Syrus, Maxim 469, First Century B.C.

Time repetitions: How do you study? Do you cram intensely the night before an exam? Research has shown that you will remember much better if you look over the material you need to memorize, then take a break before looking over it again. Next time, double the elapsed time before going back to this material. Keep doubling the off time before reviewing. This repeated exposure will fix the material in your mind much more solidly than a single, stressed time of "hard" study. When you use neural connections repeatedly, they are strengthened and make retrieval of the material easier—material in our long-term memory that is not firmly connected is very difficult to retrieve. With correct timing, the repetition method works more efficiently. As a minimum, review important new information within 10 minutes, followed by a review in 24 hours and another in one week. Most people will stop there (and most students will stop after taking an exam on the material). For long-term retrieval, review in six-month intervals—keep this timing in mind if you are facing a comprehensive exam in the future.

Pay attention: Information gathered by your senses is first recorded in the short-term or "scratch-pad" memory. Some of the information is then transferred to long-term memory. What is transferred is determined by the attention paid to the subject, as well as the context of the learning situation or experience. But when we are learning something, we are not usually involved at a highly emotional or sensory level. But we can influence what goes into our long-term memory by the degree of attention we pay to the matter to be learned. Interest, motivation, and the amount of previous knowledge already in the brain are also important to how much is transferred into long-term memory. The more we know, the easier it is to learn and connect new knowledge—the brain does this almost automatically. When we know very little about a topic, it is unlikely that we will retain casually presented information on this subject, unless we make a strong, conscious effort.

Also, if a new experience, fact, or situation is very much like something already in our memory, the mind will not pay much attention, and the new information will not be retained. Thus we must develop a habit of looking for differences, for something "odd" in the new input, to prevent forgetting. In addition, attention comes from the outside through strong stimuli. Intense experiences are unforgettable. The classic example is the John F. Kennedy assassination. People clearly remember what they were doing when they heard this shocking news: "It was a hot, sunny afternoon, and I was outside in the backyard, ironing this blue and white striped dress while listening to the radio." You probably remember birthday celebrations and other interesting happenings from years past more clearly than what you did on an uneventful Wednesday two weeks ago. Information is remembered in context. If a strong stimulus and a broad field of existing knowledge are absent, we have to make a conscious effort to *pay attention!*

Observe carefully: We not only need to be alert and pay attention (usually to verbally conveyed information), we need to sharpen our observation skills. Scan your environment. Do you notice things that are odd—things that do not belong in the particular context? Be on the lookout for things that are unusual, different, interesting. Observation differs from passive seeing; it gets you actively involved. Observation is the first step to enable you to think up useful questions. When you are curious and can ask questions on phenomena that you have observed, you will be able to remember the answers when they come. Good observation skills will enable your mind to make more and unusual connections which are helpful in remembering as well as in creative thinking.

Support your brain: Even before you make the effort to learn specific memory and visualization techniques, you can do a number of things to help your brain and your memory work better. To be good to your brain, you must be good to your body! Choose nutritious food and eat in moderation. Avoid stimulants, alcohol, and other drugs; avoid refined carbohydrates (white flour and sugar) as well as artificial sweeteners and foods high in saturated fats. Eat protein preceding work and learning to increase alertness; eat larger proportions of carbohydrates after work and study, since these tend to soothe the brain. Fish, soy, oatmeal, rice, and peanuts boost choline (which is a chemical precursor of the neurotransmitter acetyl-choline essential to memory). The folic acid in green, leafy vegetables helps improve brain function and learning. In general, food that is healthy for the heart and for cancer prevention is also good for the brain.

Exercise regularly and get enough sleep to maintain your health. Do the most difficult learning tasks during your peak mental time. Avoid distractions right after learning—sleeping on it will make recall easier. Create a supportive physical and social environment with little stress. Develop a positive (not a bored) attitude toward learning. Especially do not "program" your brain for failure by telling yourself that the subject is too difficult—instead, tell yourself that you *can* learn, because the results are solely dependent on your putting forth the necessary effort.

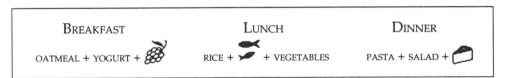

BREAKFAST	LUNCH	DINNER
OATMEAL + YOGURT + 🍇	RICE + 🐟 + VEGETABLES	PASTA + SALAD + 🍰

Use memory aids: On the other hand, who says that you have to remember everything? You can make life easier for yourself if you make use of memory aids. Go ahead, make lists; use alarm clocks and timers, appointment books, and daily "lists of things to do." Take notes. Use videos and computers as teaching and memory assistants. Use maps, charts, and other visual aids. Organize your desk so you can find things you use frequently in their assigned place. Develop a filing system and maintain it. Keep a journal or diary, and each evening jot down the most important happenings of the day. Post notes to yourself in strategic places.

Establish routines, such as a small table where you will place everything you need to take with you when you leave the house. If you frequently lose things, put your name on everything from umbrellas, caps, and gloves to books and pencils to increase your chances of getting them back. Make use of mnemonics.

Execute action tasks: There are three different categories that seem to cause the most problems with memory: (1) remembering how to do something (executing an action), (2) remembering facts, and (3) remembering to carry out an intention (going to do something). If you have to learn how to do some action in a certain sequence, you will remember it much better if you actually do it. If you think you will not be able to remember how to change a car tire in an emergency, go through a practice run. The sensory, visual, and kinesthetic information that will be involved in this action learning will reinforce your memory. Practice your piano scales. Practice the correct basketball moves or the gymnastics program. Bake the cake. Mix the paint. Grow the tomatoes. Build the book shelf. Pitch the tent; start the camp fire. Do the science experiment or lab work; use the new software. Also, you can reinforce action learning through visualization—rehearse the action step-by-step in your imagination. For example, learning how to fly an airplane is very expensive. You can cut down on the number of take-offs and landings needed by practicing in your mind the sequence of steps required at the controls. But visualization not only helps in remembering actions; it can also enhance memory for facts, and visualization is a key to remembering to carry out an intention. Thus the next section will teach several memory systems based on visualization.

🕐 *THREE-MINUTE ACTIVITY 2-3: IMPROVE YOUR MEMORY*

Among the first things first items discussed for memory improvement, select the top two that you think could really help you improve your memory. Then make a plan on how you would implement one change in study habit or life-style to reach your goal. Make a pact with a friend to help you put your plan into action during the next month—it takes a minimum of three weeks of steady practice to adopt a new habit.

VISUALIZATION—A WAY TO IMPROVE MEMORY

Visualization—or thinking in images—is very important to a good memory. This technique works because it is based on how the subconscious mind processes information. The scratch-pad memory can only remember about seven unrelated items, whereas in one visual image, the mind is able to link and store thousands of bits of information. Since the mind remembers unusual images best, we also must construct "memorable" images and "weird" linkages for best effect. Some of these techniques were already known in ancient times, when knowledge was transmitted from person to person and generation to generation through memory, not books. We will demonstrate five different methods: association, word substitution, the story link, the phonetic alphabet, and the memory peg system.

Association

In this technique we link the material we want to remember with something we already know well. Speakers in ancient Greece and Rome memorized the topics of their speeches by imagining a walk through their own homes. As they mentally walked from room to room, they associated in their minds the different points of the speech with different rooms and places in their home. However, for this approach to be effective, make sure that you make an *unusual* connection—make some unexpected change in your home environment as you make these connections in your imagination. This particular technique is known as "loci" or "places."

Association with a place—example: Three different concepts are involved for good memory. We first must understand the material, then we must file it, and finally we must be able to retrieve it. We want to use the technique of "places" to memorize the three concepts so that they will be remembered well and in sequence—say for a test several weeks later or for a speech on the subject. The house we are thinking of has a front porch with a roof supported by a post. Now make a mental picture of standing on the porch. The roof over the porch—in your mind—is a heavy slab supported by the post. On the side facing the street, in large letters, is chiseled the word MATERIAL. Now picture yourself standing under this slab of material. Do you have this image of *standing under the material* firmly in your mind? If your house does not have a front porch that can be mentally modified into a similar image, you must construct a different image, one that will have meaning for you. Now step into the entry. You want to take off your coat. But instead of hanging it in the closet as you usually do, imagine a huge filing cabinet in place of the closet. You pull out an enormous drawer and file away your coat. Can you imagine this activity—*filing* your coat? The third place in your walk through the house is the powder room. Imagine a large wash basin filled with water—a basin that grows even as you are thinking about it. On the water you see floating a large number of rubber ducks in all different shapes and sizes, representing ideas. Your task is to retrieve one particular red duck. Picture yourself in the act of stretching out your hand and reaching for this red duck. Your are *retrieving* the duck or idea.

Do you think you will now be able to remember the three concepts in order: understanding the material, filing, and retrieval? If you paid attention to what you were doing when you made the image link between the place and the item to be remembered, you will be able to "see" this association again at any time. When you are standing in front of an audience to give your speech, or if your mind suddenly goes blank during a test, all you have to do is think of walking through your house. The associated concepts will pop right into your conscious mind in the correct sequence. If you want to remember long lists of items, use a system where you will remember five items each (or ten items each) per room. Visualization is a powerful technique for retrieving information, because the visual image is like the code that will unlock the "safe" to the particular area of knowledge. If you understand some topic well, yet do not file the information effectively, you may have difficulty retrieving it later. Visualization will remind you of key words, but it will be your understanding that will fill in and let you use the information.

Remembering an intention—examples: Association works very well when you want to remember an intention, something you want to do at a later time. Let's say you wake up in the middle of the night and remember that you must take a certain book with you in the morning. You are too sleepy to get up and make a note, or you don't want to turn on the light and wake your spouse, roommate, or sibling. Instead, make a strong visual image of the particular book percolating in your pot of coffee, then go back to sleep. In the morning, when you are ready to pour the coffee from the pot into your cup, this visual image will pop into your mind. You can now place the book into your briefcase. Make sure that you build the association with an activity that you will always do, even in the case of oversleeping, and be sure to use a "crazy" image.

Another example is that during the day at work or at school, you suddenly think that you must stop at the grocery store to buy orange juice on your way home. Here you are required to remember two things—to stop at the store and to buy a specific item. In your mind, picture the road you normally take, and the critical turnoff that you must take to get to the store. Now imagine some happening that would prevent you from continuing straight home: a huge snake rearing up to hiss at you, a large wall of water sweeping you in the right direction, a bonfire in the middle of the road. Follow this image by visualizing a giant bottle of orange juice perched above the door to the store, drenching people with juice as they enter. After you have made these images, continue with your tasks of the day. Then, as you go home, your mind will get you to the store where you will remember what you are supposed to buy—because the images will pop into your mind on cue.

A word of caution: As you read through the examples in this chapter, do not try to memorize the particular images or stories. They are merely an illustration of techniques that you can use. Visualization works best when you make up your own images that are meaningful to you.

 THREE-MINUTE ACTIVITY 2-4: PRACTICE WITH ASSOCIATIONS
Teach the method of association to another person; then develop three different examples of associations together.

Word Substitution

Memory expert Jerry Lucas begins his presentation on learning and memory by saying that we first learn as young children by associating the name of an object with seeing the object. This image is then linked with the name in our memory, because we have "photographic" minds—not photographic memories which is an ability that only a few people retain into adulthood. When you hear the name of the object again, you instantly see its picture in your mind. Let's demonstrate. You are absolutely forbidden in the next moment to think of a zebra. So, what mental image popped into your mind as soon as you read or heard the word "zebra"? Of course, you "saw" a zebra. You cannot *not* think of a zebra. The difficulties come when we are asked to memorize abstract, intangible things. This is where we can use a technique called word substitution; we substitute a tangible object for the intangible word or concept. We can't pick just any word—we must select a word that will remind us of the intangible word. Thus we need to find a substitute word (or phrase) that sounds very much like the word we want to remember. Jerry Lucas gives the example of visualizing the word "pronoun" by imagining a nun playing golf—a "pro-nun." This technique is very useful for learning foreign- as well as English-language vocabulary, and examples are given in Table 2-1. Word substitution is also the key for remembering people's names.

Table 2-1
Vocabulary—Examples of the Word Substitution Method

English: *Actinoid* sounds like "act annoyed." Imagine a five-pointed star on a stage with one of its arms bent out of shape; it is acting very annoyed. The subject of the picture gives the clue to the meaning of the word—star-shaped.

French: *Trottoir* sounds like "trot war." Imagine some mounted warriors (and a canon or two) trotting down a sidewalk. You have associated images of war with the activity of trotting; in turn, these have been linked to the meaning of the word—the sidewalk.

Italian: *Prezzo ridotto* sounds like "pretzel, rid a toe." Imagine using a pretzel to rid a toe of a huge price tag attached to it. This activity causes the tag to shrink, giving you the meaning of the word—reduced price.

German: *Rolltreppe* sounds like "roll, trip, pay." Imagine a weird image of rolling down an escalator, tripping at the bottom, and then paying a fine. The "scene" gives you the meaning of the word—escalator.

Spanish: *Carne* sounds like "car neigh." Imagine a butcher having a car on his butcher block, ready to be cleaved. The car protests and neighs. The crazy subject-substitution gives the meaning of the word—meat.

Dutch: *Rok* sounds like "rock." Imagine a lady wearing a rock instead of a skirt. This crazy subject-substitution gives the meaning of the word—skirt.

Japanese: *Ahiru* sounds like "Ah hear you." Imagine a duck putting a wing up to its ears and saying the phrase to you in an accented voice. The word means duck.

Mandarin: *Wan fan* sounds like "one fun," with the vowel sound in "fun" prolonged just a bit. Imagine that the one fun thing to do in China is eating dinner, because they do not have much evening entertainment. Thus dinner is synonymous with "number one fun." This gives you the meaning of the word—dinner.

Arabic: The word for book in Arabic sounds like "key tab." Imagine a large key with a tab attached. When you pull the tab, a book emerges from the key, giving you the meaning of the word.

Russian: The word for dog in Russian sounds like "so back off" (without the f-sound). Imagine a dialogue between parent and child: "You want to pet the dog?" "What if it growls?" "So, back off!" The object of the action gives the meaning of the word; the last phrase in the scene recalls the pronunciation.

Remembering people's names—procedure: To remember names, we combine word substitution with the association technique. People who are good at remembering names have a big advantage and create much goodwill for themselves. Thus this is a useful life skill. The important steps for remembering names are:

1. You must hear and understand the name. If the name is mumbled, ask the person to repeat it slowly and perhaps spell it out. People will be flattered that you are interested enough to want to know their name and will gladly comply.

2. Next, repeat the name slowly. As you do this, pay attention to the way it sounds. Select substitute words that will help you remember the sound of the name. For example, the name Traynum may remind you of a train of M's.

3. Now look at the person's face. What is the most striking feature? Do not select eyeglasses since the person may be wearing contact lenses at other times.

4. Link the image to the feature, preferably associated with some action. If Mr. Traynum has a very high forehead, picture the train of M's chugging across his forehead. Next time when you see him, one look at his face will bring the image of the train of M's to your mind, and you will immediately remember his name.

5. Review. You will fix people's names in your memory much better, if you are able to review what you have learned about them. Within a few minutes, try to speak to the person, using the name. If you have time, jot down the name, the feature, your image, and the related story on a note card (or make a sketch). Review the name and person mentally in the evening, and again one week later.

The more practice you have doing this, the easier it will become to make up images and stories for linking names and faces so they will be remembered easily. When you are waiting somewhere, practice with magazine pictures or with the people around you. Make up the name if necessary.

The average man is more interested in his own name
than he is in all the other names on earth put together.
Remember that a man's name is to him the sweetest and most important sound in any language.
Dale Carnegie, How to Win Friends and Influence People, *1938.*

My memory is so bad that many times I forget my own name!
Miguel de Cervantes, Don Quixote de la Mancha, *1605.*

Remembering names—examples: We had two children from the same family in one of our classes, R. J. and Kristy Thayer. The boy was upset when we mistakenly called him J. R.—no wonder, names are people's most prized possessions, and who wants to be associated with the nasty J.R. of the television show "Dallas." This called for a substitute word. Thayer sounds a lot like "chair"; so we had "crispy chair" and "arm chair." Since prominent ears are a distinguishing facial feature in this family, it was easy to visualize nice little "crispy" chairs (like rice-crispies cookies) hanging from Kristy's ears, and R.J.'s ears flopping over an overstuffed armchair arranged around the back of his head. The nice thing about visualization is that you don't need to share your image with anyone—it's your own private memory aid. Now, when calling on this young man, the first sound in armchair leads our memory right to his correct name.

How would you remember a Mrs. Hoppendorfer? A substitute word for the name could be "hopping dwarfs." If this lady has very "spiky" hair, you can visualize several dwarfs hopping across these hair spikes. Make it a vivid image as you look at Mrs. Hoppendorfer. The next time you see her (and her hair), the image—and the name—will immediately come to your mind. Our name, Lumsdaine, can be remembered as a Great Dane dog snorting heavily from big "lungs" and lunging at Monika's dimples or Ed's mole next to his eye. The sound substitution does not have to be exact; it just has to remind you of the name. Make the story vivid, with action if possible; it will be remembered better this way.

 THREE-MINUTE ACTIVITY 2-5: REMEMBERING NAMES

If you are in a class, form a team of three (preferably with people that you do not know yet). Make up a word substitution and association image and story to remember one another's names. If possible, link the first name to the picture also.

The Link or Story Chain

What if you had to remember a long list of unrelated items, such as a shopping list? For a history quiz, you might need to remember the list of all the presidents of the United States or a list of technological achievements of the twentieth century.

  THREE-MINUTE ACTIVITY 2-6: REMEMBERING LISTS

Read through the list of twenty technological achievements below and try to memorize the items by going over them for a minute or two. Then cover up the list and try to reconstruct it from memory, in the correct sequence.

TECHNOLOGICAL ACHIEVEMENTS OF THE 20TH CENTURY

Henry Ford's Assembly Line	*Movies*
Alternating Current	*Pacemaker*
Airplane	*Television*
Nylon	*Credit Cards*
Plastics	*Computers*
Nuclear Energy	*Freeze-Drying*
Antibiotics	*Communications Satellites*
Transistor Radio	*Lasers*
Household Appliances	*Walk on the Moon*
Telephone	*Solar Cells*

How many of the items were you able to recall? Without visualization, the average person is usually able to remember—in the short-term scratch-pad memory—about seven bits of unrelated information. Because of the recency and primacy effects, you will probably remember most of the items at the beginning of the list, and a few at the end. The ones in the middle are the most difficult to recall, especially if they have no personal meanings attached to them.

🕐 *Five-Minute Activity 2-7: Linking Images*

Use the link method and visualization to write or sketch a wild story about the twenty technological achievements. Use your own imagination. If you have trouble doing this, read through the following example while making a strong effort to visualize the images and links.

Begin by picturing in your mind an automobile assembly line, preferably with a large number of Model T's and Model A's because Henry Ford is famous for these cars. Next, imagine wires extending out from this assembly line, with light pulsating rapidly in one direction, then in the opposite direction, to indicate alternating current. These wires are connected to an airplane. Now imagine a gigantic nylon stocking dangling from the plane. The nylon stocking is filled with all kinds of plastic objects: plastic forks, plastic toys, plastic dentures. Now the nylon stocking bursts, and the plastic items spill out all over a nuclear power plant. This makes the plant sick; it needs some treatment with antibiotics. You hear the news of this strange treatment from a huge transistor radio rolling along on wheels and passing by the nuclear plant. The radio stops in front of you; you open it up in the back and see that it is filled with household appliances: washing machines, coffee makers, mixers—the whole works. You look through the pile, and you find what you need—a shiny, purple telephone.

With the telephone, you call to reserve tickets for an exciting movie. But the show is too much excitement—you faint. When you wake up, a doctor is telling you that you have just been given a pacemaker. You decide to watch television from your hospital bed. This strange TV set can only be turned on by inserting two credit cards. But something goes wrong when you insert the credit cards—they turn into computer equipment programmed to produce freeze-dried products. These freeze-dried packets are put together to form a communications satellite being launched into space. A laser show originates from the satellite—it lights up the whole sky, replaying the first moon walk by Astronaut Neil Armstrong. But what is Neil doing? He is opening up a bag and scattering solar cells all over the lunar surface!

Close your book and notes, take a piece of paper, and jot down the list of items from memory.

How did you do this time? We think you will be surprised at the results if you have never used this technique before. Repeat your experiment in a day or even a week later. Most likely, you will remember the entire string of items forward and backward—this is how effective this technique is. If you have some problems recalling the story and some of the items, it is because you did not make a strong image or link. When you make up your own striking links and imagine your own interesting story, the association and thus the memory will be strong. You can combine the substitute word and the story link to remember lists of names such as all the U.S. presidents. The key is to imagine vivid pictures and a "weird" story that links each word-substitution picture (and associated name) to the next one.

The Phonetic Alphabet

To remember dates and numbers, a simple phonetic alphabet is used. Each numeral from 0 to 9 is assigned to a distinct sound in the English alphabet. The consonants making these sounds are given the respective numerical value. To help you memorize these pairings, you can use visualization. As you read through the list in Table 2-2, try to make a sketch of each explanation.

Table 2-2
Associating Phonetic Alphabet Sounds with Numerals

1 = T,D   Think of an umpire signaling **one** tou**ch**down at a football game. Also, the letter **T** (or **t**) has one downstroke.

2 = N   A letter **N**, when tipped over to lay on its side, looks like the numeral **2**. Also, the letter **N** (or **n**) has **two** downstrokes.

3 = M   A letter **m**, when tipped over to lay on its side, looks like the numeral **3**. Also, the letter **M** (or **m**) has **three** downstrokes.

4 = R   **Four** is a **four**-letter word ending in **r**. Remember it by emphasizing the "**r**" sound when pronouncing "four."

5 = L   **Five** fingers on the left hand, when held up with thumb out, form the letter **L**.

6 = J, SH, CH, soft G:
    A **capital G** resembles a **6**. Or imagine a **6** as a piece of pearl jewelry. "**Sh**ell to **j**ewel, a **g**iant **ch**ange" gives you the representative sounds.

7 = K, hard C, hard G, Q:
    The letter **K** looks like it is made up of two **7s**, back to back and laying horizontally. "**Q**ueen, **g**o, **k**ick a **c**ow" will remind you of the sounds.

8 = F, V, PH
    Visualize the number **8** eating **ph**ony **f**ruits and **v**egetables (or drinking **V-8** juice). Also, the script letter *f* resembles the number **8**.

9 = P,B   A **reversed** letter **P** looks like a number **9**. The **9** also looks like a baby spoon. Imagine dipping the baby spoon into a jar of **p**eanut **b**utter.

0 = S, Z, soft C, X:
    These are all "hissing" sounds. The word "**z**ero" begins with the letter **Z**. The image of a **s**nake rolled into a **z**ero and hissing at a **c**ent perched on an **x** will remind you of the sounds that go with the number **0**.

Silent letters make no sounds; thus they have no numerical value. Vowel sounds (A, E, I, O, U, W, Y) and the letter H also have no value. Repeated consonants and combinations of consonants count as one letter if they make only one distinct sound. Thus, batter = 914; elephants = 58210; recharge = 4646; wheat = 1; muck = 37. The numerals to consonant sounds relationship is very useful for remembering all kinds of numbers. As a first step, assign letter sounds to each numeral. Then use trial and error and imagination to make words and phrases out of the string of sounds. Look for words and images that can relate the meaning of the word to the event or person associated with the number. With this technique linked to the story chain, adults as well as middle-school students have memorized the number *pi* or similar random strings of numbers to a hundred digits or more.

Substituting numbers with words: Here are five examples of words that have been developed to help remember the "encoded" numbers:

+ Zip code in Toledo, Ohio = 43610. The sounds of R, M, J, T, S can be made into RAMJETS.

+ Phone number of 552-9500 (from an ad for a computer printer demonstration). The sounds of L, L, N, B, L, S, S can be made into YELLOW LINE BLAZES or WHOLE LINE BLISS.

+ Phone number of 363-8744. The sounds of M, J, M, F, K, R, R can be made into MUSHY MOVIE CRIER. Picture your sentimental friends crying every time they go and see a romantic love story at the movies.

+ Phone number of 429-5010 (from an ad for a moving company). The sounds of R, N, P, L, S, T, S can be made into RUN UP LISTS. Making lists is certainly a big part of getting ready for a move.

+ Frequent Flyer number = 074-684-724. The sounds of S, K, R, J, M, N, K, N, and R can be made into SCREECH FREE CANARY. This would be quite appropriate for someone who exhuberantly loves to fly.

Remembering historical events and years: For important events in history, you will already know to which millennium the event belongs; thus you will not usually have to remember the 1 in front of the years between 1000 and 1999.

+ To remember that Napoleon's final defeat at Waterloo happened in 1815, you want to associate the sounds for 815—F,T, L—with Waterloo. Thus you could make a sentence which says: Waterloo was FATAL to Napoleon's career. The numerical value of the word "fatal" will give you the year.

+ When and where was the traffic light invented? Picture a headline that reads: Traffic jams BEATEN in Salt Lake City! You have the place, and the odd word in the sentence is "beaten"—it's the clue for the year—1912!

+ Isaac Newton went for a walk "with JOY in the FOG" after he published his three fundamental laws of motion. When was this? The "odd" words here are "joy" and "fog." When their numerical values are joined to the millennium, the result is 1687.

+ What year was ether first used as an anesthetic for surgery? If you remember that a big FERN plant was the first patient, you will know the year—1842.

+ The novel *Cry the Beloved Country* was published in 1948 by South African writer Alan Paton. Think of it as a BRIEF to introduce you to the subject of racial persecution, and you will be able to remember the year of publication.

+ James Naismith invented the game of basketball in 1891. Think of it as an OFF-BEAT game—in the beginning, play was frequently interrupted when ladders had to be used to retrieve the ball from the peach baskets.

◔ *FIVE-MINUTE ACTIVITY 2-8: REMEMBERING NUMBERS*

With another person, practice the conversion of numbers into a memorable word or sentence, either with a historical event or a telephone number. Try different combinations, then select the best one and make a strong image that relates to the historical event (or the organization or person in case of a phone number).

Memory Pegs

This is a more complicated method that requires some preparation and training. You need to take the time to devise a structured system or pegboard on which you can "hang" the items that you want to remember. For example, you can begin by associating the image of a drum with "one," a shoe with "two," a tree with "three," a door with "four," and so on. Once you have memorized your numbering system with associated objects, you can use this pegboard for memorizing long lists of items by linking them sequentially to your pegs with a wild image. Let's illustrate this process with the list of twentieth-century technological achievements used earlier. Thus you could imagine Henry Ford's assembly line sitting on a huge rotating drum. The wires with the alternating current could form the laces of your favorite jogging shoes, and you can picture a tree growing out of a stealth fighter plane. In your mind's eye, you then see several pairs of nylon stockings wrapped around a door. We've seen this rhyming pegboard effectively illustrated in helping people memorize sermon topics.

A more complete peg system has been developed by Jerry Lucas and others. You can invent your own, or you can use or modify an existing system. These peg systems are based on the simple phonetic alphabet—words are chosen for the peg with a numerical value equal to the peg number. If you remember this rule, you can always construct a new peg if you cannot recall a particular peg.

 THREE-MINUTE ACTIVITY 2-9: PEG WORDS

A peg word for 30 could be moose, for 40 rose, for 50 whales. Alone or with another person, brainstorm pegs for the whole numbers from 31 through 49, then enter your best choice in the appropriate spaces below.

EXAMPLE OF A PEG SYSTEM

Zoo	= 0	Toes	= 10	Nose	= 20	Moose	= 30	Rose	= 40
Tie	= 1	Tot	= 11	Net	= 21	____	= 31	____	= 41
Noah	= 2	Tin	= 12	Nun	= 22	____	= 32	____	= 42
Ma	= 3	Tomb	= 13	Name	= 23	____	= 33	____	= 43
Ear	= 4	Tire	= 14	Nero	= 24	____	= 34	____	= 44
Law	= 5	Towel	= 15	Nail	= 25	____	= 35	____	= 45
Shoe	= 6	Dish	= 16	Notch	= 26	____	= 36	____	= 46
Cow	= 7	Tack	= 17	Neck	= 27	____	= 37	____	= 47
Ivy	= 8	Dive	= 18	Knife	= 28	____	= 38	____	= 48
Bee	= 9	Tub	= 19	Knob	= 29	____	= 39	____	= 49

Chemical elements—procedure: Elementary textbooks in chemistry strongly recommend that students memorize about thirty chemical elements, their symbols, and their atomic numbers. To do this effectively, we can use word substitution combined with mnemonics and the peg method to create an action image that can be vividly visualized and recalled. We have developed a three-step procedure for teams of students to use to invent memorable phrases and images:

1. *Find a tangible substitute word or phrase to identify the element.*
2. *Recall or make up a peg word for the element's atomic number.*
3. *Connect the substitute name and peg word with one or two words whose first letter consists of the letters in the element's symbol, resulting in a phrase that brings to mind an unusual image that can be sketched.*

    Chemical elements—examples: It is fun to think up substitute words for chemical elements and then link them to the peg to remember the atomic number. The following three examples were developed on the spur of the moment.

✦ MERCURY: Visualize this "weird" headline about an aristocrat: Marquis hugs green fox! This will remind you of the symbol for mercury, Hg, and its atomic number, 80.

✦ TUNGSTEN: Imagine the picture (and the phrase): Tongue whacks car. "Tongue" will remind you of tungsten; you will know that the symbol is W, and that the atomic number is 74.

✦ IRON: Imagine an ad for a super appliance—an iron: "Iron fixes every notch!" The ironing tool will remind you of the element; you will know the symbol is Fe, and the atomic number is 26.

    You can think up your own images for all the elements in the periodic table, or you can use the sketches and visualizations published by memory experts. For example, Jerry Lucas has a picture of an iron with a big notch in it. This will tell you that iron is the 26th element. For mercury, he has a sketch of the messenger in Greek mythology running toward a "smiley" face, where "face" is the peg word for 80. For potassium, he has a potato scrubbing himself in a bathtub. You probably would not have to visualize or encode all the symbols, only those that do not follow the pattern of using the first two letters of the element's name (or whatever "rule" you want to select). One class of nineteen secondary school students and six teachers in our summer math/science academy brainstormed phrases for chemical elements that were then sketched by Ken Hardy, one of the teachers. Five sketches (for B 5, Cl 17, Br 35, Ba 56, and Bi 83) are shown in Figure 2-1. What are the elements?

> *Visualization = perception + imagination + communications.*
> *We see, we imagine, we draw.*
> Walter Rodriguez, The Modeling of Design Ideas.

    After going through these examples of visualization, you may be getting frustrated. Are you thinking that this imaginative sketching is all well and good for someone who is an artist, but it is not for you because you just can't draw? In this case, *we recommend that you take the time to do Problem 2-8 at the end of the chapter,* before you get any further into this book. Even this business of imagining things may seem strange and difficult for you. This is a valid point, and that is why we will now look at some hints on how you can develop your sketching skills (and why this skill is an important aid to thinking, not just for improving your memory). Chapters 3 and 4 will show you in more detail how to develop and practice imaginative thinking and how having improved creative thinking skills (exercised with visualization) can make you a better thinker and problem solver.

BOAR BREAKS LEI.

BURY'EM BY A LODGE.

CHLORINE'S CAR LOSES DUCK.

BRO(THER)-MEAN BREAKS RED MOLE.

BEE'S MOUTH BURIED IN FOAM.

Figure 2-1 Memorizing chemical elements.

SKETCHING—A TOOL FOR
CONCEPTUAL THINKING AND VISUALIZATION

Have you ever played Pictionary®? This is a marvelous game for overcoming people's fear of sketching. The time limit on the game prevents people with drawing skills from having an advantage. Actually, what adds to the enjoyment of the game is precisely the fact that a "minimal" sketch has a better chance of being understood quickly than a detailed drawing.

Sketching and drawing: Sketching and drawing are not the same thing, and they do not have the same purpose. Sketching is above all an aid to your own thinking. Sketching is a help for developing visual ideas worth communicating. Drawing comes after this thinking and "playing" stage. Drawing is for communicating a well-formed idea to a knowledgeable audience. With computer-aided design (CAD), engineers, architects, and designers have a tool that can manipulate data and produce well-executed drawings. But these drawings cannot usually be understood by an untrained person. Because of the encoded symbols, reading blueprints depends on analysis with the left hemisphere of the brain, whereas free-hand sketching and visualization involve the right hemisphere and creative thinking. If you did not learn to sketch in school, look for a sketching and composition course at a museum or in adult education, or you can teach yourself by following the instructions in a book—references are given at the end of this chapter.

Most people can learn to sketch well by following these steps:
Learn to see. An example is given in Problem 2-8 at the end of the chapter.
Learn to handle the tools (paper and pencil) and what they can do.
Learn some specific techniques (contour drawing, shading, perspective).
Practice to develop the necessary eye-hand coordination.

Sketching objects: If you follow the steps outlined above (see Betty Edward's book *Drawing on the Right Side of the Brain* for detailed instructions), you can become proficient at sketching physical objects that you see around you. Begin by practicing your observation skills while waiting in a crowd, taking a walk, sitting on a park bench, or relaxing on a shore. Notice the texture, colors, and details of the things you see; become aware of shapes and contours; ponder relationships in perceived size by imagining yourself being a camera and visually recording the world as a two-dimensional image or "framing" it as if it were a flat canvas.

> *Still—in a way—nobody sees a flower—really—it is so small—*
> *we haven't time—and to see takes time, like to have a friend takes time.*
> *Georgia O'Keeffe, artist.*

Sketching mental images and concepts: Being able to sketch objects and landscapes directly from your observation is one thing, but you also will need to practice another kind of sketching—one that is more closely related to playing *Pictionary*. This kind of sketching deals with visualizing and communicating ideas; it involves "seeing" the object in your mind's eye or imagination. How would you sketch the concept of a cow without directly looking at one? Certainly, you can sketch a head with horns and a body with four legs and perhaps a tail. But there is one other aspect of a cow that distinguishes it from most other four-legged, horned animals—and that is its milk delivery system. If you emphasize that anatomical part, your sketch will be immediately understood without any verbal translation.

How would you visualize and sketch the relationship between learning and creativity? You could imagine learning as a linear process moving through four stages: (1) You don't know that you don't know; (2) you know what you don't know; (3) you consciously learn the missing knowledge and skill; and (4) you become so proficient in using this new knowledge that it becomes a subconscious act. Picture a cylindrical column where you are moving upward as you learn new things while progressing through the four stages of learning. As you build on what you have already learned, you are in essence adding more loops to the cylindrical spiral. Now think of creativity as enlarging your horizons, of making the radius of the learning cycles larger. Instead of a cylinder, you now have a cone opening upward, as shown in Figure 2-2.

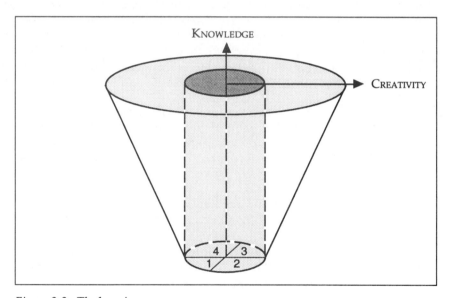

Figure 2-2   The learning cone.

During the conceptual design phase in product development, sketching enables engineers to concentrate on essentials and leave out distracting details, allowing right-brain intuitive and creative thinking and idea synthesis. Thus this type of communication is useful for brainstorming as well as for clarifying ideas. But sketches do not only visualize objects and ideas; they can represent processes and relationships through flowcharts and other types of diagrams. In many instances, making a sketch helps the mind to answer a question or "see" a solution. Thus sketching is an essential thinking tool, not just for engineers, but for anyone!

TEN-MINUTE ACTIVITY 2-10: SKETCHING A HOUSE

For this exercise, you will need a timer or a stop watch.
Part 1: Take 30 seconds to quickly sketch a house.
Part 2: Take 2 minutes to sketch a house that you see from your window or in a picture.
Part 3: Take 3 minutes to sketch your dream house.
Part 4: Take 3 minutes to discuss with another person how each of the three sketching exercises differed in the type of thinking you had to do to carry out the assignment. Which one was the easiest for you to do?

Sketching charts and diagrams: Making flow charts and diagrams is another way of using sketching as a thinking tool. When you read a textbook and try to "file" important information in your memory, first restate the ideas in your own words and make up visual images that will help you recall the information. Find applications and "create stories" to give you better visualization. As you study, summarize or connect the material by making charts and diagrams. These can be very simple, as for example the learning cone in Figure 2-2 or the little diagram shown below. Figures 10-1 and 10-2 are examples of flowcharts.

The flowchart is a diagram that connects ideas or tasks sequentially. The PERT chart (program evaluation and review technique) is a good example; this type of chart is used to plan and monitor the implementation of very large projects. It can be done by computer; it visualizes the tasks that must be done simultaneously, as well as the key bottlenecks, critical paths, and prerequisite steps that must be done for the project to proceed in a timely manner. This type of chart is also known as the precedence diagram (see Reference 2-15 at the end of the chapter). Here is a simple flowchart illustrating the creative problem-solving process.

THE CREATIVE PROBLEM-SOLVING PROCESS

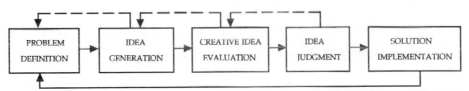

This diagram shows at a glance that creative problem solving is a sequential process with iterative loops (or flexibility) built in. This means that to achieve satisfactory results at each step, it is sometimes necessary to return to an earlier phase to get additional ideas or insight. Also, implementation is a problem in itself that requires another cycle of creative problem solving. The creative problem-solving process is diagrammed in a circular arrangement in Figure 3-9, as well as on the cover of this book where the flow path is superimposed on a metaphorical model of brain dominance and shown with the associated mindsets.

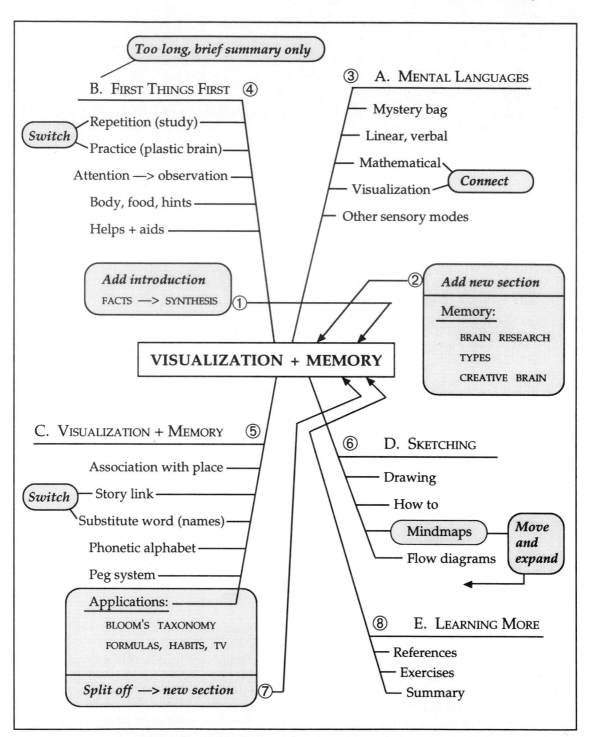

Figure 2-3   Brainstorming Chapter 2 by mindmapping.

Creating a Mindmap

Mindmapping combines aspects of brainstorming, sketching, and diagramming in the process of thinking through a subject. The term was coined by Tony Buzan, a British brain researcher who invented this method in the 1970s. The process is also known as *pattern noting* since it was originally developed as a note-taking technique which can display the relationships between facts and ideas. Contrast this to the way most people are taught to take notes, in which information is jotted down in a linear sequence. Mindmapping employs quick, intense, unrestricted, nonjudgmental thinking; it encourages creative thinking and helps generate many ideas. You can use mindmapping for brainstorming, planning, and organizing a topic before you begin a writing project—you can modify the basic format to whatever style (graphic, verbal, or a combination) works best for you. Figure 2-3 is an example of a verbal mindmap used for organizing material; the notes we jotted down for improving the outline of the material are highlighted with the shaded areas. Mindmaps are useful for final checking and summarizing when your project has been completed. More graphical mindmaps aid in communicating and memorizing information—the mindmaps given at the start of most chapters are of this type. Computer programs such as *Inspiration*™ can do mindmapping.

Procedure for "sketching" a mindmap:

1. Start your mindmap (alone or in a team) by writing the main topic in the center of a very large, blank piece of paper turned sideways. Add a sketch; then draw a box around the subject title. Background music helps you think creatively.

2. Think about what main factors, ideas, concepts, or components are directly related to your topic. Quickly jot these down on a piece of scratch paper, then write down the most important ones as main branches off the central title. Underline them and connect them to the center title box (see Figure 2-4 for a simple example). You can later assign the remaining topics as subtopics to the main ideas.

3. Now concentrate on one of these headings or main ideas. Identify the factors or issues related to this particular idea. Additional branches and details can be added if desired. Use key words, not phrases, if at all possible, to keep the map uncluttered. Capital lettering is recommended, even for the subtopics.

4. Repeat the process for each of the main ideas. During this process, associations and ideas will not always come to mind in an orderly arrangement—soon you will find yourself making extensions all over the mindmap. Continue the process for 10 minutes or so until you can no longer add ideas to the map. Use symbols and simple sketches—they constitute a direct path to your subconscious brain. If you are researching a subject, you can add items to the map as you discover them in your reading.

5. Next comes the organization and analysis phase of mindmapping. With colored markers, connect related areas, ideas, and concepts. Review, annotate, organize, revise. Figure 2-3 shows a mindmap of Chapter 2 at this stage of material organization. Edit and redraw the mindmap (on paper or the computer screen) until you are satisfied with the logic of the relationships among all the ideas.

6. Finally, you are ready to begin writing. You will be amazed how the time spent in thinking up and organizing the mindmap will make the writing task easier. The result will be a high-quality, well-organized, and well-understood product.

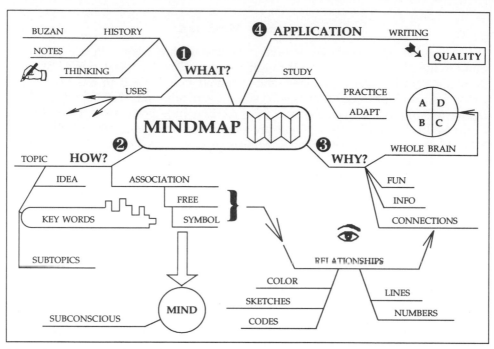

Figure 2-4   Example of a simple mindmap.

Why is mindmapping effective? It is a whole-brain thinking process that expands as well as focuses your thinking; it is logical as well as imaginative. It is a quick, flexible, spontaneous process that lets you play with ideas, yet it also encourages in-depth thinking. It lets you see the whole picture as well as a tremendous amount of detail. It is fun; it conveys much information in a small space; it lets you see connections; and it is a powerful tool for communicating ideas.

From experience we have found that mindmapping is an excellent tool for analyzing writing problems. It overcomes writer's block. It allows you to see the relationships between topics, subtopics, and ideas; it helps you identify duplications and flaws in logical progression as you develop your ideas. For example, the introductory chapter in this book was restructured three times (with major sections placed in a different order), until the resulting mindmap showed a clear and logical organization of the material without duplication and jumps. The mindmap for Chapter 4 yielded the sudden insight that each group of mental barriers to creative thinking corresponded closely to one of the brain quadrants of thinking preferences discussed in Chapter 3. Amazingly, before the mindmap, we had worked with this material for years without seeing the connection.

At times we use brief phrases along the branches, although experts on mindmapping recommend that only single words be used. Experiment with the technique and adapt it to your own preferences. When organizing and numbering the main ideas off the central topic, most people follow clockwise sequencing. We frequently use a counterclockwise arrangement roughly corresponding to the four-quadrant model of thinking preferences. The mindmap is a useful device for studying, learning, and review. As you read a chapter in a textbook, try making a mindmap of the material. Add sketches to help you remember the major points or topics. Examples of these more graphic types of mindmaps are given at the opening of each chapter in Parts 1 and 2 of this book. If you are an analytical thinker, you may experience the mindmapping process as "strange" in the beginning. But as you learn to use this efficient way of note-taking, brainstorming, and organizing material, you will become proficient and comfortable with this technique, and the quality of your writing projects will improve.

APPLICATIONS AND IMPACT OF VISUALIZATION

We tried to present an introduction to visual thinking, memory, and sketching in this chapter. This is a very broad subject, and we have merely scratched the surface. Visualization is a powerful tool, and we will close the chapter with some detailed examples, such as visualizing Bloom's taxonomy and remembering formulas. Visualization can also be used to change behavior and habits—we will demonstrate how this works. Finally, we will briefly discuss the dangers of visualization when our minds are unguarded, such as when passively watching television.

Bloom's Taxonomy

Benjamin Bloom, a professor at the University of Chicago, has developed a classification of thinking skills known as Bloom's taxonomy. We will give a summary of the main characteristics of the six levels. Visualize these levels in a pyramid arrangement, beginning with the broadest level at the bottom and building upward to higher levels, where each successive level is resting on a wider base. To help you remember the six intangible words, we will develop a visual representation. Later, in Part 2 of the book, we will use these terms in our discussion of the thinking modes involved in creative problem solving.

The first level in Bloom's taxonomy—KNOWLEDGE—refers simply to getting data and information; this is what students are primarily taught in different subjects at school. This is also what you need to know to do well on many game shows on TV. Some people have an amazing memory for bits of knowledge—they are the champions when playing the game of Trivial Pursuit®. Before the brain can do higher-level thinking, it must have a knowledge base to work with.

The second level—COMPREHENSION—is the ability to understand the meaning of the material that is being learned. Some teachers are very good at emphasizing this skill consistently, and their tests can identify students who need more help. You can do your own evaluation: review graded exam papers to gauge your comprehension of the covered material. If you diagnose a problem, take remedial action as quickly as possible. You can also check your comprehension by taking practice tests in workbooks and study guides or by doing problems on old tests. Work the problems independently first; don't just review them. Comprehension is also demonstrated through skills in giving explanations, interpretation, and summaries, as well as by drawing valid conclusions.

The third level—APPLICATION—is taught less often. It is very valuable for effective learning, especially when applied in different contexts with good examples that students can relate to. Some students learn best through hands-on application. This is the purpose of laboratory courses, and this is where computers can be valuable teaching tools. If you, as a student, are bored in a course, try to apply the principles you are learning in unusual contexts. Teach or tutor someone in the subject by coming up with a more interesting way of presenting the material, including developing special visual aids. The real test of application is the ability to use the learned concepts in new, unfamiliar situations.

The fourth level—ANALYSIS—is taught not only in mathematics and science but also in liberal arts courses such as literature, art, and music. The material is broken down into its components, and the relationships between the parts are investigated. Analysis is important in all subjects because here we look at what we are learning to find the meaning and reasons beyond the facts. Also, we look at facts and data with the intent of discerning trends. When the parts are combined in new ways, we have progressed to the next level.

The fifth level of thinking—SYNTHESIS—is rarely encouraged in school, where subjects are presented in isolation. It involves right-brain thinking as the knowledge from different areas is combined to come up with a deeper understanding and insight, which can often result in something unique. The modern (computer) word for this process is *integration*. Synthesis is an important component of creative thinking. Strategic planning as well as the development of new procedures, ideas, or products is accomplished at this level of thinking. To emphasize the importance of creative thinking skills, we would like to suggest a modification of Bloom's taxonomy by designating this step as SYNTHESIS AND CREATIVITY.

Finally, at the highest level—EVALUATION (OR JUDGMENT)—we examine the assumptions, presuppositions, and prejudices on which our thinking is based. Here, our values, our moral standards, and our social concerns enter into the picture. A manager or engineer thinking at this stage will be concerned about ethics and the short-term and long-term environmental consequences of his or her work. At this level, we are able to think beyond our personal desires to the welfare of society and the entire planet. During judgment, data, ideas, and conclusions are evaluated according to implicit or explicit criteria. If we are aware of the quality of the values on which criteria and judgment are based, we are able to better trust and evaluate our own judgment as well as that of others. In creative problem solving, our emphasis will be on positive, constructive judgment. How do you think drinking alcohol affects thinking? Recent research studies show that alcohol intake affects the highest levels of thinking first. Poor judgment then leads to risky behavior. As drinking continues, learning and memory become impaired.

The six levels of thinking in Bloom's taxonomy are all intangible concepts that are impossible to visualize, and we have found that students have not done a good job of memorizing these words or terms (and the correct sequence), even when they knew they would be tested on these items. Thus this is an opportunity to apply visualization and memory techniques. Recall that to make images of intangible words, we must use the substitute word method and find similar-sounding tangible words that can be visualized and sketched in place of the intangible terms. What we are trying to do is write a picture story, beginning with the title or subject. The pictures will be developed so that the images will remind us of the *sounds* of the intangible words. Slowly sound out the individual syllables of the intangible word and try to identify tangible words that would make a good substitution. A good substitution is a word that can be easily visualized and sketched. We will now give you one possible sequence of words, as shown in Table 2-3. Many other ways of representing the thinking skills of Bloom's taxonomy are possible, and we expect you to come up with other creative and surprising ideas. The main objective is to "see" the intangible concepts linked together. If you develop your own sketch, you will remember Bloom's taxonomy even better than if you only study our example.

Table 2-3
Story Link and Word Substitutions for Bloom's Taxonomy

Bloom's Taxonomy = blooms (with a bee, so that the image is not interpreted as flowers) + taxi + enemy. We can visualize and then sketch flowers (with a bee) growing out of a "pyramid" taxi being hailed by Public Enemy #1.

Knowledge = nail + ledge. Sketch a large nail perched on a ledge, pointing to or poking the backside of Public Enemy #1.

Comprehension = Comb + pray + hen + shine. On the right half of the ledge, imagine and then sketch a "praying" comb in front of a "shiny" hen.

Application = ape + lick + A + shine. Sketch an ape who is licking a "shiny" letter A, with the ape's hand holding on to the hen to link this image to the previous one (this is especially important here since the "direction" of the sketch is changing).

Analysis = On + alley + sis(ter). Sketch a wire from the shining letter A to an "on" switch attached to the frame of a picture. The picture shows an alley; it is held by a girl with "Sis" printed on her dress.

Synthesis = cent + ah + sis. Imagine and sketch the first sis kicking a large one-cent piece (marked 1 cent) to her twin, also wearing a "Sis" dress and exclaiming "ah." She is the "ah" sis(ter), as she is looking in "awe" at the next image. You may want to add an image of a light bulb to her dress since the light bulb is a recognized symbol for creativity—a concept you want to connect with synthesis.

Judgment = judge + mint. The second sis is looking at a judge on the bench who is pointing at a candy mint perched on top of a pyramid. The pyramid is a reminder that judgment is the apex or highest level of thinking.

Figure 2-5 A visualization of Bloom's taxonomy.

Figure 2-5 shows a possible sketch of the Bloom taxonomy "story." Your story and your artwork will most likely be very different, but they will be just as effective as a memory aid if your tangible objects remind you of the intangible concepts. This simple example illustrates a number of points. To find good substitute words, it helps to pronounce the intangible word slowly, in a variety of ways. Do not think about the spelling; listen strictly for the sounds. What other words or short phrases sound like what you have just sounded out? Sometimes you have to try different combinations to find tangible words that can easily be visualized, sketched, and linked. Finding substitute words is a skill that gets much easier with practice, because you will build up a "substitute word vocabulary" of images. Did you notice that "shine" and "sis" were used twice in Bloom's taxonomy, as parts of different words? Particular fields of knowledge have their own jargon, trade words, or technical vocabularies. Make up images for the key concepts and you will soon become fluent in sketching this language. This ability will enhance your creative thinking and memory skills tremendously.

Remembering Formulas

To visualize and remember formulas, we can use several memory techniques, since formulas are a combination of numbers, letters, and intangible concepts. Let's play around with some formulas to demonstrate the process—then it will be up to you to visualize and sketch the formulas that are important in your field of study or work. We will look at the formulas used to calculate the volume of a cone, followed by the temperature conversion from degrees C to degrees F, the ideal gas equation, and the law for critical damping of vibrations used in physics and engineering. Even if you think you will not have much use for this skill, follow along since it is a good exercise for integrating the left and right hemispheres of the brain.

To develop a story sketch for a formula, we will start with a list of substitute words. You may want to develop a system for frequently used mathematical symbols to make the process (and memorization) more efficient. For example, the formulas here will all have one thing in common—they are expressed as equalities, where something is *equal* to something else. We can use a tangible symbol for equality—the American flag. If you were of French descent, you would probably opt for the French flag and its very suitable motto of *liberté, égalité, fraternité.* For numbers, we will use the memory pegs; when numbers are used as exponents, we can visualize "power" by framing the memory peg with lightning. We do not use a special symbol for multiplication, but addition can be indicated with a saltshaker, subtraction with the outline of a submarine, and division with scissors. Another frequently-used symbol is a thermometer, which can indicate temperature or its measurement in degrees.

We hold these truths to be self-evident,
that all men are created equal.
U.S. Declaration of Independence, July 4, 1776.

FORMULA IN GEOMETRY

Volume of a right circular cone: $V = \pi r^2 h / 3$

The variables in this formula are the radius r , the height h , and the number π (pi), where π is the ratio between the circumference and the diameter of a circle (accurate to six significant figures, it is equal to 3.14159). A volume (book) is balancing on an ice cream cone stuck in an angle iron which is holding up the left edge of an American flag. The right edge of the flag is held up by the antenna of a radio serenading a pie, with lightning coming from the radio power-framing Noah. Dragging away the pie is a high kite (substitute word and image for height). Ma (to rescue the pie) is reaching up to cut the kite string with scissors.

TEMPERATURE CONVERSION FORMULA

Celsius to Fahrenheit scale: °C = (5/9) (°F - 32)

Visualize a see-saw made up of a thermometer, with an American flag draped over the ground at its lower end. An officer of the law is standing on the edge of the flag and with scissors is shooing away a bee flying along the ground toward a thermometer stuck in a fern plant. The fern is entangled in the propeller of a submarine housing a moon.

ENGINEERING FORMULAS

Ideal gas equation: $PV = mRT$

The ideal gas equation is an important equation in engineering. It expresses the relationship between temperature T, pressure P, volume V, mass m, and the gas constant R. To memorize this relationship, students may use mnemonic phrases: *pave = mart* or *Peevee = mart*. Another way is to create a visual image. This gets easier with practice, especially as you build up your catalog of images. We already have images for temperature (a thermometer) and volume (a book). We can visualize pressure with a C-clamp tightly fastened to an object; we can visualize mass with a 10-kg weight as used on an old-fashioned scale. This leaves us with the gas constant. We can use a substitute word, a "can" to stand for any constant. In the case of the gas constant, we can make this a gas(oline) can, labeled with an R.

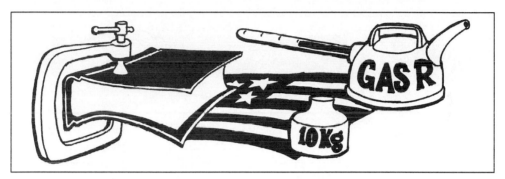

Equations in vibrations: The following equation for free vibration is important in many mechanical and even electrical engineering applications; it is used to obtain the natural frequency of a damped system with one degree of freedom, where m is the moving mass, x is the displacement, t is time, C is the damping coefficient, and k is the spring constant:

$$m\, d^2x/dt^2 + C\, dx/dt + k\, x = 0$$

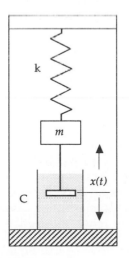

A sketch of the free-body diagram representing this equation is shown on the left. Students often have difficulty remembering the related formulas for the following three conditions: underdamped (resulting in attenuated vibration), overdamped (resulting in slow return to equilibrium without any oscillations), or critically damped (rapid return to equilibrium without vibration). Let's see if we can develop a visualization for these three relationships:

$$C < 2\sqrt{mk} \ : \ \text{underdamped}$$

$$C > 2\sqrt{mk} \ : \ \text{overdamped}$$

$$C = 2\sqrt{mk} \ : \ \text{critically damped}$$

We can imagine a critical judge at a jumping contest (jumping is a substitute word for damping). This contest is different because the objective is to come as close as possible to the target, the critical line. The jumpers who fall short of the line and have less than the required distance are judged to have "underjumped"; those getting a greater distance than the line are judged to be "overjumped"; those hitting the critical target are judged as being "equal" to the task.

Critical damping: $C = 2\sqrt{m\,k}$

To remember and sketch this equation, we have to add another item to our image catalog. The square root can be symbolized by a big carrot. C is a "jumping" can. The can is holding an American flag draped over Noah's shoulder. He in turn is holding a carrot over a weight (the symbol for mass) attached to a can with a spring wound around it (and a tag with a k).

🕐 _FIVE-MINUTE ACTIVITY 2-11: SKETCHING A FORMULA_
Working alone or in a team, select a formula that you found difficult to remember in the past. Brainstorm substitute words and images; develop a sketch and story for the formula.

Changing a Habit

So far, we have demonstrated visualization as an effective tool for enhancing memory. But visualization can improve thinking in many additional ways. We can help our mind solve problems when we encode information about the problem in visual form. The subconscious mind will work with this information while we are busy with other tasks or even while we are sleeping. This is one reason why having explicit goals and visualizing them are so important. If you frequently picture your goals in your mind in detail, your mind will help you do things that will move you toward achieving these goals. Eventually, you will become the kind of person you imagine yourself to be. Having positive role models works. Visualization also helps improve your interaction with people, as you mentally rehearse positive behavior in different situations.

We want to give you a demonstration of how you can use the power of visualization to change an undesirable habit. Let's say you want to lose weight by changing your habit of eating a junk-food breakfast. Let's try an experiment. Tell yourself three times:

Don't eat donuts for breakfast!

What is happening in your mind? What will you be thinking of all day? The mind does not "hear" the "don't"—it sees a vivid image of a donut instead, with the result that you will be thinking of donuts and most likely will give in to your craving, especially if one comes within sight of you. So—what can you do that will help your subconscious mind establish a new habit? You can give yourself a positive command; you can tell your mind to

 Eat a healthy breakfast of oatmeal and fruit!

Picture in your mind the positive command and the result! Visualize yourself preparing and enjoying a steaming bowl of cinnamon oatmeal together with a baked apple on a cold winter morning, or savoring a tasty serving of oat flakes, yogurt, and summer berries. Use sensory thinking—imagine the smells, texture, and taste! This technique of strong visualization works for many situations like giving up smoking, restraining a bad temper, or fighting illness. We also need to keep this ability of the brain in mind when we interact with small children. If we tell them, "Don't touch the vase," what we are doing is giving them an image of touching the vase—no wonder most toddlers will touch the vase within minutes of our warning. Thus, help the child by giving positive directions and opportunities for exploration, play, learning, and creative thinking. Show the child some objects and ask, "How does this object feel when you touch it? How does it smell when you rub it? How does it sound when you tap it?" Be sure you identify the object by name. Keep treasures out of reach or better yet out of sight until the child is older.

What we spend our time thinking about is important because the way we think will affect our behavior. We should not let *uncontrolled* images influence us— instead, we should use the power available through visual thinking to make us more effective thinkers. Some people fear visualization as a tool of philosophies or political programs which they oppose. But each person is the final judge who decides what to think and how to use the marvelous capabilities of the human mind. An informed, thinking mind is the best defense against unwanted subconscious influences.

Whatever is true, whatever is noble, whatever is right, whatever is pure, whatever is lovely,
whatever is admirable—if anything is excellent or praiseworthy—think about such things.
Whatever you have learned or received or heard from me, or seen in me—put it into practice.
Paul's Letter to the Philippians, Chapter 4, Verses 8-9, NIV Bible.

Watching Television

It is interesting to note that the subconscious mind cannot distinguish between real situations and "make-believe" images. What are we feeding our subconscious minds when we indiscriminately watch television without pausing to do some evaluative thinking? Lately, many voices are raising concern about the effect of television watching on young children. Watching TV for hours seems to inhibit or shut down the young brain's capacity for imagination and creativity. Experts now recommend that children's television viewing be stopped completely or severely curtailed (and then be accompanied by discussion, additional reading, and other conscious "thinking" activities). In the summer of 1989, a *Time* magazine article stated that by the time a typical American child is sixteen, he or she has watched 200,000 episodes of mostly glorified violence on television. Can this be a contributing factor to the alarming increase in the crime rate (especially in younger people) in modern society?

Television and videotapes are powerful media because they are visual, as was demonstrated vividly by the Rodney King case in Los Angeles. To most viewers watching the tape of Rodney King being beaten by Los Angeles police officers, the image spoke a clear message. The defense lawyers had to make a concerted effort to overcome this influence on the jury, since the members saw the same videotape over and over again. Thus the lawyers gave the jury intense instructions on how the videotape images should be interpreted. The second jury relied more on the evidence shown on the tape than the first jury and thus arrived at a guilty verdict.

When watching television, we must be mindful, however, that it can give a distorted, incomplete view of the world, since most of its news is packaged into brief sound bites which are accompanied by graphic and often negative visual images seen through the eye and "filters" of the reporter and editors. For example, during the Loma Prieta earthquake, the cameras focused almost exclusively on burning and heavily damaged buildings, while ignoring the large majority of structures in San Francisco that remained sound. When one of us flew to San Francisco and then drove to San Jose two days after the quake, very little damage was noted. We were in Beijing during the time of the Tiananmen Square student demonstrations and were able to observe firsthand the difference between what happened and how it was reported on CNN (with cable news available only to tourists, not the local people). Early events were exaggerated, but we believe restricted access by Western reporters to areas outside the road blocks made it impossible for them to see the whole extent of the tragedy. We gained a new appreciation for our freedom of the press—wild rumors took over during the periods when the news was blacked out in Beijing, and people were unable to find out what was really going on.

Use your imagination—broaden your horizon! Visualize—invent the future!
Laurie Bruns, engineering student, and Monika Lumsdaine.

The realms of the imagination are inhabited by creative ideas—they are waiting to be called forth.
Jodi Miller, engineering student.

The most serious danger to our minds and the minds of our children today comes not from various philosophies, but from the hours spent each day in uncritical, passive watching of television. We have to make an effort to search out those programs that encourage thought and inform us about the marvels of our world. We need to be very careful with much of the commercial programming, its values (or lack thereof), and its audience manipulation. The effect of television- and video-watching on the developing brain is being studied, and the results seem to indicate that television may be neurologically addictive by changing the frequency of electric impulses which can block normal mental processing. Frequent visual and auditory changes force the brain to pay attention in ways that overpower its natural defense mechanisms. Reading, in contrast, develops the language and reasoning skills needed for problem solving, and it encourages imagination. We must learn critical thinking skills that will enable us to evaluate the subtle messages that we are receiving from the media through visual images and the distortions caused by the focus on negative news. In the next two chapters, we will discuss other barriers, besides television, that can keep us from thinking creatively.

RESOURCES FOR FURTHER LEARNING

Material from two books recommended earlier was used in this chapter. James L. Adams, in Chapter 6 of *Conceptual Blockbusting* (Ref. 1-2), gives a very good discussion of alternate thinking languages, especially visualization and other sensory modes. Robert H. McKim, in *Experiences in Visual Thinking* (Ref. 1-11), progresses from teaching visual thinking skills and how to see and draw to the use of imagination and sketching of ideas (including visual brainstorming).

2-1   George C. Beakley, Donovan L. Evans, and John B. Keats, *Engineering—An Introduction to a Creative Profession*, 5th edition, Macmillan, New York, 1986. Chapter 7 gives a nice introduction to learning how to sketch, including many sequential exercises.

2-2   Tony Buzan, *Use Both Sides of Your Brain*, revised edition, Dutton, New York, 1983. This book includes technique and examples of mindmapping—a way to take notes, brainstorm, and connect ideas visually. The memory pegs used differ slightly from those listed by Jerry Lucas.

2-3   Jeremy Campbell, *The Improbable Machine: What the Upheavals in Artificial Intelligence Research Reveal about How the Mind Really Works*, Simon & Schuster, New York, 1989. This book describes the discoveries made by researchers trying to create a "thinking" machine. It shows that experience, not logic, is the governing characteristic of the human mind.

2-4   Betty Edwards, *Drawing on the Right Side of the Brain*, J. P. Tarcher, Los Angeles, 1979. This text has become a classic on the relationship between drawing, right-brain thinking, and creativity; it contains a lot of facts, examples, and exercises. It is excellent for letting the reader experience the difference between "verbal" drawing and right-brain processing of visual information. Also recommended is *Drawing on the Artist Within* by the same author.

2-5   Jane M. Healy, *Endangered Minds: Why Our Children Don't Think and What We Can Do About It*, Simon & Schuster, New York, 1991. Dr. Healy has investigated the influence of television on language development and thinking skills of children. When TV replaces reading, the ability to process language on a level needed for academic achievement will not develop.

2-6   Douglas J. Herrmann, *Super Memory*, Rodale Press, Emmaus, Pennsylvania, 1991. This book can help you analyze specific areas where you may want to strengthen your powers of recall. It discusses life-style changes (including nutrition and relaxation) that can help improve memory. It lists over 100 task-specific items (including social situations) that can be improved through various techniques, and it includes questionnaires that let you pinpoint your strengths and weaknesses. This book has many practical ideas but does not emphasize visual thinking.

2-7   Harry Lorayne and Jerry Lucas, *The Memory Book*, Stein and Day, New York, 1974; paper-back, Ballantine, New York, 1985. This book describes different techniques and schemes (such as the memory pegs) for improving memory.

2-8   Jerry Lucas, *How to Learn: Learning That Lasts*, Lucas Learning, Mansfield, Texas , 1988. This is a learning system for the whole family featuring Jerry Lucas and includes learning and general memory techniques to make learning fun and easy. It consists of two videotapes, six audiotapes, three workbooks, and *The Memory Book* by Harry Lorayne and Jerry Lucas, paperback edition. Several *Ready—Set—Remember* books for children and adults are available (times table, names, grammar, spelling, Bible study, Spanish, etc.) as well as seminars and college preparatory courses. These materials are based on developing and using visual thinking.

2 9   Jack Maguire, *Care and Feeding of the Brain: A Guide to Your Gray Matter*, Doubleday, New York, 1990. This book gives a fascinating tour through the functions of the mind, the myths, and the discoveries being made on the frontiers of brain science. It discusses the concepts and connections between consciousness, memory, intelligence, and emotions.

2-10   Philip Morrison and Phylis Morrison, *Powers of Ten: About the Relative Size of Things in the Universe*, Scientific American Library, Redding, Connecticut, 1982. Stunning photographs illustrate this tour on magnitudes from the atom's interior to the far reaches of the universe.

2-11   John Allen Paulos, *Innumeracy: Mathematical Illiteracy and Its Consequences*, Hill & Wang, New York, 1988. This small and easy-to-read book shows how we must and can become more comfortable with numbers, quantities, and probability—how we can overcome the mathematical ignorance so pervasive in our society.

2-12   Walter Rodriguez, *The Modeling of Design Ideas*, McGraw-Hill, New York, 1992. This texbook on computer graphics and modeling includes some freehand sketching. However, the main focus is on visualization as expressed in two- and three-dimensional computer-aided drawings in a structured, analytical approach to design.

2-13   Moshe Rubinstein, *Tools for Thinking and Problem Solving*, Prentice-Hall, Englewood Cliffs, New Jersey, 1987. This book is highly recommended by A. M. Starfield and colleagues for offering interesting and useful tools for representations.

2-14   Roger Schank (with Peter Childers), *The Creative Attitude: Learning to Ask and Answer the Right Questions*, Macmillan, New York, 1988. This book looks at various aspects of creativity, and it discusses memory as a phenomenon of "reminding." It also has a very interesting chapter on script-based thinking.

2-15   Anthony M. Starfield, Karl A. Smith, and Andrew L. Bleloch, *How to Model It—Problem Solving for the Computer Age*, McGraw-Hill, New York, 1990. This handy book has an unusual approach to teaching—it asks a lot of questions that require interaction with the material being taught. Chapter 6 takes the student through the steps needed for building a precedence diagram.

2-16   Steve Moidel, *Memory Power*, CareerTrack, Boulder, Colorado, 1989. Four audiotapes review a number of memory techniques and provide practice in different applications.

2-17   Joyce Wycoff, *Mindmapping®: Your Guide to Explaining Creativity and Problem Solving*, Berkley, New York, 1991. This paperback expands on Tony Buzan's whole-brain technique of mindmapping by presenting many applications the author has taught in workshops for creative problem solving, decision making, organizational skills, and improved memory.

EXERCISES AND ACTIVITIES

2-1   GRAPHS
Find five different visual representations of data (graph, diagram, table, histogram, etc.) or invent your own. You are allowed to copy these "charts" and make any additions or modifications that you desire to improve the presentation. The five charts should all have a different visual form and contain data about different subjects. Check that the purpose of the data and chart is clearly represented.

2-2   POP SONG
Imagine the following situation. You are in a taxi in a city in China. Your friend, who can speak Chinese, has gone into a store to do some shopping. You are tired and choose to wait in the cab for her return. The cab driver, who does not speak English, turns on a tape of pleasant-sounding Chinese pop songs. Suddenly, he becomes aware of your presence and switches to a Beethoven symphony. You want him to switch back to his songs. You sing, you gesture, but he does not understand. In desperation, you grab a pencil and note pad and draw a sketch. His face lights up in sudden understanding, and he restores the song tape. Draw a couple of sketches you think would have this kind of result.

2-3   MOUNTAIN PATH
Read through the following problem. The primary objective here is not getting the answer—your assignment is to be aware of the thinking strategies that you are employing in your attempts to solve the problem. Please jot down some notes on the different ways you are thinking about the problem and on the different mental languages that you are using to arrive at an answer.

A certain mountain in Nepal has a shrine at its peak and only one narrow path to reach it. A monk leaves his monastery at the base of the mountain at 6 a.m. one morning and ascends the mountain at a steady pace. After some hours, he tires and takes a long rest. Then he resumes his climb, albeit more slowly, and he pauses often to meditate or enjoy the view. He also takes a couple of breaks to refresh himself at a spring and to enjoy the lunch he has carried along. Finally, at sunset, he reaches the shrine where he spends the night. At sunrise, he begins his descent, quickly at first, and then more slowly as his knees begin to ache. Finally, after a couple of rest stops, he accelerates his pace again— he is hungry and does not want to miss mealtime at the monastery. Prove that there is a point in the path that the monk reached at exactly the same time of day on his ascent and descent.

2-4   SENSORY IMAGES IN CREATIVE WRITING
Instead of becoming impatient or angry when waiting in line, mentally exercise various sensory languages. The following is a sample list:

The crowing of a rooster. The bark of your dog (or your neighbor's, if you don't have one). The sound of thunder. The sound of running water (from a brook, faucet, or rain gutter). The sound of a cricket in the night. The feel of a plush carpet on bare feet; the feel of walking on wet sand. The feel of a baby's skin. The feel of a beloved's hair. The feel of cold water on your face. The feel of a sore throat. The smell of fish. The smell of burned toast. The smell of roses. The smell and feel of sunwarmed pine bark. The taste of salt. The taste of mouthwash. The taste of vanilla ice cream. The sensation of climbing a pole. The sensation of pushing a heavy wheelbarrow. The sensation of sending a rock skipping across a lake. The sensation of walking tippy-toed. The sensation of sneezing. The sensation of being very cold, then becoming uncomfortably hot. The sensation of being very relaxed and sleepy. The sensation of riding downhill on a bike. The feel of sun and a breeze on your skin.

a. *See how clearly you can imagine each of the items on the list.*
b. *What is the most beautiful place in the world for you? Why? Write a paragraph on this.*
c. *Now make a list of your favorite sounds, smells, tastes, touching sensations, and physical activities. Can you write a limerick, a short poem, or a song that uses several of these items?*
d. *Is it possible to include sensory information in a sketch? How?*

2-5 AIRPLANE SEATING

Read through the following problem and devise a seating scheme that makes the maximum number of people happy, taking the stated facts into account. Note down the steps in your thinking that help you solve this problem.

Seven passengers have just boarded a Boeing 747 aircraft for a transpacific flight. They find their assigned seats and sit down. Here is a sketch of the seats involved in this problem:

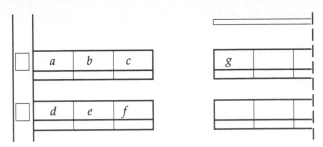

For this 14-hour flight, the people, their seat assignments, and their needs and wishes are:

a. A Korean man who speaks some English; he has the bulkhead window seat.

b. His wife who appears to be ill; she is in the bulkhead middle seat.

c. A big English-speaking Filipino carrying a large bag which she refuses to stow in an overhead luggage bin. She is in the bulkhead aisle seat but demands a seat farther back.

d. A Korean lady who does not speak English; she has the window seat in the row behind the bulkhead. She carried on a large package which does not fit under the seat in front of her. She stows it in the leg space and covers it with a blanket and her purse. Of necessity, her legs extend into the space of the middle seat.

e. A hunky U.S. serviceman; he squeezes into the middle seat.

f. A middle-aged American woman on crutches with a broken foot; she has the adjoining aisle seat. She finds that it will be impossible for her to elevate her foot from this seat using the small folding camping seat she has brought along for this purpose.

g. The woman's son, a six-foot-four-inch skinny guy with very long legs. He has the bulkhead seat across the aisle (behind the lavatory partition). He trades seats with his mother to give her more leg room. However, this is not sufficient to allow her to prop up her foot.

The stewardess has found a seat in the back of the crowded plane for the Filipino. The Korean couple is delighted at first, but then they find that the armrests in the bulkhead seats cannot be raised; thus the ill wife cannot lie down. How would you help out these six remaining passengers to achieve win-win trades (where each person ends up with an improved situation that meets their needs as well as the safety regulations on board the aircraft)? Which person do you think was the most desperate and had to think up these trades? If you are working on this problem in a group, make up a role-playing skit to better illustrate the interrelated problems involved and solution process.

2-6 MONEY

Solve the following problem and again pay attention to the mental languages that you are using: Becky and Cory together have three times as much money as Arnold. Dotty has twice as much money as Ernie. Arnold has one-and-a-half times as much money as Dotty. Cory and Dotty together have as much money as Becky plus twice the amount that Ernie has. Ernie, Dotty, Arnold, Becky, and Cory together have $60. How much money does each child have? What thinking modes would you used to solve this problem? Chapter 9 solves this problem using MATHEMATICA. Can you think of possible ways of solving this problem without algebra?

2-7   OBSERVATION (CLASSROOM ACTIVITY)
In preparation, the leader changes five to ten items in the room (or brings in some objects that are not usually found in the classroom or conference room). Then after the group or class members have entered, they must identify the "odd" items or changes.

2-8   LEARNING HOW TO SEE AND DRAW—SPECIAL SIXTY-MINUTE EXERCISE
Preparation: *Do this exercise with a group of people—it is fun to observe the results (especially if some of them are convinced that they can't draw). Set a time and place. Find a line drawing of a person in an art book (for example a portrait done by Henry Matisse or Pablo Picasso) or a line-drawn cartoon of a person's face in a newspaper. Make a copy for each group member. Also, for each person, have five sheets of blank paper, pencils, and a piece of masking tape. Have an assistant with a watch give the instructions. Note: This exercise is for adults and students high-school age on up; it does not have the same kind of results for younger students.*

Instructions: *The pencil ✏ indicates the team needs to execute the given task before continuing.*

Sheet 1:   *Draw (don't trace) your hand—you have about five minutes. ✏ When finished, sign your name and label it Drawing #1. ✏*
Sheet 2:   *Draw a profile of a human face (see the figure on page 134 in this book). ✏ Go over the line again, naming the different body parts you are tracing: forehead, nose, lips, chin, etc. This is left-brain, analytical drawing. ✏ Next, draw the horizontal line at the top and bottom to turn the profile into a vase. ✏ Now, complete the vase by drawing the mirror image of the profile. This is right-brain, holistic, spacial drawing: you are concentrating on the spacing of the line, not on the body part it represents. ✏ What differences did you notice in the types of thinking you needed to do this drawing? Discuss.*
Sheet 3:   *Turn the line drawing (artist or cartoon) upside down and copy it—your drawing will also be upside down. You will have 10 to 15 minutes. ✏*
Sheet 4:   *Fasten the sheet to your table with a piece of masking tape. Sit sideways, so you will not look at the paper. Instead, look intently at the palm of your hand. Imagine a small insect slowly following the lines of your hand. With your writing hand, copy these paths and lines onto the paper. Keep negative thoughts out of your mind. DO NOT TALK! You will be given a signal to stop in 10 minutes. Repeat lines if you have drawn all the tiniest lines you think you see. The purpose of this exercise is to bore your left brain so it will "go to sleep." ✏*
Sheet 5:   *Now we will draw the hand again. This time, form it into an interesting shape with bent fingers and hollow spaces between. The left brain does not like to deal with complexity, so it will leave this drawing task to the right brain. Look for negative space around the hand, rather than at the hand itself—negative space shares edges with the object you are to draw. Do not name the parts of the hand; look for the intersection of lines instead. Add fine details, such as shading and lines, if you wish. Closing one eye may help flatten the image. Again, you will have 10 to 15 minutes. ✏ Sign your work and label it Drawing #2. ✏ Now compare Drawing #2 with #1. Share the results with the group. Does the outcome for each person in this exercise surprise you? You have discovered how to see! Practice this new-found thinking skill. See Betty Edward's books for other exercise ideas.*

2-9   *MINDMAP*
Make a mindmap of a report or paper that you have recently written. Then analyze the process: Did this activity give you additional ideas that you could have included in the paper? Did you notice different connections that would have improved the logic presentation of ideas? Does the map help you "see" in a more holistic way what your subject is all about?

2-10   AMENDMENTS TO THE CONSTITUTION
Develop a brief story and visualization to memorize the first five amendments to the U.S. Constitution: 1. freedom of religion, speech, press; 2. the right to bear arms; 3. no soldiers quartered in homes; 4. no search and seizure without a warrant; 5. the right to a grand jury.

─────────────

*Asterisks denote advanced exercises requiring extra time but providing more in-depth learning.

2-11   PICTIONARY®
This excellent game combines visual thinking, sketching, and free association. Play the game according to the rules—or modify rules if you are playing with friends from other countries.

2-12   *TECHNOLOGY AND FOREIGN CUSTOMS PICTIONARY®*
Make up game cards with engineering, technology, and foreign business "etiquette" subjects, then play with your colleagues.

2-13   VOCABULARY SKETCHES
If you are learning a foreign language, make up five visualizations with word substitution. Otherwise, do this activity with definitions of English words. Then select your best one and teach the word to your class, group, or a friend. One week later, check to see if the word is remembered. Has teaching (and sketching the images) helped you remember the words?

2-14   *CHEMICHAL ELEMENTS—TEAM PROJECT*
Develop memorable images for learning the number, abbreviation, and name of all chemical elements in the periodic chart. Then sketch your images. If you work with a group, you may want to put together a booklet with the sketches of all the elements.

2-15   *MEMORIZING THE NUMBER PI TO 100 DIGITS*
Use a computer to print out pi to 100 digits, then develop a story that will help you memorize the chain of numbers. Hint: bunch the digits into groups of five; this will help prevent skipping a digit.

2-16   *OBSERVATION (LARGE GROUP ACTIVITY)*
Have the people in the group line up in two rows that are about five feet apart. Each person from one row directly faces a corresponding person in the other row. At a signal, all turn around and change three things on their clothing or appearance. After three minutes, all turn back to face their opposite partners. Three minutes are given for silent observation, so the pairs can try to identify the three changes. Then a time-out is taken; the participants can verify with each other how good their powers of observation were. Ask for a show of hands of those who were able to find all three changes. Next, shuffle the people and repeat the exercise, except this time ask for five changes to be made. Even though the task has been increased, most people will do better the second time around, because they are now paying attention to their partner's appearance before as well as after the changes are made.

2-17   *COMPOUND INTEREST*
The variables in the formulas used as examples in this chapter were easy to sketch with their substitute words. Other formulas are more difficult, when the variables are complicated concepts to visualize and have been assigned arbitrary symbols. For example, the formula used in economics for calculating the total amount of money A after investing a principal sum P for n years at an annual interest rate i compounded m times a year is

$$A = P \left(1 + i/m \right)^{nm}$$

We can visualize A as a \$-mountain, with a flag on its side; P as a school principal; i as an eye (that was easy). But what should we do for m and n since they are both connected with time? What about picturing M&M candies as "m" of the year on a Time magazine cover? For n you can imagine a huge # symbol, with a calendar attached (to denote years).

Sketch your own "story" for the compound interest formula. Develop a peg system for each letter of the alphabet; make a new sketch using the pegs.

2-18   WHAT HAVE YOU LEARNED?
What are the most important ideas that you have learned from this chapter? What is an important question you still have (sketch the question if possible)?

CHAPTER 2 — SUMMARY

Memory and the Brain: The human brain is the most complex arrangement of matter in the known universe. Different types of memory are located in different parts of the brain, depending on the type of learning involved in the initial acquisition of the particular item of knowledge. The brain is experience-based, not logic-based (in contrast to computers, which are logic machines).

Mental Languages: *Verbal thinking* is linear and sequential; it is not suitable for solving certain kinds of problems, although it is heavily emphasized in Western culture. *Mathematical thinking* is used in solving quantitative problems. *Visualization* helps in memorization and in generating and communicating ideas. It is essential for creativity. *Sensory input* influences creativity through the physical environment and by providing direct stimuli.

First Things First: To improve your memory, do these well-known things: correctly time repetitions; practice; pay attention; sharpen your observation skills; maintain a healthy life-style; develop good eating habits and a positive attitude; use memory aids. Learn how to visualize ideas and actions.

Visualization—A Tool for Improving Memory: Memory depends on understanding, filing, and retrieving information. Visualization can help in all three areas—it clarifies relationships for better understanding and is a key in several memorization techniques that improve filing and retrieval. *Association:* Link the item to be learned to something you already know (such as your home) in a "wild" image. *Substitute word:* A sound-alike tangible word is substituted for an intangible word or concept; the tangible word is then visualized. *Story link:* A list of unrelated items is memorized by making up an image for each item and then linking the images in a story. Remembering just one of the items will bring the entire chain into the conscious mind. *Phonetic alphabet* and *memory pegs:* These methods are more complicated because a pegboard—a system linking numerals to consonant sounds in the English language and then to corresponding images—must be developed. Items are memorized by associating them "wildly" with the pegs. Numbers can be transformed into tangible words.

Sketching: Sketching helps develop visual ideas worth communicating through drawings. It is a skill that can be learned. This process involves four steps: (1) Learn to see. (2) Learn to use the tools. (3) Learn specific techniques for representation. (4) Practice to develop eye-hand coordination. Use sketching for brainstorming and for clarifying ideas. Use sketches to visualize processes and relationships with flowcharts and diagrams. The mindmap encourages nonsequential brainstorming and presentation of ideas in order to explore their connections more intuitively and creatively. Ideas can then be connected with bubbles and lines to build a logical network. Sketching is an essential thinking tool for everyone!

Applications: Things like Bloom's taxonomy and mathematical formulas can be visualized and sketched for easy remembering. Visualization is a powerful tool to help change a habit. Caution: visual images on TV may be hazardous to your subconscious mind; thus, combine television viewing with critical thinking.

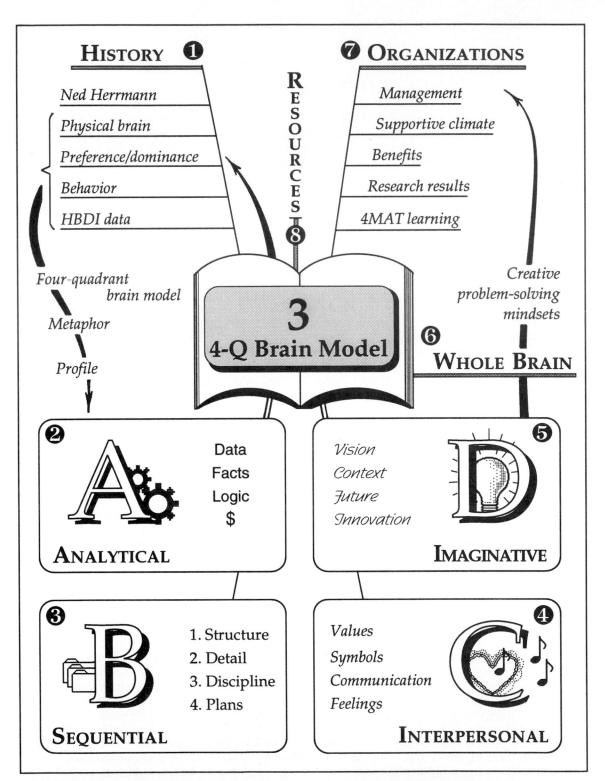

HISTORY ❶

Ned Herrmann

Physical brain

Preference/dominance

Behavior

HBDI data

Four-quadrant brain model

Metaphor

Profile

R E S O U R C E S ❽

❼ ORGANIZATIONS

Management

Supportive climate

Benefits

Research results

4MAT learning

Creative problem-solving mindsets

3
4-Q Brain Model

❻ WHOLE BRAIN

❷

Data
Facts
Logic
$

ANALYTICAL

❺

Vision
Context
Future
Innovation

IMAGINATIVE

❸

1. Structure
2. Detail
3. Discipline
4. Plans

SEQUENTIAL

❹

Values
Symbols
Communication
Feelings

INTERPERSONAL

Mindmap of Chapter 3. See Chapter 2 (pp. 55-56) to learn more about mindmapping.

3

THE FOUR-QUADRANT BRAIN MODEL OF THINKING PREFERENCES

WHAT YOU CAN LEARN FROM THIS CHAPTER:
- How the 4-quadrant model of thinking preferences was developed; the concept of dominance.
- The characteristics of analytical quadrant A thinking.
- The characteristics of sequential quadrant B thinking.
- The characteristics of interpersonal quadrant C thinking.
- The characteristics of imaginative quadrant D thinking.
- Relating whole-brain thinking to creative problem solving.
- Organizational implications of whole-brain thinking and learning; HBDI research results; comparison with the 4MAT system of learning styles.
- Further learning and insight: resources, exercises and activities, and summary.

As you compared mathematical and verbal thinking with visualization and sensory thinking in the last chapter, you may have noticed that the different mental languages required distinct and perhaps unfamiliar thinking abilities. Now we want to investigate differences in the way people think and process information in more detail. We will explore a model of thinking preferences that we have found useful for learning to become more effective thinkers and problem solvers.

HOW THE FOUR-QUADRANT MODEL OF THINKING PREFERENCES WAS DEVELOPED

It is time to introduce to you a remarkable person with a fascinating history. Ned Herrmann is the father of brain dominance technology. Through his research and life experiences, he came to recognize that the brain is specialized, not just physically, but in the way it functions—its specialized modes can be organized into four distinct quadrants, each with its own language, values, and ways of knowing.

Each person is a unique mix of these modes of thinking preferences, and these preferences result in different expressions of behavior. How Ned Herrmann came to these conclusions and insight is quite a story, as told in his book *The Creative Brain*. We are giving you here just the briefest of summaries so that you can glimpse the thinking behind his development of the four-quadrant model of brain dominance.

Already during his high school years, Ned Herrmann noticed that he had two separate sets of enjoyable friends, his math and science pals and his acting and singing buddies. His life became less balanced in college; through the family mindset he chose to major in chemical engineering. He soon found that he was "dumb" in chemistry. After military service during World War II, he returned to college and graduated in physics and music, both of which provided an enriching education. But for his professional life, he had to make a choice—a career in science *or* as an opera singer. He convinced General Electric Company to hire him because of the thinking skills he had learned through his studies in physics. Although he was treated as an engineer at first, he soon advanced through a variety of managerial positions in sales, marketing, employee relations, human resources, and management education. Then, when months of illness made it impossible for him to perform his musical hobbies, his wife brought home an oil-painting set. The results astonished him and his family—they launched his artistic and sculpting career. When he was asked to be on a panel to discuss creativity in 1976, he did a library search on the topic. He learned that creativity was related to a specific, physiological brain function and gained a sudden insight into his own behavior—who he really was and why he seemed to have a "divided" brain.

He began to investigate this area of left-brain, right-brain functioning, particularly the work of Dr. Roger W. Sperry and associates on split-brain research and testing (which earned Dr. Sperry a Nobel prize in 1981). The science of neuropsychology has found that for most people, mathematical and verbal thinking (formal speaking, reading, writing) are done primarily in the left hemisphere, while spatial, holistic, imaginative thinking is done in the right hemisphere. As briefly seen in Chapter 2, brain research now makes use of very sophisticated high-tech machines to chart brain function, and many new things are being learned about this most amazing and complex organ.

> *Within the brain of a patient, there can exist a mental duality.*
> *Hippocrates, 450 B.C.*

> *Humans exercise two modes of knowing: One through verbal argument,*
> *one through non-verbal experience.*
> *Roger Bacon, 1286.*

> *One of the two half-brains in humans takes the lead and is the dominant hemisphere.*
> *John Hughlings-Jackson, M.D., 1874.*

> *Why are some people so smart and dull at the same time?*
> *Henry Mintzberg, professor of management, McGill University, 1976.*

> *All four of these quotes are from* The Creative Brain.

The Concept of Brain Dominance

The next thing Ned Herrmann discovered during his search was the concept of dominance. The two halves of the brain are not used by people in the same way and with the same frequency. People develop dominances or—as Ned calls these cognitive (thinking) preferences—"preferred modes of knowing." Dominance has advantages: quick response time and higher skill level. We use our dominant mode when we need to solve a problem or learn something new. For example, if you solve a problem analytically by looking at facts and numbers and then plug these into a logical formula or sequential procedure, you are using a left-brain approach. If you search for patterns and images involving sensory impressions to give you an intuitive understanding of the whole, you are using a right-brain approach. Left-brain students learn by reading about a subject, whereas right-brain students learn by watching demonstrations and doing hands-on activities. The more extreme our preference for one way of thinking, the stronger is our dislike or discomfort for the opposite mode. These "opposite" people will also have great difficulty communicating and understanding each other, because they use different vocabularies and see the world through very different "filters." The question of course that comes up at this point is: "Which way is better?" Ned Herrmann found out that each brain mode is best—for those types of tasks that it was designed to perform.

How do people develop these preferences? Are we born that way and "stuck" with these preferred thinking modes? Ned Herrmann thinks that each person is born with a given genetic set of cognitive capabilities, with strengths and weaknesses. As we interact with the world, we learn to respond with our stronger abilities, because they lead to more frequent success and reward. As we have seen, the use of the brain in certain ways of thinking strengthens those structures. Ned Herrmann says that the performance-praise-preference feedback loop can turn a small difference in hemispheric specialization into a powerful preference for one cognitive mode over another. And this is not only true for individuals, it also works for entire cultures. The Industrial Revolution shifted success heavily toward analytical thinking, whereas historically in the native American culture and in agrarian societies, survival depends more on holistic, intuitive skills such as understanding animal behavior (for hunting and husbandry), understanding the ecology of the environment, the healing arts, the weather, and social interdependence. Because our school systems concentrate so heavily on sequential reasoning skills, more creative abilities have been completely overshadowed and are often actively discouraged, not only by teachers but also by well-meaning parents, family members, employers, and managers. What is needed is a better balance and an appreciation for all thinking abilities; we need to learn how to use and integrate these abilities for whole-brain thinking and problem solving.

To understand how Ned Herrmann invented the four-quadrant brain dominance model, we need to visualize the physical brain. Most people are familiar with the main hemispherical division into left-brain and right-brain. Strictly speaking, these are the cerebral hemispheres and contain about 80 percent of the brain. Primary mental processes in these hemispheres include: vision, hearing, body sensation, intentional motor control, reasoning, conscious thinking and decision making, language and nonverbal visualization, imagination, and idea synthesis. Each cerebral hemisphere has a separate structure nestled into it—one-half of the limbic system. The limbic system is a vital control center that regulates hunger, thirst, sleeping, waking, body temperature, chemical balances, heart rate, blood pressure, hormones, and emotions (pleasure, punishment, aggression, and rage). It plays a powerful role in learning since it is crucial in transferring incoming information to memory. A diagram of the brain together with its relationship to Ned Herrmann's four-quadrant concept is given in Figure 3-1.

The hemispheres are connected with fibers that carry communications within and between the hemispheres. Association fibers form a complex network connecting the different specialized areas within each hemisphere. The limbic system was long considered to be a single entity, but is now recognized as two halves or lobes linked through the hippocampal commissure. The two cerebral hemispheres are connected by the corpus callosum which contains between 200 to 300 million axonic fibers. When one part of the brain is actively thinking, the other parts are in "idle" mode in order not to interfere with the specialized thinking task. However, when solving a complex problem or doing an intricate task, more than one thinking skill is involved. The brain has the ability to switch signals back and forth very

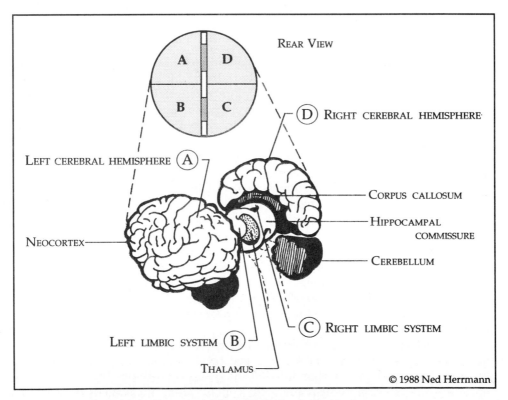

Figure 3-1 How the four-quadrant model relates to the physical brain.

rapidly between different specialized areas within and across the hemispheres through the fiber networks. Switching thinking modes within the left hemisphere or within the right hemisphere is quite easy. Switching between the two lower (limbic) or upper (cerebral) quadrants is somewhat more difficult. Switching diagonally is very difficult and stressful because no direct fiber connections exist between the diagonally opposed quadrants of the brain, and thoughts have to be translated or processed via one of the connecting quadrants.

As discussed in *The Creative Brain,* researchers have discovered a physical difference between sexes where the corpus callosum is concerned. Females on the average have 10 percent more fibers, and impulses travel 10 percent faster than in males. Thus many females can move ideas back and forth between hemispheres faster than the average male. Also, their corpus callosum matures as much as three years earlier. According to Ned Herrmann, young females have an advantage in learning to be comfortable with right-brain processing—which is due to biological as well as cultural influences. He found that women are potentially more interhemispheric in their mental processing, more whole-brain oriented, more intuitive, more open to new ideas, and more people-oriented than thing-oriented. They perceive their surroundings more sensitively, manage the innovation process more comfortably, and respond faster to changing circumstances. Thus, an organization that can cooperatively integrate male and female capabilities—in essence become whole-brained—will increase its competitive advantage.

Ned Herrmann, as the next step in his research about the brain and creativity, now wanted to find out more information about people's brain dominances. How does the brain choose which specialized part to engage? This question related directly to his job as a teacher and trainer, because he found that left-brain thinkers had to be taught differently from right-brain thinkers for optimum learning. For example, when asked to answer the question, "What is wrong with education?" a left brain thinker would typically reply, "Get back to the basics and discipline—get rid of unnecessary frills like sports and art." A right-brain thinker would rather suggest cooperative, hands-on educational activities, including integrating social and creative aspects into the whole of education. Ned Herrmann noted a frequent lack of appreciation for the different thinking modes and their contributions: how learning to be more sensitive to people could help a left-brain person get along better with coworkers, or how having set goals and keeping to a schedule could help a right-brain person be more efficient. He concluded that both types of thinkers benefit by knowing how to use the whole brain for learning, working, solving problems, and communicating with each other. Thus early in this work, he recognized this important insight: No part of the brain works as fully or creatively on its own as it does when stimulated or supported by input from the other parts.

When Ned Herrmann looked around for a questionnaire or way to diagnose thinking preferences based on brain specialization, he was surprised that he could not find any existing tools that were suitable for his purposes. So he developed his own, now known as the Herrmann Brain Dominance Instrument (HBDI). The Myers-Briggs Type Indicator by comparison is based on psychological concepts, not specialization, and thus yields different (though correlating) information. As Ned Herrmann began to teach workshops, he collected much data with early versions of his questionnaire. The data seemed to fall into four clusters, not into two cerebral hemispherical divisions. Then, one day as he was driving along in his car, the mental image of the divided brain rotated to where he realized that the limbic system was also divided into two hemispheres, in essence giving a brain divided into four quadrants, as indicated schematically in Figure 3-1. This enabled him to organize his data into the four-quadrant whole-brain model as a "descriptive metaphor." Now that we are familiar with this model, the division seems logical and obvious—it is difficult to realize that this understanding is quite recent.

Figure 3-2 shows the metaphorical model of the four-quadrant brain of thinking preferences. The brain is visualized as a circle divided into four quadrants. To emphasize the metaphorical status of the model, Ned Herrmann named the quadrants alphabetically to de-emphasize their connection with the cerebral-limbic brain. The upper left (cerebral) quadrant is designated A, followed by B, C, and D in a counterclockwise direction. Each quadrant has very distinct clusters of thinking abilities or ways of learning and knowing. The following sections in this chapter will discuss and illustrate the characteristics of each of the quadrants from the viewpoint of how people learn in these thinking modes and how they can strengthen these abilities. Keep in mind that each person is a unique "coalition" of thinking preferences and learning styles, as emphasized by Ned Herrmann. You might be interested to know that he and his coworkers, as well as independent researchers, have done a tremendous amount of work in validating the HBDI with hundreds of studies over more than a decade. Ned Herrmann now has a database exceeding 500,000 individual and organizational profiles; the HBDI survey forms

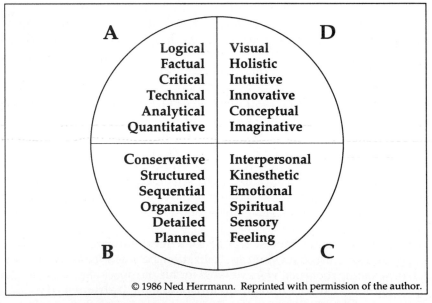

Figure 3-2 The four-quadrant Herrmann model of thinking preferences.

are scored by a computer at his headquarters, and he personally trains and certifies people in the interpretation and evaluation of the instrument to ensure the quality and reliability of the results.

Although the four-quadrant model started with the divisions in the physical brain, it is now a metaphorical model based on people's behavior because:

1.   With new brain research, it is getting more and more difficult to determine exactly which part of the brain is involved in specific thinking tasks—the brain is simply too complex, subtle, and versatile.

2.   Knowledge about the limbic system at that time was less clear, which made correlation with the data less precise. Yet Ned Herrmann strongly felt that this organization could clarify the way we think about modes of knowing.

3.   The database kept confirming that the four-quadrant model was consistent and much more useful than the left-brain, right-brain classification. The four-quadrant model also allowed for multidominance.

The Brain Dominance Profile

When thinking preferences are assessed with the HBDI, the output is a brain dominance profile. When the relative dominances are marked on axes bisecting the four quadrants, with the four scores connected by lines, the result is a four-sided figure or profile. The scale or intensity of dominance is indicated by circles dividing the quadrants into areas of preferences. The innermost circle is designated as Region 3 as shown in Figure 3-3. When given a choice, people scoring in this region for a particular quadrant will avoid thinking in this mode, but this does not mean that they cannot think in this manner. Students can even earn top grades in subjects that require thinking in modes that they tend to avoid, if they are willing to make a strong effort, since thinking preference is not correlated with I.Q. Conversely, as we shall see in Chapter 4, someone with a strong thinking preference or high I.Q. does not necessarily know how to be a good thinker. Brain scans have shown that thinking in an unfamiliar mode takes more energy and will make you feel exhausted if you have to do it over any length of time.

A score in Region 2 on the HBDI profile denotes a secondary preference—the person is comfortable with using this thinking mode. A score in Region 1 indicates a strong preference. A profile extending into the outermost band bounded by the dotted circle indicates an extremely strong preference which can usually be observed in behavior. For example, someone who uses a ruler to draw lines across the blank spaces on a check or receipt would likely score very high in B-quadrant thinking. Ned Herrmann has found that people in certain professions tend to have similar profiles. As an example, Figure 3-3 exhibits the average brain dominance profile for the engineering faculty at the University of Toledo in 1990 with a strong preference in analytical, quadrant A thinking, lesser preferences in structured as well as conceptual thinking, and least preference in interpersonal thinking. This profile is typical for engineers. However, we have seen many profiles that are very different from this typical profile, yet these individuals have been successful engineering students, faculty members, and practicing engineers.

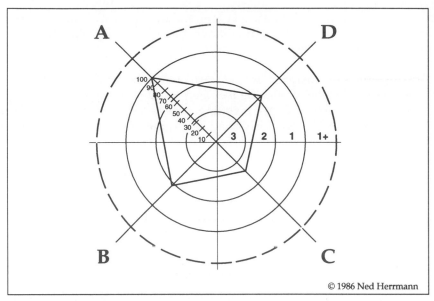

Figure 3-3   Averaged Herrmann brain dominance profile for engineering faculty.

HBDI profiles are frequently expressed in the numerical code related to the strength of preference or dominance. This code designates a *generic* profile. With this code, the profile in Figure 3-3 would be expressed as 1-1-2-1. For the two academic years from 1990 to 1992, the three most frequent profiles for engineering students at the University of Toledo were (where N is the number of profiles):

Freshmen: 1-1-2-2 (21%)   1-2-2-1 (13%)   1-1-2-1 (12%)   N = 713

Seniors: 1-1-2-2 (22%)   1-1-2-1 (20%)   1-1-3-2 (18%)   N = 135

The averaged profile for the seniors matches that of the faculty very closely. In 1990, we initiated a research project to determine if innovation in the curriculum, such as teaching creative problem solving, would change the brain dominance profiles of students individually and collectively, and if this would help retain the gifted and talented students in engineering who felt "different" and uncomfortable in the traditional classroom. We will discuss some of our early results and findings toward the end of this chapter.

🕐 *THREE-MINUTE ACTIVITY 3-1: QUADRANT C CLASSROOM*

In groups of three, briefly discuss the differences in a classroom where the majority of the students have quadrant C preferences, as compared to an environment where students and professors avoid quadrant C thinking modes. In general, would these differences be more important to male or female students? What changes in the classroom climate should be made to attract more female students to engineering? Hint: In the data given above, the 1-1-3-2 generic profiles were all from male students—subsequent studies confirmed that female engineering students very rarely score a 3 in quadrant C.

CHARACTERISTICS OF ANALYTICAL
QUADRANT A THINKING

Definition: Quadrant A thinking is factual, analytical, quantitative, technical, logical, rational, and critical. It deals with data analysis, risk assessment, statistics, financial budgets, and computation, as well as with technical hardware, analytical problem solving, and making decisions based on logic and reasoning. An A- quadrant culture is materialistic, academic, and authoritarian. It is achievement-oriented and performance-driven. An example of a quadrant A thinker is *Star Trek's* Mr. Spock; another is George Gallup, the pollster.

Preferred subjects and careers: People who prefer quadrant A thinking also have preferences for certain subjects in school and for certain careers. Preferred subject areas would be arithmetic, algebra, calculus, and accounting, as well as science and technology. Lawyers, engineers, computer scientists, analysts and technicians, bankers, and physicians show preferences in quadrant A thinking. People with quadrant A thinking talk about "the bottom line" or "getting the facts" or "critical analysis." They are talked about as "number crunchers" or "human machines" or "eggheads."

> *What I want is Facts. Teach these boys and girls nothing but Facts.*
> *Facts alone are wanted in life. Plant nothing else, and root out everything else.*
> *Mr. Gradgrind in Charles Dickens,* Hard Times, *1854.*

Preferred learning activities: If you are an A-quadrant thinker, you prefer to learn and behave in this way:

- Collecting data and information.
- Organizing information logically in a framework, not to the last detail.
- Listening to informational lectures.
- Reading textbooks (most textbooks are written for quadrant A thinkers).
- Studying example problems and solutions.
- Thinking through ideas.
- Doing library searches.
- Doing research using the scientific method.
- Making up a hypothesis, then testing it to find out if it is true.
- Judging ideas based on facts, criteria, and logical reasoning.
- Doing technical case studies.
- Doing financial case studies.
- Dealing with hardware and things, rather than people.
- Dealing with reality and the present, rather than with future possibilities.
- Traveling to other cultures to study technological artifacts.

 One-Minute Activity 3-2: Quadrant A Learning
Circle the dots in the list of learning activities above for those items that are easy for you and that you enjoy doing.

Behavior: Thinking preferences are expressed in behaviors. To demonstrate this, we used a brief exercise in one of our classes with first-year engineering students. We grouped students according to their thinking preferences (determined from an experimental brief assessment). We wanted to have a team with strong A-thinkers, a team with strong B-thinkers, a team with strong C-thinkers, a team with strong D-thinkers, and one team with multidominant thinkers, where the team members together constituted a whole brain. Each team had five members. The students at the time of this exercise did not yet know their own brain dominance profiles nor anything about the four-quadrant model of thinking preferences. The thinking profiles shown with the team results in this chapter are the actual HBDI profiles of the participating students.

The assignment was to write a definition of "What is an engineer?" in two minutes, with the answer to be written on a flip chart. The result for the A-team is shown in Figure 3-4. Generically, this team had a 1⁺-1-2-2 average profile. It is interesting to note A-quadrant words and phrases such as "technical, understanding how things work, factual information, and making big bucks." This group wrote down facts; they were not concerned with the details of correct grammar.

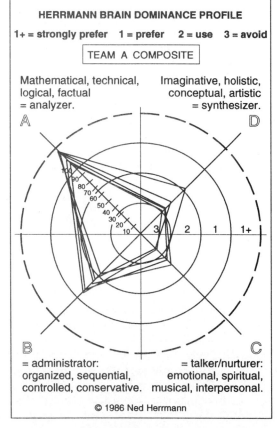

Engineers:

ᴬᴿᴱ Not Necessarily a Train Drivers.

ᴬᴿᴱ Creative Person in a Technical Way.

See and understand how things work.

Use factual info to solve problems.

Make big bucks.

Figure 3-4
Result of quadrant A thinking: definition of an engineer
(flip chart facsimile and team HBDI profile).

Exercises to strengthen quadrant A thinking: What can you do if you are not a quadrant A thinker and would like to develop this ability? Mathematics and science courses (and their homework problems) develop quadrant A thinking skills. Exercises in this book in Chapter 5 (using the detective's analytical mindset) and Chapter 7 (using the judge's mindset) will help you practice quadrant A thinking. Table 3-1 lists a variety of activities that do not specifically involve math. Many of these exercises—here as well as for the later quadrants—have been recommended by Ned Herrmann; others we have added to the list.

Table 3-1
Activities for Practicing Quadrant A Thinking

- *Collect data and information about a particular subject or problem.*
- *Organize the collected information logically into categories.*
- *Develop graphs, flowcharts, and outlines from data and information.*
- *Do a library search or patent search on a special topic of interest.*
- *Find out how a frequently used machine actually works by reading about it.*
- *Take a broken small appliance apart: find out about the function of each part.*
- *Take a current problem situation and analyze it into its main parts.*
- *Review a recent impulse decision and identify its rational, logical aspects.*
- *Analyze some politicians running for office—where do they stand on the issues?*
- *Join an investment club.*
- *Do logic puzzles or games.*
- *Play chess.*
- *Learn how to use an analytical software package or program on your computer.*
- *Play "devil's advocate" in a group decision process.*
- *Write a critical review based on logical reasoning of your favorite TV program, movie, essay, poem, book, or work of art.*

When to use quadrant A thinking in problem solving: As you will learn, the analytical aspects of quadrant A thinking are needed in the creative problem-solving process during the stage when we are trying to determine what the *real* problem is. The detective's mindset typifies quadrant A thinking (although detectives also use quadrant B procedures for problem identification). The critical aspects of quadrant A thinking are needed during the stage when we are trying to determine the best solutions, although they will be combined with creative, positive thinking that seeks to overcome the flaws that are being identified. Analytical thinking is a useful paradigm when solving routine problems that do not justify the extra investment in time and effort needed for creative problem solving. In Bloom's taxonomy, the two lowest thinking skills—knowledge and comprehension—as well as the fourth level—analysis—involve quadrant A thinking. Quadrant A thinking is also involved at the highest level—in evaluation and critical judgment —here it must be integrated with right-brain thinking for best results. The description of technological devices used in brain research given in Chapter 2, the factual discussion and the logic behind the four-quadrant brain model, as well as the statistical data about the HBDI profile included in this chapter are typical quadrant A "languages" of teaching and conveying information.

CHARACTERISTICS OF SEQUENTIAL
QUADRANT B THINKING

Definition: Quadrant B thinking is organized, sequential, controlled, planned, conservative, structured, detailed, disciplined, and persistent. It deals with administration, tactical planning, organizational form, safekeeping, solution implementation, maintaining the status quo, and the "tried-and-true." The culture is traditional, bureaucratic, and reliable. It is production-oriented and task-driven. Edgar Hoover, former Director of the FBI (Federal Bureau of Investigation), and Prince Otto von Bismarck, Prussian Chancellor of Germany (1871-1890), exemplify quadrant B thinkers.

Preferred subjects and careers: People who prefer quadrant B thinking like their subjects in school to be very structured and sequentially organized. Planners, bureaucrats, administrators, and bookkeepers exhibit preferences for quadrant B thinking. People with quadrant B preferences talk about "we have always done it this way" or "law and order" or "self-discipline" or "play it safe." They are talked about as "pedants" or "picky" or "nose to the grindstone."

> *Order and simplification are the first steps toward the mastery of a subject*
> *—the actual enemy is the unknown.*
> *Thomas Mann,* The Magic Mountain, *1924.*

Preferred learning activities: If you are a B-quadrant thinker, you prefer to learn and act in this way:

- Following directions instead of trying to do something in a different way.
- Doing repetitive, detailed homework problems.
- Testing theories and procedures to find out what is wrong with them.
- Doing lab work, step by step.
- Writing a sequential report on the results of experiments.
- Using programmed learning and tutoring.
- Finding practical uses for knowledge learned—theory is not enough.
- Planning projects; doing schedules, then executing according to plan.
- Listening to detailed lectures.
- Taking detailed notes.
- Making time management schedules—the schedule is important, not people.
- Making up a detailed budget.
- Practicing new skills through frequent repetition.
- Taking a field trip to learn about organizations and procedures.
- Writing a "how-to" manual about a project.

🕐 *ONE-MINUTE ACTIVITY 3-3: QUADRANT B LEARNING*
Circle the dots in the list of learning activities above for those items that are easy for you and that you enjoy doing.

Behavior: B-quadrant behavior is one of the easiest to notice in the area of time—B-thinkers stick to schedules, and they get very annoyed when others do not have the same kind of discipline! Since many D-quadrant thinkers have no sense of time, the potential for conflict exists between strong B-quadrant people and strong D-quadrant people. Strong B-quadrant thinking can be a barrier to strengthening other thinking skills, because quadrant B thinkers can be quite inflexible.

The result for the B-team in the exercise with engineering students is shown in Figure 3-5. The generic profile is 1-1-2-2. Note the obvious structure and consistency, including the punctuation. This group was the only one who elected a leader and a scribe. This team reviewed their work and indicated a change in the sequence. The team members were careful with grammar and spelling because quadrant B thinkers pay attention to detail. Here, the interesting words are "breaking the rules" and "leader"—yes, quadrant B thinkers notice when people do not follow procedures, and they are aware of proper leadership roles. The team results exhibit clear B-characteristics, even though this team had an equally strong preference for A-thinking. For these students, quadrant B behavior took precedence over quadrant A behavior in this unfamiliar problem-solving situation.

A WHOLE-BRAINED THINKER;

A COMMUNICATOR;

A CREATIVE PROBLEM SOLVER;

A DESIGNER & INVENTOR OF
    NEW & INNOVATIVE THINGS;

A PERSON WITH GOOD JUDGMENT;

A PERSON WHO BREAKS THE RULES;

A LEADER;

A CORRELATOR OF ABSTRACT IDEAS.

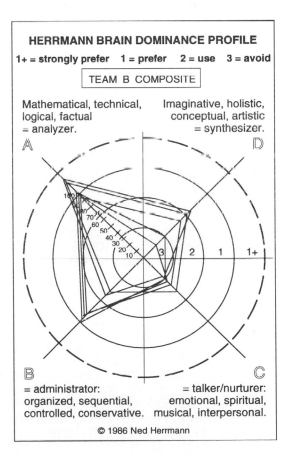

HERRMANN BRAIN DOMINANCE PROFILE

1+ = strongly prefer   1 = prefer   2 = use   3 = avoid

TEAM B COMPOSITE

Mathematical, technical, logical, factual = analyzer.

Imaginative, holistic, conceptual, artistic = synthesizer.

= administrator: organized, sequential, controlled, conservative.

= talker/nurturer: emotional, spiritual, musical, interpersonal.

© 1986 Ned Herrmann

Figure 3-5
Result of quadrant B thinking: definition of an engineer (flip chart facsimile and team HBDI profile).

Exercises to strengthen quadrant B thinking: What can you do if you are not a quadrant B thinker and would like to develop this ability? A course in bookkeeping would be good training for quadrant B thinking. Exercises in Chapter 8 later in this book (using the producer's mindset for tactical planning) will sharpen quadrant B thinking skills. Table 3-2 lists additional activities.

Table 3-2
Activities for Practicing Quadrant B Thinking

- *Learn a new habit through planning and self-discipline.*
- *Cook a new dish by following the instructions in a complicated recipe.*
- *Use a "programmed learning" software package to learn something new.*
- *Plan a project by writing down each step in detail; then do it.*
- *Assemble a model kit by instruction (or a piece of modular furniture).*
- *Develop a personal budget, then keep it for two weeks.*
- *Prepare a personal property list; then put it into a safe-deposit box.*
- *Set up a filing system for your paperwork and correspondence.*
- *Organize your desk drawer or clothes closet.*
- *Organize your records, disks, books, photographs, or collection.*
- *Prepare a family tree, or play* Scrabble.
- *Find a mistake in your bank statement or monthly bills.*
- *Be exactly on time all day.*
- *Visit a hands-on science museum; follow the directions for all the activities.*
- *Learn time management skills—read a self-help book and then* do *what it says.*

When to use quadrant B thinking: As you will learn, the organizational aspects of quadrant B thinking are needed in the creative problem-solving process when we are planning how to put an idea into action and when we actually execute the prepared work plan. The fault-finding aspects of quadrant B thinking are needed by the judge ahead of implementation, in order to prevent errors and minimize risk. The organizational, sequential, disciplined aspects of quadrant B thinking will make the implementation possible. In Bloom's taxonomy, quadrant B thinking is not mentioned explicitly; it is used when students do routine practice problems, and it is a preferred teaching style. Highly talented people who lack quadrant B thinking skills may not be successful simply because they are unable to get their good ideas to the implementation stage. Artists who learn how to be organized in their business dealings (and in scheduling their time) find that this enhances their creativity because they are no longer distracted by dunning notices for unpaid bills and irate customers complaining about missed deadlines. Thus developing a comfortable level of quadrant B thinking—and the judgmental ability of choosing when to apply it most profitably—can enhance the effectiveness of all the other thinking quadrants. Some quadrant B thinking is essential in order to have a balanced life. But if it is used exclusively, it can be a strong barrier to creative thinking. In this chapter, you can see a quadrant B approach in the sequential description of the development of the four-quadrant brain model, in the parallel structure of the chapter sections describing each quadrant, as well as in the detailed "lists of things to do" to practice a particular thinking mode.

CHARACTERISTICS OF INTERPERSONAL QUADRANT C THINKING

Definition: Quadrant C thinking is sensory, kinesthetic, emotional, interpersonal (people-oriented), and symbolic. It deals with awareness of feelings, body sensations, values, music, and communications; it is needed for teaching and training. A quadrant C culture is humanistic, cooperative, and spiritual. It is value-driven and feelings-oriented. Mahatma Gandhi, the Hindu social reformer, typifies a strong quadrant C person.

Preferred subjects and careers: People who prefer quadrant C thinking have preferences for certain subjects in school, such as social sciences, music, dance, drama, and highly-skilled sports, and they participate in group activities rather than work alone. Teachers, nurses, social workers, and musicians have strong preferences for quadrant C thinking, although musicians and composers involve quadrant A thinking when they analyze musical scores or evaluate a performance. People with quadrant C thinking talk about "the family" or "teamwork" or "personal growth" and "values." Stereotypically, they are viewed as "bleeding hearts" or "soft touch" or "talk, talk, talk."

> *With the sense of sight, the idea communicates the emotion, whereas, with sound,*
> *the emotion communicates the idea, which is more direct and therefore more powerful.*
> *Alfred North Whitehead, 1943.*

Preferred learning activities: If you are a C-quadrant thinker, you prefer to learn and act in this way:

- Listening to and sharing ideas.
- Motivating yourself by asking "why"—looking for personal meaning.
- Experiencing sensory input—moving, feeling, touching, smelling, tasting.
- Using group-study opportunities and group discussions.
- Keeping a journal to record feelings and spiritual values, not details.
- Doing dramatics—the physical acting out is important, not imagination.
- Taking people-oriented field trips.
- Traveling to other cultures to meet people; hosting a foreign student.
- Studying with classical background music; making up rap songs.
- Using people-oriented case studies.
- Respecting others' rights and views; people are important, not things.
- Learning by teaching others.
- Learning by touching, feeling, and using a tool, object, or machinery.
- Reading the preface of a book to get clues on the author's purpose.
- Preferring video to audio to make use of body language clues.

ONE-MINUTE ACTIVITY 3-4: QUADRANT C LEARNING
Circle the dots in the list of learning activities above for those items that are easy for you and that you enjoy doing.

Behavior: Quadrant C thinkers are nurturing and reach out to others; they are interested in teamwork, sharing ideas, feelings, and cooperation, not competition. Faith, values, and religious beliefs can have a strong influence on this thinking. Nurses, social workers, teachers, and trainers usually exhibit profiles with strong C-quadrant preferences.

The result for the C-team in the exercise with the engineering students is shown in Figure 3-6; the generic profile is 1-2-1-1. This was the only team that used such personal, emotionally-loaded terms as "overworked, underpaid, underappreciated." Their focus was on people—social interaction was important. This team did a lot of talking and consequently had little time to actually write down ideas. Despite a strong preference for D-quadrant thinking by some of the team members, the C-behavioral characteristics dominated, possibly due to strong interpersonal skills present, as well as the mutual reinforcement the C-thinkers found in working together—students with strong C-dominance are rare in engineering. Even though this was definitely a right-brain thinking team, the presence of a strong double dominance in B and C is seen in the nice presentation and grammatical symmetry of the team's results on the page.

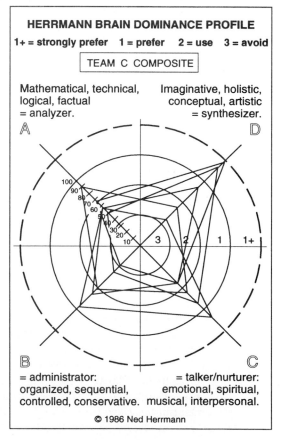

Figure 3-6
Result of quadrant C thinking: definition of an engineer
(flip chart facsimile and team HBDI profile).

Exercises to strengthen quadrant C thinking: What can you do if you are not a quadrant C thinker and would like to develop this ability? Teaching, nursing, and child care all require quadrant C thinking and attitudes. Problem 2-5 in the preceding chapter is an example of a situation that needs quadrant C thinking. Activities that enhance good communications as well as all the team activities in the book will give training in quadrant C thinking skills. Table 3-3 lists additional ideas.

Table 3-3
Activities for Practicing Quadrant C Thinking

+ *Get together with a friend; share your feelings about a topic or issue.*
+ *When in a conversation, spend most of the time listening to the other person.*
+ *Study in a group, or do a group project.*
+ *Get involved in a play or musical, or do charades at a party.*
+ *Compose a song, then get someone to sing it.*
+ *Get involved in a new sport or exercise activity.*
+ *Play with a small child the way he or she wants to play.*
+ *Adopt a pet from the local animal shelter.*
+ *Allow tears to come to your eyes without feeling shame or guilt.*
+ *Think about what other people have done for you and find a way to thank them.*
+ *Become a volunteer in your community on an environmental issue.*
+ *Get involved in a program that teaches adults to read.*
+ *Get involved in a Big Brother/Big Sister program or in scouting.*
+ *"Adopt" an elderly person, or help with "Meals on Wheels."*
+ *Become a tutor or mentor to a disadvantaged child or a fellow student.*
+ *Get to know your neighbors—get together and have a block party.*
+ *Explore your spirituality. Read the religious documents of the major faiths.*
+ *Join a church choir, a barbershop quartet, a square dancing club.*
+ *Savor a vegetable or fruit that you have never tasted before; grow and use herbs.*
+ *Grow flowers; make artistic bouquets and cheer up someone who is lonely.*
+ *Enjoy a walk in nature: pay attention to sounds, smells, and other sensory input.*
+ *Use artwork, colors, and accessories to create a specific "mood" in a room.*
+ *Take a seminar on how to communicate or express your feelings better.*
+ *Find a pen pal from another country or a different culture.*
+ *Make time for family meals—think up a reason to have a special celebration.*
+ *Play a musical instrument "playfully"; learn to enjoy a different style of music.*

When to use quadrant C thinking: The interpersonal aspects of quadrant C thinking are needed in creative problem-solving for teamwork, when collecting data on the "customer's voice," and for meeting the needs of the customer. It is required during implementation when you are trying to build support for your ideas and at any time when good communication with others is required. Bloom's taxonomy does not explicitly include quadrant C thinking. We believe that group discussions can help in acquiring knowledge, in comprehension, application, analysis, and synthesis. An awareness of values and bias will improve judgment—thus quadrant C thinking can improve learning at each level of Bloom's taxonomy. The quotes throughout the book, the periodic group discussions, and explanations of "why" are quadrant C ways of getting you to interact with the ideas being taught.

CHARACTERISTICS OF IMAGINATIVE QUADRANT D THINKING

Definition: Quadrant D thinking is visual, holistic, innovative, metaphorical, creative, imaginative, conceptual, spacial, flexible, and intuitive. It deals with futures, possibilities, synthesis, play, dreams, vision, strategic planning, the broader context, entrepreneurship, change, and innovation. A D-quadrant culture is explorative, entrepreneurial, inventive, and future-oriented; it is playful, risk-driven, and independent. Pablo Picasso, the modern painter, and Leonardo da Vinci, the Renaissance painter, sculptor, architect, and scientist, had strong quadrant D thinking preferences.

Preferred subjects and careers: People who prefer quadrant D thinking prefer subjects such as the arts (painting, sculpture), as well as geometry, design, poetry, and architecture. Entrepreneurs, explorers, artists, and playwrights have strong preferences for quadrant D thinking, as do scientists involved in research and development (R&D) in medicine, physics, and engineering. People with quadrant D thinking talk about "playing with an idea" or "the big picture" or "the cutting edge" and "innovation." They are talked about as "having their heads in the clouds" or as being "undisciplined" or "unrealistic dreamers."

> *As a rule, indeed, grown-up people are fairly correct on matters of fact;*
> *it is in the higher gift of imagination that they are so sadly to seek.*
> Kenneth Grahame, The Golden Age, 1895.

> *Without this playing with fantasy no creative work has ever yet come to birth.*
> *The debt we owe to the play of imagination is incalculable.*
> Carl Gustav Jung, Psychological Types, 1923.

Preferred learning activities: If you are a D-quadrant thinker, you prefer to learn and act in this way:

- Looking for the big picture and context, not the details, of a new topic.
- Taking the initiative—getting actively involved.
- Doing simulations—asking what-if questions.
- Making use of the visual aids in lectures.
- Doing problems with many possible answers.
- Appreciating the beauty in the problem (and in the solution).
- Leading a brainstorming session—wild ideas, not the team, are important.
- Experimenting; playing with ideas.
- Exploring hidden possibilities.
- Thinking about trends.
- Thinking about the future.
- Relying on intuition, not facts or logic.
- Synthesizing ideas and information to come up with something new.
- Using future-oriented case discussions.
- Trying a different way of doing something just for the fun of it.

☉ ONE-MINUTE ACTIVITY 3-5: QUADRANT D LEARNING
Circle the dots in the list of learning activities on the preceding page for those items that are easy for you and that you enjoy doing.

Behavior: Students with strong quadrant D thinking preferences have persisted despite everything the educational system may have done to discourage them. These students may feel like outsiders—different, odd, weird, or even crazy. In self-defense, they may have developed a chip-on-the-shoulder attitude. On the other hand, they may have learned to enjoy independence; they are self-motivated; they truly march to a different drummer. A strong quadrant D person may not be able to understand the language and "tribal" bonding between members of a left-brain-dominant group. But since the world and society need the intuitive insight and innovative ideas of the creative quadrant D mind, we must develop more nurturing environments for quadrant D thinking that can be expressed in many acceptable ways—quadrant D behavior can be more successful if balanced with self-discipline, caring, and logic.

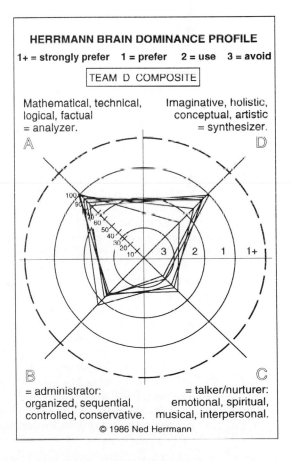

AN ENGINEER IS SOMEONE THAT

solve problems   generates ideas

find solutions   synthesizes

implement solutions
                    defines
translates solutions   problems

synthesizes ideas

new + unique ideas

makes a lot of money

works hard, (sometimes)

destroys bridges

is creative

Figure 3-7
Result of quadrant D thinking: definition of an engineer (flip chart facsimile and team HBDI profile).

HERRMANN BRAIN DOMINANCE PROFILE
1+ = strongly prefer   1 = prefer   2 = use   3 = avoid

TEAM D COMPOSITE

Mathematical, technical, logical, factual = analyzer.

Imaginative, holistic, conceptual, artistic = synthesizer.

A   D

B   C

= administrator: organized, sequential, controlled, conservative.

= talker/nurturer: emotional, spiritual, musical, interpersonal.

© 1986 Ned Herrmann

The result for the D-team in the exercise with the engineering students is shown in Figure 3-7. Typical of quadrant D thinkers, this team had little sense of timing—the students continued working long after the set time had elapsed, and they had to be reminded several times to stop and rejoin the class. The generic profile of this group was 1-2-2-1. Even a quick glance at their page shows that this team operated very differently. They had many ideas all over the page; they changed their mind, they made connections with arrows; they underlined, and they had to give an explanation of what they had written, because it did not make sense to their classmates—the "sometimes" did not belong to "working hard" but to "destroying bridges." In brief, this team brainstormed ideas without thinking about any organization or structure—thus they too had problems with grammar. They had probably the most unusual idea: "An engineer sometimes destroys bridges." This is a part of creativity—we know intuitively that sometimes we must break something or some "rule" before we can put ideas or parts together in a new way. The influence of the joint dominance with A is shown by the phrase "making big bucks." The informal exercise format and setting enabled this group to express their D-quadrant thinking instead of their A-quadrant thinking preference.

Exercises to strengthen quadrant D thinking: What can you do if you are not a quadrant D thinker and would like to develop this ability? Exercises in Part 1, in Chapter 5 (explorer's mindset), Chapter 6, and Chapter 7 (engineer's mindset), will give you training in quadrant D thinking. Table 3-4 lists additional activities.

Table 3-4
Activities for Practicing Quadrant D Thinking

- *Look at the big picture, not the details, of a problem or issue.*
- *Make a study of a trend; then predict at least three different future developments.*
- *Ask what-if questions and come up with a lot of different answers.*
- *Allow yourself to daydream.*
- *Make sketches to help you memorize material that you are learning. Create a logo.*
- *When solving problems, find two or three different ways to do them.*
- *Do problems that require brainstorming; find at least ten possible answers.*
- *Appreciate the "beauty" of a design: building, appliance, object.*
- *Play with* Tinkertoys, Skill Sticks, Legos.
- *Learn to paint, sketch, draw; play with modeling clay. Take an art class.*
- *Attend a "story-telling" session; read a book of folktales or myths; participate in role-playing games.*
- *Design and build a kite. Fly the kite the way it is meant to be flown.*
- *Invent a gourmet dish and then prepare it.*
- *What time of day are you the most creative—when first waking up, exercising, or taking a shower? Use this time to think up and jot down ideas, then take the next afternoon off to further explore one of these creative ideas.*
- *Take a drive (or walk) to nowhere in particular without feeling guilty.*
- *Take 200 photographs without worrying about cost; try unusual shots.*
- *Imagine yourself in the year 2000, 2020, 2040.*
- *Investigate how a particular subject can be connected to other things you know.*
- *Use analogies and metaphors in writing or when explaining a concept or idea.*

When to use quadrant D thinking: The imaginative, wishful aspects of quadrant D thinking are needed in the creative problem-solving process during the brainstorming phase. Exploratory, holistic, contextual thinking is also needed for problem definition. Innovative thinking is needed when you are trying to find the best way to implement the solution. Even during evaluation and judgment, creative thinking is needed to overcome flaws and difficulties. And creative thinking is required when dealing with difficult people, when dealing with change, when making plans and developing goals for the future. In Bloom's taxonomy, creative thinking is involved in the process of idea synthesis—yet quadrant D thinking, just like quadrant C thinking, can improve each of the other levels of thinking and learning. Figure 2-5 and the mindmaps with sketches illustrate D-quadrant ways of presenting material.

> *No matter how different you are, there are other normal people like you*
> *somewhere in the world. Celebrate your uniqueness!*
> *Ned Herrmann.*

WHOLE-BRAIN THINKING
AND CREATIVE PROBLEM SOLVING

Whole-brain thinking: In the exercise with the teams of engineering students, we have one more result to consider—the definition from the multi-dominant team. It is shown in Figure 3-8 with the individual profiles of the five students in the team. The generic profile of their composite is 1-1-2-1, with almost equally strong dominances in the A, B, and D quadrants and the preference in C only slightly smaller. This team wrote a definition that presents a more complete appearance than the work of the other teams: it has whole sentences; it is well-balanced on the page; it gives two options, and it even includes a sketch (and a stab at D-quadrant humor). The team's lack of quadrant C thinking shows in the terminology of "guy" and "individual." As you compare these five team results, keep in mind that this was a very brief exercise where the teams did not have time to revise their first draft. None of these students had any previous training in teamwork. But the results with these inexperienced teams do show that differences in thinking are expressed in different languages and problem-solving outcomes.

In assessing over half a million people, Ned Herrmann has found that 7 percent have a single dominance; 60 percent have a double dominance; 30 percent have a triple dominance, and only 3 percent have a quadruple (1-1-1-1) dominance. He has also determined the thinking preference (pro-forma) profiles of many famous people by analyzing their writings and life work. These people are distributed all across the four quadrants—no single profile is more prominent or more valuable than any of the others. People are happier and usually will do well when their activities and job requirements match their thinking preferences. Evaluating the HBDI questionnaire is not a routine, analytical task; it requires in-depth training, a thorough understanding of the characteristics of each mode, as well as intuitive judgment based on the information provided. Even though two people

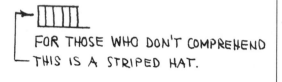

AN ENGINEER IS:

1. A GUY WHO DRIVES TRAINS...

—OR—

2. AN INDIVIDUAL WHO USES EXISTING KNOWLEDGE CREATIVELY TO SOLVE CURRENT PROBLEMS.

FOR THOSE WHO DON'T COMPREHEND
THIS IS A STRIPED HAT.

Figure 3-8
Result from a whole-brain team: definition of an engineer (flip chart facsimile and team HBDI profile).

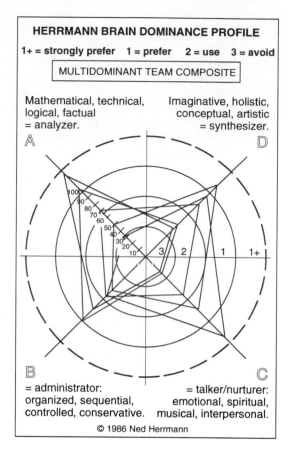

HERRMANN BRAIN DOMINANCE PROFILE

1+ = strongly prefer   1 = prefer   2 = use   3 = avoid

MULTIDOMINANT TEAM COMPOSITE

Mathematical, technical, logical, factual = analyzer.
A

Imaginative, holistic, conceptual, artistic = synthesizer.
D

B
= administrator: organized, sequential, controlled, conservative.

C
= talker/nurturer: emotional, spiritual, musical, interpersonal.

© 1986 Ned Herrmann

can have almost identical profiles, they will be "different" thinkers with differing abilities and competencies, since they will have different clusters of preferences within each quadrant. However, they will most likely find it easier to communicate and understand each other than people who have opposite profiles.

Brain preference does not equal competence. Competency is achieved through training, motivation, and practice—we can learn to use our primary and secondary preferred modes more effectively. Ned Herrmann says that operating in a less preferred mode is akin to traveling to a foreign country and experiencing culture shock. We might struggle initially with the language and the customs, but we can become comfortable with the culture *if we are motivated and if we practice*. This is true particularly for the quadrant C and D modes which are not encouraged in our educational system. If you have a strong thinking preference in these modes, but your daily life and work are devoid of opportunities to express these modes, you can change and learn to "claim your space." If you have an avoidance in one of the quadrants, our advice is to make a career choice that will not require you to have to function in this mode on a day-to-day basis—the frustration and energy level required would be too great.

All four thinking quadrants are involved in learning, as shown in Figure 3-9. We have what is called *external* learning taught from authority through lectures and textbooks —> quadrant A learning. Then we have creative thinking, which is *internal* learning through a flash, an insight, a visualization, an idea synthesis, or a sudden understanding of a concept holistically and intuitively —> quadrant D learning. Then we have *interactive* learning through discussions and hands-on, sensory-based experiments where we try, fail, try again with an opportunity for verbal feedback and encouragement —> quadrant C learning. And finally, we have *procedural* learning through a methodical, step-by-step testing of what is being taught, as well as practice and repetition to improve skills —> quadrant B learning. Effective teachers have discovered ways of incorporating each one of these learning modes into their teaching strategies. This goal is not always easy to achieve when the instructor may have strong thinking preferences in only one or two quadrants.

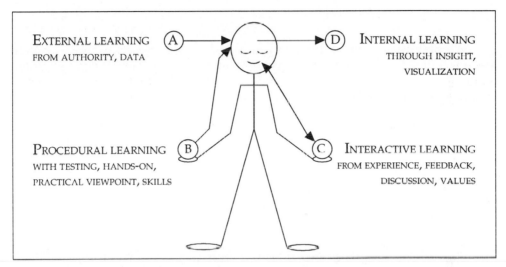

Figure 3-9   Four modes of how students learn.

⏱ *Three-Minute Team Activity 3-6: Whole-Brain Teaching*
In groups of four people, select a concept in some subject (science, mathematics, politics, economics, the arts, etc.), then make up examples of how to teach the concept in the quadrant A, quadrant B, quadrant C, and quadrant D thinking modes. Share your ideas with a larger group, or develop additional examples for several different fields.

A balanced view between wholeness and specialization is the key:
The brain is designed to be whole, but at the same time we can and must
learn to appreciate our brain's uniqueness and that of others.
Ned Herrmann.

Marriage is something you have to give your whole mind to.
Chinese fortune cookie.

Creative problem-solving mindsets: So far, we have looked at creative problem solving as a process. We will now look at the thinking skills or mindsets that are required at each step of the creative problem-solving process.

MINDSETS REQUIRED FOR CREATIVE PROBLEM SOLVING

Problem Definition — Detective and Explorer
Idea Generation — Artist
Creative Idea Evaluation — Engineer
Idea Judgment — Judge
Solution Implementation — Producer

We first learned about different problem-solving mindsets from Roger Von Oech, but then we added the detective for data analysis, since problem definition requires divergent and convergent thinking. We also invented the "engineer" since creative idea evaluation is a very important step in creative problem solving—a step that differs from the artist or the judge, as we shall see. And we changed the warrior into the producer in response to requests for a change from students and teachers. These metaphors make it easier to remember the type of thinking that we need to use in each step of the creative problem-solving process.

Let's visualize these concepts or metaphors in some detail. If you are struggling with a problem that involves some difficulty, imagine seeing Hercule Poirot (or the other endearing detective created by Agatha Christie—Miss Jane Marple) or Sherlock Holmes, or any of the many different and interesting detectives in recent movies or TV shows, such as J. B. Fletcher in "Murder She Wrote" or Virgil Tibbs of "In the Heat of the Night." Can you see these detectives walking around in the dark with a flashlight or matches? Can you see them using their minds to evaluate clues? Can you hear them asking questions? For problem definition—when we are looking for opportunities—we need to think like an explorer. Imagine being armed with field glasses, a compass, and a large notebook, keeping a sharp eye out for ideas. Imagine a character like Indiana Jones. Can you think of other explorers? What about Dian Fossey, Christopher Columbus, David Livingstone, Jean-Jacques Cousteau, Thor Heyerdahl, Roald Amundsen, or the astronauts and cosmonauts? During the problem definition phase, we find out as much as possible about the problem. As explorers and detectives we search for opportunities and information, we analyze data, and then come up with a positive problem definition statement.

Now picture yourself as an artist, a Picasso, a Michelangelo, a Steve Martin, a Walt Disney, a Georgia O'Keeffe, a Diego Rivera, a Grandma Moses, a Jim Henson, or a Margot Fonteyn. This is the stage where we do brainstorming to come up with a multitude of creative ideas—here we need to think like an artist—the wilder and crazier, the better. Imagine standing in front of a large sketch pad, furiously drawing a slew of sketches, noting down all kinds of ideas and fragments of ideas. Perhaps you can imagine a team of artists collaborating and contributing to an idea "collage." Next, conjure up a new image in your mind—that of an engineer, a design engineer, a tinkerer. She sits in front of a drawing board or a computer screen. She examines and plays around with all kinds of wild ideas with a view toward combining them to develop more complete, more practical ideas. She calls in a team to improve ideas through this second round of brainstorming activity. The team tries to borrow ideas from nature and force-fits them into innovative solutions to problems. It is during this process of idea synthesis that our minds can come up with "better" and more practical ideas! Now, in your mind's eye, enter into a courtroom. In front of you sits the judge, gavel in hand and ready to render a judgment as to which ideas and solutions are best. As judges we set up criteria and make decisions about which solution or solutions will be implemented. Judges look for flaws, but then try to overcome them with additional creative thinking.

Finally, it is time for the producer. A producer is a jack-(or jane)-of-all trades: maker, creator, nurturer, mover, parent, organizer, inventor, builder, executor, entrepreneur, director, practitioner, actor, planter-grower-harvester, seller, coach, quarterback, general, leader, manager. The producer is responsible for putting out a quality product, for fund-raising, and for budgeting. Producers need courage; they take risks; they deal with creative ideas and innovation. They need good communication skills, be it in producing a movie, in manufacturing a consumer product, or in growing a crop—to be successful, the teamwork of many must be managed, as well as budgets, organization, technology, and strategic planning. In essence, the producer is in charge of a new round of creative problem solving, where the problem is the implementation process. Another image that is important here is the picture of a person who is ready to fight or stand up for an idea. The producer must be persistent to carry the project to a successful conclusion while working to overcome opposition and other adversity.

Now let's play around with these six roles or mindsets a bit. What would happen if we left out the explorer or the detective? We can still come up with many good ideas and get a good solution except that the solution may not fit the original problem, because we did not take the time to find out exactly what the problem was, or we did not look at the problem in its entire context. So we may still have a problem. What would happen if we left out the artist? Actually, this happens quite frequently. This happens all the time when you get an idea and right away tell yourself: "No, this won't work; this is a crazy, dumb idea." This happens each time you take the very first idea that comes into your head and rush to implement it without looking for alternatives. As we will see later, it is very important to get a lot of wild ideas before judging any of them. This is one of the basic principles of brainstorming. This deferred judgment means that many creative ideas are generated first—in the artist's mindset—before the engineer and judge come into the picture to determine how good the ideas are. That is why we must have the artist, and that is why we must give this mindset a sufficient amount of time.

What would happen if we left out the engineer? Without the engineer, the judge may be getting only half-baked ideas to choose from. The engineer is needed to make good ideas better, to make wild ideas more practical, and to develop optimum solutions. Most analytical problem-solving methods leave out the artist and the engineer. And how important is the judge? Without the judge we would probably not be able to select the best idea. What would happen if we only had the producer? Do you think that could work? It might in very rare cases, just because the enthusiasm and energy of producers could carry it off—if they are lucky enough to pick a solution out of the blue that actually solves the problem. You can probably think of times when you and your friends took this approach—young people have a marvelous knack for getting excited about ideas. But most of the time, ideas must be evaluated by the engineer and judge to prevent producers from taking reckless risks.

But to create the best conditions for coming up with a good solution to the problem, it is best to follow the process in this sequence: detective and explorer, artist, engineer, judge, and producer. As you will see, even during the different steps we may have to pause and backtrack before we can continue. Sometimes we can separate the different steps by doing them with different groups of people on different days; sometimes we have to do them all ourselves. Sometimes we have a lot of time (weeks, or even months), sometimes only days, hours, or just minutes, but even then we should try to do problem solving in steps, with the different mindsets in the correct order, to get the best results.

Combining the whole-brain model with creative problem solving: The six problem-solving mindsets can be related to the Ned Herrmann whole-brain model as shown in the diagram of Figure 3-10. In analytical problem solving, only the left modes are used, where we move immediately from quadrant A thinking for problem definition to quadrant B for solution implementation. In creative problem solving, we also start with quadrant A and quadrant B thinking as detectives and analyzers, but then we move to quadrant D and quadrant C as explorers who investigate the larger possibilities, ecological aspects, and social ramifications of the problem. As artists, we make use of the capabilities of the right modes—we employ visual as well as sensory thinking, feelings, and intuition.

To integrate ideas with the mindset of the engineer, we must move rapidly back and forth between cerebral quadrant D and quadrant A thinking, as we seek to make ideas more practical through creative evaluation. As judges, we use the critical ability of the entire left hemisphere, beginning with quadrant A reasoning based on criteria and logic, followed by quadrant B procedures to further evaluate ideas and their associated risks. An iteration back to quadrant D may be needed to overcome flaws before making the final decision in quadrant A. As producers, we begin in quadrant B with preliminary tactical planning for solution implementation. We use quadrant C thinking for the selling plan where the communications aspects of solution implementation are considered, such as overcoming opposition and the fear of change to gain idea acceptance, meeting the customer needs, and assessing the staff educational requirements in the organization. Then we use

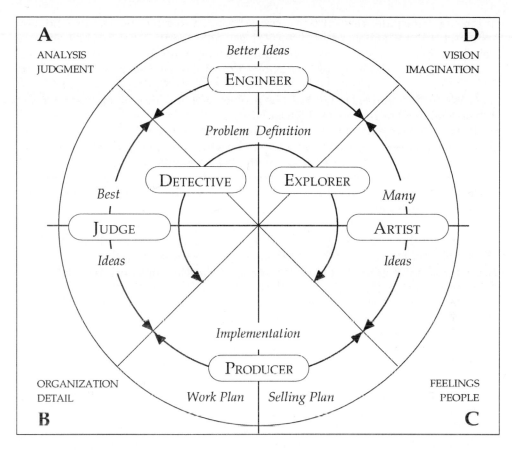

Figure 3-10   The Herrmann model and the creative problem-solving mindsets.

quadrant B thinking to prepare a detailed work plan. Since solution implementation is a new, unstructured problem, we as producers must do a new round of creative problem solving, using all four quadrants. Creative problem solving ends with B-quadrant thinking for checking the work plan, implementing the solution, monitoring the results, and conducting a final evaluation of the entire process.

The color graphic on the book cover is an enhanced visualization of the two models, and it includes the directional path and iterative loops of the creative problem-solving process. Each color is associated with one of the mindsets—the location and range of the colored band identifies the primary thinking quadrants used at that step in creative problem solving. As you study the remainder of the book, you may want to use highlighters in these colors for material that relates to the respective mindsets. When reviewing later, green markings on clippings, notes, mindmaps, and information in the book would indicate to you at a glance that this matter has to do with the engineering mindset (where the artist's brainstormed ideas are synthesized to result in fewer but better ideas). If you are preparing transparencies to teach creative problem solving, you can use the same color scheme for enhanced learning. Please note that this color scheme has no relationship to the colors Ned Herrmann associates with the four quadrants of his brain dominance model in his workshops and HBDI data sheets.

When do you need to use creative problem solving, or when is it appropriate to use analytical or routine problem solving?

1. When we have a repetitive, routine, well-structured problem (such as building an ordinary bookcase, or getting to work expediently under normal traffic conditions), we can use our usual methods and standard procedures.
2. When we have an unstructured problem which is very <u>elusive</u>, <u>ambiguous</u>, and poorly understood and which involves changing conditions (such as dissatisfied employees or customers), we need creative problem solving.
3. When we have a strategic problem which is important and long-term so that it can affect the existence of the entire organization (for example, when new paradigms are pioneered in our field), we need creative thinking.
4. Operational, tactical problems in our day-to-day dealings must be solved analytically (to deal with the crisis aspect) as well as creatively (to deal with the opportunity aspect), or they can grow into strategic problems. Creative problem solving will prevent us from finding superficial solutions to deep-seated problems and from finding long-term solutions to short-term problems (such as hiring a permanent employee for a temporary work overload).
5. When we have an emergency or crisis, we need to react quickly—we must use authoritarian procedures. This immediate attention does not involve creative thinking. We do not have time to identify ten alternate ways of fighting a fire: we grab the extinguisher, call the fire department, and evacuate the building—we follow a predetermined procedure.
6. But when a problem represents an opportunity, we can be proactive; this requires visual, imaginative, creative thinking and problem solving and can result in true innovation—we can prevent future crises.

ORGANIZATIONAL IMPLICATIONS OF WHOLE-BRAIN THINKING AND LEARNING

Creative problem solving and four-quadrant thinking are not only important for individuals—they have serious implications for organizations and teamwork. Many organizations, especially as they get older, tend to get more entrenched in quadrant B thinking—after all, this is the mode that minimizes risk and preserves the status quo, the existing paradigm of success. However, many companies can benefit when the staff's thinking preferences match the work requirements. According to Ned Herrmann, executives having triple or quadruple dominances have an advantage—they can communicate and understand a wider range of employees and solve problems more easily, but the particular mix of thinking preferences that would be optimum in an organization depends on the nature of its business.

The example I set is important, so it's critical that I encourage ideas.
Bill Gates, president of Microsoft.

The unit within any system that has the most behavioral responses available to it controls the system.
First Law of Cybernetics.

You can provide an environment in which creativity will flourish.
Allen F. Jacobson, CEO, 3M Company.

Organizations, to be successful, must have all four brain quadrants available for effective problem solving and response to changing situations. These organizations can be anything from a small family unit, a school, a club, or a small business, to large universities or multinational corporations, and they must create a supportive climate for creative problem solving and innovation. As we are learning in this book, it is really not that difficult to have creative ideas—implementing a creative idea is the difficult part. An idea becomes an innovation when it is a new paradigm that is being pioneered and adopted, leading to permanent change in people's lives. Innovation can be a thing—a new product or a new technology; innovation can also be a service or a procedure—a new way of doing something.

Two kinds of training are needed to support a whole-brain approach: first, creative thinking skills must be encouraged and developed. Then these skills have to be followed up with integration into the creative problem-solving process, particularly as they apply to working in teams. It has been found that creativity training of employees will not be effective without the second step. It is crucial that creative problem solving be used at all levels of the organization in order to create a supportive climate for change and innovation.

What constitutes a supportive environment for creativity? Many aspects fall into this: not being satisfied with things as they are, but always looking for improvement and better ways of doing things; recognizing that technology is a competitive tool that can make difficult products achievable. Calculated risk is acceptable, and bureaucracy is at a minimum. Creative thinkers and innovators are rewarded, not only financially, but even more importantly by seeing their ideas developed into tangible products and services that succeed in the marketplace. In highly innovative companies, people from all departments are involved in creative problem-solving teams and in the evaluation and implementation process, while management develops policies and procedures that encourage this process of never-ending innovation and improvement. These organizations go out of their way to patiently followthrough on behalf of new ideas.

An environment that encourages creativity is a protective environment—the innovator is allowed to learn and work by playing. How does a tiger cub become a successful hunter? It stalks and pounces around playfully while being protected by the strength and watchful eye of the parent tiger. Creating a supportive environment is not as easy as it sounds, because innovation is often chaotic, and change is seen as threatening to the well-oiled gears of an organization. Thus it is not surprising to see that innovators often work best as parts of small teams or "skunk works" operating outside regular channels. Management for creativity is much more than paying lip service by providing a "suggestion box." The organizational culture unwittingly sabotages innovation unless it establishes definite mechanisms to develop and fund creative ideas. "If it ain't broke, don't fix it" is an attitude that blocks creative thinking, continuous improvement, and innovation—a major factor in the decline of U.S. industry. We may disagree with this assessment, but it should give us food for thought about our own problem-solving habits—analytical approaches and a hierarchal chain of command may feel more comfortable to us than the ambiguity introduced by right-brain thinking or by cooperative activities.

An organization that trains and encourages people in whole-brain thinking and creative problem solving can expect to reap the following benefits:

1. People will be able to better understand and communicate with each other; they will not only have a common vocabulary but also common meanings.
2. People will be able to work together better in teams, with common goals. These whole-brain teams can usually achieve better-quality solutions.
3. People's thinking preferences can be better matched to their job requirements, resulting in increased commitment, enthusiasm, energy, and productivity. Ned Herrmann believes the potential productivity gain is around 30 percent (*The Creative Brain*, p. 157).
4. Managers will be more effective if they can recognize and appropriately match or complement their employees' thinking skills with the tasks that need to be accomplished. People will be hired because they think "differently," not because they think just like everyone else in the department.
5. Learning and teaching styles within the organization can be matched more effectively, as people's training and skills are continuously upgraded.
6. Whole-brain people act as "translators" to coordinate the work of groups with very diverse thinking preferences. How would you manage a team when the A-thinkers are looking for "ways to count," the B-thinkers for "ways to save," the C-thinkers for "ways to help," and the D-thinkers for "ways to spend"?
7. When all the thinking preferences are used, the organization gains a powerful advantage in problem solving, innovation, and long-term success.

An article in October 1990 over the *Times-Post News Service*[1] reported on a study conducted at the University of California-Irvine on women in management. According to these findings, female executives tend to lead in nontraditional ways: by sharing information and power. They inspire good work by interacting with others, by encouraging employee participation, and by showing how employees' personal goals can be reached as they meet organizational goals. This style of leadership may be especially appropriate for the future, for the global economy of multinational companies, for service industries, and for businesses in fields of fast-changing technology, where having only a few top people giving orders does not work well. This study thus supports the idea that the most effective leaders and problem solvers are people who can use the whole brain.

---

[1]"Study Says Women Manage Differently," *Toledo Blade*, October 26, 1990, p. 22.

The discussion so far has taken the view of supporting creativity from the top down, through managerial leadership and providing the right kind of training in creative thinking and problem solving. But organizations can also grow creative—at least in the first two stages—from the bottom up. Most organizations are at the first stage in which a few creative individuals work quietly in isolation, so they will not be noticed and get into trouble. In the second stage, two or more of these individuals discover each other—they begin to collaborate and mutually enhance each other's creativity. This process may start a chain reaction and involve more and more people in teams—creativity has become contagious and acceptable! Finally, in the third stage, management will come to support these efforts with creative challenges and assignments; it will evaluate, appreciate, and implement good solutions—with the outcome that the organization has become a creative community because creative individuals have initiated the process. As the following discussion will show, support and encouragement are crucial if people are to become (or remain) creative thinkers.

Thinking Preferences of Engineering Students: Research Results

Questions: What happens when an educational system introduces creative problem solving into the curriculum? The thinking preferences of engineering students at the University of Toledo were assessed in a longitudinal study using the HBDI. In the spring of 1994, we evaluated four academic years of HBDI data going back to the fall of 1990—each year, all students in the fall creativity classes completed the HBDI; seniors were invited during the spring term before graduation to voluntarily participate. The study was to answer a number of questions:

1. Can the HBDI be used to evaluate curriculum changes (for instance the effect of teaching creative problem solving and other whole-brain approaches)?
2. How easily can a student's thinking profile be changed?
3. What happens to students who think "differently" from the prevailing organizational, analytical mindset?
4. Are we graduating students with the thinking skills required by industry to succeed in the global environment of the twenty-first century?

Context: Industry is increasingly looking for employees who can think holistically, who can innovate, who can work in teams, who can synthesize and who can integrate environmental and societal values and ethics into their work—all activities that demand right-brain thinking skills. Just by interacting closely with industry, we arrived at the same conclusions as Ned Herrmann did about the paradigm shift in thinking preferences needed for success. The progression of this paradigm shift over the last three decades is shown in Figure 3-11. Compare this figure with the average profile of the typical engineering faculty (from Figure 3-3) and with the fact that the graduating seniors exhibit a profile very much like that of the faculty. We believe more engineers with the 1990s profile are needed—engineers with a 1970s mindset will no longer find a ready acceptance in the marketplace. The following discussion of our research results is most meaningful when we divide the seniors into three different groups: the "traditional" students, the "different" students, and the creativity assistants.

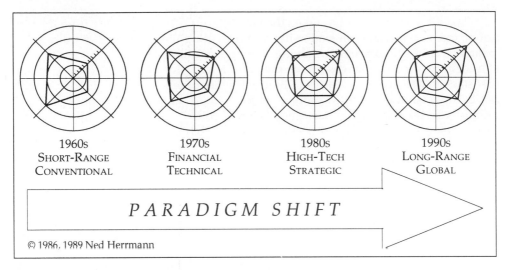

Figure 3-11   Thinking skills required for success.

Results: For the traditionally educated group of eighteen students, the drop in quadrant C thinking shown in Figure 3-12 is of deep concern since quadrant C skills are needed in industry for effective teamwork. As freshmen, this group has a whole had a score of 60 in quadrant C; this dropped to a low of 46 (slightly below that of the faculty average). Their quadrant D preference also dropped.

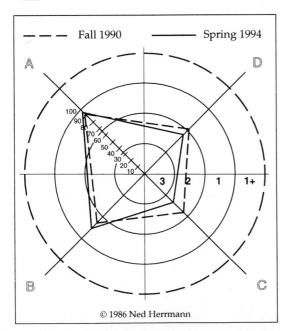

Figure 3-12   Changes in average profile for eighteen traditional engineering students.

The creativity class assistants were students who helped in creative problem solving classes or in our math/science Saturday academy where we taught the creative problem solving process to middle school students and their parents. The six creativity assistants—without exception—had profiles which moved to the right. If they were left-brained as freshmen, they dramatically increased creative or interpersonal thinking to become more whole-brained; if they were right-brained, they strengthened these preferences. They were the only group that increased in the quadrant C mode (which is beneficial for student-centered teaching), as shown in Figure 3-13.

The six "different" students were students whose HBDI results at first glance did not fit the pattern; thus we attempted to interview them by telephone.

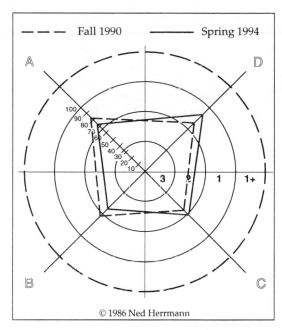

Figure 3-13 *Changes in average profile for six creativity class assistants.*

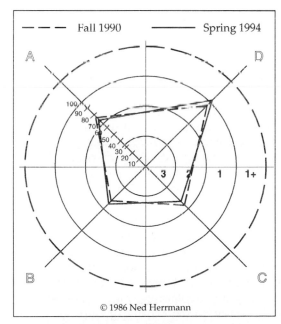

Figure 3-14 *Changes in average profile for six "different" engineering students.*

If they could not be reached, we talked to close friends or family members instead to explore the reasons for the unusual results. These students were "different" in that they all had strong quadrant D preferences as freshmen, and they remained relatively unchanged, as shown in Figure 3-14. What made them able to resist the pressures of a strongly left-tilted curriculum? We found that the creative problem- solving course provided an affirmation that it was OK to be different—the course validated their independent outlook. These students also built quadrant C and quadrant D activities into their lives: they studied in groups or they immersed themselves in music and other creative activities in their free time.

Conclusions: We found that the HBDI is a very useful tool for evaluating the effect of curriculum changes. The noticeable rise in the overall quadrant D scores of the seniors in the four years of the study (from a low of 57 to a high of 75) is due solely to the influence of the creativity students and the higher retention of the "different" students.

We were surprised to find how much individual students had changed in their thinking preferences, some making almost complete reversals from left- to right-brain thinking or vice versa, depending on the type of teaching they received and the subjects that they studied. For example, a majority of students in computer science developed an avoidance of quadrant C thinking, and it is not surprising that the particular department did not have any females in group. Because thinking preference can be changed, the freshmen profiles are not a good predictor of success in engineering. The profile is best used to gain insight on how to match a student's thinking preference with optimum learning strategies.

Students who think "differently" can be identified with the HBDI and given special encouragement, mentoring, and support to persist to graduation—they will have excellent career opportunities once they make it through the left-brain curriculum. Without a nurturing classroom environment, many will switch to different fields or drop out of college. When HBDI results are used to put together whole-brain teams for senior design projects and other team activities, students with quadrant C and D preferences are recognized and appreciated. We are convinced that we can design the curriculum and train the faculty to teach in such a way as to assure that we will graduate students who have the whole-brain thinking skills required to succeed in the diverse and rapidly changing world of the new century.

In addition to creative problem solving, another model of whole-brain teaching exists—the 4MAT system of learning styles. It is beginning to be used in many U.S. schools (including universities) to improve teaching and learning. We want to give a brief overview of the four-quadrant 4MAT system since it is occasionally mistaken for the Herrmann four-quadrant model of thinking preferences.

The 4MAT System of Learning Styles

The 4MAT system of learning styles was developed by Bernice McCarthy and is derived from the Kolb learning cycle which is based on two sets of dichotomies in the way students learn. Some learners are most comfortable in evaluating experiences through their senses while others prefer to analyze and logically think about what is happening in an abstract, more detached way. To fully understand a problem or situation, learners need to employ both approaches. But experience alone is not sufficient for learning—information must be processed and evaluated to become part of the learner. Some people are watchers; they reflect and slowly create meaning through deliberate choice of perspective. Doers try to work with the new information immediately; reflection comes only after experiment and experience. All four ways of learning are valuable and need to be used and encouraged since they are complementary. When the four ways of learning are arranged in a circle, with feeling at the top, thinking at the bottom, doing on the left, and watching on the right, four quadrants are formed that describe four major learning styles. Kolb and McCarthy have shown that the four styles are distributed nearly equally in the general population. A schematic of the four poles and the location of the associated quadrants or learning styles is shown in Figure 3-15.

Type 1—Divergers: To learn, these students (in the upper right-hand quadrant of the model) prefer to listen and to share ideas. Personal experiences and involvement are important to them. They are interested in people, harmony, values, culture, context (and all sides) of a problem. They find the traditional school curriculum too fragmented, disconnected, and unable to provide them with a holistic understanding of their world. Teachers must explain the meaning—the "why"—of the material they are about to teach. These students thrive on questioning, brainstorming, and class discussions; their focus is on relationships. In the Herrmann model, these students have quadrant C and D thinking preferences.

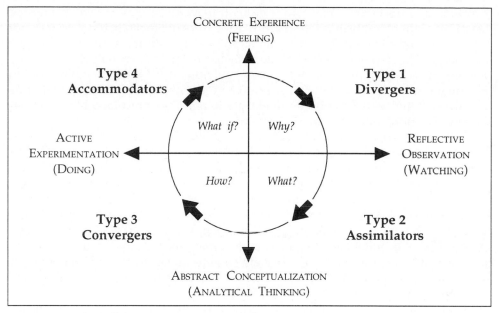

Figure 3-15   The Kolb learning cycle and types of learners, together with the 4MAT questions.

Type 2—Assimilators: These students are integrators of observation with existing knowledge—they have the engineer's mindset in the creative problem-solving model. They conceptualize ideas; they like to play around with new thoughts, concepts, and models. They use logical, deductive reasoning. They prefer personal intellectual achievement to teamwork and social interaction. They do quite well in the traditional, lecture-based, textbook-driven classroom and look to teachers to provide knowledge in answer to the "what" question. The focus is on facts, data, and knowledge. In the Herrmann model, these students have strong quadrant A and some quadrant D thinking preferences.

Type 3—Convergers: These students (in the bottom left-hand quadrant) have a practical, hands-on outlook. They use both abstract knowledge and common sense. They like to test theories and believe, "If something works, use it." They are pragmatic problem solvers; they want to find out how things work. School is often frustrating to them, because many subjects are too theoretical and do not include immediate application and use of what is being taught. Teachers must emphasize the "how"—and demonstrate the usefulness of the material. Field trips and lab experiences are preferred ways of teaching these learners. The focus is on developing skills. In the Herrmann model, these learners have strong quadrant B thinking preferences, supplemented with quadrant A abstract, deductive reasoning.

Type 4—Accommodators: These students (in the upper left-hand quadrant of the model) often learn by trial and error. They find good solutions intuitively but may be unable to provide a logical explanation of how they got the answer. They are comfortable with people, and they frequently are natural leaders and very expressive performers. They find it difficult to sit still—they thrive on physical activities. School for them is too structured and does not meet their need for a wide variety of experiences. Teachers can support these students' interests in idea synthesis and creative thinking by asking "what-if" questions and giving them independent learning projects, not routine assignments. When compared to the Herrmann model, these students seem to be the most whole-brained, with a tilt toward preferences for quadrant C and D thinking. They are adaptable and thrive in new situations since they are risk takers.

Each of the four learning style quadrants contains right-mode, left-mode, and whole-brain learners, although for Types 2 and 3, left-brain processing dominates and for Types 1 and 4, right-brain modes seem to dominate. Bernice McCarthy at this point made the decision to incorporate right-brain and left-brain teaching techniques into the 4MAT cycle in order to fully address each one of the four learning styles. All four styles are to be taught to all learners, with each style equally valued. As a result, each student will be comfortable some of the time and stretched to think in different ways at other times. For effective teaching, instructors must use all segments of the cycle in the 4MAT system, even though they themselves may have preferences for particular modes. The cycle moves from concrete experience to reflective observation to abstract conceptualization to active experimentation, and then back to concrete experience.

The eight steps of the 4MAT learning cycle: Specifically, the 4MAT system has the following steps (two steps per learning style), as discussed in the October 1990 article in *Educational Leadership.* The quadrants indicated by the arrow at each step make a connection to the thinking preferences of the Herrmann model that we believe are being addressed.

TYPE 1: The key concept here is *motivation.* The subject is introduced in a way that provides the big picture. Meaning and relevance to the student are provided through positive class discussion, stories, simulations, journal writing, interactive lectures, field trips, and team activities. The steps are:
1. Creating an experience —> right mode, quadrants C and D.
2. Reflecting: analyzing experience —> left mode, quadrant A.

TYPE 2: *Expert knowledge* is emphasized here. Well-organized information is provided and integrated with previous knowledge in a way that invites reflection. Lectures use visual aids, problem examples worked by the instructor, and demonstrations. Students are encouraged to do library searches, as well as collect and analyze data. The instructor may hand out notes; textbook problems are assigned. The steps are:
3. Integrating reflective analysis into concepts —> right mode, quadrant D.
4. Developing knowledge, concepts, skills —> left mode, quadrants A and B.

TYPE 3: The focus during this stage in the cycle is on *applications.* The instructors coach students in the use of heuristics, algorithms, and procedures for problem solving through demonstrations and lectures. Example problems are worked by the students. Homework, lab work, computer work, field trips, and report writing provide opportunities for applying the knowledge being learned.   The steps are:

    5. Practicing defined "givens"—> left mode, quadrant B.

    6. Practicing and adding something of oneself—> right mode, quadrant C.

TYPE 4: *Discovery,* as well as sharing and performance evaluation, is the key concept here. Open-ended problems, labs, and research are used, as for example in so-called capstone courses.   Teamwork is important.   Group discussions, role playing, field trips, think tanks, quality circles, simulations, and group project reports (oral and written presentations) are additional techniques that are effective, as are optimization, brainstorming, and asking what-if questions. This quadrant can also include simulation, team design projects, and inventing, and students may do some of the teaching. The steps are:

    7. Analyzing applications for usefulness —> left mode, quadrants A and B.

    8. Applying to new, complex experience —> right mode, quadrants C and D.

If teachers or instructors are not teaching to all four learning types, students should try to find activities on their own or with fellow students that will teach important concepts through the different steps of the cycle. Alternatively, students can use the Herrmann model to design learning activities that will use all four thinking quadrants. Either way, the whole-brain approach will make for better thinking and learning. A major difference between the 4MAT system and creative problem solving is the placement of the "what if" phase. In creative problem solving, brainstorming and playing with ideas come early in the process (before evaluation); in the 4MAT system, they come as the last step.

Example of an application: How can the topic of centripetal force be taught using the eight-step sequence outlined above? 1. Let students ride on a merry-go-round. 2. Discuss what the students have experienced—what have they felt or sensed about the forces involved? 3. Connect the analysis to the theory and relevant general equations in physics. 4. Derive the particular equations and help students define and solve centripetal-force problems. 5. Reinforce the new problem-solving skills with homework problems. 6. Assign further projects—for instance, let students experiment with throwing a ball while riding the merry-go-round, or have them write a report. 7. Students are encouraged to find or explore practical applications for what they have learned (for example, the spin cycle in a washing machine or the motion of a weather satellite).   8. Students can play with the equations and simulate further applications on the computer by asking what-if questions, or they can do an invention project.

A four-foot diameter "centripetal rider" was invented by a group of mostly junior high school students for this type of classroom use in response to the request from a high school physics teacher, and a prototype was constructed in the College of Engineering at the University of Toledo as part of the math/science summer academy. The invention team of six students, two teachers, and one engineering professor followed the creative problem-solving process throughout the project. Their final report is included at the end of Chapter 12 in this book.

RESOURCES FOR FURTHER LEARNING

These books and articles are just a sampling; check your library or bookstore for additional titles and periodicals on these subjects—results from brain research are being reported all the time in scientific as well as popular journals.

3-1   Teresa M. Amabile, *The Social Psychology of Creativity*, Springer Verlag, New York, 1972. This book presents technical discussions of relevant psychological studies; it also investigates the role of motivation in creativity.

3-2   Pietro Corsi, ed., *The Enchanted Loom: Chapters in the History of Neuroscience*, Oxford University Press, 1991. This is just one example of a publication containing reports on brain research.

3-3   J. N. Harb et al., *Teaching through the Cycle; Application of Learning Style to Engineering Education at Brigham Young University*. This monograph includes sample lesson plans for courses in civil, chemical, construction, electrical, computer, and manufacturing engineering used at Brigham Young University. Their 1989/1990 faculty development program is also described.

3-4   David Keirsey and Marilyn Bates, *Please Understand Me*, Prometheus, Buffalo, 1978. This book is recommended by James L. Adams for discussing the Myers-Briggs test, which measures problem-solving preferences according to the theory of Carl Jung.

3-5   Bernice McCarthy, *The 4-MAT in Action, Creative Lesson Plans for Teaching to Learning Styles with Right/Left Mode Techniques*, Excel, Barrington, Illinois, 1983. Teachers using the 4MAT system in their lesson plans will cover all four quadrants of thinking preferences, though not in the same sequence as used in creative problem solving (for example, asking "what-if" comes after evaluation). The system will enrich classroom experiences for students.

3-6   Robert Ornstein and Richard F. Thomas, *The Amazing Brain*, Houghton Mifflin, Boston, 1984. This book is easy to read and its illustrations help convey the marvels of the human brain.

3-7   Michael Ray and Rochelle Myers, *Creativity in Business*, Doubleday, Garden City, New York, 1986. This book is based on material taught at Stanford University. It focuses on personal creativity, with applications in the business world. Its approach contains insights both from Western culture and Eastern philosophies.

The book *The Creative Brain* by Ned Herrmann (Ref. 1-7) was used as the main resource for this chapter. It not only explains the theory and development of the four-quadrant model of brain dominance and thinking abilities, it also contains applications to many different areas of life, including personal relationships, organizational management, and education. The strength of this book is in the way it demonstrates the process of creative thinking and the advantages of a whole-brain approach to problem solving and living. We recommend this work very highly; it is a must for anyone who is involved in teaching or management. The book itself is a beautiful example of whole-brain teaching. In addition, we have attended two Ned Herrmann workshops which have given us a more in-depth understanding of the four-quadrant model and different thinking preferences. Two additional books referenced in Chapter 1 contributed to this chapter. *The Care and Feeding of Ideas* by James L. Adams (Ref. 1-1) gives many interesting facts about the workings of the human brain and its limitations. *A Kick in the Seat of the Pants* by Roger Von Oech (Ref. 1-16) discusses four roles of the creative process: explorer, artist, judge, and warrior and provides interesting stories and exercises.

ASSESSING YOUR THINKING PREFERENCES

The Herrmann Brain Dominance Instrument (HBDI): Knowing one's thinking preferences is very useful—the HBDI is a very powerful tool. We gain insight into why we do things the way we do and why we have problems in communicating with people who think differently from us. We can set goals, practice specific thinking skills, and expand the range of our thinking repertoire to become more whole-brain thinkers and effective problem solvers. Thus we strongly recommend that you complete the HBDI, particularly if you are a teacher, trainer, or instructor. The instrument is included (with the evaluation prepaid) in the hardcover edition of *The Creative Brain.* Or it can be obtained separately from Applied Creative Services, Ltd., 2075 Buffalo Creek Road, Lake Lure, NC 28746; phone 704/625-9153, fax 704/625-2198. The cost in 1994 is $35 for the form and evaluation.

We have abandoned our experiments with versions of a brief assessment for two reasons—the results were unreliable and Ned Herrmann was very concerned about the issue of validity and the danger of inappropriate applications. We concur with his judgment which is based on fifteen years of experience. In his soon-to-be-published new work *What Will I Be When I Grow Up?* he has developed a new technology to provide individuals with a way of discovering the direction of their preferences through the use of "locator maps" on which they can indicate their successes and failures in high school subjects, vocational courses, college courses, and occupations. When accumulated, these indications form a pattern which seems to correlate well with the HBDI profile. However, these maps do not result in an individual's brain dominance profile; only the HBDI can provide these results. By the end of 1994, Ned Herrmann expects to have the validation work completed on an instrument suitable for assessing the thinking preferences of children.

Differences in learning styles: Classroom team activities are more beneficial when students with different learning styles work together. You may obtain a preliminary idea about your learning style preferences from Activity 3-7.

ACTIVITY 3-7: LEARNING PREFERENCE DISTRIBUTION

Tabulate the total number of circles for each learning preference from One-Minute Activities 3-2, 3-3, 3-4, and 3-5 in this chapter. Then add up the total for all the responses and calculate the percentage contribution for each quadrant:

Quadrant A = _____ = _____ %

Quadrant B = _____ = _____ %

Quadrant C = _____ = _____ %

Quadrant D = _____ = _____ %

Number of Responses = _____ = 100 %

Evaluation: The ranking according to the percentages above gives you a set of data to determine where you will have to make a special effort and where you may have unique abilities and interests to contribute to your team's class project or other group activity. If you have taken the Myers-Briggs Type Indicator or the HBDI, you may want to compare those results with this ranking. Note that this brief assessment is not a substitute for the HBDI, the only instrument available for obtaining an accurate brain dominance profile.

Example: Number of Circles: A = 12; B = 3; C = 7; D = 8. Total = 30.
Percentages: A = 40%; B = 10%; C = 23%; D = 27%. Total = 100%.

Conclusion: From these results, it can be conjectured that this student is an analytical learner. For creative problem-solving exercises and projects, this student should be grouped with someone who has different learning preferences to form a heterogeneous, whole-brain team. People with similar preferences can be grouped into a team when quick consensus and quick decisions are needed, but most likely they will not be able to develop optimum solutions to complex problems.

EXERCISES AND ACTIVITIES

3-1   FOUR-QUADRANT TEAM EXERCISE
Divide a class into teams of five students each, grouping students with similar learning preferences together (from the results of the self-assessment). Select one of the topics below and give the students 3 to 5 minutes to brainstorm the assigned definition on a flip chart.
a. What makes an ideal breakfast?
b. What is a teacher?
c. What is an ideal parent?
d. What is your idea of a good vacation?
e. What is your idea of a great bathroom?
f.   What makes a good toy for a two-year-old child?
Then discuss the results in view of the behavioral characteristics of the Herrmann model. What is the dominant thinking preference of each team?

3-2   INTERVIEW
Interview an engineer, teacher, and person in business about their experiences with creativity at work. Do they have a supportive climate? What would they change? Write up a summary and discuss the factors that you think are important to encourage creative thinking and innovation on the job. What kind of questions would you ask during a job interview to gauge the creative climate of a prospective employer?

3-3   OPTIONAL STUDY: EXODUS
Chapter 3 in the Book of Exodus (Bible, Old Testament) makes an interesting study in four-quadrant thinking. Moses exhibits strong quadrant B thinking—very appropriate when one's business is safeguarding sheep. The chapter shows two phases of God trying to move Moses out of quadrant B thinking toward creative thinking. First, diagram or sketch the action of verses 1 through 6 as they relate to the four thinking quadrants (along the lines of a mindmap). Next, diagram or sketch the action of verses 7 through 22 as they relate to the four-quadrant model (again use an approach similar to constructing a mindmap).

CHAPTER 3 — SUMMARY

Development of the Four-Quadrant Model of Thinking Preferences: Ned Herrmann, the father of brain dominance technology, experienced very distinct thinking abilities in his education, hobbies, and professional life. These led him to research the origin of creativity in the brain. He studied the development of preferences and their relationship to the physical brain and neural connections. He developed an instrument to assess thinking dominances, and he then synthesized the information into a four-quadrant metaphorical model of thinking preferences or "ways of knowing." The data from his instrument yield a profile which can be interpreted generically, and different occupational groups exhibit characteristic generic profiles. Each person is a unique coalition of all four thinking modes.

Quadrant A: Quadrant A thinking is logical, critical, factual, technical, analytical, and quantitative. In creative problem solving, it is used to analyze clues and data about the real problem. It is also used to evaluate and judge ideas (though with a positive attitude that seeks to overcome flaws).

Quadrant B: Quadrant B thinking is conservative, structured, sequential, organized, detailed, and planned. In creative problem solving, it is used during idea implementation for planning, execution, follow-up, and final process evaluation.

Quadrant C: Quadrant C thinking is interpersonal, people-oriented, kinesthetic, emotional, spiritual, sensory, and feeling-based. In creative problem solving, it is used during problem definition (to take the customer's needs into account), during idea generation (when we use feeling-based ideas and intuition), and during implementation (when we are working for acceptance of new solutions).

Quadrant D: Quadrant D thinking is visual, holistic, intuitive, innovative, imaginative, future-oriented, and conceptual. In creative problem solving, it is needed during problem definition (to explore trends and context), during idea generation (in brainstorming), and during idea evaluation and judgment (when ideas and solutions are being improved).

Whole-Brain Thinking and Creative Problem Solving: Appreciate your unique thinking profile, but also learn to use the whole brain to respond well to many different situations. Creative problem solving uses mindset metaphors for each step: the explorer and detective for problem definition, the artist for idea generation, the engineer for creative idea evaluation, the judge for critical judgment and decision making, and the producer for solution implementation. Whole-brain thinking can benefit organizations; productivity is improved when people's thinking skills are matched with the task requirements. Three areas need strengthening in organizations: (1) training in creative thinking; (2) applying creative thinking in the creative problem-solving process; (3) management support for an environment conducive to creativy. Management styles for future success are seen as less hierarchal and more cooperative, with power and information shared at different levels. Research results with engineering students have shown that the thinking preferences of college students can be changed; instructors should learn to use a system that will ensure that they teach to all thinking preferences or learning styles.

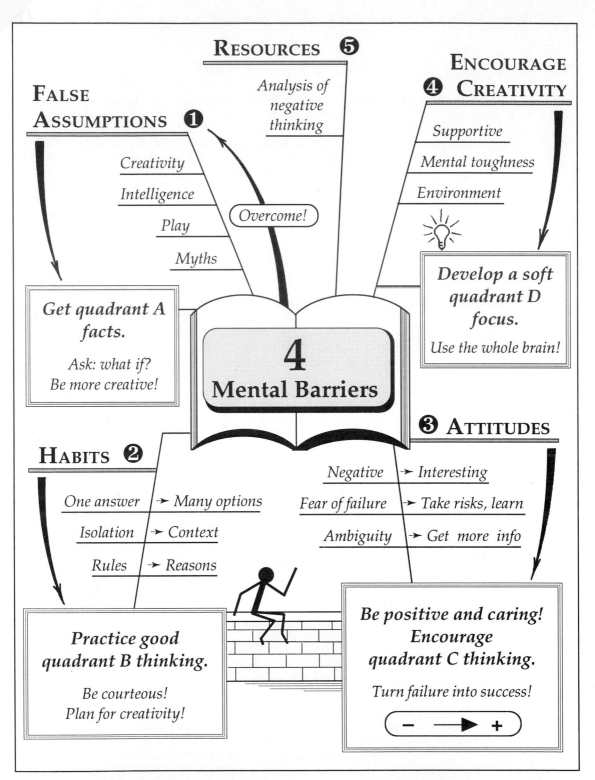

RESOURCES ❺

FALSE ASSUMPTIONS ❶

Analysis of negative thinking

Creativity

Intelligence

Overcome!

Play

Myths

ENCOURAGE CREATIVITY ❹

Supportive

Mental toughness

Environment

Develop a soft quadrant D focus.

Use the whole brain!

Get quadrant A facts.

Ask: what if? Be more creative!

4 Mental Barriers

HABITS ❷

One answer → *Many options*

Isolation → *Context*

Rules → *Reasons*

Practice good quadrant B thinking.

Be courteous! Plan for creativity!

❸ **ATTITUDES**

Negative → *Interesting*

Fear of failure → *Take risks, learn*

Ambiguity → *Get more info*

Be positive and caring! Encourage quadrant C thinking.

Turn failure into success!

− ➔ *+*

Mindmap of Chapter 4. Refer to Chapter 2 (pp. 55-56) to learn more about mindmapping.

4

OVERCOMING MENTAL BARRIERS TO CREATIVE THINKING

WHAT YOU CAN LEARN FROM THIS CHAPTER:
- *Barriers to creative thinking: False assumptions. "I am not creative." An intelligent mind is a good thinker. Play is frivolous.*
- *Barriers to creative thinking: Habits. There is only one right answer. Looking at a problem in isolation. Following the rules.*
- *Barriers to creative thinking: attitudes and emotions; negative thinking; fear of failure or risk avoidance; ambiguity.*
- *How to encourage creative thinking; boundaries and discipline.*
- *Further learning: resources, exercises, Negativism Score Sheet, and summary.*
  Review of Part 1—questions and thinking report.

In this chapter, we will explore how our quadrant A, quadrant B, and quadrant C thinking can be improved to make us more creative. We need factual knowledge to counter myths and false assumptions; we need to be aware of habits that we have been taught; and we need to overcome mental blocks due to attitudes and emotions. We will outline strategies that can help us overcome these mental blocks to creative thinking in ourselves and others. We will also show how the unstructured process of creativity can be enhanced and balanced through boundaries and discipline to result in productive problem solving. Our creativity is progressively enhanced when we improve thinking in all four quadrants (as illustrated on the mindmap facing this page) and as we learn to freely iterate between all four brain quadrants.

> *Creativity to me means actually becoming a child again*
> *and seeing things from that perspective.*
> *Herbert Ford, engineering student.*

> *Creativity is looking at the same thing as everyone else and thinking something different.*
> *Albert Szent-Györgyi, Nobel Prize-winning physicist.*

BARRIERS TO CREATIVE THINKING: FALSE ASSUMPTIONS

What we believe about creativity has a major impact on how creative we become, on how much creative thinking we do, on how we encourage others to express their creativity. Right at the beginning of the book we stated that we believe everyone is creative and can learn to be more creative—we can learn to use the D-quadrant thinking abilities of our brain. Believing otherwise is a major barrier to creativity, with serious consequences. Let's illustrate. If a state agency is seeking proposals on testing and assessments that can better identify talented and gifted students, the underlying assumption is that only some students have (or are born with) these exceptional talents. But if the agency is seeking proposals on how to improve classrooms and teaching to encourage creative thinking in problem solving, the underlying assumption is quite different: creativity can be nurtured and developed. The outcome or benefit of these projects to the state's children will be very different, too. We will now look at some false assumptions people may have about creativity and about thinking in general.

ⓧ One-Minute Activity 4-1: Group Problem
Circle the group you think is the most creative.

NASA Engineers	High School Teachers	Homemakers
College Students	First Graders	Journalists
Movie Producers	Abstract Painters	Auto Mechanics

Here are some statistics that will help you evaluate your answer. When individuals at various ages were tested for creativity, the results were as follows: At age 40, 2 percent were creative. At age 30, 2 percent were creative. At age 25, 2 percent were creative. At age 17, 10 percent were creative, but at age 5, over 90 percent were creative. These were all people who had never been taught how to nurture their creativity. Thus as a group, first graders are the most creative, because they have not yet learned the mental blocks to creative thinking; they can still let their imaginations run wild. When shown a sketch of two circles, one inside the other, they come up with imaginative answers, whereas adults usually only see the geometric figures. Homemakers were also found to be very creative because the job requires much flexibility and improvisation in handling many different tasks—often simultaneously (although homemakers are not receiving encouragement and recognition). With proper use, creative ability is independent of age! False assumptions can be likened to prejudice. Have you ever thought to yourself: "I am not creative"? This is a false assumption, because we have an astounding potential to be creative and can learn to use our creative talents and our whole mind better, as we have already seen in Chapters 2 and 3.

> ## MENTAL BLOCK 1: False assumptions.

Here is another example of a false assumption: "An intelligent mind is a good thinker." According to Edward de Bono, a highly intelligent person who is not properly trained may be a poor thinker for a number of reasons listed in Table 4-1. Can you think of an example from your own experience where "quick thinking" led you to wrong conclusions? We often take verbal fluency or quick responses as signs of good thinking—such false assumptions can keep us from thinking creatively.

THREE-MINUTE TEAM ACTIVITY 4-2: POOR/GOOD THINKING
Look over the list in Table 4-1. Write a short paragraph (with specific examples) on one or two of the items listed and how they relate to your experiences. Then share your insight with two other people.

Another false assumption that is prevalent not only in the business environment but also in our schools and sometimes even among parents is the attitude that "play is frivolous." Play is very important to our mental well-being. Play is as beneficial to our creative minds as physical exercise is to our bodies, and for maximum benefit we must do it regularly. Play with your family members, especially with young children. Play pretend games, play pretend ball. Play word games. Play around with words by yourself or in groups; playing around with words will lead you to play around with ideas. The Moebius strip is an example of a very practical idea that was considered to be only a "plaything"—an abstract mathematical concept—for many years, but now is used to reduce wear on continuously moving tapes and conveyor belts since the configuration has only one side and thus only one edge for even wear. Make yourself a Moebius strip. Play around with it. What would happen if you cut it into two strips lengthwise? What would happen if you cut it into thirds (three strips) lengthwise?

Two additional items need to be mentioned in connection with playfulness. One attitude related to play that is very beneficial to creative thinking is humor. Humor helps creative thinking because it turns the mind from the usual, expected track. Thus funny ideas may lead to unusual combinations—they can be stepping stones to creative solutions. Let's look at some examples. Are you familiar with the *Far Side* cartoons? Here, a technique of turning familiar things around is often used, such as switching the roles of people and animals. There is one that asks: "Do you know why dinosaurs became extinct?" The cartoon shows a bunch of dinosaurs puffing away at cigarettes. Or, perhaps you have seen the orange "smiley" faces used in Michigan highway renovation projects in the last few years. *USA TODAY* had a story on these signs in February 1987. We think someone with a sense of humor as well as a good portion of quadrant C thinking was behind the idea, asking how funny signs could be used to cheer people through construction areas. Humor helps in creative thinking because it relieves stress, tension, and monotony—it switches the mind to unexpected tracks.

Table 4-1 ,
The Intelligence Trap

A highly intelligent person who is not properly trained may be a poor thinker for the following reasons. *Our comments are given in italics.*

1. A highly intelligent person can construct a rational and well-argued case for virtually any point of view. The more coherent this support for a particular idea, the less the thinker sees any need to accurately explore the situation and look for alternatives. *Quadrant A thinking needs to be improved with input from quadrant D.*

2. Verbal fluency is often mistaken, in school and after, for thinking. An intelligent person learns this and may substitute one for the other. *Quadrant C "talk" and Quadrant B "lectures" need to be backed up with quadrant A reasoning and good judgment.*

3. The ego, self-image, and peer status of a highly intelligent person are often based on that intelligence. From this arises the need to always be right, clever, and orthodox. *Quadrant B thinking needs to be enlarged with some quadrant D risk-taking and innovation.*

4. The critical use of intelligence is always more immediately satisfying than the constructive use. To prove someone else wrong gives instant achievement and superiority, but does not lead to creative thinking; it destroys it in the critical individual as well as in all within "hearing" distance. *The judgmental attitudes of quadrants A and B need to be tempered with quadrant C compassion or nurturing and quadrant D ability to see flaws or "mistakes" as stepping stones to ultimate success.*

5. Highly intelligent minds often seem to prefer the certainty of reactive thinking (solving puzzles, sorting data) where a mass of material is placed before them and they are asked to react to it—*typical quadrant A thinking.* In projective thinking, it is the thinker who has to create the context, the concepts, the objectives. The thinking has to be expansive and speculative—*typical quadrant D thinking.* Through inclination or previous training, the highly intelligent mind seems to prefer reactive type thinking, although real life most usually demands projective thinking. *Thus intelligent minds can benefit from learning quadrant D thinking skills.*

6. The sheer physical quickness of the highly intelligent mind leads it to jump to conclusions from only a few signals. The slower mind has to wait longer and take in more data, thus often reaching a more appropriate conclusion. *It takes time to use all four quadrants to get the best results in problem solving (which is applied creative thinking).*

7. Highly intelligent minds confuse quick understanding with quick thinking and slowness with being dull-witted. If "leisurely" or "exploratory" are substituted for "slow," the benefits of slower thinking can be appreciated. *Taking the time to understand the real problem is an important first step for good thinking and problem solving.*

8. The highly intelligent mind is frequently encouraged to place a higher value on cleverness than on wisdom (where wisdom can be defined as good judgment derived from meditation on inner resources, truth, and experience). *Quadrant A judgmental and critical thinking needs input from other thinking modes, especially quadrant C values.*

9. Arrogance is the major sin of intelligent minds. *This attitude stems from not appreciating the specializations, differences, and contributions that the unique minds of others can bring to problem solving and community.*

Adapted from *De Bono's Thinking Course* by Edward de Bono. Copyright © 1986 by Pentacor B.V. Reprinted with permission by Facts On File, Inc., New York.

Another useful technique for playing with ideas is asking "what-if" questions. Roger Von Oech, in his book *A Whack on the Side of the Head*, tells the following story:

> A few years ago, a Dutch city had a trash problem. A once-clean section of town had become an eyesore because people stopped using the trash cans. Cigarette butts, candy wrappers, newspapers, bottles, and other garbage littered the streets. The sanitation department became concerned. One idea was to double the littering fine from 25 to 50 guilders for each offense. This didn't work. Increased patrolling didn't work. Then someone had an idea: What if trash cans paid people money for putting in trash? This idea, to say the least, whacked everyone's thinking. The what-if question changed the situation from a "punish the litterer" problem to a "reward the law-abider" problem. The idea, however, had one major fault—if implemented, the city would quickly go bankrupt. But the people did not reject the idea but used it as a stepping-stone instead. They came up with the following "reward": The sanitation department developed electronic trash cans which had a sensor on the top that detected when a piece of trash was deposited. This activated a tape recorder that would play a recording of a joke. Different trash cans told different jokes. Some developed quite a reputation for their shaggy dog stories; others told puns or elephant jokes. The jokes were changed periodically. As a result, people went out of their way to put their trash in the cans, and the town was soon clean again.

Is this a true story? We haven't seen this particular city, but when we were in Holland in 1971, our children were searching for trash all over one park. Why? Because the trash cans there said *"dank u zeer"* when something was deposited.

Why don't we ask what-if questions more often? First of all, according to Roger Von Oech, we're not taught to do it; we are not in the habit of doing it. Then it is a low-probability technique—you have to ask many what-if questions and follow many different stepping stones before you come up with a truly practical idea. You can practice asking what-if questions at least once a day. Do it; once you get the hang of it, it will be quite a bit of fun, and it may lead to some unexpected, useful ideas. Here is a question to get you started: What if people were only two feet tall? Play around with this idea for a while. What would be the effect on energy consumption? What would be the effect on overcrowded cities, on the world's food supplies? What would our homes look like? This exercise is not as preposterous as it may appear at first. We may get an appreciation for the idea that small is beautiful and that "the bigger the better" is not necessarily true. Also, we may gain insight into what the world looks like to a small child or to a person in a wheelchair.

We have another tool that we can use to ask what-if questions—the computer. We can simulate many situations on the computer; it is a handy tool for investigating many different conditions instead of running actual experiments in engineering, physics, biology, mathematics, and the social sciences. The computer really lets us be inquisitive; here we can take risks and explore situations that would be too dangerous (or too expensive) to do in real life. The purpose of this—besides learning—is to find the best way, the optimum solution, to a given problem. Take a computer lab, especially one using software with declarative languages such as MATHEMATICA, MATHCAD, MAPLE, or MACSYMA. Or take a field trip to see a large simulator in action. Some courses are beginning to be taught in which a software package is integrated with the textbook. When solving problems, students can ask many what-if questions about changing parameters; this enables them to get a much better understanding of the principles and theory of the subject.

QUIZ: FACTS AND MYTHS ABOUT CREATIVITY

Complete the following quiz, circling true (T) or false (F) for each statement. Then rewrite each false statement to make it into a true statement. (For more information on these myths, see Ref. 1-10.)

1. To be creative means imagining something completely new.   T   F

2. Only an expert knows enough to create something meaningful.   T   F

3. Only a gifted minority of people are creative.   T   F

4. Creative people are weird or insane.   T   F

5. If you really have a creative talent, someone will discover you.   T   F

6. Ideas are like magic; you do not have to work for them.   T   F

7. Creative thinking is nice but impractical.   T   F

8. Creativity means complexity or high technology.   T   F

9. The best ways and inventions have already been found.   T   F

10. People over 40 are too old to be creative.   T   F

11. Only men are creative geniuses.   T   F

12. Only people with a high I.Q. can be creative.   T   F

Answers: Turn to the bottom of page 150 to check which statements are true.

Team Activity 4-3: Writing True Statements
In teams of two or three, complete the Myth Quiz on the facing page. The quiz contains many false assumptions, so go over each statement very thoughtfully. Try to rewrite each sentence as a more accurate statement.

Cure for False Assumptions

Get the facts—use quadrant A thinking about creativity.
Learn to play with ideas—practice quadrant D thinking.
Implement creativity and whole-brain thinking into your life.

Next, we are going to investigate mental blocks that we have been taught—habits we have acquired that keep us from thinking creatively.

BARRIERS TO CREATIVE THINKING: HABITS

  Two-Minute Activity 4-4: Symbols Problem
1. Circle the figure that is different from all the others. Explain the reason for your choice.

(a) ◯   *(b)* △   *(c)* ◡   *(d)* ◖   *(e)* ‡

Reason:

2. What is one-half of twelve? Give five different answers.

The two problems above illustrate the same mental block. Were you able to find reasons to justify each of the figures as "different from all the others"? The first problem illustrates that different answers can be "correct" or appropriate, depending on the questions being asked or the criteria being used. Most people stop after finding one answer, because we have not been trained to look for alternatives. The second problem does ask for different answers. Answers here can be strictly mathematical, or they can be imaginative (TWE or 2) or visual (such as writing 12 in Roman numerals and then cutting them in half horizontally, which yields a 7). Even getting alternative mathematical answers is not easy for some people because of a mental block that we have learned in school at an early age:

MENTAL BLOCK 2: There is only one right answer.

When we solve problems, we must not assume that a problem will have only one right answer. This is a serious mental block when we are dealing with other than purely mathematical problems. Therefore, don't stop after the first answer—investigate to see if other answers would be better depending on the circumstances. This is especially important when dealing with ideas!

> *Nothing is more dangerous than an idea when it is the only one you have.*
> Émile Chartier, French philosopher.

How do we know that the idea that we have is best if we have nothing else to compare it with? We have an expression for this type of thinker—a person with a one-track mind. This brings us to a related mental block:

MENTAL BLOCK 3: Looking at a problem in isolation.

⏱ ONE-MINUTE ACTIVITY 4-5: GRID PROBLEM
How many squares do you see?

How many squares did you "see" in the figure above? You can use a mathematical formula to find the total number of squares in a grid with n squares along a side:

$$\Sigma \;=\; 1^2 \;+\; 2^2 \;+\; 3^2 \;\ldots\; +\; n^2$$

Here we have an $n = 3$ square with a smaller, heavier square superimposed; this adds three additional squares to the figure (the new square as well as two smaller squares cut from the existing squares). But this is still not a complete answer, because you have at least one other square in the figure, and that is the spelled-out word in "squares." Did you use another way to look at the problem? Did you determine that you are really looking at a town planning map, where the darker square indicates the location of *one* city square? Or perhaps you saw an infinite number of squares in your mind, if you imagined that you were looking at the top view of a three-dimensional figure—a cube or column.

How many squares did you see, not just on the sketch on the opposite page, but in your immediate surroundings: on your desk or clothing, on the ceiling, or on the floor? What if you looked outside? Before we can find answers to a problem, we must first determine what we are dealing with. We must also find out if the problem is part of a larger problem. We must look at the whole situation; we must look at the entire *context* of the problem. Having a very narrow view of a problem is a mental block.

Especially when people have become experts in their work, they are naturally more narrowly focused on what they know so well. They tend to forget to look beyond the familiar to new horizons. Thus people who can take a multidisciplinary approach (or a wider, "softer" focus) are usually very valuable to their organizations. To use another picture or metaphor, we need to get into a habit of looking not just at a leaf or a branch of a tree—we need to look at the whole tree and the whole forest, and perhaps even beyond. We need to take the time for the long-term view—the wide-angle lens (Figure 4-1). Some decisions we make may have consequences that last beyond our lifetime. When dealing with problems, we need divergent as well as convergent thinking. This brings up an activity to illustrate another mental block.

Figure 4-1 The context is never irrelevant, unless you're dead.

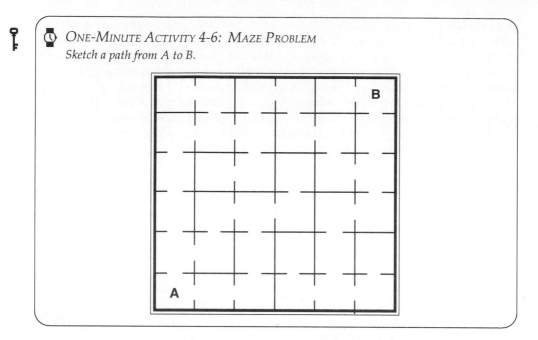

ONE-MINUTE ACTIVITY 4-6: MAZE PROBLEM
Sketch a path from A to B.

In trying to find a path through the maze, did you get to a dead end and have to backtrack? That's one problem-solving strategy: sometimes we need to know what doesn't work. Or, did you start at B and then work backward? This also is good strategy: we look at our goal and then find the best way to get there. Did you trace a path through the maze? Most people, when given this type of problem, think this is what they are *supposed* to do, even though the assignment did not specify a path through the maze. But wait—let's look at the problem again: What did it ask you to do, exactly? Why not draw a line straight from A to B? Or could you go around the outside of the figure from A to B? What about folding the page to put A right on top of B? Another possibility is drawing a line from A to B in the instructions. These last four options illustrate overcoming another mental block:

MENTAL BLOCK 4: Following "The Rules."

We have to make sure we do not make up our own rules and barriers where none exist. Before we can come up with novel ideas, we must question existing barriers. Especially managers and administrators need to develop a habit of looking at the purpose of paperwork and procedures. Are all the "rules" we put on others and ourselves really necessary or helpful? How often are we encouraging our friends or family members to look for unusual solutions to problems, even in a combative situation? "Following the rules" can develop into a judgmental, critical attitude which is a mental block discussed in the next section. Insistence on conformity is a type of following unspoken rules. When we are afraid of questioning convention or arbitrary criteria, we may miss opportunities for creative thinking and improvement.

Sometimes we follow rules when the original reason for the rules no longer exists. An example of this is the computer keyboard. Have you ever wondered why the letters are arranged the way they are? When the typewriter was first invented, it soon developed a problem: the keys were jamming frequently because the typists worked too fast. The obvious solution was to slow down the typist. This is the reason some of the most frequently used letters (such as the A, S, L) are typed with the weaker fingers, and E, N, or T have been placed above or below the main level. But why are we still using this inefficient keyboard when we now have equipment that works faster than any operator can type? Actually, better keyboards have been invented, but only a few people take the initial effort to install the new system and to overcome the old habit by learning how to use it. Today, we need to examine many of our work routines and our ways of teaching and learning in light of the power now available in calculators and computers. Do our traditional ways still make sense? Change is possible: just think a few years back when Great Britain switched to the metric system of measurements.

At this point we need to pause and clear up a false assumption: "Creative thinking and breaking the rules lets you act in an undisciplined manner." Yes, creativity is an unstructured activity, and yes, we need to break "rules" when playing with ideas. However, creativity blossoms better when it has some boundaries and direction. This is why we have the steps in creative problem solving; thus we also follow rules of etiquette and acceptable behavior as we interact with people! Rude behavior creates stress, and stress creates chemicals in the brain that keep it from thinking creatively. Courtesy and consideration for others are important parts of a creative environment. Thus we need judgment to discern which "rules" in quadrant B thinking have to be suspended to let quadrant D flourish and which rules are helping to enhance creativity.

The last three mental blocks that we have looked at—there is only one right answer, looking at a problem in isolation, and following the rules—are habits that we have been taught. Here is a case study that illustrates the process and result of not "following the rules."

Case Study: Kitchen Design

Problem: A young couple in California had a comfortable home they liked very much, except for the kitchen, which was old-fashioned and had an impractical and potentially dangerous floor plan (with the cooking area and wall oven impinging into the main traffic lane) as shown in Figure 4-2a. When the house was destroyed by an act of nature, this presented an opportunity for improvement.

Conventional Solution: The house was redesigned along the lines of the original plan, since it suited the couple's life-style. The improved layout for the kitchen and laundry/garden room is shown in Figure 4-2b. At this point, all the people involved (the designer, the couple, the interior decorator, and the architect) were still bound by the old paradigm—the way the house was originally configured. The new plan was acceptable since it was better than the original design.

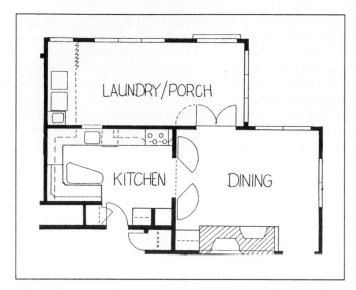

*Figure 4-2a
Original kitchen/laundry
layout.*

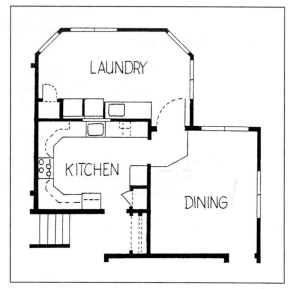

*Figure 4-2b
Improved kitchen/
laundry design.*

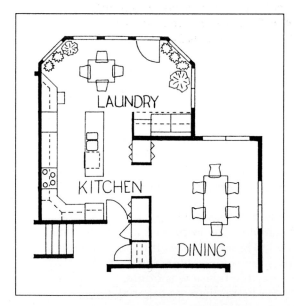

*Figure 4-2c
Creative garden kitchen
design.*

Creative Solution: The couple showed off the plans to friends who immediately questioned the kitchen design and suggested some "wild" ideas—they had no investment in the old plan and were not bound by "the rules." Thus they were able to see different solutions. When these changes were described to the designer over the telephone, she was unable to visualize them. However, just the idea that other, better solutions were possible sent her back to the drawing board. She tried different layouts, but nothing seemed to click until she removed the dividing wall between the kitchen and garden room to create a large, open space. Immediately, the most logical place for the laundry equipment became clear. When the kitchen island was turned perpendicular to the garden room (not a "logical" but an "intuitive" solution), the entire plan suddenly fell into place. It was easy to accommodate the food and dish pantries, the desk, the wall ovens, the large sink and dishwasher, the cook top, and the refrigerator for an efficient work triangle easily accessible from the breakfast area as well as from the formal dining room. In less than two hours, the new plan was designed, drawn, and faxed to the couple, who loved it (see Figure 4-2c). When the architect saw the new design, he was surprised and then commented, "Why didn't we see this solution earlier?" He incorporated it into the house plan with some additional improvements. The couple thinks that this new kitchen has increased the resale value of the house by $20,000 without adding to the construction costs. Even though this design is clearly better than the other plans, this solution was not obvious in the early design stages, just because the old paradigm or rules blinded those involved with the project to a "different" way of seeing and thinking, despite their experience in creative design. It was the outsiders who were able to see other, more imaginative possibilities.

CURE FOR HABIT BARRIERS

Consciously broaden your outlook to make divergent thinking a habit.
Develop an adventurous mind; look for "different" alternatives!
But hone your quadrant B rules of courteous behavior.
Follow the steps in the creative problem-solving process.

ONE-MINUTE ACTIVITY 4-7: DOT PROBLEM

What do you see in the figure below?

This little exercise has more than one purpose. Have you learned to think of alternate answers? With your imagination, were you able to "see" other things besides just a small black dot? The second purpose of this exercise is to show that it is our human nature to notice small negatives more easily than large positives. Do you realize that the figure is 99.9 percent white? Why do we focus on the black dot that covers only one one-thousandth of the area shown? An unexpected proof of this happened with this illustration when the printer of the second edition removed the dot, thinking that it was a blemish. Actually, the discriminating ability of the human mind is very important to survival, but often we misuse this critical thinking mode by concentrating only on negatives because of our attitude.

BARRIERS TO CREATIVE THINKING:
ATTITUDES AND EMOTIONS

The group of mental blocks that we want to consider next is more difficult to deal with because they involve our attitudes and emotions. They require improvements in our quadrant C thinking. Here is the first of these barriers:

> **MENTAL BLOCK 5: Negative thinking.**

Negativism, criticism, sarcasm, and put-downs are mental blocks that not only inhibit creative thinking in the person using these blocks; they have the same effect on all those coming in contact with the negative thinker and are thus doubly destructive. It is so easy for us to focus on small shortcomings of an idea—or of people—rather than appreciate or recognize or compliment the good features or qualities. When we are presented with new ideas, we should make a real effort to react positively. A judgmental attitude, including your own inner "critical voice of judgment" can be powerful barriers against expressing creative thoughts. Learn to appreciate your own creative ideas and nurture the creativity in others!

We must be especially careful when we are dealing with children. Do you know that typically a child receives about 150 negative reactions for each positive reinforcement, within the family as well as outside the home? If you must reprimand a child, try to give encouragement and praise, both before and after the negative statement. Even then, keep the negative statement neutral; do not attack the person, only the undesirable action. Also, focus on positive goals, not prohibitions. Edward de Bono recommends a useful technique for overcoming a spirit of criticism and negative thinking:

 Look at things as being different or interesting— not good or bad!

"Different" and "interesting" have positive or neutral connotations and thus will not activate our emotional mental blocks. These words can stir up our curiosity and thus direct us toward further investigation. Quick judgments tend to be negative judgments; thus, take time to make thoughtful, creative evaluations when judging ideas. Another negative thought that inhibits creativity is the fear of sharing ideas because others might "steal" them. This is especially true in an overly competitive environment. The opposite attitude is fostered when collaborative teams brainstorm—the members are encouraged to borrow one another's ideas and use them as stepping stones for further creative thinking and idea synthesis.

Negative thinking is not the same thing as constructive discontent. Discontent can be used as a stepping stone and motivator to find a better way to solve the problems that are bothering us. To do this, we do not have to be a genius; anybody can do it. All we need is an attitude that looks at problems as merely being temporary inconveniences. We need an attitude that considers a problem a challenge. In essence, negative thinking is an attitude that says: "Things are bad, but we are powerless to change the situation." It seeks to place blame: "It's someone else's fault," or it refuses to take responsibility: "It's not my problem." Constructive discontent also acknowledges that things are bad, but then has the attitude: "Let's get to work to change and improve the situation!" We are probably not conscious of how much negative thinking we do routinely and the effect it has on our success. The "investigation of negative thinking" exercise at the end of this chapter is a tool to help diagnose the frequency and types of negative thinking.

When dealing with an unsatisfactory situation as a team, try to use the PMI approach also developed by Edward de Bono. First, make a list of positive statements and solutions about the problem or idea—look for the PLUSES. Then, switch to the opposite viewpoint and find what is negative about *not* making changes or search out the pitfalls of the new idea—look at the MINUSES. Finally, concentrate on what is INTERESTING about the whole "mess." This technique is fun and can be used by young and old. For example, when a group of bright seventh-graders brainstormed the question, "Why is it important to use the whole brain?" they came up with 40 different answers. Here are some excerpts:

PLUSES:
- We would be more effective and competitive economically in the world
- You can grasp the whole picture of an idea rather than just one point of view.
- You can learn better and be more creative.
- By using all of your given equipment, you can get better answers to any problem.

MINUSES:
- With only right-brain thinkers, the world would be a mess, with only left-brain thinkers, sad and dreary—thus life would be neither fun nor safe.
- If you don't use your brain, you will get it from your parents and the principal.
- You could end up wearing a paper hat and flipping hamburgers.
- You may have a disability, or you may just need tutoring.

INTERESTING:
- I like my brain just the way it is.
- You may get better grades to get into the school you want.

The last two statements certainly call for more data and discussion on the relationship between thinking skills, grades, and learning, or an assessment of the speaker's thinking profile and skills.

> *No pessimist has ever won a battle.*
> *General Dwight D. Eisenhower.*

Here is another important barrier that involves our emotions; it prevents creative thinking as well as action:

MENTAL BLOCK 6: Risk-avoidance or fear of failure.

The turtle only makes progress when it sticks its neck out!

Lack of risk taking is also expressed when we are overly pedantic, nitpicking, fussy, or anxious. The risk involved in ideas is not the same thing as physical risk taking. We all know from personal experience that teenagers especially do like to take physical risks, for example, when driving, experimenting with things (including drugs), and other activities. You have to use good judgment to decide when to take a risk. You would not jump from a ten-story building—you know what the physical law of gravity would do to you. The risks that we are thinking about here are things like speaking out in a group when you have an idea (even though it may be hooted down), learning something new where you may fail at first until you become good at it, or standing up against peer pressure to get involved in serious studying and excellence, because it is your future that is at stake. Yes, you have to stick your neck out when you are championing a creative idea; you also need a thick shell, and you have to be persistent in getting to your goal—you can expect your critics to make "turtle soup" out of you and your idea.

What do you think? Is a person who misses in two out of three tries very successful? Well, a 0.333 batting average is among the best. What of one out of 6000 tries? This is success also—that's about how many different experiments it took Thomas Edison to invent the incandescent light bulb. Mistakes can be triggers or stepping stones for creative ideas. We can learn from mistakes—thus we should not be afraid to make them. Japanese companies that are very quality-conscious "applaud" the appearance of a flaw in their assembly lines instead of hiding the defect. They recognize these occurrences are good learning experiences and real opportunities to make improvements.

We can learn from and use mistakes. Here is an illustration from industry. The 3M Company encourages creativity in its employees, and its researchers are allowed to spend about 15 percent of their time exploring creative ideas and projects of interest. Some years ago, a scientist by the name of Art Fry decided to make use of this time to deal with a small irritation in his life. He sang in the church choir and used small bits of paper to mark his pages in the hymnal. Invariably, these pieces of paper would fall out and end up on the floor. He remembered that a colleague had developed an adhesive everyone thought was a failure because it did not stick very well. Art Fry played around with this adhesive and found that it made not only a good bookmark but was great for writing notes because it would stay in place as long as needed, yet could be removed without damage. The resulting product's trade name is Post-it™. It has become one of 3M's most successful office products, although it failed at first to generate sales in its test markets. It is interesting to note that Art Fry had to invent and build a machine in his basement to produce the blocks of sheets since the traditional 3M products came in rolls. The story of how these problems were overcome creatively is told in more detail in Section 3 of Chapter 8. To get a creative idea implemented takes persistence; we may have to turn early failures into success at several points when moving an idea from the original dream to the marketplace.

> *Failure is a necessary and productive part of the innovation process.*
> *Jack V. Matson, director, Leonard Center, Pennsylvania State University.*

Jack Matson was an engineering professor at the University of Houston when he did research in the area of encouraging student creativity. He found that those students who made more mistakes initially in a project ended up being the most successful. As a student, you may have to make a decision—will you play it safe and follow conventional paths in your projects, or will you risk failure for the chance to really come up with some especially creative ideas? When we try a lot of different approaches, three things happen: we increase our chance of failure, we increase our chance of winning, and either way we learn! Our task as instructors (or parents) is to find ways of evaluating the progress of our students (or children) so that they are not being penalized for early failure but only assessed on the total of their learning at the end. This raises the question of how the quality of teaching affects learning. How is teaching to be evaluated—does "early failure" apply to the instructional process as well? How (or should) we give teachers and school systems the opportunity to experiment to become better educators?

Do you realize that a very popular model of diving board used in the Olympics and at 50,000 other pools all over the world was a "failed" airplane wing? You may have heard of it by its trade name, *Duraflex*. Keep looking for better ideas; take some risks; be persistent! Use failures as stepping stones to more creative thinking and success! The fear of failure leads to other associated mental blocks: not being willing to take the responsibility for independent thinking or being passive and uncurious—letting life simply pass by—in essence leading to mental "laziness" that refuses to ask questions and does not want to get involved in finding solutions to problems. Such a person may learn to become a chronic complainer and negative thinker without the motivation to become part of the solution or to develop mental flexibility—we could use the image of a "couch potato mindset" to describe such a thinker.

> *Creativity is a learned response to a situation, drawing from within the necessary energy, information, and other resources necessary to solve a problem.*
> *William J. Riffe, professor and director of Manufacturing Systems Engineering, GMI.*

Here is another mental block that acts on our emotions:

> **MENTAL BLOCK 7: Discomfort with ambiguity.**

We know from experience that ambiguous situations may lead to serious misunderstandings and conflict, and thus we try to avoid them. Visual images as well as verbal information can contain ambiguity. Let's quickly look at some examples. What do you see in the simple sketch on the left?

Do you see a vase or goblet, or mirror images of a face in profile? What if you turned the drawing around? Can you imagine a bell hanging from a rafter, or do you see a candlestick? A-quadrant and especially B-quadrant thinkers are uncomfortable with ambiguity; these minds prefer things to be black or white, not both, and not various shades of gray. You may be familiar with the fascinating drawings of M. C. Escher; they illustrate the conflict that can arise between visual cues and the brain's interpretation of the situation.

Below are three ambiguous statements from job references. Can you discern two different meanings for each statement? If you are in an ambiguous situation, use it to get more information!

> AMBIGUOUS STATEMENTS
>
> I am pleased to say that this person is a former friend of mine.
>
> In my opinion, you will be fortunate to get this person to work for you.
>
> I can assure you that no person would be better for this job.

Robert J. Thornton, Lehigh University, "Lexicon of Internationally Ambiguous Recommendations (LIAR)," *Detroit News*, February 8, 1988, pp. D 1-2.

 THREE-MINUTE ACTIVITY 4-8: AMBIGUITY
In groups of three, discuss the following statements—what are some possible interpretations? Then rewrite them in at least two different ways to remove the ambiguity.
Thank you very much for your memo—I will waste no time in reading it.
The longer he is away, the more I appreciate him.

It's O.K. to have "strange" ideas—don't be afraid to stick your neck out!
Laurie Bruns, engineering student.

High school and college students are often uncomfortable with ambiguity. They do not mind so much having a lengthy homework assignment, as long as it comes with detailed, specific instructions and information. They dislike having to make assumptions or dig up information on their own in order to solve problems. They do not realize that they are getting a great opportunity to cope with real-life ambiguous situations. Even though all of us may be uncomfortable with ambiguity, we sometimes need to explore ambiguous situations or ideas because they can be a source of especially creative trains of thought. Therefore, don't be in a hurry to resolve an ambiguous situation too quickly; look at the situation from many different angles. Ask more questions, and especially give your subconscious mind time to "think" about the ambiguity.

Paradoxes serve a similar purpose; they can be "whacks" that can get us to think in a new direction by putting things into a different context. Jesus Christ used many paradoxical statements in His teachings to get people to think. For example, He told His disciples that "whoever wants to save his life will lose it, but whoever loses his life for me will find it" (Matthew 16:25, NIV Bible). Paradoxical thinking can lead to creative thinking and inventions. For example, *Corelle* ™ dishes are "unbreakable" china. Could the idea of a "water-repellent sponge" lead someone to invent a new way of separating chemical solutions?

The three mental blocks that we discussed in this section—negative thinking, the fear of failure, and ambiguity—involve our attitudes and emotions. They, too, can be overcome with practice.

CURE FOR EMOTIONAL BARRIERS

Cultivate a positive outlook.
Use quadrant C thinking to encourage the people around you.
Practice quadrant D thinking—take risks with learning.
Use failure as a stepping-stone to success.
Use ambiguity as an opportunity for creativity.

Creativity is being able to solve problems in the mind that don't even exist yet.
Brian Webb, engineering student.

RECOGNIZING AND ENCOURAGING CREATIVE THINKING

In addition to overcoming the mental barriers by improving our quadrant A, quadrant B, and quadrant C thinking, we can encourage creativity with improvements in quadrant D thinking in several ways. First of all, we can be an example: we need to practice and exhibit creative thinking in our life and in our work, so others can see it in action. You have been presented with many ideas on how to do this in Part 1 of this book. You also must balance imaginative thinking with mental toughness to get results from your creativity. Second, we can recognize and encourage creative thinking in others—remember, Ned Herrmann found that praise increases the use and preference for specific thinking modes. And third, we can build a favorable physical environment for creative expression, for children, for adults, and for brainstorming in teams.

Be an example: To check up on yourself and your own creativity, Table 4-2 lists seven important points that Harold McMaster, a noted inventor in the field of glass-making from Toledo, mentioned when he spoke to students and engineers. You will note that several of these items relate to the mental blocks or to steps in the creative problem-solving process.

Table 4-2
Requirements for Creative Thinking

- Be curious—look at the frontiers of knowledge.
- Obtain a solid foundation in the field you're working in.
- Invent to satisfy a need.
- Look for new ways of doing things; take the familiar and look at it in another way.
- Question conventional wisdom.
- Observe trends, look for opportunities, then work hard.
- Realize that most progress is made in small steps.

We need to discuss one other factor needed to achieve the goals in creativity and thinking abilities that you are setting for yourself, and that is *mental toughness.* This, too, is learned, not inherited. Top athletes have learned this well. Personality style is unrelated to mental toughness: you can be introverted or extroverted, energetic or low-key—this has no bearing on your success. The characteristics of mental toughness in competitive sports are described in *Mental Toughness Training for Sports: Achieving Athletic Excellence* (Ref. 4-15). We believe that they apply equally to success in learning, in thinking, in problem solving. Mental toughness provides a context of discipline for the expression and implementation of creativity. The characteristics are listed in Table 4-3. The lack of discipline and motivation and the resulting chaos in one's life can be a serious barrier to applied creativity.

When you adopt a disciplined mental attitude about creativity, you will be calm and relaxed, you will have fun and enjoy your activities, you will have energy, and you will be in control. Developing mental toughness and discipline is hard work, because you have to build different structures in your mind and form new habits, but it will require less effort as you practice and as these attitudes become more automatic. We encourage you to build this structure to shelter your inner creative environment, because with this type of discipline, the mind is sharpened for its tasks, and you will be encouraged by the results you will be able to achieve!

Table 4-3
Characteristics of Mental Toughness

1. You are self-motivated and self-directed—you are doing *your* thing.

2. You are positive but realistic—you build up, you praise, you are optimistic; your eye is on the goal, not on possible failure.

3. You are in control of negative emotions—so what if the environment or your co-workers are not perfect or make mistakes. You may need their forgiveness at times, too. Reacting with anger does not solve problems; it makes solutions more difficult.

4. You are calm and relaxed under pressure—you deliberately see the opportunity, not the crisis.

5. You are energetic and ready for action—you are determined to give your best performance and do the best job that can be done.

6. You are persistent—you have a vision. You know what you want to achieve; temporary setbacks do not faze you.

7. You are mentally alert and focused—you are in control of your concentration; you can use divergent or convergent thinking in response to what the situation requires.

8. You are self-confident—you believe in yourself and know that you can perform well. You are well-prepared; you have past successes; you can do it again, even in a new, unfamiliar situation.

9. You are responsible for your own actions and behavior—you take responsibility for your own thinking skills and ideas (or when working in a team, for the results of the group effort). You will see the project through, and you are accountable for the outcome and consequences.

 FIVE-MINUTE ACTIVITY 4-9: BASKETBALL

In the following story (condensed from Guideposts, *May 1992), how many aspects of creative thinking can you identify? Discuss your findings with two others in your group.*

James A. Naismith, a physician, teacher, and minister, invented the game of basketball because he wanted to give the bored young men at the YMCA something more exciting to do than calisthenics. He remembered a game he played as a child, where players tried to knock a stone off a rock by lobbing other stones at it. He also recalled that rugby players practiced during the winter by tossing a ball into a box while team members tried to block the shots. James Naismith combined the two ideas, divided players into two teams, with boxes to be fastened to the railing 10 feet above the gym floor, and he added a soccer ball. The janitor at the "Y" however, did not have any boxes—what he had were two peach baskets. The players loved the new game (most of the original rules are still being used today). The only annoyance was that someone frequently had to climb a ladder to retrieve the ball from the baskets. With time, the baskets were replaced by nets, and eventually someone thought of cutting off the bottom of the net so the ball could pass through.

Encourage others: How do you recognize creative thinking in someone else? It is possible to look for creative expressions in people who do not appear to be "gifted" or "talented" in special areas. How many traits of creative thinking were you able to identify in the story told in Activity 4-9? Table 4-4 is an example of a list of creative thinking traits.

Table 4-4
Recognizing Traits of Creative Thinking

1. Look for the unexpected—is the idea something different and original? Did it overcome some of the barriers discussed in this chapter? Does it show mental flexibility?

2. Is the idea, product, or solution an unusual or new combination or synthesis of ideas? Is it an improvement of an existing idea or product?

3. Does it have a potential for further creative development, even though the idea appears to be "useless" in its present form?

4. Does the idea or product "feel" right? Is it a logical answer to a problem?

5. Does it respond to the situation or context; does it solve the need well?

6. Does the solution have a sense of wholeness, beauty, or elegance?

7. Does the solution work on several levels? Does it stimulate thinking and other applications?

8. Did the person listen to "something inside the head"?

9. Was the solution a result of brainstorming (alone or with others)?

We can encourage the creativity of others by giving positive feedback and "playing" with ideas together—in essence providing a nurturing, supportive social environment. We can also encourage creativity when we create a stimulating physical environment for people of all ages.

Build a creative environment: What is a creative physical environment? The idea of protection was already mentioned in Chapter 3 in connection with creative organizations. Protection from harm while having the freedom to explore plays an important part in encouraging the creativity and mental development of young children. Babies and toddlers need opportunities to touch, to feel (with hands and mouth), to taste, to smell, and to manipulate many different objects, so that the young brain can make many neural connections and thus develop to its full potential. Parents must provide a stimulating environment for their young children that is free from physical hazards but also free from unnecessary restraints. Such an environment does not have to be expensive—it only has to invite imagination! Provide cardboard boxes of all sizes, wood blocks, paper, crayons, rags, all kinds of odds and ends—children will come up with amazing ways to invent and play with these things. Shut off the television; speaking dolls and automated teddy bears are not necessary. Building blocks, *Tinkertoys*, simple (home-made) stuffed animals encourage imagination. Read to children, frequently, and from a wide variety of subjects. Let the children help with cooking and other household projects. Take them to the public library; attend the story hour. Take them to the zoo (local residents may have certain hours when admittance is free or greatly reduced). Take them to parks, to nature programs, to sandboxes, to the beach or a mountain creek . . . let the children get their hands and feet wet and muddy (and join the fun). Then talk about all these fascinating experiences.

> *The person who teaches your child to talk teaches your child to think.*
> Jane M. Healy, Endangered Minds, 1991.

> *Think creatively: Turn an idea from stupid into stupendous.*
> Adam Macklin, engineering student.

For older children and adults, a relaxed atmosphere, too, is best for creative thinking. If you must come up with creative ideas to solve a problem, select a location that is not familiar to the person or the team, because familiar surroundings encourage old thought patterns. By the way, did you know that the opposite is true also: it helps to study in the same room where you will be taking a test later. Thus, to help children study for a test at home, fix up a quiet place that is similar to their space or desk at school and have them study the material in this spot. It will come back into their mind easier than if they were to study in a lounge chair, on a bed, or on a rug in front of the TV set. So, when studying to remember material later, use a familiar environment; if you want to think creatively, use a new, different, stimulating environment.

When brainstorming in a group, use a circular arrangement for the chairs—do not face each other across a narrow table (which implies confrontation). Have lots of wall space where ideas can be posted. Have a flip chart and colored markers. Minimize distraction (like a telephone)—have comfortable lighting and room climate control. Don't rush people—give them time for social interaction before getting to the task. Provide a protein snack, some humor, and some beauty (interesting pictures, flowers, or a nice view)—and most importantly, a positive attitude! A positive attitude is very important, not just for the leader, but for everyone involved in brainstorming or the entire creative problem-solving process. Creative people (as individuals and as a team) have no false pride; they believe in their ability to succeed. They have a sense of wholeness and balance, as well as moral and ethical standards. They are self-controlled, persistent, honest, unselfish, and willing to work hard. A truly creative person or group is disciplined, good-tempered, unselfish, well-behaved, responsible, unprejudiced, and caring. We will talk more about what it takes to be a good team member in Chapter 6.

Now that we have examined the mental blocks that keep us from thinking creatively, we can develop and practice habits and attitudes that encourage creative thinking. Do the exercise problems in the first four chapters of this book. Get together with a friend and review your mutual progress periodically. In Part 2 of the book you will learn how to apply the four different thinking modes most effectively in the creative problem-solving process. In summary, whether you are looking at your organization, your job, or your personal life, this is true:

CREATIVITY DOESN'T JUST HAPPEN—PLAN FOR IT!

Practice positive, exploratory thinking—take risks with learning.
Cultivate an attitude that looks for alternatives.
Develop flexibility and fluency in all four modes of thinking abilities.

The basic aim of education is not to accumulate knowledge,
but rather to learn to think creatively, teach oneself,
and "seek answers to questions as yet unexplored."
Jim Killian, former president of MIT, as quoted by Charles M. Vest, president of MIT.

RESOURCES FOR FURTHER LEARNING

4-1   Edward de Bono, *Lateral Thinking*, Harper and Row, 1970. This is a well-known book and highly recommended for teaching flexible thinking skills. A more recent book by the author, *Serious Creativity*, Harper Business, New York, 1992, is now available in paperback; it builds on twenty-five years of practical experience with lateral thinking and the deliberate use of creativity.

4-2   Kent G. Burtt, *Smart Times: A Parent's Guide to Quality Time with Preschoolers*, Harper and Row, New York, 1984. This book contains a wide variety of creative activities, learning games, and interactive experiences including kitchen play, arts and crafts, let's-pretend scenarios, rough-housing, family fun, and bedtime talk (including gentle hints to impart values).

4-3   William Coleman, *You Can Be Creative: Making Your Dreams Come True*, Harvest House, Eugene, Oregon. 1983. This easy-to-read paperback book puts creativity and its applications to life into the Christian frame of reference.

4-4   Dorothy Einon, *Play with a Purpose: Learning Games for Children Six Weeks to Ten Years*, Pantheon Books, New York, 1985. This book includes many activities that encourage verbal skills as well as exploration and sensory learning. Check your public library—it will have a collection of similar books for children of all ages.

4-5   Siegfried Englemann and Therese Englemann, *Give Your Child a Superior Mind*, Simon & Schuster, New York, 1966. Although this book is out of print, it can still be found in libraries and used-book sales. We found it to be an excellent book when raising our children; it has excellent insight on how children's minds can be influenced by the environment and parent feedback.

4-6   Robert Fritz, *Creating*, Fawcett, New York, 1991. This author clearly differentiates between creativity (or thinking creatively) and creating what one really wants. Creating is a skill that can be mastered. When this skill is used in music or painting, the results are artistic. When creativity is used in technology, the results are inventions. When creativity is used in business, the results are production, and when creativity is used with people, the results are improved relationships. Although this book contains many practical ideas, it also presents some challenging questions and unconventional philosophical views.

4-7   Robert Fulghum, *All I Really Need to Know I Learned in Kindergarten: Uncommon Thoughts on Common Things*, Villard Books, New York, 1989. This warmhearted best-seller contains many low-key stories of positive thinking—fine recreational reading!

4-8   Martin Gardner, *aha! Gotcha—Paradoxes to Puzzle and Delight*, W. H. Freeman, San Francisco, 1982. This softcover book presents a humorous collection of puzzlers from logic, probability, numbers, geometry, time, and statistics.

4-9   Kurt Hanks and Jay A. Parry, *Wake Up Your Creative Genius*, Kaufmann, Los Altos, California, 1983. This softcover book is a compilation of many different ideas, techniques, and examples that can stimulate creative thinking.

4-10   Peter Jacoby, *Unlocking Your Creative Power*, Ramsey Press, San Diego, 1993. This small volume is a fun guide for leading you to discover your own creativity.

4-11   Loyal Jones and Billy Edd Wheeler, *Laughter in Appalachia: A Festival of Southern Mountain Humor*, Ivy Books, New York, 1987. This small paperback is typical of many joke books that are available, but it also contains a discussion on humor and its context.

4-12   John M. Keil, *How to Zig in a Zagging World—Unleashing Your Hidden Creativity*, Wiley, New York, 1987 (also available as a cassette, 1988). Our students have liked this small book.

4-13   Arthur Koestler, *The Act of Creation*, Dell, New York, 1967. This is a very comprehensive work on the nature of creativity—it is, however, not easy to read.

4-14   Herbert Kohl, *A Book of Puzzlements: Play and Invention with Language*, Schocken Books, New York, 1981. This is a nice source for ideas and examples of word games.

4-15   James E. Loehr, *Mental Toughness Training for Sports: Achieving Athletic Excellence*, Greene Press, New York, 1982. This book shows how winning athletes develop the mind to do their best. These principles and exercises can be applied to learning self-discipline for thinking tasks.

4-16   Ruth Stafford Peale, ed., *Guideposts* magazine. This monthly publication presents tested methods for developing courage, strength, and positive attitudes. (For example, see "The Choir Singer's Bookmark," by Arthur L. Fry, January 1989, pp. 7-9.) *The Power of Positive Thinking*, is also recommended; it was authored by her late husband, Dr. Norman Vincent Peale.

4-17   *Project XL—The Inventive Thinking Curriculum Project: An Outreach Program of the United States Patent and Trademark Office*, and *The Inventive Thinking Resource Directory*, bicentennial edition, 1990. Project XL is an outreach program of the U.S. Patent and Trademark Office that will help teachers teach all their students analytical and creative thinking daily without the student labels of "gifted," "special," or "high-risk." The program is infectious. The *Resource Directory* contains material to assist teachers in developing and implementing their own programs: it is a comprehensive guide to programs, materials, literature, organizations, and other sources that promote thinking across all disciplines. The *Curriculum* is suitable for students of all ages.

Four previously cited references contain information pertinent to this chapter. *Conceptual Blockbusting* by James L. Adams (Ref. 1-2) gives a different viewpoint on recognizing and overcoming mental blocks to improve creative thinking. *De Bono's Thinking Course* by Edward de Bono (Ref. 1-5) outlines tools and techniques of good thinking in a practical, easy-to-follow manner. Roger Schank, in *The Creative Attitude: Learning to Ask and Answer the Right Questions* (Ref. 2-14) especially addresses the problems of how to overcome the fear of failure and the mental barriers that have been taught, such as following the rules and looking for only one answer. *A Whack on the Side of the Head* by Roger Von Oech (Ref. 1-17) teaches how to overcome ten mental blocks to creative thinking. The American Creativity Association, St. Paul, Minnesota (phone/fax 612/784-8375) has a list of interesting materials available that encourage creative thinking. Two examples are: *Are We Creative Yet?* written by DuPont employees based on the "Frank and Ernest"humor of cartoonist Bob Thaves, and two cassette tapes by Marilyn Schoeman Dow: "Creative Innovative Thinking" and "Positive Parenting: Nurturing Your Child's Creativity." Most of the books listed in Part 1 of this textbook list additional references about the nature, discovery, development, and application of creativity.

CREATIVE THINKING EXERCISES

4-1   DEFINITIONS AND SLOGANS
Make up some slogans to encourage creative thinking. Pick one that you especially like; write it on a notecard and place it where you will see it frequently—such as on the dashboard of your car, your bathroom mirror, or over your TV or stereo set.

4-2   MEANING OF PROVERB
There is an old Jewish saying: "If there are two courses of action, always take the third." Explain what this means to you and include an application. Then try to come up with an entirely different explanation.

4-3   AMBIGUITY
"I am here because I have nothing better to do." Rewrite this statement in a least two different ways to remove the ambiguity.

4-4   PARADOXES
Take a few minutes to make up some paradoxes. A paradox is a contradictory statement. See if you can use your paradoxes as "jazzy" book titles. Examples of paradoxes are: warm ice, a soothing rock concert, unbreakable glass, soft stones, dry rain, a calm tornado, a timid hero, dreadful happiness. If the last three were book titles—what would these books be about? Can you see where writing paradoxes is useful? Wouldn't it be wonderful to have "warm" ice for treating a sprain, so it would have the benefit of coldness without making the skin numb?

4-5   ANALOGIES AND METAPHORS
a. You can play around with words by writing metaphors. A metaphor is a figure of speech in which one thing is talked about as if it were something else. Example: "Break the cocoon—fly with creative thinking" (contributed by Dawn Rinehart). Kim Steger developed an essay using the growth of a rose to illustrate creativity. Can you make up five different metaphors for creative thinking?

b. An analogy is a comparison to something that is similar. For example, you could say: "My college dorm is like a maze with no exit, with occupants always hungry and scrounging for food." Or, "My university is like a supertanker—huge and powerful, with a set course that's tough to change." Or, "My student organization is like a beacon of light, helping people avoid rocks and shipwrecks." Can you think of five additional analogies along the lines of the examples? Next, repeat this exercise but use only positive statements.

4-6   HUMOR
Humor is good for creative thinking because it gets the mind off the expected, usual track. Find some examples of jokes, cartoons, or comic strips that are good illustrations of the surprise ending. Or even better, make up your own funny story or pun.

4-7   YOUR MIND MATTERS
Use the words YOU, MIND, MATTER in different combinations to come up with ten different slogans. (You may add one or two additional words to the sentence, if desired.) Which one of your slogans could be used by a professor lecturing students, by a young man whispering to his girlfriend, or by a child "educating" a parent?

4-8   NEW WORD
Make up a new word and its definition.
EXAMPLES: Fluffle—a new whipped apple dessert served on a flat thin pancake.
Shrumble—an alarm clock that shakes the bed (contributed by Brandon Slotterbeck).
Spadles—the round dirt marks left on your car after a light rain (contributed by Brad Cramer).
Haphazardous—casually, by chance, but with risky and even dangerous consequences.

4-9   LEARNING FROM FAILURE
Describe an incident in which you made a mistake "with class" or when an initial failure was used as a stepping-stone to success.

4-10   GOOD NEWS BULLETIN
You are in charge of writing an "all good news" bulletin. List at least ten items that you would feature in the bulletin and make up appropriate headlines. Do you find that this is more difficult to do than thinking of "bad" news? Why?

4-11   GOOD NEWS/BAD NEWS/GOOD NEWS
Make up four different Good News/Bad News/Good News "stories" with three statements.
EXAMPLE: (Good News) Last summer, I found a $50 bill in the grass. (Bad News) When I picked it up, a poisonous snake bit me. (Good News) The ambulance attendant was so kind and caring—we are now married.

4-12   IT'S HOPELESS—SO WHAT
Make up five different sets of sentence pairs for "It's hopeless—so what."
EXAMPLE: My car's not working. So what—I can jog to the store.

4-13   *PLAYING WITH WORDS*
The last three problems above have been adapted from A Book of Puzzlements, Play and Invention with Language *by Herbert Kohl. Look up a "playing with words" book in the library. Do some of the games with your family or friends next time you have a party or when taking a long road trip by car.*

4-14   WHAT-IF QUESTIONS
Make up some interesting what-if questions.

EXAMPLES: What if everybody decided to be perfectly considerate for just one day? What if computer viruses got out of hand and made computers unreliable to use? What if transportation became so advanced that cars and highways became obsolete—what kind of transportation could this be? What if the moon suddenly blew up? This last question has all kinds of interesting implications, from dealing with the debris of the catastrophe to the changes in the tidewater ecology if the tides disappeared. Use the PMI approach on this question.

By yourself or with a friend, play around with the example problems or use some of your own what-if questions. Can you come up with twenty ideas of what might happen for each question?

4-15   *PROJECTS*
a. Do a project that will demonstrate creative thinking.
b. Make up example problems that demonstrate that the context is never irrelevant.
c. Describe situations where more than just one right answer was needed.
d. Develop a plan and describe the steps you can take to overcome the "follow the rules" barrier.

4-16   *WHAT IF WE HAD INSTANTANEOUS TRANSPORTATION?*
Travel-Holiday *magazine in June 1990 had an interesting article by Arthur C. Clarke, the science fiction writer. In his characteristically imaginative way, he wrote about some possibilities to the scenario: What if transportation were instantaneous—people could "dial" themselves to anywhere they wanted to be, as easily as sending a voice or fax message today. With two or three other people, brainstorm and discuss what society would be like if this type of transportation were possible. Then compare your answers to Arthur Clarke's ideas. We think that your group will come up with ideas that are just as creative as this famous writer's ideas. Remember, playing with what-if questions is good exercise for the quadrant D part of your brain.*

---

*The asterisk denotes exercises that require extra time but provide more in-depth learning.

Homework Project

INVESTIGATION OF NEGATIVE THINKING

Week 1—General Data collection and Preliminary Analysis

Instructions: Duplicate the following two pages; then use the Negativism Score Sheet to increase awareness of your own negative thinking. This investigation can also be a useful tool if you have to work with people who have a habitually negative outlook or who are very critical. Keep the tally sheet with you at all times, together with a pencil. Each time you catch yourself being negative or experiencing one of the items listed in the tabulation below, make a hatchmark on the sheet. Add up the total for each day for 7 days.

Example Results: We did this assignment with a class of 14-year-old students from the inner city of Detroit—for two weeks. The results showed these students doing better with less incidences of negative thinking during the second week (as expected) since their outlook improved as they became aware of their own negative thinking. We also found—quite surprisingly—a trend of lowest scores on Sunday and Monday, with highest scores (more frequent negative thinking) in midweek.

Preliminary Analysis: After Week 1, write down your conclusions about your data on the Negativism Score Sheet What do your results show about your thinking patterns? What do you think are the major causes of your negativism? Write down the six most important categories in 1(b) of the Analysis Worksheet and on the bottom of the Negativism Score Sheet (see the table below for ideas). Use the seventh category for "Miscellaneous."

Week 2—Detailed Data Collection, Analysis, and Application

Instructions: To investigate your negative thinking habits in detail, keep detailed scores on your negative thinking according to the categories—for one week. Add up the totals for each category as well as the daily totals.

Analysis: Compare your Week 1 and Week 2 totals—what do you conclude from these results? Then construct a Pareto diagram using the data from Week 2 (see Chapter 5 for examples of Pareto diagrams).

Application: Describe two things that you want to change to become a more positive thinker and supportive person. Do Problem 5-4 at the end of the next chapter. Also, describe a situation where you turned a negative into a positive.

Examples of Types of Negative Thinking

1. Judging others: being critical, nitpicking, looking for "wrongs."
2. Avoiding people or situations; procrastinating.
3. General complaining, whining, moaning (from habit).
4. Lack of self-discipline.
5. Angry and spiteful; looking for trouble and revenge.
6. Expressing sarcasm, scolding, intolerance.
7. Using abusive language and profanity; lack of respect for others.
8. Down on self: dejected, pessimistic, sad, fearful.
9. Receiving put-downs, sarcasm, negative feedback from others.
10. Imagining slights; having a "self-pity party"—seeing life as unfair.
11. Lack of vision, positive goals, meaning, hope; seeing "no way out."

NEGATIVISM SCORE SHEET (Data Collection)

Lumsdaine and Lumsdaine, *Creative Problem Solving*, p. 145. © 1994 McGraw-Hill.

For each day, make a hatchmark every time you catch yourself doing negative thinking, using sarcasm, or putting yourself or someone else down.

Starting Date: _____

Day 1							Total =
Day 2							Total =
Day 3							Total =
Day 4							Total =
Day 5							Total =
Day 6							Total =
Day 7							Total =
Total							**Week 1:**
Day 8							Total =
Day 9							Total =
Day 10							Total =
Day 11							Total =
Day 12							Total =
Day 13							Total =
Day 14							Total =
Total							**Week 2:**
	A	B	C	D	E	F	Misc.

ANALYSIS WORKSHEET

1. ANALYSIS OF WEEK 1 RESULTS:

a. My conclusions about the Week 1 scores are:

b. The six most important types of negative thinking that I seem to be doing are:

A =

B =

C =

D =

E =

F =

2. ANALYSIS OF WEEK 2 RESULTS:

a. My conclusions from comparing the Week 2 totals with the Week 1 results are:

b. Construction of a Pareto analysis and diagram to identify the most frequent causes of my negative thinking:

3. APPLICATION

a. I will become a more positive thinker by:

b. I will become a more supportive person by:

CHAPTER 4 — SUMMARY

Barriers to Creative Thinking —> False Assumptions: *"I am not creative"* is a false assumption. Everybody is creative and with some training can learn to be even more creative. We have marvelous brains. *"Intelligent minds are good thinkers"* is a false assumption. For example, quick thinking can lead to false conclusions, whereas leisurely or exploratory thinking can give more creative results. We need to use the whole brain for good thinking. *"Play is frivolous"* is another false assumption. Play is very important to creative thinking. Play around with your imagination; play around with words. Ask what-if questions frequently, alone or with others, verbally, in writing, or with the computer. Many myths exist about creativity—remember that anyone can be more creative!

Cure: Learn quadrant A facts about creativity; learn good thinking skills; practice and play! Use fantasy and wishful thinking; exercise your imagination.

Three Mental Blocks That We Have Been Taught —> Habits: *There is only one right answer*—to be creative, look for several answers; develop a habit of asking many different questions. *Looking at a problem in isolation*— to be creative, look at the whole picture; the context is never irrelevant. Develop a habit of divergent thinking. *Following the rules*—to be creative, don't follow rules that are not there; find out the reason for the rules (maybe they are no longer valid or needed).

Cure: Explore the entire problem and its connections; look for different and alternative answers; break the rules and mental boundaries—use contextual thinking. But keep quadrant B rules of courteous behavior and self-discipline.

Three Mental Blocks That Involve Our Emotions —> Attitudes: *Negative Thinking*—to be creative, look at things as being different or interesting, not good or bad. Be positive; look for the good points. Praise yourself and others for all these interesting ideas. Negative comments are contagious and thus doubly harmful. Use the PMI approach when investigating problems. *Avoiding risk/fear of failure*—to be creative, do not be too fussy. Also, do not let the fear of possible ridicule stop you from having and expressing creative ideas. Failing in two out of three tries is actually a very good batting average. Mistakes are valuable because we can learn from them—so don't be afraid to make them. Mistakes are opportunities for making improvements. *Ambiguity*—to be creative, explore ambiguous situations. Look at the problem from many different angles; look for motives and reasons to learn why people do things or what the problems are all about. Do not rush to a quick answer, even though ambiguous situations may make you uncomfortable; instead, ask questions.

Cure: Nurture a positive quadrant C attitude in yourself and others; take risks with learning—use failure as a stepping-stone to success. Take time to explore ambiguities—these are opportunities to practice creative thinking.

How to Encourage Creativity: Be a role model—let others see how you practice and express your creativity within the bounds of mental toughness. Then establish a support network—encourage others to be creative. Know how to recognize creativity. Build a creative environment for children and adults. Use a creative physical and attitudinal environment, especially when brainstorming. Practice quadrant D thinking supported by good thinking in quadrants A, B, and C. **Plan for creativity!**

REVIEW OF PART 1—CREATIVE THINKING

The questions below will help you check up on your learning—on your own or as a midterm review or exam to gauge your progress if you are using this book in a formal class.

QUADRANT A AND QUADRANT B TYPE QUESTIONS

1. *Write down the simple phonetic alphabet and how the sounds correspond to the ten numerals. How did you visualize (memorize) the relationships?*

2. *Describe the thinking characteristics of each brain quadrant of the Herrmann model. Give an example of behavior for each quadrant.*

3. *Discuss two reasons why intelligence does not necessarily result in good thinking.*

4. *List five benefits of having creative problem-solving skills.*

5. *List the steps and associated mindsets of the creative problem-solving process.*

6. *List the higher-level thinking skills in Bloom's taxonomy in the right sequence.*

7. *You are designing a toy for a small child. Four different mental languages are needed to do this task well. Explain how and why each language is used.*

8. *How would you overcome the mental block, "There is only one right answer"?*

9. *Describe a myth about creativity that you have heard people express, and write down the facts to counter this myth.*

10. *Describe the steps of how you have applied something that you have learned so far from this book.*

11. *Briefly describe and analyze your own learning style.*

12. *Describe the characteristics of an organizational environment that encourages creativity. Then select an organization you are familiar with and use your list of characteristics to evaluate its creative climate and performance.*

QUADRANT C AND QUADRANT D TYPE QUESTIONS

1. *Brainstorm a mindmap on one of the following topics: The brain. The future. An object of technology. The ideal breakfast. Continuous learning. What makes a good teacher? What makes a good student?*

2. *Select an intangible word, foreign word, name, number, or date; develop a story or an image that can be visualized or sketched to help remember the item.*

3. *Brainstorm the problem: In what ways can people have fun at a party without using alcohol? List at least twenty different, positive ideas not involving risky behavior.*

4. *Write a brief explanation of "paradigm shift" and include your own illustration.*

5. *Answer the paradigm shift question about one of the following: (a) elementary school, (b) secondary education (junior high or high school), (c) college. What is impossible to do today but if it could be done would fundamentally change these educational systems in the United States? EXAMPLE: Replace the traditional eighth grade with a year of community service and just-in-time teaching. Students will learn hands-on skills, project management, cooperation, communication skills, and success in implementing creative problem solving as they renovate neighborhoods and are involved in environmental projects and other constructive, real-life activities.*

6. *Design a collage (or quilt) which illustrates and summarizes the characteristics of the Herrmann Four-Quadrant Model of Thinking Preferences, using symbols and imagery rather than verbal language.*

7. *Describe a home environment that would foster creativity in people of all ages.*

8. *How could you best teach Pareto analysis to a quadrant C person?*

9. *Ask an imaginative what-if question and give five different answers.*

10. *Why is a positive attitude important when you are solving a problem?*

11. *Imagine yourself seven years in the future. Write a short paragraph (in the present tense) of the kind of work you are doing. Describe your workday, your environment, your tools. Think of some trends in today's technology and incorporate new developments into your picture. Your verbal sketch must be more than just describing yourself in a work situation that exists today—you must invent something new.*

12. *Write a sound bite or television commercial to remind people to think creatively. Start by brainstorming several ideas; then try to combine them into one "best" idea.*

13. *Make a contour drawing of your left hand (holding the pencil in your right hand). Then switch hands and repeat. In both cases, do not look at your drawing but keep your eyes firmly on the hand being drawn. Then discuss the feelings you experienced while doing this activity, as well as the artistic results.*

14. *What are the five most important concepts that you have learned in Part 1 of this book? Select one concept overall and one from each chapter; then try to connect them visually. Write down one important question that you still have.*

THINKING REPORT (WHOLE-BRAIN PROJECT)

Another way of applying what you have learned by doing a thinking report. This "different" approach to report writing will help you improve your writing and communication skills (and it should be fun).

Assignment

Demonstrate your understanding of the broad subject of creative thinking covered in Part 1 of this book by doing a piece of writing that makes connections between what you have learned and material elsewhere. If you are short of ideas, check Section 4 in Chapter 12 or think about some important questions you still have that you may want to investigate further. Make sure that you use all four quadrants of thinking preferences. Other suggestions and requirements are outlined in the steps below.

Report Writing Procedure

1. *Brainstorm a mindmap—choose your topic and main ideas. Research the subject and select your sources: books, journal articles, speeches, TV programs, artworks or cartoons, audiotapes, technological artifacts, etc. Elaborate, edit, redraw the mindmap.*

2. *Decide on your objectives; address the "why"; explain your choice of topic and the purpose of your project.*

3. *Select the approach and format for the written "report": essay, poem, short story, case study, etc. Required length: one to three pages on a word processor.*

4. *Do your writing: include facts and examples to support your views and conclusions.*

5. *Use a spell-checking software program. Submit your work to two people (one who has some knowledge of your topic and one who does not) for proofreading your final draft. Ask your editors to make suggestions on how your content and your writing style could be improved for increased clarity of the ideas being communicated.*

6. *Revise your work. Make a mindmap representation of your final report to check the logical flow and connections of the ideas presented.*

7. *Experiment with various features of your word processing program to achieve a pleasing page layout for the final "best" version of your project. Include the last mindmap as an appendix (done neatly by hand or on the computer). You may also want to append a brief evaluation of your experience with this assignment: How will it help you in future writing assignments? Which thinking preferences do you need to develop to produce top-quality written communications?*

Answers to the Myth Quiz on page 122: All statements express false assumptions or myths.

Part Two

Creative Problem Solving

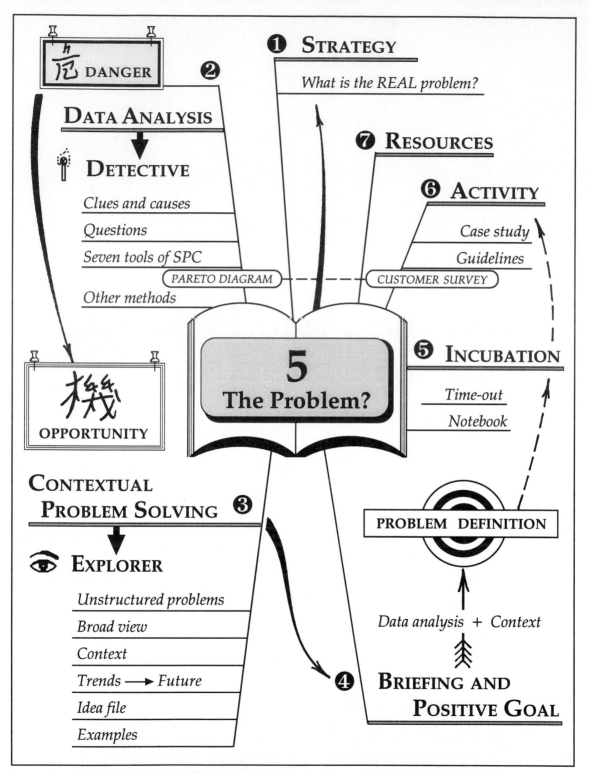

Mindmap of Chapter 5. Refer to Chapter 2 (pp. 55-56) to learn more about mindmapping.

5

PROBLEM DEFINITION

WHAT YOU CAN LEARN FROM THIS CHAPTER:
- *Strategies for finding the real problem—overview.*
- *Data collection and problem analysis—the detective searches for clues and specific causes. Tools (including the customer survey and Pareto analysis) are used to identify major aspects and causes of a problem.*
- *Contextual problem solving—the explorer looks at the bigger picture (including trends and opportunities).*
- *Convergence—the briefing document and the positive problem definition statement.*
- *Problem incubation—introspection and purging: keeping a notebook.*
- *Hands-on activity for problem definition—case study and guidelines.*
- *Resources for further learning: reference books; activities and exercises for detectives and for explorers; summary.*

In Part 1 of this book we focused on sharpening different thinking skills. Since school systems in our industrial society put a heavy emphasis on analytical and sequential (left-brain) thinking, we took time to introduce you to visualization and its connection to memory since this thinking mode helps to access the right side of the brain. The discussion of the four-quadrant model of thinking preferences purposed to build an appreciation of different thinking abilities and presented the advantages of whole-brain thinking by individuals and teams. Chapter 4 demonstrated how to improve creativity by overcoming mental blocks and honing thinking skills in each quadrant. The emphasis was on making you more comfortable as well as more disciplined with quadrant C (interpersonal, emotional) thinking and with quadrant D (imaginative, holistic, intuitive) thinking. But since we are not truly creative unless our creative ideas are applied or put into action, Part 2 of the book will focus on the creative problem-solving process—where we will use our whole brain to find good solutions to problems. The format of each chapter in Part 2 is similar to that used in Part 1, except for the addition of instructions for doing a hands-on team activity to immediately practice each step in the creative problem-solving process. Also, when there is a definite switch from analytical (left-brain) to imaginative (right-brain) thinking, we have inserted a creative thinking warm-up exercise and recommend that you take the time for this brief but necessary activity. So let's start problem definition at the beginning—with what we already know.

Problem: [pro-, forward + ballein, to throw]
1. a question or matter to be thought about or worked out.
2. a matter, person, etc. that is perplexing or difficult.

Definition: [de-, from + finis, boundary]
1. a defining or being defined; a determination of the boundaries, extent, or nature.
2. a statement of the meaning of a word, phrase, etc.
3. (a) a putting or being in clear, sharp outline (b) a making or being definite or explicit.
4. the power of a lens to show (an object) in clear, sharp outline.
5. radio and television: the clearness with which sounds or images are reproduced.
Webster's New World Dictionary.

WHAT IS THE REAL PROBLEM?

Remember the Chinese character *wei ji* ? It means *problem* or *crisis* and is composed of two words: *danger* and *opportunity.* To remember this, visualize a little scenario: You have a crisis; you are on board a leaky boat. Your sister is crying (– *cry, sis*). People are crowding the gangplank. You yell, "Make way!" "Way" is the pronunciation for *wei* and you have associated it with the idea of danger. You reach the pier safely; you look back and exclaim: "Gee!" (which sounds like the pronunciation of *ji*) because you see an astounding sight—the tide has gone out and the boat is perched high and dry on a rock. You are thinking up opportunities for how to turn this into a profitable business venture—a hotel or a locale for a mystery movie. Because a problem can have these two aspects, problem definition requires two mindsets: the detective for dealing with the crisis and the analytical parts of problem definition, and the explorer for dealing with the context and the opportunity aspects of the problem. Table 5-1 summarizes and compares the different thinking tasks and inputs of the detective and the explorer. The "bulleted" tasks in bold print will be discussed in detail in the subsequent sections: data collection and analysis in the detective's mindset, contextual problem search and trend spotting in the explorer's mindset, with these activities converging to the problem definition statement as the objective or target of the creative problem-solving process. For comprehensive problem definition, we need facts and details—the quadrant A (and some quadrant B) thinking of the detective—as well as a view of the whole picture, the contextual connections, and the systems approach—the quadrant D (and some quadrant C) thinking of the explorer.

In essence, the main assignment for the detective and the explorer is to find out what the real problem is. It is surprising to realize how much effort and how many arguments are wasted in families, groups, and entire organizations because no time is taken to carefully define a problem. The very first step here is to agree or accept that a problem exists. If you or other people concerned in the situation deny that there is a problem, nothing will happen to improve the situation or find a solution. Problem definition makes sure that everyone involved understands the situation and works on solving the same problem, because what may appear to be the problem may not be the real problem. Let's look at some examples. It is Monday morning, and some child you know does not want to get out of bed. The child moans and complains about a stomachache. Is the problem an upset stomach? If

Table 5-1
The Problem Definition Process—The Detective and the Explorer

DETECTIVE	FOCUS	EXPLORER
Assigned a problem or crisis: something is not working right.	**Type of Problem**	Find or identify a "mess." Uncover a problem or opportunity.
List facts already known. Determine what data are needed.	**Facts**	Look into the context; set goals. Imagine the ideal situation.
Determine constraints and limits: time, budget, staff, other resources.	**Boundaries**	Keep limits in the back of the mind; seek to overcome the boundaries.
What is so terrible about the situation?	**Feelings**	What would be nice … if it could be done?
Autocratic chain-of-command: Who is responsible? Who is the expert?	**Communications and People**	Cooperative teamwork: Include people not too close to the problem, and from related fields, not experts.
Use existing tools and methods; traditional approach. Analytical and sequential thinking.	**Problem-Solving Paradigms**	Look for new paradigms, trends, and alternatives; use divergent, intuitive, sensory, flexible thinking.
Search for causes and clues. • **Collect and analyze data.**	**Tasks**	• **Seek out the context and trends;** see connections; look to the future.
Problem: bikes are stolen. Experiments: test chains/cables with hacksaw and bolt cutters. Conduct a customer survey. Conclusion: Need better locks.	**Example**	See the stolen bikes in context— a systems problem in bike security. Consider changes in bike design, bike parking, bike registration, and other uses for security systems.
Narrow scope—focus on task. If a solution is identified that will solve the problem, adopt it. Then terminate problem solving.	**Scope**	Initial wide scope converges to goal and most important factors; explore resistance to change. • **Write the problem definition statement.**
If no good solution appears, shift to explorer's mindset and continue with contextual problem definition.	**Action**	No clear solution. • **Observe an incubation period.** Continue the creative problem-solving process.

you take the time to investigate the situation, you may find that the real problem is the bully that is tormenting the younger kids at school. Here is another example—a bit more scientific. Let's say you live in a small town, and your house is located close to a dairy farm. You and your neighbors are having quite a problem with too many houseflies. How are you going to solve this problem? Actually, what is the real problem? Do you want to design a more effective fly trap, or do you want to look at the broader picture and improve the health of the community by eliminating flies altogether by preventing them from breeding? You can see that the solution that you will be seeking will depend on what you see as being the real problem. You need a clearly defined objective! Or look at a mouse or mole trap. What is the purpose of such a trap—to trap these rodents or to kill them? If you only want to trap these "critters," you will design a different trap than if you want to kill the "vermin."

DATA COLLECTION AND PROBLEM ANALYSIS— THE DETECTIVE'S MINDSET

Figure 5-1   The detective.

An important objective guides our activities during the problem definition phase—to collect as much information as possible that is related to the problem. Collect this data even if you think it is not very important or if you think you already know what the problem is. But you are not to be a judge at this point, so do not decide too early whether something is or is not important. We have a whole toolbox of methods available to collect data, depending on the type of problem, your organization's problem-solving culture, and your expertise. In creative problem solving, we begin with the detective's mindset—we are either given a problem-solving assignment or are faced with a crisis. If we are in the fortunate position where we can look around for a problem to solve, we can start with the explorer's mindset and switch to the analytical detective's mindset as needed. Why do you suppose we associated the detective's mindset with the color blue?

What attitudes must a detective have? Detectives are looking for information that is hidden—to find it, they must be persistent, and they must think logically about where to look and how to go about finding the desired information and clues. Figure 5-1 is a humorous illustration of this mindset. A methodical, quadrant B approach is helpful; thus the quantitative information that is collected is best assembled in a notebook or file. Detectives ask questions about who, what, where, when, why, and how much. Some people have made up long **lists of questions** to help in this process of data collection. These questions should be answered in as much detail as possible. By going through a list of questions, we may be able to find enough relevant data and information to define the real problem. Table 5-2 gives some sample questions that are useful for problem definition.

Table 5-2
The List of Questions

- How big is the problem?
- What is distinct about the problem?
- What makes this a problem?
- What makes this problem different from other problems?
- What events caused this problem?
- How long has it existed?
- Why is it a problem?
- How did the problem get started?
- Who has been involved; in what way were they involved, and why?
- When and how was the problem discovered?
- Where is it located?
- What changes (in surroundings, equipment, procedures, personnel) occurred that could possibly be related to the problem?

- What are the specific causes of the problem—what is your evidence? How are these causes related?
- Does the problem pose a threat to people, your organization, or your community? In what way is it a threat?
- Does the problem have long-term or only short-term effects on individual people, on the community, or on the environment? How?
- How complex is the problem?
- How are the different parts related?
- Is the problem connected to other problems? In what way?
- Can some of the factors be dealt with separately? How would this affect the overall problem?

This approach of using a list of questions is used by companies and businesses in their efforts to solve problems. The **Kepner-Tregoe method** of problem solving uses this type of analysis. The problem is defined as the extent of change from a former satisfactory state to the present unsatisfactory state, and finding the causes of the deviation should help solve the problem. This requires specific, quantitative data about the entire problem area. It also helps to describe the problem in terms of what it is *not*. Thus the Kepner-Tregoe method is very good for finding the boundaries of a problem. This is a useful approach. As you collect data and information about your problem, be sure to also include what it is not or things that were already tried and did not work. For example, we would have liked to write a book in the early 1970s about all the things that people tried and that didn't work in solar energy, especially in passive solar home design and construction. This would have been very helpful and could have prevented many other people from making the same mistakes all over again. Unfortunately, we found it just too difficult to get this kind of information. People do not like to talk about their failures, only their successes. The Kepner-Tregoe analysis clarifies the factors that are irrelevant and focuses on the areas that need to be examined and tested. Companies, especially manufacturing companies, use a number of analytical methods for collecting specific data about problems. One of these approaches is called **Statistical Process Control or SPC**. SPC uses a number of different tools that we will describe below. Japanese companies train all their employees (from top management to shop floor workers) in statistical process control. SPC techniques are useful for analyzing many different types of problems, not just those in manufacturing. Managers in U.S. companies now expect their college-educated personnel to be familar with SPC concepts—thus we have included a brief description here.

Statistical Process Control (SPC)

Statistical process control (or SPC) is a tool that uses statistical data and comparisons to monitor processes; its primary objective is to prevent problems, like a physical exam. A process is simply a sequence of actions or activities which result in a product or service. One of the primary goals of quality control is to assure that the results of a process remain uniform—or in other words, that variations and the resulting decrease in quality are kept to a minimum. When a process is controlled, this means that it is monitored to assure the product meets the quality targets. SPC is also a valuable tool for problem analysis: by making graphs of the data and then analyzing the results, the causes of problems can be identified. The originators of SPC were two American quality experts: Dr. Walter A. Shewhart of Bell Laboratories in the 1920s and Dr. W. Edwards Deming after World War II.

The first step in using SPC to analyze a problem is to collect data. This data must be relevant; it must be usable; and it must be accurate and reliable. Always check the calibration of the instruments that are used for measurements. Carefully think about data collection to avoid creating a data swamp with too much information. SPC has seven tools: checksheet, histogram, fishbone diagram, Pareto diagram, scatter diagram, control charts, and additional documentation.

1. Checksheets: Basically, there are three different types of checksheets. The **checklist** is a memory aid; it helps us to follow procedures correctly. A simple example is the shopping list you take to the grocery store. Following a checklist scrupulously can be very important—airline pilots use checklists before takeoff. In a manufacturing plant, checklists are frequently used for maintenance procedures and schedules of various sorts. Your car dealer or service station may use a checksheet for your vehicle's 20,000-mile maintenance work. **Recording checksheets** are for collecting data on frequency. Instructors may use a recording checksheet for class evaluation at the end of a term to get a statistical look for identifying areas for improvement. A **location checksheet** is a drawing of a product part or map; the location of defects is marked so that the investigation can zero in on the most critical areas. Another example is a floor plan of the plant, where all locations of accidents are noted that occurred within a specified time period. Or a state's highway department marks accidents and fatalities on roads and highways on a large map to identify danger spots and areas for improvement. Data visualized in this form can be very helpful in pinpointing critical areas for an investigation and corrective action; then this baseline data simplifies comparison once the improvements have been made. Checksheets are important in problem solving because they are a tool for collecting necessary data in easily used form. If set up carefully, they make data analysis and troubleshooting easier.

2. Histogram: A histogram is a bar graph depicting frequency versus variation of some product parameter, as shown in Figure 5-2. It is often used in charting the precision of machines or in process capability studies to determine the relationship between target values and actual production values. Its aim is to eliminate defects and improve yield and product quality; it is an essential tool in the factory to chart and monitor "continuous improvement."

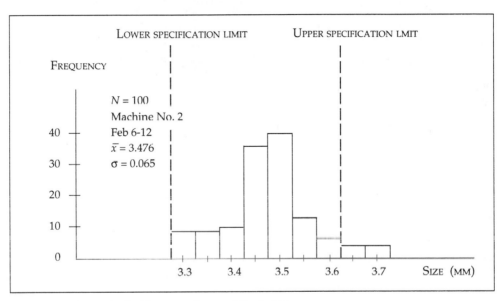

Figure 5-2   Example of a histogram for metal block thickness.

3. Cause-and-effect (fishbone) diagram: Raw materials, work methods, equipment, and measurement can be the cause of quality variation. Through brainstorming, a cause-and-effect diagram can be built-up in detail by asking the question: "Why does this variation or dispersion occur?" Such diagrams clearly illustrate the various causes affecting product quality, and they can aid record-keeping of production deviations. The cause-and-effects diagram was invented by Kaoru Ishikawa in 1943. The process for making a fishbone diagram is similar to creating a mindmap, except that the main characteristic that needs improvement is placed along the main horizontal arrow (or "fish vertebra") instead of in a central circle. The main factors which may be causing the problem are determined and entered as main branches (or bones) to the diagram. To each of these branches, the detailed factors which may be regarded as a possible cause are added as branchlets. For example, for the effect "wobble during machine rotation," a main factor would be material, which has two possible related factors or causes: "G" axle bearing or central axle. "G" axle bearing then branches into material quality and variations in the dimensions (which can in turn be broken down into their possible causes).

4. **Pareto diagram:** The Pareto diagram is a specialized bar graph used to identify and separate the vital, the most important, causes of trouble from the more trivial items. The Pareto diagram arranges classification items by order of importance and thus points out which factors or causes need to be addressed first for improvement. The vertical axis can be expressed in numbers or percentage of cases, but the most useful parameter is money lost by the defect. This diagram was "invented" by Vilfredo Pareto, an Italian economist who was struck by the fact that 20 percent of the population in a country control 80 percent of the wealth in the country. This "80/20 principle" makes it possible to concentrate resources on removing the top 20 percent of the causes and thus cure 80 percent of the problems. The Pareto diagram is very useful for assigning priorities for continuous improvement; an example is given in Figure 5-3. Among the five operations causing defects, caulking should be the initial target to improve quality.

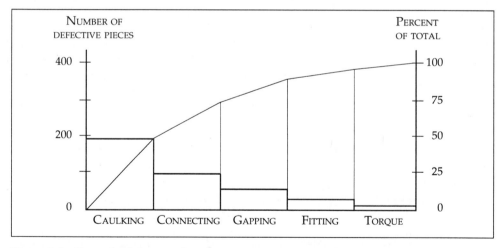

Figure 5-3   Example of a Pareto diagram.

5. **Scatter diagram:** Scatter diagrams are most frequently used to study the relationship between a cause and its effect, with cause values commonly plotted on the horizontal axis and the corresponding effect on the vertical axis (as shown in Figure 5-4). The resulting scatter patterns will indicate positive correlation if both axes show an increasing trend, negative correlation if the effect decreases for increasing cause values, or no correlation if the plotted dots are randomly scattered. If the pattern is crescent-shaped, more than one cause may be involved, and further analysis is needed to clarify these effects.

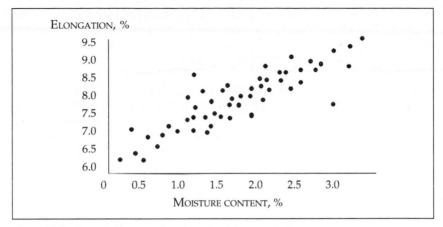

Figure 5-4   Scatter diagram showing a positive correlation.

6. Control charts: Control charts present data plotted in a chronological sequence to show how the influence of various factors in the production process (materials, workers, methods, equipment) changes over a period of time. Data points falling outside the limit lines signal an abnormal situation that needs appropriate action. The control limits are calculated mathematically—they are not specifications. The daily data, for example, are averaged in order to obtain an average value for that day. Each of these values then becomes a point on the control chart which represents the characteristics of that day. Both the mean value x and the range R are plotted; the x portion of the chart shows changes in the mean value of the process, while the R portion shows changes in the dispersion of the process. Control charts are powerful tools for preventing the fabrication of defective products; they also chart the progress of continuous quality improvement and identify when a process has achieved a decrease in variation. Quality circles or teams use this data to brainstorm ways of improving processes and quality.

7. Additional graphs and documentation: Other types of graphs can be used to document defects or identified causes. Among these are flow diagrams, stem-and-leaf plots, various line graphs, circle diagrams, spaghetti charts, and matrices. If you have an opportunity to take courses in statistical methods or SPC, we recommend that you take this training; it will give you a valuable advantage in being a problem detective. For the purposes of creative problem solving, we will primarily use the Pareto analysis for evaluating customer survey data, and you will see an example toward the end of the chapter. We will also use checksheets.

Other Tools for Data Collection and Analysis

In addition to SPC, manufacturers, researchers, designers, and service companies use many other tools for collecting and analyzing data. Why should so much care and time be spent on defining the problem? Many people and even cultures have recognized that "a problem well defined is half solved." Companies have their own training workshops, manuals, and procedures to teach their employees the analytical methods that they prefer using. We are briefly describing some of these methods and their purposes to give you the vocabulary and an understanding of where they fit into the problem-solving process.

Experiments and surveys: Sometimes, experiments are conducted to get the data needed to answer the list of questions and define the problem accurately. Many manufacturing and service companies depend on surveys to collect data on "the voice of the customer." The data collected from surveys can then be analyzed and visualized with a Pareto diagram. This usually makes an interesting exercise for students since the causes (or aspects) of a problem identified through the survey are surprisingly different from what the students originally think is the "real" problem. For example, when middle-school students surveyed their classmates to find out what the most important problems with their schools were, the top answer was the lunch period—too short, no social interaction possible, unsatisfactory food, and an unpleasant room. Thus by making changes in this one area, students' attitudes toward school could be greatly improved at little cost. A survey on bicycle locks found that the biggest problem was not the theft of bikes but that the bike owners could not open their own locks. Not surprisingly, having this particular insight affected the direction of the problem-solving effort. A survey on toasters found several instances of kitchen fires caused by toasters. Although the frequency was low here, the costs and potential dangers were high, and thus fire prevention became one of the important design criteria in the design of an improved toaster.

FMEA, FTA, and benchmarking: Ford Motor Company, for example, uses two specific methods to analyze causes of failures. Failure mode and effects analysis (FMEA) explores all possible failure modes for a product or a process, whereas fault tree analysis (FTA) is restricted to the identification of the system elements and events that could lead to or have led to a single, particular failure. The FMEA allows engineers to assess the probability of a failure as well as the effect of a failure. By identifying potential problem areas, an FMEA conducted early in the design process can aid in preventing defects and in planning appropriate test programs. Identified causes of failures are ranked according to frequency of occurrence, severity, and ease of detection. An FMEA can be used for services as well as for products—see Appendix B for more details. On a simple level, you use this type of analysis when you are thinking about how to prevent your suitcase from being lost by an airline. You know that you need to have identification inside and on the tag and that you must verify the destination tag. But if you absolutely require the bag on arrival, even the small probability of a "failure" will make you decide to pack lightly and carry on the bag instead. **Weibull Analysis** is a technique used by manufacturing companies where the results of testing products to failure are plotted on a log-log paper. Cumulative failures (in percent) are graphed versus a product life parameter such as hours of operation or miles driven.

Fault tree analysis was invented by H. A. Watson of Bell Laboratories to evaluate the safety of the Minuteman launch control system; this deductive analysis requires considerable information about the system. It graphically represents

Boolean logic associated with the development of a particular system failure (see Appendix C). The FTA considers a single undesirable event and directs activities toward eliminating the event by controlling all the factors that could contribute to the failure. On a simple level, you use this kind of thinking, for example, when the lights frequently go out in your house. First you might check to see if your house is the only one that is dark or if the entire neighborhood is affected. If you are the only one, you would seek for causes, such as a faulty circuit breaker or frayed powerline to your house; if the entire neighborhood is dark, you would probably try to contact your utility company. If you find that the problem is in your fuse, you will need to investigate what caused it to blow—do you have an appliance that is leaking current, for example? Both the FMEA and FTA result in recommendations and corrective actions. These techniques make valuable contributions in problem definition because they help in differentiating and identifying causes and effects.

In addition, warranty claims and complaints about a product need to be analyzed, and products and services need to be evaluated against the competition—this is known as *benchmarking*. The "House of Quality"—the first step in a very structured procedure called **quality function deployment** or **QFD**—is useful for collecting warranty data and comparing critical product quality characteristics against the competition in a benchmarking process (see Appendix A). The purpose is to improve the quality of components of a product above the level of the best competing product for those areas identified as most crucial to the customers. Benchmarking sets design criteria and targets for continuous improvement. Another technique which uses benchmarking is the Pugh method of creative design concept evaluation—it is discussed in more detail in Chapters 7 and 10. We use benchmarking when we compare ourselves to a role model, whether in sports, in a musical or scholastic achievement, in growing flowers, or in being a good parent. When we set goals or benchmarks of what we want to achieve, we have taken the first step that will help us reach the goal. Collecting the necessary data for benchmarking is a challenging and often time-consuming task for detectives.

Morphological creativity and Synectics: Morphological creativity is very useful for problem definition since all elements of the problem are presented in all possible relationships, together with the values sought. The problem is then synthesized to at most seven parameters, with seven components each in a matrix format with movable columns. Selection of the primary objective will make it possible to identify the specific elements of the problem and will then focus the problem down to a manageable level. It is a complicated method that involves brainstorming all the different factors—but the concept of this method can be used by explorers in trying to get new views about a problem. A somewhat different approach is used with Synectics. This approach starts by considering how each group member understands the problem submitted by the client/expert. The brainstorming panel begins to generate a large number of possible goal-wishes or objectives, and the client/expert selects the alternative which is closest to a plausible solution to the real problem. Then the panel concentrates its problem-solving activities in that particular direction. Synectics is a very complicated method that requires special training and employs analogy to stimulate creative thinking.

Resource assessment: We began this section by saying that we need to collect *all* the information or data connected with the problem. Thus as detectives, we need to consider these factors: **Time**—is the problem an emergency, or do we have time to find causes and good solutions? **People**—should we try to solve the problem ourselves or can we find people to help us solve the problem as a team? **Resources**—do you think the problem will take a lot of money to fix? Do you have this money? If you don't, you will have to concentrate on solutions that do not need money or that include raising funds as part of the problem.

Force field analysis: The resource assessment factors can be incorporated into a force field analysis in which the situation or problem is analyzed in terms of supporting and hindering forces on the way toward achieving a satisfactory state or solution. In the next step of brainstorming, ways are sought to strengthen support and eliminate or minimize the obstacles.

Detectives thus have a wide array of tools available for data collection and analysis. You can learn these methods by studying books, attending short courses or the continuing education courses offered by professional engineering societies, or taking an elective in the traditional engineering curriculum. When such courses are offered by your employer, use this opportunity to sharpen your skills.

THE CONTEXT OF THE PROBLEM— THE EXPLORER'S MINDSET

When you analyze problems as a detective, two outcomes are possible, as was illustrated in Table 5-1. You may be able to identify the specific causes of the problem; this may suggest an obvious solution—either by removing these causes or by using a remedy to repair the damage and prevent future problems. Sometimes, this is a satisfactory solution = end of problem solving. The problem definition process may need to be continued if the problem is complicated and needs contextual input, or if it is a case that requires creative problem solving because of its unstructured nature. Then it is time to adopt the holistic mindset of the explorer.

Holistic, imaginative thinking can lead us to more thoroughly understand a problem than is possible with a strictly analytical approach. An unusual technique for problem exploration has been developed by Ned Herrmann. We have not yet tried it ourselves because it requires a wide variety of materials. It is called **problem modeling** and involves the team members' constructing a model—a physical representation—of the problem. A plethora of arts and crafts materials, construction toys, machine parts, tools, and objects from nature are invitingly displayed in a workroom. The participants select the materials they want—then they begin constructing a model for an unstructured problem that needs a creative solution. The assignment is to visualize all aspects of the problem. This right-brain activity gives the participants a surprising amount of insight into the problem as well as ideas about potential solutions.

We believe that being a successful explorer requires us to have a sense of adventure. We can develop an adventuresome spirit; we can make it a habit. How can we be explorers in our ordinary, conventional surroundings? Whereas it is possible to learn how to use specific tools for detectives through special training, being explorers is more a matter of developing and practicing quadrant D thinking skills and attitudes. Here are some practical suggestions:

TO DEVELOP THE MINDSET OF AN EXPLORER

1. *Take one afternoon a month to look around in a subject you don't know anything about, by reading, speaking to people, visiting exhibits, etc.*

2. *Read regularly outside your own field; develop a new interest or hobby.*

3. *As you read or listen to the daily news, look for trends that are developing in many areas, not just in your own community, in your circle of friends, or in your own area of expertise or study.*

Figure 5-5   The explorer.

The information we gather through this process will prepare us to recognize and solve problems. These exploratory habits have a tremendous benefit for our minds by keeping us mentally fit, because the new neural connections will prevent a decline in our brain's functioning as we get older. In the explorer's mindset (see Figure 5-5), we ponder futures and possibilities with quadrant D thinking; we envision what might happen in the problem area and its context; we think about long-term effects and anticipate developments that might impact us. Explorers use quadrant C thinking to investigate how the problem affects people. Does the primary problem have a communications problem connected with it? Do people need special training to solve the problem? In essence, in this part of problem definition, we look at the broader context of the problem. Why do you suppose we have linked the explorer's mindset with the color yellow?

> *Breakthrough ideas are most likely to occur when you are actively, confidently*
> *searching for new opportunities. They occur to those who are prepared.*
> Denis E. Waitley and Robert B. Tucker, Winning the Innovation Game.

Trend Watching—How to Anticipate the Future

Studying trends can help us see the development of problems in a wider context and time frame. With this information, we may be better able to devise appropriate solutions. Studying trends lets us identify areas for action, developing markets, and future products or services. Studying trends is also important in career development. Where are the opportunities? What new technology—if you get into it quickly—will give you a competitive advantage? Students need to watch trends to make wise choices in the courses they select. Be on the lookout for areas of rapid development to prevent becoming quickly obsolete; learn as much about the newest technology as you can. Here we can share a personal experience. Midstream in Monika's undergraduate mathematics program, the university decided that two computer courses would now be required for graduation. Students could elect to graduate under the "old" or the "new" rules. But because of our growing family (and because no one counseled us otherwise or taught us to look at trends), she decided to take the easy way out—she graduated without a single computer course. When she interviewed for jobs, she was in for a jolt. Invariably the first question asked by the interviewers was: "What computer courses have you taken?" Look into courses being offered by newly hired assistant professors; attend seminars, colloquia, or other types of lectures by guest speakers—these may give you a glimpse of coming new paradigms. Watching for trends is very important!

In an article on "How to Think Like an Innovator" in the May-June 1988 issue of *The Futurist* magazine, authors Denis E. Waitley and Robert B. Tucker, two California consultants on personal and executive development, offer some ideas on how to become a good trend watcher (see Table 5-3). They also have this to say:

> In studying America's leading inventors, we were constantly struck by how well informed they were on a broad range of current events, issues, and trends, both within and outside their particular fields. A knack for trend watching is one of the inventor's secret skills. It is one of the things innovators do to make their own luck. Innovators ride the wave of change because they constantly study the wave. ... Successful information gathering is not something we are born with; it is a skill that can be developed.

Keeping an idea file: When you are an explorer, it is a good habit to keep a small notebook or a stack of note cards handy, in your briefcase, on your desk, in your purse or backpack, in your pocket, and even beside your bed. Good ideas about anything may pop into your mind at any time, and if you record and file them, you will have a gold mine available when you need some "thought starters." Steve Allen, a very creative writer, performer, and comedian, assembled shelves and walls full of notebooks with ideas. The notebook is a tool that will help you become even more creative. If you are doubtful about your own creative potential, keeping a record of your own good ideas will soon convince you that you are very capable and creative. Try exploring many different subjects—you will be surprised

Table 5-3
How to Become a Good Trend Spotter

1. READ AND AUDIT YOUR INFORMATION INTAKE.

   Make informed choices about what you currently read—cut down on mental "junk food." What potentially useful new sources should you add instead? Innovators may spend as much as a third of their day reading. Read actively, rather than passively. Look for the point the writer is trying to make. Read articles that contain ideas; take notes as you read. Read intuitively—look for what is different, incongruous, new, worrisome, exciting, unexpected. Seek to broaden your worldview.

2. DEVELOP FRONTLINE OBSERVATIONAL SKILLS.

   It is important to remain actively involved when absorbing information; then draw your own conclusions. Also, become a people watcher; listen to other people's conversations to find out how they think and feel. What are people complaining about? How do their main topics of conversation change over time? Take the initiative to ask questions, even of perfect strangers. Asking questions from customers about what they want has been perfected by the Japanese and is a key to their success in manufacturing quality products.

3. ADOPT THE METHODS OF PROFESSIONAL TREND WATCHERS.

   One of these is John Naisbitt, author of *Megatrends,* whose organization does content analyses of 300 daily U.S. newspapers. You can adopt this method for your incoming mail. Scan the junk mail before you toss it; observe how it differs from last year's. You can find valuable clues about developing trends. The same goes for the popular culture (movies, pop music, MTV, videos). Monitor the media—listen to the news and to call-in talk shows on the radio or on television when you have a chance. What perspectives do you pick up? Who advertises on these stations and why? Make use of cassettes such as the *Hines Report,* or attend seminars and lectures by experts.

4. FIND OPPORTUNITIES.

   The purpose of trend watching is to discover opportunities and problems to solve. Watch for patterns that can tip you off to new opportunities. Opportunities abound. Search for solutions to negative trends and offer a means of prevention. Look at current activities and interests for ideas that may appeal to others. Even when a present trend is against us, we can use this to inspire us to come up with a new, breakthrough idea to counteract it. (An example of this thinking is a manager from Lansdale Semiconductor, Inc., a company which was not doing well in the overcrowded chip industry. He decided to go against the trend of developing more and more sophisticated technology and instead concentrated on making outdated lines of integrated circuits for the military. It did not take long for this company to take the lead in after-market sales of obsolete chips.) And finally, we need to watch what our competition is doing and do it better, with added value.

at how some will turn you on and lead you to discoveries, new interests, and increased creativity. Also, scan a news or business magazine periodically for trends and jot down any ideas that come to your mind as you do this. Looking to the future is not easy, especially for young people—but it is crucial that you develop and practice this ability because your survival (economically, socially, environmentally, and physically) is at stake.

 FIVE-MINUTE ACTIVITY 5-1: TRENDS

In groups of three, discuss in what ways can studying trends and acting upon this information help young people today prevent future economic, medical, social, or environmental calamities?

Contextual Problem Solving

How do most people learn to solve problems? One way, of course, is through experience, but the other way is in school. We have learned in Part 1 that our schools teach mostly analytical problem solving, which works well when problems are well defined. The difficulty arises once students move out into the real world where many problems are no longer well defined. Thus to do good problem solving, we need to use additional information and additional thinking skills—we need a different point of view. Using the analytical tools described in the detective's mindset is essential, but it is not enough if we want to determine what the real problem is as a first step to finding the best solution. What happens when problems are solved without considering the context? This may be one of the reasons why many technological solutions to problems are causing new and bigger problems— because the larger context is not defined and analyzed properly. As an example, the construction of the Aswan High Dam in Egypt comes to mind—an overpowering and inappropriate solution to the country's lack of energy sources and annual flood control. Electricity is being generated (although distribution is a problem); floods are being controlled. But now the lake behind the dam is disappearing rapidly due to silt buildup, and the interrupted flows of the Nile waters and nutrients have had a profound and negative impact on Egypt's agriculture and the health of the *fellahin* living along the Nile delta.

Another related concept is systems thinking, which is difficult for most of us, because we are not trained to do it—we develop habits of considering problems in isolation. For example, physicians in Western medicine as a whole tend to look at their patients as made up of separate parts: "Here is the enlarged heart." Or, "Today I had four tennis elbows." Note, it is not, "Today I treated a person with an enlarged heart" or "I saw four patients with tennis elbows." With this mindset, the symptom or the body part is treated as a separate entity, not as part of the whole person made up of body, mind, and emotions. Eastern cultures, in comparison, have a different, more holistic approach to medicine. To illustrate what we mean by contextual problem solving (perhaps best contrasted to plug-and-chug problem solving), we have collected some test examples.

EXAMPLE PROBLEM 1—HOW MANY BUSES?

> An army has to move some soldiers to a different location. If a maximum of 39 soldiers and their gear fit safely into one bus, how many buses are needed to move 1261 soldiers?
>
> (a) 31   (b) 32   (c) 32.33   (d) 33   (e) 34

When this word problem is presented without multiple-choice answers, most students will do the long-division problem of dividing 1261 by 39 to get the answer of 32.33 or 32-$\frac{1}{3}$ (or they will say, 32 buses, remainder of 13 soldiers). When the students think about what the mathematical solution means, they will give 33 buses as the answer (since it is impossible to have one-third of a bus). In multiple-choice tests, some students will either guess at (c) or "deduce" that (d) is the right answer, just because of the way the answers are structured; thus this test will not show if they understand the word problem, if they can do long division correctly, or if they can make sense of the answer.

This problem was used in an article by *Newsweek* magazine to illustrate that U.S. students do not do as well in contextual problem solving as students from other countries. To us, the problem also illustrates another inadequacy of multiple-choice testing as compared to contextual problem solving. We have as many as one-third of our students say that the answer is one bus (if you have a lot of time), or "Let the soldiers walk and use the bus for the gear" (if the distance is not too far). The best answer very much depends on the context or situation, and students should be encouraged to ask questions about the problem or put down the reasoning behind their answers.

EXAMPLE PROBLEM 2—CALCULATING PERCENTAGES

How would you test students to see if they understand the concept of percentage and can do the calculations? Here is a problem as given in two countries:

> UNITED STATES
> What percent of 500 is 30?
> (a) 6%   (b) 16%   (c) 60%   (d) 166.67%   (e) None of the above
>
> NETHERLANDS
> *The facts:*
> In 1980, the defense budget of a certain country was $30 million out of a total budget of $500 million. In 1981, the defense budget of that same country was $35 million out of a total budget of $605 million. The country's inflation rate for that one-year period was 10%.
>
> *The tasks:*
> 1. Use the facts to argue that the defense budget increased from 1980 to 1981.
> 2. Use the facts to argue that the defense budget declined from 1980 to 1981.

When we compare the thinking skills involved in solving this example, do we need to wonder why American students do poorly when competing in math and science contests against students from other countries? Which way of learning do you think is more effective? Which way do you think is easier? Which way do you think is more interesting?

EXAMPLE PROBLEM 3—ENGINEERING ANALYSIS

But what does contextual problem solving mean in more advanced courses, let's say in engineering? What are the majority of problems like that engineering students are trained to solve? Here is a typical example at the first-year level:

> A 16-ft beam that weighs 8 lb_f per foot is resting horizontally. The left end of the beam is pinned to a vertical wall; the right end is supported by a cable that is attached to the vertical wall 8 ft above the left end of the beam. There is a 250-lb_f concentrated load acting vertically downward 4 ft from the right end of the beam. Determine the tension in the cable and the amount and direction of the reaction at the left end of the beam.

Students need to be taught to solve these types of problems—that is not the question. The problem is that they are *only* taught these well-defined problems. What is the context of this particular problem—why is the beam 16 feet long? Why does it have to be supported with a cable? Could there possibly be a better engineering solution for accomplishing the same purpose? The problems of course increase in complexity as the students advance, but the basic lack of attention to the overall context, to problem definition and to looking for alternate solutions, persists into graduate school.

EXAMPLE PROBLEM 4—TRUCK ECONOMICS

Here is another example to illustrate the difference between plug-and-chug problem solving and contextual problem solving. Consider this problem:

> The cost of gas, oil, maintenance, and depreciation for running a certain truck is $[50 + S/8]$ ¢/mile when it travels at a speed of S mph. A truck driver earns $10/hour. What is the most economical speed at which to operate the truck?

To solve this problem, the "cookbook" approach is to find the derivative of cost with respect to speed, set it equal to zero, and thus determine the speed in miles per hour (mph) that yields the minimum cost. You can see the equation and its solution in Chapter 9, where the problem is solved using *Mathematica*. When the computation is carried out (whether by computer or calculator), the numerical output for S is 89.4427 miles per hour. This is the answer that most students will give; rarely will they round off to 89 or 90 miles per hour. But how could a driver keep the speedometer at 89.4427 mph? Even though these calculations are mathematically "correct," they give an inadequate answer to the contextual problem.

In contextual problem solving, students examine the problem and whether the answer makes sense in the context of the whole situation. With the availability of computers and software or with a hand-held calculator with graphics output, it is a simple matter to explore the problem in more depth by looking at a graph of cost versus speed. *Turn for a moment to Example 4 in Chapter 9 and look at the graph.*

From this curve you can see that the mathematical minimum is around 90 miles per hour. This is an unusual speed for a truck, and you might want to ask if the assumptions made in the original formula still hold. At such a high speed, the possibility of an accident increases tremendously, as would the certainty of collecting a stack of speeding tickets and fines (and corresponding delays), not to speak of added wear and tear on the truck, all of which are not included in the cost formula. Another thing you may want to explore with this graph is the changes in costs as the speed increases. Thus at 30 mph, the cost is 87 ¢/mile, but at 60 mph, the cost is 74 ¢/mile. Thus by doubling the speed, the cost has dropped by 13 ¢/mile. Now what happens if we increase the speed by another 30 mph? At 90 mph, the cost drops to 72 ¢/mile, or a differential of only 2 ¢/mile. Thus an increase from 30 to 60 mph makes a noticeable difference in cost and should be made whenever traffic conditions allow, but the extra 2 ¢/mile increase that would result in going from 60 to 90 mph is not justifiable. With this kind of analysis, there is no question that the most economical speed would be at the legal speed limit of 55 to 65 mph. Yet a student who will give this as an answer may very well be marked "wrong" in a typical class, even though this student would have had a better understanding of the "real" problem and its context.

Example Problem 5—Solar Energy

It is possible to have a heat transfer problem that asks for the determination of the optimum tilt angle and orientation for a solar panel used for water heating, given monthly solar gains data for a particular latitude. However, additional data may be needed to find the optimum solution for a specific location: What shading is present on the site? What is the cloud pattern for this location? For the new solar-heated engineering building on the campus of New Mexico State University, the optimum orientation was determined to be slightly west of south because steam from a cooling tower located to the east would affect heat gains negatively into the late morning hours, especially during the colder seasons. And what about air pollution—would a slightly steeper slope allow more heat gain over an average year because dirt buildup would be prevented? What would be the optimum solution, taking the economics of periodic collector cleaning into account? What about fixed versus movable-tilt?

Basically, in engineering analysis, students are trained to concentrate on defining the physical system so that the fundamental physical laws can be applied to find a numerical solution. Thus in Example Problem 3, a free-body diagram (an idealized representation or visualization of the system) would be the first step in problem analysis, followed by mathematical modeling, calculations, and a discussion of errors introduced by the assumptions made. But even when the problem is solved within its context, this is not all that students have to learn—contextual problem solving draws a wider circle. Students must know how to present their results in the form of a technical document; they need good writing skills and familiarity with acceptable and effective formats; and they must know how to summarize data in appropriate and easily understood tables, graphs, or histograms. And finally, students need good verbal communication skills, because ideas and solutions have to be "sold" to others in the process of getting them implemented.

Does this kind of learning require changes in the way we teach? Many projects and coalitions between universities are being funded by the National Science Foundation with the objective of effecting major changes in the way we educate engineers. Similar efforts are under way in mathematics and science education. Table 5-4 shows how Ed Lumsdaine has changed the teaching of one course, Heat Transfer for Mechanical Engineers, from analytical problem solving to creative, contextual problem solving. As a result, students are learning more; they perform better on tests (with a class average shift from C to B); they gain self-confidence; they don't drop out or fail; they participate in class; and they develop an improved understanding of the subject and its connections to other fields. He team-teaches the course with colleagues, and the team uses the Pugh method on the course syllabus to identify the topics that are of greatest benefit to the students (see p 367).

Table 5-4
Two Ways of Teaching Heat Transfer

ANALYTICAL APPROACH	CONTEXTUAL APPROACH
• Students must know the fundamentals.	• Students must know the fundamentals.
• Minimal computer use.	• Extensive computer use.
• Only one "correct" solution expected.	• Multiple solutions/alternatives expected.
• Right-or-wrong answers.	• Contextual problem solving.
• Narrow focus on course or discipline.	• Multidisciplinary focus.
• Pure analysis—no design content.	• Application to design is central.
• Students work alone.	• Students work alone and in teams.
• Problems are fully defined.	• Problems are open-ended (less defined).
• Students spend much time substituting in equations (plug-and-chug).	• Students spend much time thinking critically and asking what-if questions.
• Learning is teacher-centered.	• Learning is student-centered.
• Students fear risk; failure is punished. Learning from failure does not occur.	• Students are encouraged to examine causes of failure for continuous improvement.
• Quick idea judgment.	• Deferred idea judgment.
• Artificial, neat problems.	• Real-life, "messy" problems.
• Isolated, disconnected learning; students learn no communications skills.	• Students are required to make a verbal presentation and a written project report.
• Left-brain thinking only; the creative problem-solving process is not used.	• Creative problem-solving approach and mindsets are emphasized.

THE BRIEFING DOCUMENT AND
THE PROBLEM DEFINITION STATEMENT

During the process of problem definition, especially when a problem is very complicated, people may get so involved with the details of the problem that they become discouraged. By looking at all the data, it may appear that a small hill has grown into a huge mountain. This is why it is important to use the explorer's mindset—it is needed to get a divergent view and better perspective on the context to balance the narrow, convergent, and often negative thinking of the detective. If you are involved with a difficult problem, you must take steps to overcome a negative attitude. First of all, you are in charge of your life, and you can make decisions to make your life better. You can ask yourself: "How does the problem relate to my life, to the goals I have for my life? Is it my responsibility to do something about the problem? Do I have talents and abilities that will help me find a solution?" Sometimes, the answer here needs to be "no." You may need to turn a problem over to people who are trained to treat it. For example, you cannot solve the problem of a friend who is suicidal—you need to get the help of others. But many times you will find that you are able to do more than you give yourself credit for. Go for it! A positive attitude helps your mind be creative. Such an attitude is very important if the same team members that were involved in problem analysis will go on to do the brainstorming and subsequent steps in problem solving.

As the explorer, you now have two final tasks to do (as shown in Table 5-5)—you must assemble the information about the problem in a briefing document, and you must converge the problem down to a problem definition statement in terms of a positive goal! In an organization, the facilitator commonly assembles the briefing document from the data collected by the problem definition team (both in the detective and explorer modes). The problem definition statement is important since it will direct the thoughts of the brainstorming team toward solutions. The goal can be quite specific and even "impossible"—a big dream or wishful thinking. "How can we serve our customers better?" most likely will result in mundane ideas, but "How can we provide *instant* service?" will force the mind to seek unusual or innovative ways to reach the goal. In your team (or alone), play around with several versions of the statement before selecting the best one. Use a dictionary and thesaurus for concise or alternate meanings of words and to find synonyms. Here is another example of a positive goal statement. In the search for a suitable problem topic for their class exercise, a group of high school students had a lively discussion about various problems they were facing. They zeroed in on peer pressure. Instead of phrasing the problem as "Peer pressure can get teenagers into a lot of trouble with drugs, sex, spending too much money on clothes, and not doing well in school," they focused on the specific goal of "How to cope with peer pressure." In this statement, the focus is on coping, not on the problems. Why is it important to express the problem in words that focus on a vision or goal? By doing this, we are working with the capabilities designed into our brain—when we visualize the ideal situation or the goal, our subconscious mind will synthesize ideas that will help us achieve the ideal condition and thus solve the problem.

Table 5-5
The Problem Definition Procedure

DETECTIVE'S JOB: DATA COLLECTION AND ANALYSIS

Keep a notebook with the following:
a. General information about the problem.
b. Specific data collected about the problem.
c. Results from data analysis.
d. Things that were tried but didn't work.
e. Thoughts on possible solutions that come to mind during this phase.

EXPLORER'S JOB: PROBLEM CONTEXT AND BRIEFING DOCUMENT

Prepare a document with the following:
a. A summary of the detective's data, information, and analysis.
b. The context of the problem, including a view on trends.
c. Conclusions from the data and context: What is the *real* problem?
d. The problem definition statement expressed as a positive goal!

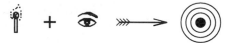

INCUBATION—INTROSPECTION AND PURGING

Before going to the second step in creative problem solving—idea genera-tion—you need to have a briefing and then a time-out. Do not move directly from problem definition to idea generation if at all possible, either when working alone or when brainstorming with a team. It is necessary to take the time for a briefing about the problem and a discussion of the problem definition statement; otherwise this activity will surface during the brainstorming session, thus interrupting the creative thinking process. During the briefing, the team members can ask questions to be sure they understand the problem. They may also want to contribute some additional insight into the problem. Then the problem definition statement can be paraphrased until all team members clearly understand the goal of the problem solving. Then a time-out period is provided for incubating the information and ideas about the problem in order to prime the subconscious mind to be at its best for creative thinking. An overnight period makes a good time-out, if that can be conveniently scheduled. Otherwise, organize a refreshment period with some relaxing or creativity-stimulating activities (such as an interlude with music).

Albert Einstein made good use of the incubation period to help his mind think through a problem. When he was stuck on a problem, he took a time-out and occupied himself with improvising on his violin or the piano. He was not a good musician, but this activity gave his subconscious mind a chance to process the information about the problem. Then when he returned to work, he found that new ideas would come to him (particularly since music engages the right hemisphere of the brain). Thomas Edison also took this approach. He would play the organ he had in his laboratory; then he and his researchers returned to their tasks mentally refreshed and ready to think of new ideas. We all can recount experiences in which our subconscious mind worked out a problem while we were busy with something else (a different work assignment, housekeeping tasks, exercise, creative hobbies, or social interaction with friends or family). If you are a student, do not wait with your longer homework projects until the last minute. Instead, leave yourself enough time so that you will be able to "sleep on it" if stuck. Make things easier on yourself by giving your subconscious mind a chance to work out the problem, particularly if you have primed it by visualizing the relevant information.

During this time-out or transition period you must try to put the problem out of your conscious mind. To avoid thinking about the problem, get busy with some other absorbing task—the subconscious mind cannot work on a problem if you are consciously thinking about it. Thus the time-out gives your subconscious mind a chance to incubate the problem. But keep a notebook at hand—sometimes, your subconscious mind will suddenly pop up an idea on the problem when you least expect it. Such unexpected illuminations about the problem must be written down immediately; they can be shared with the team before the brainstorming session begins. These "aha" ideas are usually quite creative and very easily forgotten if they are not written down while fresh in your mind. You may also have some intuitive insight into the problem—write this down in your notebook also. If you are thinking of some well-known ideas, jot even these down; this process is called *purging.* Purging is an important activity that should not be skipped during incubation, because the mind has to be cleared of these mundane solutions before it will be able to come up with truly novel ideas.

Some people consider the transition period very important in group creative problem solving and may stretch this period over several weeks. The **collective notebook method** places a great emphasis on collecting the individual ideas of the team members during the incubation phase that precedes the group session. After receiving the briefing document and the problem definition, the participants are instructed to daily jot down all ideas and thoughts that come to mind on the problem, typically for an entire month. The notebooks are then collected by the team leader and a summary of the results is prepared, with the most interesting ideas selected for further exploration and brainstorming during the group session. Actually what happens during such a long incubation period with this method is that each person is doing some individual brainstorming. If you are facing a nagging problem that you just don't know how to handle, try this notebook approach, alone or with a concerned family member or friend.

When people get together spontaneously for brainstorming, or when little time for an in-depth problem analysis is available because the problem is very urgent, we can engage in *introspection,* in which we dig into our memory to bring up any information that we already have about the problem. This information should be jotted down and shared, perhaps by developing a mindmap about the problem on a blackboard or large posterboard.

HANDS-ON ACTIVITY FOR PROBLEM DEFINITION

To learn the material in this chapter, you must conduct an exercise that will let you practice this step in the creative problem-solving process. We will first present a simple case study as an example. Then we will discuss some guidelines on how you can organize and conduct a team activity in problem definition.

Case Study: Curling Iron

Problem finding: A group of four students in a heat transfer class had to pick a topic for their team project. They discussed some ideas. When one mentioned that she was dissatisfied with the performance of her hair curling iron, the group decided to investigate the problem to see if it would make a suitable design project.

Data collection and problem context: Information on customer needs was a requirement in the project. Thus the students developed a survey form to collect data for problem definition. They also did a patent search, looked at a popular curling iron for benchmarking data, and investigated merchandising journal articles for trends in curling irons and other developments in personal-care products.

Data analysis and Pareto diagram: The results of the customer survey showed that price was the major determinant in the purchase decision of a curling iron, followed by features and necessity. Brand name ranked a distant fourth. Almost all respondents used the iron at a high setting, about 15 percent at a medium setting, and only a very few at a low setting. Close to 80 percent are willing to spend between $6 to $15 on an iron. About 50 percent would use the iron at home, followed by gym, school, and work. The problem areas with curling irons that were identified are shown in the Pareto diagram.

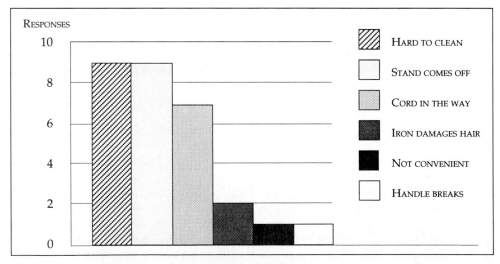

Figure 5-6   Pareto diagram for problems with curling irons.

CURLING IRON SURVEY

1. How old are you?
    5-13   14-17   18-25   over 25

2. Which of the following apply to your hair?
    _____ fine   _____ permed   _____ short
    _____ coarse   _____ color-treated   _____ shoulder length
    _____ thick   _____ long

3. How many curling irons have you bought or been given?
    1   2   3   4   5   6   more than 6

4. Do you buy a curling iron for its
    features   brand name   price   necessity?

5. How long does a curling iron usually last?
    < 6 months   < 1 year   1-2 years   2-5 years   > 5 years

6. What length of hair have you used an iron on?
    short   shoulder length   long

7. How much would you pay for a good curling iron? $ _____

8. Do you use anything else to curl your hair?
    hot rollers   hot sticks   permanent   other _____

9. Where do you use or would you use a curling iron if you could?
    home   school   car   bus   gym   work

10. How often do you use your curling iron?
    1/month   1/week   2-5/week   1/day   more

11. How much time do you spend curling your hair?   _____ minutes/day

12. Approximately how much time is spent on each curl that curls the way you want?   _____ seconds

13. How much time *total* do you spend getting ready in the morning?
    _____

14. How important is it that your hair looks good ? (5 = very important)
    1   2   3   4   5

15. How many temperature settings do you use?
    low   medium   high   more

16. What do you like least about your curling iron?
    _____

17. What problems (if any) do you have with your curling iron?
    _____

Briefing: Through the consumer survey, consumer needs and several design flaws in today's curling irons were brought to light. Problems with existing curling irons were plotted on the Pareto diagram. Quick warm-up time was mentioned as a desirable feature, as was portability. The survey showed that there is much room for improvement; thus the team decided to go ahead with the project.

Problem definition statement —> positive goal:

> *Design an improved curling iron that meets the customer's needs (while using heat transfer analysis for optimum performance).*

Guidelines for the Team Activity in Problem Definition

Before you are ready to start the team activity, you must select your team members. Use one of the methods discussed in Chapter 3 to identify thinking modes; then assemble a team of three to six people with complementary thinking preferences—all four thinking modes need to be represented in the team. The team members should get along well together—the presence of conflict causes stress and will prevent creative thinking. As mentioned earlier, cultivate a positive attitude and expect the team to do well. Expertise in the problem area is not a prerequisite for team membership—having people with different backgrounds is an advantage. Facilitating this type of team activity will give you practice in leadership skills.

How to select a problem topic: You have several options about choosing a problem topic. You can select one of the problems given in the exercises at the end of this chapter. You can brainstorm problems with your team—for one week pay attention to items at work, at school, or around the house that are not working well and could benefit from redesign. Or you may catch yourself saying something like this: "I wish someone would invent a gadget to do this task." Or, "Why hasn't anybody thought of doing this a different way?" These are problem areas that can give you ideas about a good topic for your creative problem-solving practice. The objective is to take the chosen problem through the entire process as you study each chapter. Assign a problem topic if you are facilitating an inexperienced team or very diverse group of people; this saves time and argument and still serves the purpose of practicing the particular skill. From our experience, we have found that people learn more about the process from a problem in which they are not too closely involved—otherwise they get carried away by the results and lose sight of the learning objectives. If your team is made up of technically-oriented people, we recommend that you select a simple design project—this will make it easy later on to practice the Pugh method of creative design concept evaluation. However, Phase I of the Pugh method can be used for "soft" problems also.

How to focus a problem topic: We recommend that you do not select a problem that is too large for your first project. On the other hand, do not narrow down the topic too much in the early stages or you will limit the creative possibilities in the solutions that will be generated. With practice, you will be able to select a problem definition that is the right size for your team exercise. If you have a good topic but need to expand the problem, you can ask a series of "what is this about" questions. This technique invites divergent and contextual thinking.

EXAMPLE OF A DIVERGING CHAIN OF QUESTIONS

"What is this problem about?"
Answer: *"Housing."*
"What is housing about?"
Answer: *"Being warm and cosy."*
"What is being warm and cosy about?"
Answer: *"Feeling loved, cared for, and safe."*

Can you see that this chain has brought out some aspect of the problem that involves not only a physical need but also emotional needs? It has helped us get the bigger picture. We must encourage our customers and other people involved with the problem to express the needs that are important but often remain unspoken.

At other times, in order to obtain a solution, we have to break problems down into smaller parts through convergent thinking. If we want to "squeeze" a problem, we can use a chain question process by asking "why?" Why questions can bring out the real reasons why people have a problem or what is important about the problem. We can ask ourselves questions, or two people can have fun asking the questions and giving quick answers.

EXAMPLE OF A CONVERGING CHAIN OF QUESTIONS

"Why do you want to improve your budgeting procedure?"
Answer: *"Bcause I'm always late in paying my bills."*
"Why are you always late?"
Answer: *"Because I do not keep track of schedules."*
"Why don't you have a sense of time?"
Answer: *"Because I have a habit of procrastination."*
"Why do you procrastinate?"
Answer: *"Because I really don't like to do paperwork—I wish I could give the job to someone else."*

Chain questions let us eliminate all kinds of superficial motives and rationalizations; the process zeroes in on the real motivation underlying a problem. Now we have more specific information—the real problem is a mismatch between the task and the person's thinking preference, not the budgeting procedure, the routine task, or the paperwork. Have you noticed how good children are at asking chain questions? Don't get impatient with this game: play along—it is good training for all the participants. Develop the mental habit and flexibility to be able to change your focus on a problem all the way from a close-up to a bird's-eye view.

Table 5-6 shows how to focus or expand a problem topic. If you move upward, you are moving from a specific task to the wider problem; if you move downward, you are moving from a general problem to a narrower task. If we ask: "In what way can we make the world a better place to live?" that is a job that is probably outside the scope of our life right now. It is probably outside of what most people can handle. A manager of a company could ask: "In what way can we help our employees do their jobs better?" As an individual you could ask: "In what way can I help my family or my friends solve one problem?" Then if you complete the problem solving and implement a creative solution, you will have been successful in making your area of the world a better place to live. By concentrating on specific groups, on individuals, or especially on tasks, we will be able to cut a problem down to a size that is suitable for brainstorming and come up with solutions that can be implemented and will make a real difference. However, there is a place for doing exploratory brainstorming on large problems. Through such a divergent activity, many factors can be identified that are contributing to the overall problem. Each of the important subareas can then be brainstormed separately in more detail.

Table 5-6 Diverging or Converging a Problem	
FOCUS	PROBLEM DEFINITION
	In what way can we
SOCIETY	— make the world a better place to live? — help stimulate the national economy?
COMPANY	— help the company make more money? — contribute to company productivity?
GROUP	— accomplish more as a group? — use group activity time more wisely? — make group meetings more efficient? — work together better as a group?
INDIVIDUAL	— help others do their job better?
TASK	— improve access to job-related data?

The problem can be expanded or contracted as the team plays around with the problem definition. However, do not be concerned about perfection at this stage. If the problem is too narrow, the team will most likely add divergent ideas during brainstorming. If the topic is too broad, narrower subtopics can be selected during the idea evaluation stage. But just to cut down on the amount of work and the length of time required for the team exercise project, we recommend that a reasonably narrow topic be selected—one that may generate around 40 brainstorming ideas as compared to 300. If the team has several good topics, you (as the leader) can select one, or you can let the team vote for its choice by secret ballot.

Data collection—the customer survey: The next task is to collect data about the problem. The notebook method is useful. If appropriate, we divide the task into a library search and a "customer survey." The problem is brainstormed (perhaps in the format of a mindmap). From this preliminary information, survey questions are developed. Through discussion, the team should gain a feel for the important aspects of the problem which can then be addressed in the survey. The survey can collect different types of information, either purely quantitative data, or "weighted" data—in which people can indicate, for example, not only if they have a problem but how severe the problem is by ranking it as 0 for no problem, 1 for a small problem, 2 for a moderate problem, and 3 for a severe problem. In this case, the replies can be tabulated as total points, or they can be stratified into the number of answers in each category. Stratified data collection can be more useful—in the survey on problems with schools, it was interesting to see that even though quite a number of students had a severe problem with teachers, this category was highest in the number of respondents *not* having any problems. The team can have quite an interesting discussion on how data should be analyzed and interpreted.

Data analysis and the Pareto diagram: The data collected with the survey can then be summarized. The ranked frequency of identified problems and causes can be plotted and displayed in the format of a Pareto diagram. Sketch the diagram on a large chart and post it in a prominent place in the room to guide idea generation and further phases of creative problem solving, because it will continue to highlight those areas that need to be addressed as being the most important factors or causes of the problem—those areas that are of greatest concern to the customers.

Briefing document and problem definition statement: As a final step, the team members are now ready to prepare a summary paragraph or briefing on what they think is the real problem. This can be brainstormed and then presented as a handout or written on a large poster to be displayed in the room. Each team member can suggest a problem definition statement as a positive goal or objective. Spend about 10 minutes playing around with various ideas for the problem definition statement, then converge this activity to a single, best, synthesized statement—one that has the general agreement of the team. Each person in the team should now have a clear understanding of what the real problem is. Close the meeting by posting the final problem definition statement in the room, and make sure that each team member writes down this statement in the problem-solving notebook. Now the team is prepared for incubation. Remind each person to keep the notebook handy for jotting down any thoughts about the problem (or early solutions) that come to mind before the team meets again for idea generation.

Incubation period: Be sure to allow for an incubation period before proceeding to the next step in the creative problem-solving process. With the problem definition statement, the briefing, and the incubation period, the mental "pump" (the subconscious mind) is primed for generating creative ideas. Various methods and variations on brainstorming have been devised to help different groups come up with creative ideas. In the next chapter we will learn about verbal brainstorming and some alternative techniques that help people generate creative ideas.

A Final Thought on Incubation

Relax! Stop working on the problem.
Give your subconscious mind a chance to work. Go have fun.
Listen to your favorite music; strum the guitar.
Pay ball with your friends; pull some weeds, rake your leaves.
Do aerobic exercises, swim, or go for a walk in the woods—
physical activity is very good for your creative mind!

RESOURCES FOR FURTHER LEARNING

Large corporations may have their own manuals for the specific procedures they have developed for problem definition and data analysis within their particular business culture. These procedures are usually part of companywide quality control efforts. For example, Ford Motor Company publishes its own manuals on FMEA and FTA. Since national professional engineering societies are beginning to offer workshops for training in FMEA and FTA, reference material should become more widely available on these two analytical methods. If you work for a company that is introducing new methods, take every opportunity to learn about these techniques by reading and by attending workshops and seminars. If you need to develop problem definition skills, we highly recommend Reference 5-7; it will give you a good starting point for the detective's mindset and for learning to ask the kinds of questions that can give you data and insight into different aspects of a problem. Appendix F has an additional list of interesting books on the topics of quality, innovation, and manufacturing that include in-depth discussions on problems and solutions.

> *A problem is an imbalance between what should be and what actually is.*
> *Paraphrased from Kepner-Tregoe definition, Ref. 5-10.*

Reference Books

5-1   George Ainsworth-Land and Vaune Ainsworth-Land, *Forward to Basics*, DOK Publishers, New York, 1982. This teacher's manual discusses how forthcoming changes in education and society can be an exciting challenge rather than a devastating experience.

5-2   Myron S. Allen, *Morphological Creativity: The Miracle of Your Hidden Brain Power*, Prentice-Hall, Englewood Cliffs, New Jersey, 1962. This book presents the principles of morphological creativity; the technique is demonstrated by the organization of the material in the book.

5-3   Don P. Clausing, *Total Quality Development: A Step-by-Step Guide to World-Class Concurrent Engineering*, ASME Press, Fairfield, New Jersey, 1994. This book addresses the problem of quality in engineering and is written for technical readers.

5-4   W. Edwards Deming, *Out of the Crisis*, MIT Center for Advanced Engineering Study, Cambridge, Massachusetts, 1982. Quality in manufacturing is discussed by one of the early leaders of the quality movement in Japan.

5-5   Andrew Fluegelman and Jeremy Joan Hewes, *Writing in the Computer Age: Word Processing Skills and Style for Every Writer*, Doubleday, Garden City, New York, 1983. This softcover book is just one example of books that cover a skill that should be learned by anyone who does even occasional writing with a word processor. Depending on your access and familiarity with particular computers, also learn to use desktop publishing software—these programs can add immensely to the appearance of your reports.

5-6   Eliyahu M. Goldratt and Jeff Cox, *The Goal: Excellence in Manufacturing*, Creative Output, Milford, Connecticut, 1984. This book is in the form of an easy-to-read novel; it presents steps and concepts of problem solving in the context of manufacturing.

5-7   Kaoru Ishikawa, *Guide to Quality Control*, available from Unipub, Box 433, Murray Hill Station, New York, NY 10157. This book, translated from the Japanese, gives a nice overview of the tools of statistical process control.

5-8   Herman Kahn and Anthony J. Wiener, *The Year 2000*, Macmillan, New York, 1967. This book provides a detailed example of the "science" of forecasting and the interpretation of trends. It is interesting to observe how recent history and developments agree or disagree with the various scenarios that were proposed more than twenty years ago.

5-9   Thomas S. Kane, *The New Oxford Guide to Writing*, Oxford University Press, New York, 1988. This book takes the writing process from searching and brainstorming for ideas to statement of purpose, outline, and draft revision. In addition, it covers grammar and style—in general, it provides guidance on how to write well.

5-10   Charles H. Kepner and Benjamin B. Tregoe, *The Rational Manager*, McGraw-Hill, New York, 1965. This book thoroughly explains the Kepner-Tregoe method of problem solving. Even if you or your organization want to use more creative methods, the Kepner-Tregoe approach is excellent for initial problem definition and data analysis as well as for identifying potential problems during the solution implementation phase.

5-11   Herbert B. Michaelson, *How to Write and Publish Engineering Papers and Reports*, second edition, iSi Press, Philadelphia, 1986. This small softcover book covers a broad range of topics and practical ideas connected with the writing chores of students and engineers.

5-12   John Naisbitt, *Megatrends: Ten New Directions Transforming Our Lives*, Warner Books, New York, 1982. This is required reading for learning more about the "science" of trend watching.

5-13   John Naisbitt and Patricia Aburdene, *Megatrends 2000: Ten New Directions for the 1990's*, Morrow, New York, 1990. The social forecasters focus on ten new forces shaping our world and our global future.

5-14   P. Ranganath Nayak and John M. Ketteringham, *Breakthroughs!* Rawson Associates, New York, 1986. Written by two consultants of Arthur D. Little, this book examines the creativity and ingenuity, vision and persistence of innovators in sixteen companies who created commercial breakthroughs that swept the world.

5-15   Thomas J. Peters and Robert H. Waterman, Jr., *In Search of Excellence: Lessons from America's Best-Run Companies*, Warner Books, New York, 1982. It would make an interesting exercise to analyze how well the lessons described in this best-seller have been learned and how these companies are meeting the challenges of the approaching twenty-first century.

5-16   George M. Prince, *Practice of Creativity*, Macmillan, New York, 1970. Although the main topic is Synectics, this book has useful comments for anyone who has to attend committee meetings. It also discusses the importance of the briefing document.

5-17   Denis E. Waitley and Robert B. Tucker, *Winning the Innovation Game*, Fleming N. Revell, Old Tappan, New Jersey, 1986. This book emphasizes creative thinking, innovation, and managing change; it shows how to obtain possible breakthrough ideas from observing trends.

> ### REMINDER
> *Hands-on practice is important for learning and experiencing the creative problem-solving process.*
> *Be sure to take the time to apply your theoretical knowledge in a team project.*

ACTIVITIES AND EXERCISES FOR DETECTIVES

5-1   PROBLEM DEFINITION
Make up a positive problem definition statement and then paraphrase it several times. Do the paraphrases help you improve on the original definition? Try several ways of making the problem more divergent or more convergent. Look up the precise meaning of each word used in the definition in a dictionary.

5-2   TIME USE ANALYSIS
Over a period of three days, complete a detailed log on how you are using your time (in 15- minute chunks). Then do an analysis to determine which activities waste the most time. Make a Pareto diagram to find "the 20 percent that cause 80 percent of the trouble." Make a plan to eliminate the top three time wasters (one at a time over a three-month period).

5-3   IDENTIFYING PROBLEMS
Make up a list of five problems that you would like to see a university brainstorm. Make up a list of three problems engineers should brainstorm. Make up a list of five problems that would be fun for students (and their family or friends) to brainstorm. Make up a list of five topics that teachers should brainstorm. Use divergent thinking—make up a list that different groups of people should brainstorm (in their own groups or together).

5-4   *CHANGING NEGATIVE THINKING TO POSITIVE ACTION*
Using the results from the negative thinking project in Chapter 4, determine which negative thinking habit you should change first to gain the greatest results with the least amount of effort. Remember that it takes three weeks to establish a new habit. Concentrate on changing one habit at a time to increase your chance of success. To motivate yourself, write up a list of benefits that you will gain by reducing the amount of negative thinking that you do—then post the list in a prominent place (like your refrigerator door). But do not just stop doing the negative activity—substitute a positive action in its place!

5-5   *A PROBLEM FOR THE GOVERNMENT*
With two friends, make up a problem definition for a problem you think the government should solve. For a week, keep a notebook each and collect as much information and ideas as you can about the problem. Then get together and look at the combined results. Has this process helped you understand the problem better? Write a summary; if you have a good problem, send it to an appropriate government official with a request for action.

5-6   *FOLLOW-UP TO THE GOVERNMENT PROBLEM*
As a follow-up to Problem 5-5, if you and your friends became really enthusiastic about a problem involving your local government, in your letter express your desire of wanting to be involved in the problem-solving process, particularly if there will be brainstorming in the search for a good solution.

---

*Asterisks denote advanced exercises for more in-depth learning.

5-7   *BRIEFING DOCUMENT*

As a team project, obtain samples of briefing documents from three different organizations. How was the data collected and presented? In what way could you improve the problem definition statement?

5-8   *CAUSE-AND-EFFECTS ANALYSIS*

Select an item that you are using in your daily life that is not functioning properly. Examples: the front door "howls" when the wind blows above 10 mph, your bicycle's kickstand sticks in one position, your alarm clock fails to ring at least once a week, your computer has developed a strange quirk, or your car is pulling to the right when you are driving down a straight road. Make up a cause-and-effects chart (fishbone diagram) that identifies all the factors that could possibly be involved in causing the problem.

5-9   *SAMM*

Do a library search and write a summary report (including an example) that explains the sequence attribute/modification matrix. How is it related to Dr. Osborn's nine thought-starter questions and the method of attribute listing (see pp. 210-211 in this book)? How are these techniques related to the storyboard approach (see p. 208)?

ACTIVITIES AND EXERCISES FOR EXPLORERS

5-10   EXPLORING A TOASTER

Think about designing a better toaster. What problem does a toaster solve? Imagine being a toaster yourself. Make statements such as: "I have to take bread slices into myself." "I have to heat bread uniformly, without burning." "I have to kick the toast out at the right time and then shut off." "I have to keep a cool skin." The purpose here is to really identify with the problem. This helps for problem definition. Write five more statements like these.

5-11   ENGINEERING ACTIVITIES OF CHILDREN

Children are natural explorers and engineers. With some friends, dig back into your memories and brainstorm some things you did as a child that were elementary forms of engineering or that exhibited a true spirit of exploration and adventure. Did you build a tree house or construct a snow fort? Did you dam a small creek, launch a boat, fly a plane, hike an unfamiliar trail, explore a cave, eat a bug? See if you can get a list of fifty items.

5-12   CREATIVE THINKING WARM-UP

When you do not have much time to incubate a problem (or to take a short break between the detective's and the explorer's mindset during problem definition), you must take a five-minute time-out and do a creative thinking warm-up exercise. Here are some suggestions:
a. Play around with a metaphor related to the problem.
b. Play around with an idea, concept, or symbol not at all related to the problem and try to make it fit the problem.
c. Make some creative, crazy, humorous sketches about the problem or the ideal state.
d. Briefly brainstorm other uses for an object involved in the problem. Try to come up with twenty ideas, but you can stop when answers are getting the team to smile or laugh—the group has been warmed up!

5-13   *LANDFILLS*

If you are not yet recycling your garbage, you may need to make some major changes within the next ten years, as most of the industrialized world will run out of space for landfills. What trends are you predicting—in government regulation, in business opportunities, in many people taking personal responsibility?

5-14 THE GREENHOUSE EFFECT

Many scientists predict that the earth's climate will keep getting warmer. Brainstorm some positive outcomes or opportunities. For example, more air conditioners will be in demand, yet they will not be allowed to use freon. New cosmetics providing better protection from the sun will be needed. See if you can think of other markets, products, or scientific breakthroughs and new paradigms that may result from the greenhouse problem.

5-15 *TECHNICAL KNOWLEDGE*

It has been predicted that all the technological knowledge we have today will represent only about 1 percent of the knowledge that will be available by the year 2050. What are the implications of this (a) for education and schools, (b) for the workplace, (c) for libraries, (d) for book publishers, (e) for authors, (f) for business? Brainstorm one of these topics and see if you can come up with an opportunity that you can use.

5-16 *READ ABOUT EXPLORATION*

Read a biography of an explorer or a book written by an explorer. What made this person be an explorer? What are some of the most striking personal characteristics? What were his or her goals and rewards? Or read about a team exploration effort, such as the Voyager space program. How did the project grow and change from the original idea to final execution? What are some areas that are being explored that are currently in the news? What might be a hot topic for exploration (including medical research) five years or ten years in the future?

> *How are you going to see the sun if you lie on your stomach?*
> Ashanti Proverb.

> *Where the telescope ends, the microscope begins. Which of the two has the grander view?*
> Victor Hugo, Les Misérables.

5-17 TRENDS AND BUSINESS OPPORTUNITIES

Do an exercise in studying trends. Scan a newsmagazine or newspaper and note any trends that are mentioned, discussed, or analyzed. How will these trends affect you?

EXAMPLE: People are living longer. Who used to take care of elderly parents? A recent report on television mentioned that women now spend more time taking care of an elderly parent than raising a child. Daughters or daughters-in-law are still expected to take care of elderly parents. Now put this fact together with another trend—more and more women are working and don't have time to look after elderly parents. Can you see some business opportunities from these facts? What about "Daughters for Rent"? Would you be willing to visit, drive, run errands, help older people in other ways? Who would be your customers — where would you need to advertise and who would pay for your services? (Answer: the daughters, not the elderly.) Males are not excluded from finding opportunities in this situation. What about a company called "Friends Behind the Wheel, Inc."? You could advertise your willingness to drive older people anywhere, anytime, in your car or theirs. (We heard these suggestions from a bank manager who indicated that her bank certainly would be very willing to finance this type of enterprise.)

Brainstorm some ideas for a new business suggested by your investigation of trends. Secelct the idea you think has the greatest potential, prepare a short business proposal, and make up an ad or a business card for your new business.

5-18 WHAT HAVE YOU LEARNED?

a. What are the three most important concepts that you have learned from this chapter?
b. Relate one of these concepts to something in your life; explain or sketch the connection.
c. What is an important question you still have?

CHAPTER 5 — SUMMARY

What Is the Real Problem? First agree and accept that a problem exists; then ask questions to find the real problem. Use the detective's mindset for data collection and analysis; use the explorer's mindset to probe the problem's context and trends.

Methods for Data Collection and Analysis—Your Role as Detective: Many tools have been developed to analyze problems. Among these are:
1. The list of questions (developed by Alex Osborn, inventor of brainstorming).
2. The Kepner-Tregoe method (which asks very detailed questions).
3. Statistical process control (histogram, cause-and-effect diagram, checksheet, Pareto analysis, control chart, scatter diagram, and other graphical analyses).
4. Experiments (to obtain or verify data) and customer surveys.
5. Failure mode and effects analysis (to identify potential areas of malfunction).
6. Fault tree analysis (to trace the events that have led to a particular failure).
7. Benchmarking (to set targets/design criteria to outperform the competition).
8. Morphological creativity (a very structured approach that looks at all possible combinations of factors involved in a problem).
9. Synectics (a very complicated method of problem solving using analogies).

Your Role as Explorer: Search for information about the context of the problem. Be adventuresome. Develop a habit of reading about and exploring new ideas, hobbies, fields. Watch for trends. Be on the lookout for opportunities. Look for improvements in procedures, services, and products. Develop future goals. Keep note cards or a small notebook handy to record ideas that come to mind unexpectedly. You will be surprised at how quickly you accumulate many creative ideas.

Becoming a Good Trend Spotter: Search for material with ideas that stimulate active thinking. Look for what is different, incongruous, new, worrisome, exciting, unexpected. Ask questions of anyone (including customers and clients). Listen for trends and ideas on radio programs, television, tapes, lectures. Find opportunities. Observe a trend and exploit it; search for solutions to negative trends or trends that are running against you; look at your current activities for ideas that may appeal to others; watch what the competition is doing—then do it better.

Data Collection and Context (Notebook): (a) General information about the problem. (b) Specific data collected about the problem. (c) Results from data analysis. (d) Things that were tried but did not work. (e) Thoughts on possible solutions that come to mind during the problem definition and data analysis phase.

The Briefing Document: (a) Summary of the collected data and notebook information assembled in the detective's mindset. (b) The context of the problem, including trends. (c) Conclusions from the data and context: What is the *real* problem? (d) The problem definition statement expressed as a positive goal.

Practice Problem and Incubation: CONDUCT A TEAM EXERCISE! Jot down all ideas that come to mind during the incubation period—well-known ideas as well as sudden, creative ideas. For a complicated problem, the team can apply the collective notebook method extending over a period of several weeks.

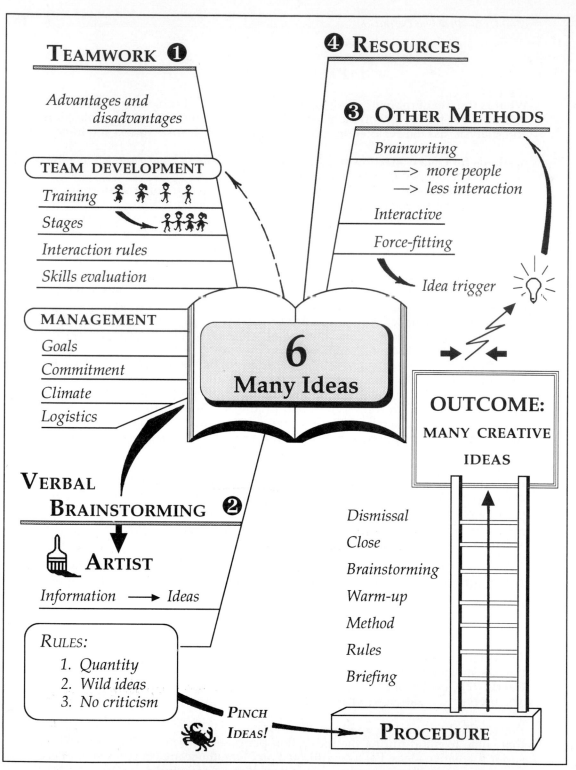

TEAMWORK ❶

Advantages and disadvantages

TEAM DEVELOPMENT

Training

Stages

Interaction rules

Skills evaluation

MANAGEMENT

Goals

Commitment

Climate

Logistics

VERBAL BRAINSTORMING ❷

ARTIST

Information ⟶ *Ideas*

RULES:
1. Quantity
2. Wild ideas
3. No criticism

PINCH IDEAS!

❹ **RESOURCES**

❸ **OTHER METHODS**

Brainwriting
—> *more people*
—> *less interaction*

Interactive

Force-fitting

Idea trigger

6
Many Ideas

OUTCOME:

MANY CREATIVE IDEAS

Dismissal

Close

Brainstorming

Warm-up

Method

Rules

Briefing

PROCEDURE

Mindmap of Chapter 6. Refer to Chapter 2 (pp. 55-56) to learn more about mindmapping.

6

IDEA GENERATION, BRAINSTORMING, AND TEAMWORK

WHAT YOU CAN LEARN FROM THIS CHAPTER:
- *Principles of teamwork and team management.*
- *Procedure for verbal brainstorming.*
- *Other methods for idea generation, including force-fitting techniques.*
- *Resources for further learning: reference books and idea-generator tools; exercises and activities for artists; and summary.*

Before we discuss specific idea generation procedures, we want to summarize some important principles of teamwork and team management. Teamwork is beneficial in creative problem solving at each step, but nowhere more so than for idea generation. Although brainstorming can be done alone—and it can be productive if the rules are observed—it should be done as a team activity for best results. However, teams do not just happen; developing effective teams takes time and effort. This is a key concept when total quality management (TQM) is introduced into an organization.

TEAMWORK AND TEAM MANAGEMENT

We have learned that everyone is creative; we can develop our creativity by overcoming mental barriers, and with practice we can learn to be more creative thinkers and doers. Now we will see that teams collectively can be developed to be creative. We have already discussed the first step—teamwork requires quadrant C thinking skills. Here we will investigate some of the factors involved in team development beyond selecting members with complementary thinking preferences, and we will discuss some principles of team management.

Before we begin, we need to conduct a small demonstration. This activity requires a group of five or more people.

> ⏱ *TEN-MINUTE GROUP ACTIVITY 7-1: IDEA GENERATION*
>
> *First, each person in the group works alone. Take a sheet of paper and write down as many uses for a telephone book as you can think of. Stop after three minutes. Next, the group prepares a common list of all the ideas on a board or large chart. Each person gets a turn to quickly share ideas. The others can cross duplicates off their lists. If you get a new thought as you hear the ideas presented by the others, jot it down on the back of your sheet. After all ideas from the first round have been recorded, repeat the process with the new ideas, but now go through the group in reverse order. Finally, give the group an opportunity to express additional ideas that come to mind as each person contemplates the group list.*

Now let's review what usually happens during this exercise. At the end of the first three minutes, the average person will have generated at least ten ideas. For the group as a whole, many of these ideas will be duplicates, but many will be unique. But what is even more interesting is to note the ideas from the second list and general brainstorming—seeing or hearing other people's ideas helps your mind think of additional new ideas. Thus this activity demonstrates one of the benefits of teamwork—the interaction between minds increases the output of ideas.

Leonardo da Vinci is an example of a person who knew almost everything that was known in his time. Today, it is rare for an individual to thoroughly know everything in an entire field, such as chemistry or literature or electrical engineering or even gourmet cooking. We know that Eli Whitney invented the cotton gin, Alexander Graham Bell the telephone, and Jonas Salk a vaccine for polio. Why don't we talk about the inventor of the Boeing 747 airplane? One reason is that it was developed by many teams of engineers. Also, about 90 percent of patents for inventions are not for completely new products but for improvements of existing patents. Most inventors do not work in isolation—they build on the ideas of others. The invention of the cotton gin is an example. The original idea came from Catherine Littlefield Greene who supported Whitney financially on the project; she also improved the prototype and shared in the royalties from the patent.

With today's knowledge explosion and technology, it is no longer possible for a single person to know all the data connected to a problem. This is why teams are often used for problem solving. Consider three individuals with brain dominances A+D, B, and C, as shown in Table 6-1. Each person differs in background and experiences and thus has a different knowledge base as well as different ideas and biases on particular subjects and problems. When the three people work together, team W now contains a large background of which only a small amount is in common. Because a team has a broader background of information available, the possibility is increased that new, creative combinations of ideas will occur—the ideas and suggestions made by one person can stimulate the imagination of the other team members. Also, they compensate for biases to achieve better judgment.

Table 6-1
Team Approach to Problem Solving

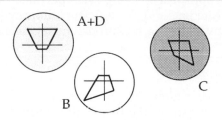 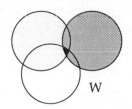

| The three circles represent three different persons with varied backgrounds, experiences, and thinking preferences (indicated by a generic HBDI profile). | Team W represents a large background of which only a small amount is shared experience. Together, the team constitutes a "whole brain" (circle on right). |

RESULT: The larger amount of information available to the team increases the possibility of new, creative combinations, since ideas and suggestions offered by one can trigger associations in the other minds. Because all thinking modes are present, the team is able to develop more complete solutions.

Thus teams have advantages when used in the creative problem-solving process:

1. More information and knowledge is available to help solve the problem; people with different expertise and thinking skills can be brought together from within the organization. This results in cost savings over hiring outside "experts."

2. People interact with one another; ideas are used as stepping-stones to more creative, better solutions. The team members are encouraged to build on one another's ideas. When this process really clicks, productivity increases and we have an example of the whole being greater than the sum of its parts.

3. If there is one "best" solution for a particular problem, the team has a good chance of finding it. Teams have an advantage in identifying opportunities and in taking greater risks (and thus increasing the chance of innovation).

4. People who take part in the problem-solving and decision-making process are usually more willing to accept the solution than if the solution were developed by only one person and imposed by the "voice of authority."

5. The team members learn from each other, and the team provides an encouraging environment for developing leadership skills.

The use of teams for problem solving may also have disadvantages:

1. A greater investment in effort and total personnel time is needed, not just for solving the problem but also for team development.

2. In general, the team process has a low efficiency—a large number of ideas may be generated, but only a few of these will be truly good solutions.

3. The people making up the group or team may not get along with one another. When there is conflict and hostility, the creative idea output of the team will be lessened because of these negative emotions.

4. Teams can suffer from the "group think" phenomenon. When this is the case, the group exhibits extreme conformity and peer pressure; independent ideas are not allowed; and group members may also be intimidated by the leader or a vocal minority; or they may want to gain approval by being "yes" persons.

The negative group interactions can be overcome and the efficiency improved through team development, training in creative thinking, and proper management.

Team Development

Our school systems train students to work alone by rewarding individual achievement. Public education since the early part of the twentieth century has mainly been geared to producing docile assembly-line workers who will "check in their brains at the factory gate upon entering." However, U.S. companies are finding that higher-level thinking skills and teamwork are needed to increase productivity and achieve competitive quality for the global marketplace. Because teamwork is not yet practiced widely in schools, businesses and industry are spending large amounts of money and effort to train their employees in how to work in teams. Teamwork requires cooperative relationships at all levels in an organization—in essence it demands a cultural paradigm shift between the traditionally more adversarial roles of management, employees, and labor unions. This change is not easy, because it must happen on many different levels, from cultural values to organizational structure to the attitude of each individual person. The broader issues can be investigated in a study of TQM (total quality management—see Appendix F); what we want to explore here is how to develop good teams.

Let's begin with the individual. What traits in individuals should we look for when we make up problem-solving teams? Table 6-2 lists some ideas. Not everyone in the team will have all the traits, but these traits should be present in the team as a whole. Note how many of these characteristics are involved in quadrant D thinking. Creative thinkers are also open to feelings and emotions. Macho males—influenced by the social culture around them—are usually less open to their feelings and thus may be embarrassed about expressing creative ideas. People with aesthetic interests tend to be concerned with form and beauty in their surroundings. Because they want to go beyond just a practical solution to an elegant solution, they often achieve a higher-quality final product.

Let's say we take several individuals who together have most of the traits listed in Table 6-2. Will we then automatically have a good problem-solving team? Most likely not—this is only the first step. The team members will need training in creative problem solving, as well as in developing specific team skills such as good communication and listening; asking questions; borrowing and building on other people's ideas; learning from one another and from self-education; being considerate and polite; having a spirit of patience and humility; and being willing to help, contribute, share ideas, and "walk in the other person's moccasins."

In a well-organized system, all of the components work together to support each other.
W. Edwards Deming, quality expert for manufacturing industries (Japan and United States).

Table 6-2
What Makes a Creative Team?

• Intelligence	• Openness to new ideas
• Expertise in problem area/related field	• Originality
• Analytical, logical thinking	• Flexibility
• Independent judgment	• Ability to toy with ideas
• Willingness to test assumptions	• Tolerance for ambiguity
	• Ability to defer judgment
• Self-confidence	• Curiosity
• Self-esteem	• Imagination
• Commitment	• Humor
• Optimism	• Willingness to take risks
• Enthusiasm	• Divergent thinking skills
• Verbal fluency	• Knack of elaboration
	• Ability to see the whole picture
• Self-discipline	• Aesthetic interests
• Perseverance	• Impulsiveness
• Concentration	• Intuitive viewpoint

It may take as long as two years before a dozen people working together will become an effective, productive, well-functioning team. In the early stages, when the group is first formed, the group members still act as individuals; they do not contribute effectively to the group as a whole but look out for themselves. They are merely an assembly under a "boss" or manager. When team objectives are worked out collectively, the common problems or goals begin to draw the individuals together into a group, although the sense of individual responsibility and autonomy is still very strong. Conversations among group members now extend beyond neutral subjects to organizational and budget matters. After achieving some successes in problem solving, the group begins to realize that team development is important, and the individuals as well as management participate in sharpening communications and other team skills. Team responsibility begins to develop, and the team presents a united front to outsiders. Value questions and motivation are discussed. Finally, the group truly melds into a team; the team becomes involved in problem solving in a wider area within the organization. Team members feel united and strongly bonded; they are now open to outsiders and seek contact with the wider community to extend their influence. The purpose of the team is seen in the context of the organization's goals and connected to its broader tasks and responsibilities. Dreams and visions are shared; new ideas and personal differences are evaluated and worked out in light of the common vision. The team is self-directed and no longer managed from above. The team members fully share accountability for the team's actions, and they operate from a basis of trust and mutual respect.

In a successful team, all members are equal. All parties affected by the problem are represented, and all members agree to the common objective. This agreement can be strengthened if the entire team participates in the problem definition process, although at times teams are specifically assembled for the steps of brainstorming and "idea engineering" only. The briefing document and prob-lem discussion in this case serve to supply the common objective. Clear ground rules for interaction must be established—no sarcasm or put-downs are allowed; the team knows the idea generation procedure that will be used; they agree to mutually respect each member's contributions and thinking skills. If the idea gen-eration team continues problem solving to the judgment phase, it will be able to establish agreement on proposed actions. Finally, the team must have assurance that its recommendations will be seriously considered for implementation.

It is not usually the case that teams are kept together for years to solve problems. Most problem-solving teams are assembled for short-term projects. However, the principles of good teamwork outlined above still apply. Attention will need to be paid by the team leader or facilitator at the outset to establish mutual trust and understanding through some team-building activities, such as taking out time for leisurely introductions of all team members, discussing the team rules, and providing the "big picture" and motivation.

In an academic setting where team activities and output are expected to be graded, each student's team skills can be evaluated by an observer (such as a class assistant), by self-evaluation (completing a checksheet), and by anonymous feed-back on each team member by the others on the same team. Characteristics to be graded would be items such as: Did the student participate fully? Did the student contribute to the maximum of his or her ability? Were good listening skills observed? Were good communication skills observed? How were conflicts resolved? Did the student exhibit a cooperative attitude? What was the level of support for the team's efforts? Was leadership used to help each team member? Students and instructors can cooperatively brainstorm such a list of evaluation criteria.

The project team grade can then be combined with the team skills evaluation to arrive at each student's grade. This will encourage students to become active, contributing members in their teams instead of letting a few carry the entire load. The use of cooperative learning strategies and teamwork is avoided by many instructors because of the difficulty of fairly evaluating a group project. However, the problems can be overcome when students are required to subdivide responsi-bilities according to their thinking preferences; the specific contributions are ac-knowledged in the written project report. Each student can be required to make a brief oral presentation on his or her area of expertise as well as on the overall understanding of the entire project. This approach will again emphasize individual as well as team responsibility for the final result and will provide incentives for developing good team skills.

> *Groups reach consensus on the best solution no one can disagree with.*
> *Teams reach consensus on the best solution everyone can agree with.*
> *Charles Henning, president of Innovation-Productivity-Quality (IPQ).*

> *Heterogeneous teams consisting of differences in mental preferences are capable of higher and more effective*
> *creative output than homogeneous teams consisting of similar mental preferences and same gender.*
> *Ned Herrmann.*

Management is an intellectual process that provides leadership and an environment
in which people are willing to work together toward an end purpose.
Paraphrased from David I. Cleland and Harold Kerzner, Engineering Team Management, *p. 1.*

Management and Team Responsibilities

To be successful, teams must be given special support and responsibilities by the management of their organization. Pay attention to the following items:

Goal: Give the team an assignment with a clear, achievable, elevated goal, mission, or purpose, as well as a specific objectives or tasks.

Structure: The objectives should be customer-driven and will determine the structure and scope of the team, preferably across departmental lines. Experienced teams can be self-directed. New teams need to be guided by a facilitator. The leadership is governed by mutually-developed criteria and principles, in effect moving toward shared leadership. The facilitator is a coach and listener, not an authoritarian manager. All team members are treated as equals; facilitators are to serve the team's interests, not their own personal agenda. It is essential that facilitators have a positive attitude toward the team and its competence.

Team members: The members of the team are selected for the thinking skills, personal characteristics, expertise, and other abilities they will be able to contribute to the team and the problem-solving task. If necessary, provide training in communication skills, including "constructive differing." Habitually negative thinkers and people with a hostile attitude should not be selected to serve on the team. Remember that diversity can add important strengths to a team.

Commitment: Management and the team share a unified commitment to the teamwork concept as well as to the problem-solving process and results. The team is allowed to focus on the agreed-upon task. The team members are committed to work hard and to nurture a positive team spirit. Both the team and management are committed to maintaining standards of excellence and to dedicating their efforts to the good of the organization. Management must be open to new ideas.

Climate: A collaborative, not competitive, climate is maintained within the team, with implicit and explicit support from management. The team results—not individual glory and self-advancement by the members—are important; credit for the problem-solving results will only be given to the team as a whole. However, the team members know that they are accountable for making good judgments to achieve team success, helping other team members when needed, and supporting team decisions. Team members support, build, clarify—they give and accept positive feedback as well as constructive criticism for continuous improvement.

Support: Management supports the team's effort with the needed resources (time, financial support, facilities, networking) as well as with recognition for the team's important role and contributions in the organization and the community.

William Golomski, who is a senior lecturer in business policy and quality management at the University of Chicago, made a presentation on teamwork and team development at the Fifth Annual Teaching Excellence Seminar in the College of Engineering at the University of Toledo. He presented the following list of tasks, roles, and maintenance responsibilities of a team:

+ The initiator/contributor proposes new ideas and methods.
+ The information seeker tests the factual accuracy of suggestions.
+ The information giver offers information, facts, and data.
+ The opinion seeker asks for clarification of values behind issues and tests for agreement.
+ The opinion giver states opinions or beliefs.
+ The elaborator diagnoses problems and presents details and embellishments.
+ The coordinator clarifies relationships among tasks, ideas, and suggestions.
+ The orienter summarizes, raises questions about the team's direction, and defines the position of the team in relation to other teams and the organization.
+ The evaluator/critic examines team accomplishments in light of standards.
+ The energizer prods the team toward action and decisiveness.
+ The procedural technician distributes materials and obtains equipment.
+ The recorder records suggestions, ideas, decisions and outlines discussions.
+ The encourager accepts, praises, and agrees with the contributions from others to build group warmth and solidarity.
+ The harmonizer attempts to reconcile differences, relieves tension, and helps team members explore differences.
+ The gatekeeper regulates the flow of discussion and encourages participation by directing conversational traffic, minimizing simultaneous conversations, quieting dominant members, and eliciting participation from quiet members.
+ The team observer provides feedback on group dynamics.

In essence, the entire team shares the responsibility for the management of the team. Shared leadership occurs naturally, as topics and the stages in the creative problem-solving process demand, and as different members take over appropriate roles for short or longer time periods.

The facilitator is responsible for preparing the team for brainstorming and for providing a relaxed setting, preferably in an unfamiliar location. The necessary equipment is set up (easels, flip charts, markers, visual aids, note cards, tape recorders) as well as creature comforts such as beverages and snacks for longer sessions. The facilitator selects the brainstorming method most appropriate for the problem and the size of the team. Brainstorming can be very exhausting. It has been found that in general morning sessions scheduled for the middle of the week are best. The team members need clear minds (from a good night's sleep and a healthy breakfast); they should avoid listening to negative news on the day of the brainstorming. After the brainstorming session, the facilitator may also have the responsibility for guiding the process through further steps such as idea evaluation, idea judgment, and presenting the best solution to management for implementation. At the minimum, the facilitator must collect all the ideas generated during the brainstorming session, thank the team members for their participation, and later inform them of the final results, making sure that the team's contribution is acknowledged.

A positive and supportive attitude toward the team members is essential for team leaders
—disapproval can be communicated through body language.
Brigit Koenig, team facilitator.

⌚ THREE-MINUTE ACTIVITY 6-2: TEAM NAME

Brainstorm a name for your team or class. Jot down all brainstormed ideas on a sheet of flip-chart paper, then save the list for Critical Thinking Activity 7-11.

Two matters of logistics need to be considered: dealing with negative comments and ways of splitting a large group into teams. First, how should a team or facilitator manage critical or negative comments made by group members? One practical approach is to use the baseball metaphor: "Three strikes and you're out!" This policy is explained to the team at the beginning of the brainstorming session; if a group member (usually an inexperienced, new person on the team) makes another negative remark after two reminders, he or she will be asked to leave the team. Negative thinking early in a creative problem-solving session will completely destroy any possibility of success; thus the facilitator must be prepared to eject anyone who continues to be critical of another member's ideas. Positive comments, praise, and humorous feedback for encouragement are allowed; sarcasm, criticism, and put-downs are out! The metaphor and peer pressure make a powerful combination; only very rarely will a "strike three" actually happen.

There are a number of ways to split up a large group into smaller teams. We have used the numbering system in workshops, where the members consecutively count off; then all the odd-numbered people go into one team and all the even-numbered people into a second team. To make three teams, the counting off can go through the entire group by repeating "one, two, three." The advantage of these simple approaches is that people who chose to sit next to each other will be on different teams—thus separating "cliques" and people who may tend to think alike. We have used notepads of different colors, randomly distributed in the room; then all those with pink pads were assigned to the "Pink Platoon," those with green pads to the "Green Team," and those with gray pads to the "Gray Company" (until they devised their own, better team names). Random grouping can also be accomplished with folders or name tags (or sticky dots) in different colors.

For longer projects (such as classes or summer institutes), we use brain dominance data to form teams. Students with similar thinking preferences are given the same color dot or tag; then they are allowed to form their own teams of five or six people (depending on class size)—the only requirement is that each team can have only one person with the same color. This way, students will become more aware of their different thinking styles and will learn to appreciate the contribution each can make at different stages in the creative problem-solving process. Responsibility for team "management" can be assigned according to the colors and corresponding mindsets: blue for the detective (quadrant A dominance), yellow for the explorer (quadrant D or D+C dominance), orange for the artist (quadrant C or C+D dominance), green for the engineer (quadrant D+A dominance), purple for the judge (quadrant A+B dominance), and red for the producer (quadrant B, B+C, A+B+C+D). These students are then responsible for making sure that the team stays on task and turns in the required output—charts, reports, etc.

VERBAL BRAINSTORMING—THE ARTIST'S MINDSET

Generating innovative ideas is the heart of the creative problem-solving process. The best-known method for idea generation is verbal (or classic) brainstorming. We will present the simple rules and procedure for how to conduct a verbal brainstorming session. Then we will discuss other methods that have been developed to help people get creative ideas. Verbal brainstorming is a team activity, but some of the alternate methods are suitable for people working alone.

Figure 6-1   The artist.

The artist's mindset or role is required for creative idea generation (Figure 6-1). If you are picturing a "wild and crazy guy" like Eddie Murphy or Steve Martin—good! This is the time when you can break out of your usual mold. Go to town with your quadrant D imagination and your quadrant C feelings! Welcome eccentric ideas when you brainstorm a problem. What do artists do? Artists create something new, something that first existed only in their minds. With the artist's mindset, our task in creative problem solving is to transform information into new ideas. In brainstorming, this process of using the imagination—this mental activity of coming up with wild, anything-but-mundane ideas—is called "freewheeling." This means we impose few restrictions on ourselves or our team members on the types of ideas that can be expressed. Why do you suppose the color orange has been chosen to represent the artist's mindset?

Now, what exactly is brainstorming? What today is called classic brainstorming was developed in 1938 by Alex Osborn. It is a group method of creative idea generation. It has been found that the best number of people in a verbal brainstorming team is from three to nine. Brainstorming does not work for all types of problems all the time, but its successes have made it a valuable problem-solving tool. It is easy to learn, and it gets more productive with practice. People frequently mistake committee discussions or meetings with brainstorming. These undirected, repetitive, critical, or routine discussions of problems and old solutions have little to do with brainstorming. As you will see, brainstorming requires careful mental preparation, including a problem briefing and a creative thinking warm-up. Athough it is a creative, freewheeling activity, definite rules and procedures are followed.

THE THREE RULES OF BRAINSTORMING

Generate as many solutions as possible—quantity counts.
Wild ideas are encouraged.
No criticism is allowed—judgment is deferred until later.

Brainstorming is easy to learn because it only has three rules:

Rule 1: Generate as many solutions as possible. Quantity counts! Don't give long explanations along with your ideas, just toss them out using key words only. Be brief. Ideas do not have to be completely new; it is perfectly fine to "pinch" or expand or "hitchhike" or build on other people's ideas. The more ideas you generate individually or collectively, the better the chance that you come up with an innovative solution.

Rule 2: Wild ideas are welcome. This point cannot be overemphasized; the more odd, weird, impossible, or crazy ideas are generated, the better are the chances of coming up with a truly original solution in the end. The only limit here is to avoid words and ideas that could be hurtful or offensive to your team members because the stress that is caused will inhibit creative thinking in addition to undermining the team spirit.

Rule 3: Do not judge ideas; do not put down ideas or the people who express them (including yourself!). Humor, favorable comments, laughter, and applause are O.K. There is no such thing as a dumb idea. There is also no such thing as a right or wrong answer in brainstorming. Brainstorming is a deferred-judgment activity—idea evaluation and critical judgment come later in the creative problem-solving process.

Since brainstorming is a team activity, you as an individual cannot hog your own ideas or take credit for your output. The interaction that occurs between the minds of the team members is important. Share all your ideas—someone else may use your idea as a stepping-stone to another idea, which in turn is used by a third person to come up with something new—and you may just use that idea to think of something even better. But don't wait for the perfect idea; look for successive steps forward! It may not be easy at first to get used to this concept of idea sharing. We do not have much training in this type of thinking because it is strongly discouraged in our schools, where teachers insist that students work without help or input from classmates. Brainstorming is different—it is teamwork, and you are supposed to make use of the ideas of the other team members. Idea pinching is allowed!

Here are a few thoughts on wild ideas. Wild ideas are valuable because the normal forces of life will tend to make the ideas more practical. This will occur especially during the "engineering" and judgment phases, as we shall see. If you are not used to welcoming weird ideas or generating wild ideas, this process can be quite uncomfortable at first. It will take a conscious effort—or force—to make this process of having and expressing wild ideas enjoyable. Later, we will look at some techniques we can use to help us get creative ideas. These tools are called forced-association techniques because they force the brain into making new connections between unrelated ideas. Having or sharing wild ideas may make you feel ridiculous, or you may feel that others will laugh at your ideas. Please do not be self-conscious; everyone on the team is in the same boat. Especially if you are a person with strong quadrant A and quadrant B thinking preferences, you may find it difficult to play with wild ideas. As you learn and practice creative thinking through a conscious effort, it will become easier to express wild ideas and overcome the "business as usual" mindset. Establishing new thinking habits takes hard work. It will help to remember that the information you have for solving a problem is not complete and not identical to that of your team members. In cooperating with the others, you will be surprised to find what the "team mind" can achieve.

We are going to present the procedure for brainstorming from the facilitator's point of view. Since you are studying this subject, it is quite likely that you will be the best-trained person in a group and thus will be "elected" for team leadership in your organization, in your circle of friends, or in your family group. We will go through the general procedure of conducting a brainstorming session step-by-step. Then we recommend that you practice with a team. The seven steps of the procedure are listed in Table 6-3, and we will briefly go through each step .

Table 6-3
Guidelines for Leading a Brainstorming Session

1. *Brief the team on the problem's background and post the problem definition statement.*
2. *Review the three brainstorming rules.*
3. *Explain the brainstorming procedure that will be used.*
4. *Do a creative thinking warm-up exercise.*
5. *Conduct the brainstorming.*
6. *End the session; collect all ideas.*
7. *Thank and dismiss the team.*

Problem briefing: After everyone has had a chance for some social interaction with the team and has comfortably settled down around the table, you, as the team facilitator, will turn on the tape recorder. If your team is large, you may appoint one or two of the members as recording assistants. Briefly review the background of the problem and important data such as the results from a Pareto analysis and customer survey. Then post the problem definition statement and jot down a few sample solutions generated during the explorer's phase. If team members have additional information, it can be presented. Did anyone get some ideas during the incubation period? These ideas can be shared and jotted down. Make sure that anything distracting, either on people's minds or in the room's environment, is taken care of before starting the session.

Brainstorming rules: Review the three brainstorming rules and also the "three strikes and you're out" procedure for preventing negative thinking.

Verbal brainstorming method: In small teams, say three to five members, ideas can just be called out as fast as they can be written down on a flip chart (or on large sheets of paper on a wall). In larger teams with a dozen members or so, people can take turns speaking. The other participants need to jot down any ideas that flash into their minds on pieces of notepaper, so that they won't forget these while they wait for their turn to share. Arrange a signa—such as a raised hand or snapped finger—to be used when someone has a modification or addition to an idea that has just been presented. Such "hitchhiking" is encouraged and given priority. The combination of two ideas that have previously been mentioned is also counted as a new idea. Do not allow long explanations; that will come during the idea evaluation phase. All ideas need to be recorded on the flip chart or paper.

Warm-up exercise: Conduct a 5-minute warm-up exercise in creative thinking using a simple, familiar object (brick, pencil, popped corn, newspaper, ruler, large goose feather, etc.). Bring a visual aid if you are not using an item present in the room. Some experts recommend that background music be turned on at this time and played until the end of the brainstorming to facilitate use of the right hemisphere of the brain.

Brainstorming: Set a time limit (say 20 to 30 minutes) and ask the team members to start sharing ideas. They can begin by bringing out obvious, well-known ideas—these have to be purged first before the mind will be able to bring out some really new, creative ideas. This process is also known as "load dumping." If the flow of ideas slows down, you as the facilitator can encourage the process by throwing out an outrageous idea that can serve as a stepping-stone. Or the team can start on a spree of wishful thinking by asking what-if questions. If things still are not rolling, the brainstorming session can be interrupted for a brief period of relaxation, including some humor, then started again by using a force-fitting technique that will force the team to combine ideas mentioned previously in an unrelated context. This should start the flow of creative ideas again.

Close: Once the flow of ideas has slowed down to a trickle and the previously announced time limit is coming up, give a 3-minute warning. Some of the best ideas can be generated during the extra time period at the end.

Dismissal: Thank the team members for their participation. Let them know what will happen next. Will the same team continue the problem-solving process with the ideas? If not, make sure the team will be informed about the results of its work. Collect all the ideas that were written down, as well as the tape recording, for later processing and evaluation.

The first of our senses which we should take care never to let rust
through disuse is that sixth sense—the imagination.
Christopher Fry, English actor and playwright.

Table 6-4
Brainstorming Exercise Topics

---

1. What can children do to show they appreciate their parents, without spending any money or using verbal expressions? Or, what can parents and children do to improve communications with each other? Or, what can bosses do to show appreciation for their staff?

2. What can parents do to encourage and help their children read, study, and do well in school? Or, how would you change the school system so more students would go on to study math, science, and engineering?

3. How can teachers and students develop a more caring attitude and mutual respect toward each other? Or, how can a schoolteacher change the classroom environment to encourage creative thinking by the students? Or, how can principals and school boards change the school environment to encourage teachers to be more creative? Or, what can a math teacher do to make an algebra or calculus class more interesting?

4. What can a person do to get more time in the daily schedule for reading?

5. How can a school library raise funds to get more books and resources?

6. Imagine that you are in the plumbing business. You want to design and sell a better bathtub. What features would you include?

7. Design a bicycle (or tricycle) that is completely recyclable.

8. How can engineers keep up to date with the latest developments in their fields?

9. Imagine that you are an engineer who is transferred to a job site in the Far East. You will have management responsibilities. What can you do to facilitate communications and minimize culture shock for yourself and your family?

10. How would you help employees who are threatened with layoffs?

11. Identify an environmental problem in your community. Do you need a recycling program? Do you have a polluted river that needs cleaning up? Brainstorm how such an effort could be organized.

12. The energy crisis has not disappeared—it is just on hold. Brainstorm one of the following energy conservation topics:
    a. How can houses be built to be more energy efficient?
    b. How can people save more energy in their homes?
    c. Design an energy-efficient transportation system for your community.

13. What changes could we expect in our economy if a superbattery were invented that would allow a car to go 1000 miles on one charge? Or, find other uses for the superbattery.

14. Think up some business ideas (or creative activities) that high school students can do during the summer.

15. How can team activities be made more attractive and pleasant for people who "hate" group activities?

16. How could paperwork and red tape be reduced in your organization? Or, identify a particularly burdensome procedure—how could it be improved?

> ⌚ *TEAM ACTIVITY 6-3: BRAINSTORMING EXERCISE*
> *First, put together a small team. Select one of the topics listed in Table 6-4; do a creative thinking warm-up; then have your team brainstorm ideas and solutions for the selected topic. This will give you practice in being a facilitator. Next, brainstorm the team problem that you explored and defined in the hands-on activity of Chapter 5. You may use the same team or a different team. Again, do not forget to do the creative thinking warm-up!*

How was the brainstorming exercise? Were you pleased with the results and the variety of ideas that were generated? Would you have been able to think them all up yourself? Did you and your team members get tired? That should not surprise you—brainstorming is mentally exhausting. This is why it is usually done in the morning, when people are well-rested and have fresh minds. That is also why it is not usually done for more than one hour at a time or for more than two problems per day. Under optimum conditions and with an experienced team, your output will become even more productive and creative. What do you think—is it necessary to be an expert in the problem area to have creative ideas? Brainstorming consultants recommend that people not familiar with the subject be included in the team because they will not be bound by previous training and thus are able to come up with really new ideas. In other words, they will not have any preconceived ideas about what cannot work, and they will have experience in other fields that may give them sudden leaps of imagination.

> *Imagination is more important than knowledge, for knowledge is limited, while imagination embraces the entire world.*
> *Albert Einstein.*

> *To become flexible, quality-conscious, and thence competitive, the modest sized, task oriented, semi autonomous, mainly self managing team should be the basic organization building block.*
> *Tom Peters, Thriving on Chaos.*

> *Without this playing with fantasy no creative work has ever yet come to birth. The debt we owe to the play of imagination is incalculable.*
> *Carl Gustav Jung, Psychological Types, 1923.*

Think back to your experiences with verbal brainstorming. Did you run into a problem or notice some shortcomings? Verbal brainstorming usually works well in all-female or all-male groups of up to a dozen members, especially with people who are comfortable with each other and like to express themselves verbally. When people know each other and like to work together, verbal brainstorming can be a very successful technique among equals, especially in a positive, interactive climate. But what can you do if you don't have these perfect conditions—what if you have shy (or domineering) team members? What if you have a group of 20, or of 100 or more, people? What if there is little leadership to keep the team focused? What if the team members are inexperienced? What if there is open conflict between the parties involved? Written and interactive versions of brainstorming were developed to overcome these types of problems and circumstances.

OTHER BRAINSTORMING METHODS

Brainwriting

Written brainstorming (which is sometimes called brainwriting) has been found to work well for engineers and for mixed-gender groups. It allows for teams larger than a dozen members, and it works well for shy people. The disadvantage of written brainstorming is the lack of direct verbal interaction between the team members; and the quantity of ideas may thus be reduced. Some of the written brainstorming methods can be used by individuals working alone on a project, or they can be done sequentially (by letter, bulletin board, or E-mail) by a group of people who cannot or do not want to meet in the same place at the same time.

WRITTEN BRAINSTORMING METHODS

Gallery Method
Pin Card Method
Nominal Group Technique
Collective Notebook Method
Cranford Slip Writing
"Ringii" Process
Method 6-3-5
Brainwriting Pool
Electronic brainstorming
Bulletin Board
Delphi Method
Mindmapping

For the **gallery method**, an easel and flip chart are provided for each participant. After the problem definition has been given out, each group member writes down on his or her chart all the ideas that come to mind within a limited period, say 20 minutes. Then a time-out is called. The group members silently circulate among the easels, reading the ideas written down by the others (as people do in an art gallery, looking at the paintings on exhibit). After this period of strolling around, the group members return to their own boards and make changes and additions to their own ideas, in essence carrying out idea hitchhiking. Finally, all the notes and ideas are collected for later evaluation by a different team. With this method, a shy person may feel freer to share ideas by writing than by speaking in front of a group.

With the **pin card method**, people sit around a large table. Ideas are written down on note cards—one idea per card. These cards are then passed to the left around the table, and group members are asked to add their related ideas and improvements to the original idea on the card. Several levels of additions can be made to the original idea in this way. Since this process is somewhat anonymous, it can get people involved who otherwise may feel too intimidated to contribute creative ideas. When the process of sending new cards around has slowed down to a trickle, the session is terminated, and the cards are collected for later evaluation by a different team. One application for this method could be in a family circle with several teenagers, because the people involved can concentrate on ideas and will not be influenced by a confrontational tone of voice.

With the **nominal group technique**, the problem is presented and the individual members silently write down their ideas during a 5- to 10-minute period. The ideas are then pooled, discussed, and voted on by the group, which has the task of selecting and ranking the top five to nine ideas or so. The method has the advantage of reducing the pressure to conform, but because of a lack of group interaction during the idea generation phase, it has a lowered chance of generating truly unique ideas. Also note that this technique combines the idea generation and idea evaluation phases into one session. It may be useful in cases where time is very short and a solution with consensus must be found quickly.

In the **collective notebook method** (which was already discussed in connection with problem definition), each person in the group receives a briefing on the problem and then over the span of several weeks brainstorms and records ideas on how to solve the problem. The notebooks are then collected by the facilitator, and a summary of the results is prepared, with the most promising ideas selected for further evaluation by a team. This method, like the nominal group technique, suffers from a lack of group interaction, but it may be useful for people who feel uncomfortable in groups and who like to take their time pondering a subject.

A method called **Cranford slip writing** is used to collect ideas when large groups of people want to be involved in the idea generation process. After the problem definition has been presented, each participant is asked to write down 20 to 30 ideas on slips of paper, with each idea on a separate slip. The slips are collected quickly, before the people have time to make corrections or deletions to the ideas. A different task force is then used for sorting the ideas into categories and taking them further through the evaluation process to arrive at a workable solution. This method can be used by large organizations—thousands of people could be involved in this way. Sometimes the number of ideas can be cut down if the people are asked to do a bit of prejudging and only submit their top two or three ideas. But then the most unusual, crazy idea may be thrown out too soon—thus prejudging is not usually a good approach. Group interaction can be inserted into the process by having small groups of two or three people brainstorm ideas for submission.

Another interesting method with minimal face-to-face interaction is the Japanese **"Ringii" process**. Here, an idea is submitted on paper to others in an organization. These people may make any modification or addition to the idea. The original proposer can then use these suggestions to rework the original idea, or a synthesized solution can be worked out by an independent panel. This second approach can be used in cases when the original proposer wants to remain anonymous. This process is beneficial in large and small organizations (even families) when there is some problem with communications or with coordinating schedules. With this method, you can get your idea across in writing; the others can add their ideas or reaction without confrontation or at a convenient time or location. You can then think about their input and rework your original idea. If you do this in several rounds, you may just come up with an idea or solution that pleases everyone, all without a single verbal argument or scheduling conflict.

The first brainwriting technique, proposed in 1970, was **Method 6-3-5**. Six people are instructed to produce three ideas in 5 minutes. The ideas are written down on a sheet with three columns (one idea per column). The sheet is passed on to the next person in the circle, and three new ideas related to the ideas already on the sheet are added on the second line. Thus 108 ideas can be produced within 30 minutes, with about half of these being useful. This method is especially appropriate when the best variation of some concept has to be found (such as an advertising slogan). Not everyone is comfortable with the restrictive rules, which can generate stress, impatience, or nervousness. For this reason, the method was modified. In the **brainwriting pool**, ideas to a given problem are written on sheets of paper (one original idea per sheet). The first four sheets are placed face-up in the middle of the table and form the pool. When group members run out of ideas, they can exchange one of their sheets for one of the sheets in the pool and then work on making modifications to the idea on the exchanged sheet. This process of exchange continues for about 30 minutes. The advantage of this method is that people can work at their own pace and control the rate of interaction with other ideas.

There is now a high-tech approach to brainstorming. **Electronic brainstorming** is used by Bill Gates of Microsoft in his company. People are connected via electronic mail; when someone has a creative thought, it can be sent to other computers where a signal will flash on. Thus instant feedback and hitchhiking ideas can be obtained. The low-tech equivalent is the **bulletin board**. The problem definition (with a short briefing about the problem's background) is posted in a prominent place for several weeks; anyone can enter new ideas as well as hitchhiking ideas at any time. These ideas are then collected and evaluated by a team or a single judge. The bulletin board is a method that is very appropriate for children.

The **Delphi method** is a technique that begins with written brainstorming but then continues the process until consensus has been obtained on the best ideas. It is used in organizations as a communications tool for large groups. The participants can remain anonymous; no direct interaction occurs since the idea collection is done by questionnaire or on-line computer. After the first batch of ideas has been evaluated, categorized, and ranked according to importance, this list is returned to the group members for another round of rankings and "debate." The process can be repeated until consensus is achieved. An advantage of this process is that it focuses on areas of agreements from what can originally be widely differing views. The process is lead by a judge or "jury." The method is applied especially in the area of planning the future direction of organizations.

Written brainstorming is a good technique to use when you want to or have to brainstorm ideas for an individual project. Be sure you follow the brainstorming guidelines and defer judgment. This means that you quickly write down each idea on a slip of paper or on a large sheet, just as it comes to mind. Don't worry about the ideas being practical, crazy, good, dumb, or bad (and especially do not worry about grammar or spelling). This can be done over a longer time period with a note-book—spend a few minutes each day thinking about the problem and add all thoughts and ideas to the notebook. Another way of doing individual brainstorm-ing is by making a **mindmap** as discussed in Chapter 2. The problem definition is centered on a large sheet of paper and circled; then ideas related to solving the problem are added as branches similar to those in a fishbone diagram.

Interactive Brainstorming

Verbal and written brainstorming techniques can be combined to take advan-tage of the best features of each approach in order to collect the largest quantity as well as better-quality ideas. Some interactive brainstorming techniques have been developed to accommodate special circumstances. Groups of up to twenty people can participate in interactive brainstorming, and these methods feature alternate periods of silent idea-writing with verbal sharing of ideas.

INTERACTIVE BRAINSTORMING TECHNIQUES

Idea Trigger Method
Panel Format
Integrated Problem Solving
Morphological Creativity
Storyboard
Synectics

With the **idea trigger method**, the problem definition is given. The brain-storming session begins with a brief period of silence so the group members can write down their ideas on a notepad in a column format. Then, going around the table in one direction, each member briefly describes his or her ideas while the others cross any duplicates off their own lists and add new or hitchhiking ideas in a second column. A second round of verbal sharing is then conducted, with the turn-taking now in the opposite direction. Sometimes the verbal and writing periods can be alternated quickly (each lasting about 3 to 5 minutes) until no further ideas are forthcoming. Then all the written ideas are collected for later evaluation.

If a larger group is present, say from 20 to 30 people, and it is not possible to separate them into smaller groups for brainstorming, the **panel format** can be used. Seven volunteers are chosen from the group and formed into a panel. The problem definition is presented to the entire group, then the panel verbally brainstorms the problem for 15 to 20 minutes, with these ideas being posted on a chart. The other group members write down their own original or hitchhiking ideas as they listen to the panel and try to synthesize or hitchhike on the posted ideas. The posted ideas of the panel are collected for later evaluation together with the written ideas of the group members in the audience. We have found this variation suitable for the classroom, where a second panel of students can get a turn after the first 10 minutes and where students get rewarded for turning in additional unique ideas.

The next four methods are more complicated. All use some form of forced relationships. We will talk about some methods of forcing ideas in the next section. In **integrated problem solving**, a list of ideas is collected quickly by either verbal or written brainstorming. From this collection, two different ideas are presented to the group, and a solution is discussed that combines the two ideas. A third idea is then added to the discussion and integrated with the previous combination, until all differing ideas have been considered and used. This technique resembles the synthesis that is normally done in the separate creative idea evaluation phase that follows the first brainstorming session in our approach to creative problem solving. This method is most useful with the type of problem—either very simple or very difficult—for which not too many ideas and solutions are expected.

Morphological creativity is a very structured approach to the search for the solution to a complex or vague problem. It generates a very large number of relationships between related factors and elements of the problem, which helps in defining the problem and in coming up with solution ideas. An interesting application of morphological creativity is the story plot matrix. An example is given in the exercises at the end of this unit. This method depends heavily on forced association between the different important elements of the problem. Groups are usually used in the initial stages of developing the morphological table; groups or individuals may then be involved in the process of finding solutions by using the elements in the table. The method is easier to visualize when arranged as a two-dimensional matrix, where each column can be moved up and down independently. Sometimes problems are set up and considered as a three-dimensional matrix; the matrix can be worked out first through discussion and writing, and then the most important relationships are brainstormed verbally with a group.

The **storyboard** is a matrix that visually displays ideas in several categories. This method can be used for brainstorming, planning, or idea evaluation. Title cards (headers) are made up for the important factors involved in a problem (or in implementation planning, these could be the words *who, what, where, when, why,* and *how*). These headers are posted across a large bulletin board so that a logical relationship exists among the categories that will lead to a solution of the problem. The first category is always "purpose." Then each category is brainstormed, and the ideas are posted on index cards below the appropriate header. Through this visual arrangement of ideas, additional creative ideas and solutions can be triggered, because the items in the different category columns on the board can be combined in different, unexpected ways. After the brainstorming, the group conducts a critical thinking session to eliminate the idea cards that do not meet the objectives. Then the remaining ideas are creatively improved. Many organizations outside advertising and filmmaking use the storyboard for planning, communication, implementation, and follow-up. In addition to being posted on the large wall board, the ideas are also recorded on a smaller, portable board.

Synectics is a highly developed brainstorming technique that emphasizes imagination; analogies and paradoxes are used to stimulate creative thinking and generate original ideas or insight into the problem. This complicated method requires a skillful leader. Even though this method works best with participants who are unfamiliar with the subject of the problem, these people should have had some experience with the more basic idea generation techniques. Books and workshops are available to those who are interested in learning more about Synectics and derivative techniques.

Force-Fitting Techniques

The idea generation process can be encouraged through the use of force-fitting techniques. Let us briefly look at different methods that can be used to "force" the mind to make creative leaps. These techniques are especially useful when a group is "stuck" and not able to bring forth many creative ideas.

FORCE-FITTING TECHNIQUES

Imagine Success
Force-Fitting Two Unrelated Ideas
Free Association
Big Dream/Wishful Thinking
Forced Relationship Matrix
Thought-Starter Questions or Charts
Idea Generator Tools
Attribute Listing
Bionics
Force-Fit Game

One of the easiest ways to free a mind that is stuck on a problem is to reverse the direction of the problem-solving process in the imagination. Turn the problem around—instead of focusing on the problem and trying to think of solutions, a time-out is taken to concentrate on the ideal state or on the "what should be." **Imagine success!** The mind will fill in the steps on how to get to this ideal state. Record all ideas; usually the team will quickly move to verbal brainstorming after starting with this change in viewpoint.

 THREE-MINUTE ACTIVITY 6-4: FORCE-FITTING IDEAS

In a group of five or more people, quickly brainstorm this concept: How could you use the idea of caged white rats to improve the food and atmosphere in a school cafeteria?

EXAMPLES: *Have a wild animal decorating scheme. Serve pizza in the shape of white rats. Use a squirrel cage for students to let off steam. Have a magician perform in the cafeteria during lunchtime. Have students do a research project using white rats to test the nutritional value of typical cafeteria meals. Can you see how different aspects of the two unrelated ideas lead to creative as well as practical ideas?*

This exercise can be used as a creative thinking warm-up for a brainstorming session. Invent additional pairs of unrelated ideas that would make a good warm-up exercise.

Activity 6-4 is an illustration of **force-fitting two unrelated ideas**. The technique is not only useful to get a sluggish brainstorming session going, it is also useful for improving and hitchhiking on ideas that have already been posted. When brainstorming has slowed down, team members select two very different ideas that were generated earlier and attempt to fit them together. This process should result in additional creative ideas.

In the **free association** technique of force-fitting to stimulate the imagination, the process is started by jotting down—on the blackboard for instance—a symbol which may or may not be related to the problem. This can be a picture, a word, a sketch, a numeral, or a relationship. The process is continued by jotting down a new symbol suggested by the first. This chain is continued until creative ideas related to the problem emerge. These ideas then become part of the brainstorming process and must be recorded. You might already be familiar with this method from children's games and psychology.

Imagination for brainstorming with the technique of **big dream/wishful thinking** is started by having the group members (or the individual) think of the biggest dream solution possible to the original problem. This is the place for really far-out ideas. Then the big-dream idea is further developed by wishful thinking and by asking related what-if questions. All the ideas coming out during this process are recorded. When these ideas begin to be more closely related to the problem at hand, the process continues with regular verbal brainstorming. You can really have some fun with this approach. This technique helps to "loosen up" a group that is too analytical and practical-minded.

The **forced relationship matrix** is similar to creative morphology. From definitions of possible forms and elements of the original problem, relationships are determined between them—such as similarities, differences, causes, and effects. These relationships are recorded and then analyzed to find new ideas and patterns. Especially when opposed and absurd concepts are combined in different ways, creative ideas may suddenly emerge. This technique can be practiced by re-arranging the words in a short sentence. For the example, different combinations of the two ideas of PAPER and SOAP give us paper soap and soap paper (both nouns), soapy paper and papery soap (adjective/noun combination), or papered soap, soaped paper, soap "wets" paper, or soap "cleans" paper (verb forms). Then each of these concepts is used as a "trigger" for creative ideas, depending on the original problem. In the example, this approach could lead to some good ideas if you are looking to develop washable wallpaper or a new way of packaging soap.

Dr. Alex Osborn, the inventor of verbal brainstorming, developed a **thought-starter chart** as a tool for helping people generate creative ideas. **Idea generator tools** based on Dr. Osborn's approach have been developed by other inventors and are available commercially, either as small tables, hand-held tools, decks of cards, large wall charts, or software packages. In essence, they are just different ways of asking "what if?" and "what else?" (Refer to the list of resources for some of these products.) The nine thought-starter questions are listed in Table 6-5.

Table 6-5
Dr. Osborn's Nine Thought-Starter Questions

1. **Put to other uses?** New ways to use object as is? Other uses if modified?

2. **Adapt?** What else is like this? What other idea does this suggest? Any idea in the past that could be copied or adapted?

3. **Modify?** Change meaning, color, motion, sound, odor, taste, form, shape? Other changes? New twist?

4. **Magnify?** What to add? Greater frequency? Stronger? Larger? Higher? Longer? Thicker? Extra value? Plus ingredient? Multiply? Exaggerate?

5. **Minify?** What to subtract? Eliminate? Smaller? Lighter? Slower? Split up? Less frequent? Condense? Miniaturize? Streamline? Understate?

6. **Substitute?** Who else instead? What else instead? Other place? Other time? Other ingredient? Other material? Other process? Other power source? Other approach? Other tone of voice?

7. **Rearrange?** Other layout? Other sequence? Change pace? Other pattern? Change schedule? Transpose cause and effect?

8. **Reverse?** Opposites? Turn it backward? Turn it upside down? Turn it inside out? Mirror-reverse it? Transpose positive and negative?

9. **Combine?** How about a blend, an assortment, an alloy, an ensemble? Combine purposes? Combine units? Combine ideas? Combine functions? Combine appeals?

An acrostic has been developed from this list: SCAMPER—substitute, combine, adapt, modify-magnify-minify, put to other uses, eliminate, rearrange-reverse.

Attribute listing is a technique that can be used as a checklist or as a matrix. All important attributes, parts, elements, or functions of the problem or object under consideration are listed; then the team focuses on each part in turn for new ideas. The questions to ask are "Why does it have to be this way?" or "Could it be done differently?" When an attribute listing is combined with Osborn's thought-starter questions, it is called the **sequence-attribute/modifications matrix (SAMM)**. Although its major application is for identifying promising areas for brainstorming, it also can be used as a tool to get idea generation started in a particular area.

Bionics is a simple technique that is useful for starting creative thinking. It employs analogy to nature and living organisms by asking: "How is the problem solved in nature?" Are there similar problems or phenomena in nature? For example, people in a Synectics brainstorming session thought up the idea of using the pressure distribution in a camel's foot on sand to design a new tire for a dune buggy. And the wings for a superlight aircraft were designed using the wing of sea gulls as a model in order to make the aircraft maneuverable yet stable in high winds. Other engineers have used the structure of a moth's eye (which is perhaps the most antireflective surface known) as a basis for improving the performance of optical-disc storage systems. Now they are investigating the utility of this idea in the development of nonglare surfaces in many other applications.

Here is another interesting example: Some years ago, a group of engineers from the University of Tennessee at the Tulahoma Space Institute were studying how to reduce aircraft noise. Manufacturers had been successful in reducing the jet noise and the compressor noise. However, airplanes still made a lot of vortex noise created by the frame of the plane moving through the air. To get ideas on how to approach this problem, the engineers looked to nature. Was there a flying creature whose success depended on silent flight? Yes, there was—the owl! If the owl wants to catch a mouse, its glide must be undetectable by the rodent's hearing. The engineers began using owls in flight tests in their anechoic chamber and discovered that the owl's wing had a serrated instead of a smooth leading edge. This minimizes vortex shedding by keeping the flow of air attached to the wing during flight, thereby reducing the noise level.

Force-fitting is not always serious business—a game can be used as a creative thinking warm-up. To play the **force-fit game**, the group is divided into two teams and given the problem definition statement. Team 1 shoots out an idea that is completely unrelated to the problem. Team 2 tries to turn the idea into a practical solution for the original problem. If they succeed, the second team earns a point; if not, the point goes to the first team. The two teams alternate in posing crazy questions and finding good applications. All ideas and solutions are recorded. The game combines imaginative thinking/wild ideas with the process of force-fitting two unrelated ideas and is thus a good creative exercise in its own right. Young people especially enjoy playing the game since it involves some team competition.

EXAMPLE OF THE FORCE-FIT GAME

Problem: How can study habits be improved to get better grades?

Team 1 challenge: To improve your study habits, eat a lot of ice cream.

Team 2 rebuttal: Reward yourself with a scoop of ice cream for every hour of study.

Scoring: Team 2 gets a point because they came up with a good solution for the problem using the idea of Team 1.

Team 2 challenge: To improve your study habits, wear purple boots.

Team 1 rebuttal: Also paint your face purple.

Scoring: Team 1 does not get the point because its answer is not a valid solution for the original problem—Team 2 scores instead. But, "Set up a routine—use the purple table for studying," or "Wear a stylish but comfortable outfit when studying" would have been legitimate solutions. The game is continued for a few rounds, until answers come easily; regular brainstorming is then started.

RESOURCES FOR FURTHER LEARNING

Creative Thinking Warm-Up Example

⏱ *FIVE-MINUTE ACTIVITY 6-5: WARM-UP EXERCISE*

Brainstorm "as many uses as you can think of" for a one-foot square piece of aluminum foil. Then compare your list with the list below. Now can you think of five additional ideas?

EIGHTY USES FOR A SQUARE OF ALUMINUM FOIL

1. Wrap food.
2. Cook (bake) food.
3. Conductor.
4. Ball.
5. Sun reflector.
6. UHF antenna.
7. Frost hair.
8. Christmas decoration.
9. Drip pan liner.
10. Boat for mouse.
11. Wrap package.
12. Shred for tinsel.
13. Scarecrow.
14. Stencil.
15. To make a relief print.
16. Hold hot or sticky pan.
17. Wrap pop for freezer.
18. Wrap sandwich.
19. Use as lid.
20. Mirror.
21. Crinkle and make a texture.
22. Imprint (rubbing).
23. Punch holes for filtering sand.
24. Cover vent.
25. Silver confetti.
26. Distress signal.
27. Start fire.
28. Get rust off other metals.
29. Put in shoe for temporary repair.
30. Demonstrate static electricity.
31. Jewelry.
32. Funnel.
33. Cake decorating tool.
34. Bookmark.
35. Temporary fuse.
36. Mouse suit.
37. Deflector.
38. Angel halo in Christmas pageant.
39. Flag for an alien country.
40. Wrap candies that you can eat in church.
41. Cigarette lighter.
42. Use as fan.
43. Little table mat.
44. Eye mask.
45. Stuffing for drafts and holes.
46. Melt to use as filling.
47. Make little toy animal.
48. Make little toy dishes.
49. Make "emergency" wedding ring.
50. Wrap for a small bouquet of flowers.
51. Beautify a flower pot.
52. Catch water under flower pot.
53. Book covers for "silver" library.
54. Make emergency drinking cup.
55. Make windmill toy.
56. Crease every inch, then use as ruler.
57. Make play money.
58. Make wall decoration to cover defect.
59. Make a butterfly mobile.
60. Roll up, use to blow soap bubbles.
61. Shower cap.
62. Shade for a transplanted plant.
63. Fountain for architectural model.
64. Garbage bag for "yucky" stuff.
65. Recycle.
66. Grill cover.
67. Shelf liner.
68. Candy mold.
69. Gift wrap.
70. Pie pan.
71. Emergency gas cap for car.
72. Picture frame.
73. Bird cage liner.
74. Window shade.
75. Party streamers.
76. Hair "spikes."
77. "Tin man" costume.
78. Emergency purse to carry small stuff.
79. Creative art material.
80. Shoe shield for walking through mud.

Reference Books

Several references listed earlier were used in writing this chapter. *Imagineering—How to Profit from Your Creative Powers* by Michael LeBoeuf (Ref. 1-10) describes brainstorming techniques useful for individuals working alone. *Applied Imagination—The Principles and Problems of Creative Problem-Solving* by the inventor of brainstorming, Alex F. Osborn (Ref. 1-13), is well worth reading; it explains the technique and its applications. We recommend this basic text particularly for team leaders. *Practice of Creativity* by George M. Prince (Ref. 5-16) is required reading for anyone who wishes to study Synectics. *Managing Group Creativity: A Modular Approach to Problem Solving* by Arthur B. Van Gundy, Jr. (Ref. 1-15) gives a very technical presentation on various brainstorming techniques and their justification.

6-1   Harvey J. Brightman, *Group Problem Solving: An Improved Managerial Approach*, College of Business Administration, Georgia State University, Atlanta, 1988. This book discusses strategies for improving team success and cooperative decision-making styles which foster learning.

6-2   Charles Clark, *How to Brainstorm for Profitable Ideas*, Creative Education Foundation, 1050 Union Road, Buffalo, New York 14224. This is just one of the fine books available from this organization for helping people conduct brainstorming sessions. This organization was founded by Dr. Alex Osborn in 1945 and teaches workshops on creative thinking and problem solving.

6-3   David I. Cleland and Harold Kerzner, *Engineering Team Management*, Krieger Publishing, Malabar, Florida, 1990. This book includes discussions on the ambience of team management, communications, leadership, motivation, planning and organizing, and decision making, especially as applied to developing high-performing technical teams.

6-4   Stephen R. Covey, *The 7 Habits of Highly Effective People: Powerful Lessons in Personal Change*, Simon & Schuster, 1989. The importance of values and character development are central to this book. It outlines a pathway for living with integrity. The principles provide the security which encourages change and gives wisdom for utilizing the opportunities brought about by change.

6-5   V. A. Howard and J. H. Barton, *Thinking on Paper*, Morrow, New York, 1986. This small hardback book teaches how to generate ideas by writing. The focus is on writing as a thinking tool and thus goes beyond the traditional view which considers writing as communication.

6-6   Stanley Krippner and Joseph Dillard, *Dreamworking: How to Use Your Dreams for Creative Problem Solving*, Bearly, Buffalo, New York, 1988. This textbook-workbook combination would be of interest to students wanting to find out more about the value of dreams in creativity and idea generation; it gives many examples from history.

6-7   H. A. Linstone, *The Delphi Method: Techniques and Application*, Addison-Wesley, Reading, Massachusetts, 1975. This group idea generation technique is designed for futures forecasting.

6-8   Tom Peters, *Thriving on Chaos: Handbook for a Management Revolution*, Knopf, New York, 1987. Managers today confront accelerating change with constant innovations in computer and telecommunications technology. This book gives practical guidelines for management for survival and flexibility by empowering people and teams.

6-9   Arthur B. Van Gundy, Jr., *Techniques of Structured Problem Solving*, second edition, Van Nostrand Reinhold, New York, 1988. Over one hundred proven problem-solving techniques are explained and evaluated.

6-10   Fritz Zwicky, *Entdecken, Erfinden, Forschen im morphologischen Weltbild*, Droemer Knaur, 1966. Available in an English translation: *Discovery, Invention, Research through the Morphological Approach*, Macmillan, New York. This is a good reference on the uses of morphological creativity.

Idea-Generator Tools

Software:
• Roy A. Nierenberg's THE IDEA GENERATOR® PLUS for DOS 2.0 or higher is available from Experience in Software, 2000 Hearst Avenue, Suite 202, Berkeley, California 94709-2176. The software package includes a manual and Gerard I. Nierenberg's book *The Art of Creative Thinking.*

• IDEA FISHER is a creative idea generator program developed by Marsh Fisher (co-founder of the Century 21 real estate chain); it is available from Fisher Idea Systems, Inc., 2222 Martin Street, Suite 110, Irvine, California 92715.

• MINDLINK PROBLEM SOLVER is a program for idea generation as well as problem solving. The package comes with a manual and a copy of Vincent Nolan's *The Innovator's Handbook.* It is available from MindLink, Box 247, North Pomfret, Vermont 05053.

• BRAINSTORMER is based on morphological creativity and is available from Soft Path Systems, 105 N. Adams, Eugene, Oregon 97402. Obtain a demonstration disk or find out from your computer vendor where you could experiment with some of this software to check out its suitability for your purposes: classroom, meetings, or brainstorming alone.

• INSPIRATION is a program for brainstorming, diagramming, and outlining and is available from Inspiration Software, Inc., P.O. Box 1629, Portland, Oregon 97207. It can be used for visualization and storyboarding as well as for sequential organization and planning, since it has a diagram (graphic) view and an outline (text) view.

Other Tools:
• THE POCKET INNOVATOR is a colorful hand-held idea generator by Creative Learning International, P.O. Box 160, Neenah, Wisconsin 54957, phone 1-800-955-IDEA, fax (414) 739-8595.

• The IDEA GENERATOR I is a wall chart developed by Marilyn Schoeman Dow; it can be purchased from The Write Stuff, 2515 39th Avenue SW, Seattle, Washington 98116.

• THE WHACK PACK is a deck of topical cards which can be used to stimulate thinking and trigger ideas during brainstorming. It is an accompaniment to Roger Von Oech's book *A Whack on the Side of the Head* (Ref. 1-17).

ACTIVITIES AND EXERCISES FOR ARTISTS

6-1   WHAT IF CREATIVE THINKING WARM-UP
Pose a what-if question and play around with it for a while, preferably in a group (but this activity can be done alone also). The what-if questions do not have to be practical; the exercise is even more valuable if you practice it with a wild or impossible idea. If you cannot think of a what-if question, select one from the following list.

a. *What if gravity were suspended for 10 minutes each day—how would bedrooms have to be redesigned?*
b. *What if people were color blind—how would you design traffic signals?*
c. *What if people were color and gender blind—what would a truly integrated society be like?*
d. *What if one country were suddenly occupied by aliens from outer space—how would (or should) people react?*
e. *What if trash could be made desirable—what would be the effects? How could it be made so?*
f. *What if people all looked identical—how would one be identified as an individual?*
g. *What if insects worldwide suddenly quadrupled in size—would we get a new food supply or be worse off?*
h. *What if you were stranded on a desert island with the three people you most dislike— what would you do to make this a pleasant experience?*

6-2 "LOOSEN UP"

Get a group of people to stand in a circle facing each other. One person pretends to throw a ball to someone else; that person "catches" the imaginary ball and passes it on to another person. Call out the name of the person before throwing the ball. This is a good exercise to do when people don't know each other well yet. Then use a sound instead of a ball to pass around. The person who receives a "rooster's crow" repeats it, then makes up a new sound, for example a "hog call" to pass on. Children especially love to play this game, but it is also an excellent tool to "loosen up" analytical-type persons toward right-brain thinking.

6-3 WARM-UP EXERCISE

Find different uses for one of the following "fun" objects as a warm-up exercise immediately preceding a brainstorming session—a worn sock or old blanket, a small mirror, a shoe box, a bucket of sawdust, a peanut, an old sneaker, a Frisbee, a jack-o'-lantern.

6-4 *COURTROOM DRAMA**

Make up a miniplay. The setting is a country where creativity is absolutely forbidden. One person is selected as the accused, and the prosecutor is attempting to prove that the defendant is guilty by presenting damaging evidence. Others in the acting group can be witnesses, judge/jury, detectives, "experts," etc.

6-5 IMAGINING THE FUTURE

Stretch your imagination with the following three situations (jot down your thoughts):

a. Picture a perfect day for yourself six months from now. Where will you be? At a picnic? In front of a group of your peers, giving a great speech? Running a race and winning? Playing the lead part in a performance? What will you look like? How will you feel? Will you feel great about yourself because you've broken a bad habit or made some long-planned improvement?

b. Now picture yourself five years from now. What have you accomplished in the past five years? Did you graduate from high school, from college? What new things might you have learned? Are you into some continuing education program? What have you accomplished in your personal life? Have you changed as a person? Have you grown, have you matured, not just physically, but also spiritually? What kind of friends do you think you will have, and what family relationships?

c. From where you are today, write as many endings as you can for this sentence: One of the things I'd really like to do during the next ten years is... Now look over your list of ideas and mark three or four that are most important to you. Then ask yourself: What can I begin doing now to make one or more of these dreams a reality? This could mean broadening your mind (and your skills) with additional formal or informal education, eating the right foods and exercising to make yourself physically fit, spending time doing something for others in your family, neighborhood, or community. It's your life, and you can make decisions that will make it better. Make a plan to include one activity in your weekly schedule that will move you in the direction of achieving your dream.

6-6 WISHFUL THINKING

The big dream/wishful thinking approach can give us practical ideas.

EXAMPLE (for teenagers): "I need to make $50 to buy a new sweater this week." Big dream: What if I won the lottery? Wishful thinking: What if I helped an older person and she left me all her money in her will? Hitchhiking idea: What if I go around the neighborhood and asked older people if they need help with odd jobs? Notice how wishful thinking led to an idea that could be explored further and made into a practical solution to the original problem.

Continue wishful thinking on the same problem (alone or with a team). Can you think of some other useful ideas? Now make up your own problem definition and use wishful thinking until you get ideas that lead to practical solutions.

*Asterisks denote advanced exercises leading to more in-depth learning.

6-7   *TEACH SOMEONE HOW TO BRAINSTORM*
Brainstorming is more fun and usually more productive if done in a group of at least three people. Get two friends together and teach them how to brainstorm.

6-8   PIN CARD METHOD
Brainstorm an idea with the pin card method, either at home or with a circle of friends. Pick a topic that is of some concern to the participating group members (and where you had some problems previously in getting a positive discussion going). Remember, this method is to be done silently, in writing. After idea generation has been completed, you—as the leader— can now be the judge and select the three best solutions. A day later, submit these to the group for a vote. Write a brief summary of the process and results.

6-9   *CRANFORD SLIP WRITING*
If you belong to a club where more than ten people attend a meeting, look for an opportunity to apply the Cranford slip writing method. Are you looking for ideas for some club activity or fund-raiser? Get together with the club officers to make up a problem definition statement and arrange for a brief period of idea generation (10 minutes or so). Then collect all ideas. You may want to enlist the help of a committee to evaluate the ideas and find the best solution.

6-10   "RINGII" PROCESS
Find an application for the "Ringii" brainstorming method. Go through the procedure, then write up a summary of your results.

6-11   BULLETIN BOARD
Find an application for the bulletin board method. Explain the procedure and the problem to be brainstormed to your targeted group; then post the problem definition on the bulletin board. Encourage the people to check the bulletin board frequently and add hitchhiking ideas. After two weeks, evaluate the ideas (either alone or with a small committee), then write a brief summary and report the results back to the group.

6-12   IDEA TRIGGER METHOD
Practice the idea trigger method with one of the ideas or objects commonly used for the warm-up creative thinking exercises (see Problem 6-3 for examples).

6-13   PANEL FORMAT
In a group of three people, attend a meeting that has a scheduled panel discussion, or watch a panel discussion on television. As you listen to the people speak, write down your own ideas that come to mind. After the session, discuss and compare your ideas. Were you able to come up with creative ideas and insight that improved upon the panel's views?

6-14   FREE ASSSOCIATION
Play a game of free association with a friend. You want to brainstorm the topic: Things teenagers can do to help older people. Assume that you do not have any ideas in mind. Start by drawing a triangle. The friend looks at the triangle and then draws or says whatever comes to mind. You respond in turn. Continue the chain (write everything down) until you get some ideas that are related to the problem—then continue brainstorming the topic (also record all these ideas). Write up a brief summary of the process—how did free association help?

6-15   *FORCE-FITTING EXAMPLE FROM TECHNOLOGY*
Think about the development of the razor in the course of history. Now draw an analogy to the design of a lawn mower. What kind of a mower would you design using each type of razor? Now reverse the process—can you think of an improved shaver by drawing an analogy to advanced lawn mower technology? Now extend the analogy to other types of cutters and hair-grooming tools—can you think up some wild as well as practical modifications?

6-16 *MORPHOLOGICAL CREATIVITY*

In morphological creativity, the elements or parameters of a problem are arranged in columns; each column may have a number of components (each with a maximum of seven items). In the morphological table below are 5x5x5x5x5 = 3125 different story combinations. Interesting stories can be made up with a "forced" plot—where the item in each column is selected by chance (a throw of a die, with 6 being a "wild card").

		Who?	Goal?	Obstacle?	Where?	How?
	1.	Explorer	Love	Lack of Courage	Big City	Cooperation
	2.	Grandma	Riches	Bad Reputation	Country	Sacrifice
	3.	Teenager	Fame	Poverty	Boat	Honesty
	4.	Fox	Survival	No Education	Arena	Hockey Team
	5.	Armadillo	Relief	Injury/Weakness	Mall	Food Item

MORPHOLOGICAL TABLE FOR STORY PLOTS (EXAMPLE)

(circled items: Explorer, Love, Bad Reputation, Sacrifice, Mall)

EXAMPLE PLOT: *An explorer wants the love of an attractive college professor. But he does not have much education; thus she ignores him. One day, when she is shopping in the mall with her mother, a terrorist begins shooting in front of the explorer's exhibits. The explorer throws a very valuable statue at the man, stunning and disarming him. The statue is broken, but the terrorist is captured and the women's lives are saved. Through this sacrifice, the professor begins to appreciate the explorer's fine character; soon she'll fall in love with him.*

a. How would the story in the example change if just one of the items were altered? Write up two alternate endings for the example plot above by selecting a different "how" item.

b. Use a die to determine the main components of the plot and make up your own story. Then brainstorm some imaginative titles.

c. Make up a 4x4 plot matrix (who, what, where, when) suitable for children, then write a poem or a song with one of these plots.

6-17 ADVERTISING

Think about TV commercials that you have seen recently and pick one that annoys you. Then, with a small group of friends, brainstorm ideas on how to make a better, more creative ad for this product.

6-18 WHAT HAVE YOU LEARNED SO FAR?

Write a brief paragraph describing the three most important, interesting, useful, or surprising things you have learned in this chapter, giving the reasons for your choice. What is an important question you still have?

To survive in a difficult environment (such as war), teamwork is essential.
It takes old-fashioned hard work to grow into a good team.
True leadership is looking out for the welfare of the team's members; it is not self-serving.
Jonathan C. Henkel, Lieutenant Colonel (retired), U.S. Army, Vietnam Conflict Veteran.

CHAPTER 6 — SUMMARY

Advantages of Teamwork: A team has a mix of knowledge and different thinking skills available to solve problems. Minds can interact beneficially; people can build on each other's ideas. Teams have a good chance of finding the best solution and of identifying opportunities. People who help find a solution will be more willing to accept the solution. With an investment in time and effort, the team can synthesize innovative solutions from the large number of brainstormed ideas.

Team Development: Groups meld into productive teams with support, training, and practice in working together and in having opportunities to participate in management (to overcome "group think" and other negative group dynamics). Cooperative relationships at all levels of the organization are required, as are patience, consideration, willingness to share ideas, communications, and good listening.

Team Management and Responsibilities: Teams need specific assignments and goals. The structure is customer-driven; members and facilitators are equal. The team members are selected for their thinking skills and expertise in the problem area. The team is committed to the teamwork concept and team excellence. The organization is maintaining a collaborative climate, and the team receives recognition from management for its efforts. The team members share and rotate the many roles and responsibilities for the functioning of the team.

Your Role as Artist: Use your imagination to think up many "wild and crazy" ideas related to the defined problem. As an artist, you can dream big and do a lot of wishful thinking. Elaborate on other people's ideas (idea pinching is allowed); use ideas as stepping-stones to get many more creative ideas.

Verbal (Classical) Brainstorming: This method for getting creative ideas with a group of three to nine people was developed in 1938 by Alex Osborn. Its three rules are easy to learn: (1) Generate as many ideas as possible–quantity counts. (2) Wild ideas are encouraged; be as odd, weird, and impossible as you can. (3) Do not be critical (not even of your own ideas)— judgment is deferred until later.

Procedure for Leading a Brainstorming Session:
1. Brief the team on the problem's context; give the problem definition statement.
2. Review the three brainstorming rules.
3. Explain the brainstorming procedure that will be used.
4. Conduct a creative-thinking warm-up exercise.
5. Do the brainstorming for 30 to 45 minutes.
6. End the session by allowing a few extra minutes at the end; collect all ideas.
7. Thank and dismiss the team; inform them of the next step in problem solving.

Other Methods for Idea Generation: Written and interactive brainstorming methods have been developed to accommodate different circumstances, such as large, heterogeneous groups, shy persons, interpersonal conflict, different schedules, and different levels of participation, interaction, and technology. Force-fitting techniques and idea generator tools can be used to "force" the mind to make creative leaps. They are also useful for moving people into a right-brain thinking mode.

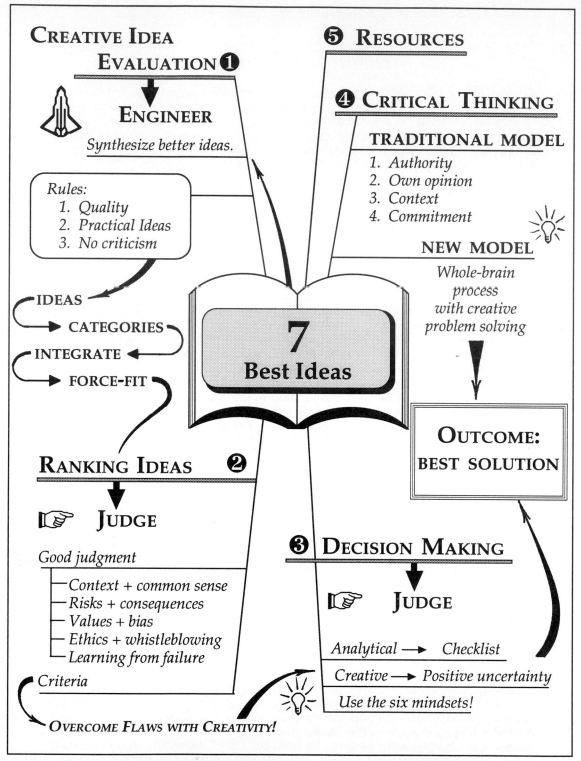

CREATIVE IDEA
EVALUATION ❶

❺ RESOURCES

❹ CRITICAL THINKING

ENGINEER

Synthesize better ideas.

TRADITIONAL MODEL

1. *Authority*
2. *Own opinion*
3. *Context*
4. *Commitment*

Rules:
1. *Quality*
2. *Practical Ideas*
3. *No criticism*

NEW MODEL

*Whole-brain
process
with creative
problem solving*

IDEAS

CATEGORIES

INTEGRATE

FORCE-FIT

**7
Best Ideas**

OUTCOME:
BEST SOLUTION

RANKING IDEAS ❷

JUDGE

❸ DECISION MAKING

JUDGE

Good judgment

— *Context + common sense*
— *Risks + consequences*
— *Values + bias*
— *Ethics + whistleblowing*
— *Learning from failure*

Analytical ⟶ *Checklist*

Creative ⟶ *Positive uncertainty*

Use the six mindsets!

Criteria

OVERCOME FLAWS WITH CREATIVITY!

Mindmap of Chapter 7. Refer to Chapter 2 (pp. 55-56) to learn more about mindmapping.

7

CREATIVE EVALUATION, JUDGMENT, AND CRITICAL THINKING

WHAT YOU CAN LEARN FROM THIS CHAPTER:
- *Creative idea evaluation: the engineer synthesizes better, more practical ideas.*
- *Ranking ideas: the judge uses context and common sense, valid criteria, and an appropriate judgment technique. Risk and consequences are assessed. Underlying values, presuppositions, and bias are questioned.*
- *Decision making: analytical and more intuitive, integrated processes.*
- *How is critical thinking related to creative problem solving?*
- *Resources for further learning: case studies, references, activities and exercises for engineers and for judges, and summary.*

Do you remember the five steps of creative problem solving and associated mindsets? So far, we have covered the first two steps: problem definition (with the detective's and explorer's mindsets) and idea generation (with the artist's mindset). Now we will look at the next two steps: creative idea evaluation, with the engineer's mindset, and critical idea evaluation (more simply known as idea judgment) using the judge's mindset. Much is heard today about schools needing to do a better job in teaching students to think critically. Thus we will discuss critical thinking and how it relates to the creative problem-solving process. Evaluation (or judgment) is the highest level in Bloom's taxonomy of thinking skills.

CREATIVE IDEA EVALUATION—THE ENGINEER'S MINDSET

Creative idea evaluation is in essence a second round of brainstorming. It is more focused than the divergent thinking process used during the idea generation phase, because we now want to use convergent thinking to clarify concepts and arrive at practical ideas that can be implemented to solve the problem. This step is

221

the key process involved in "engineering" wild ideas into more practical concepts and product designs. The process can be applied by any team or person who wants to improve the quality of the original output of brainstorming ideas. Idea synthesis is the heart of the process and characterizes the engineer's mindset in creative idea evaluation.

What do engineers do? Engineers think, plan, design, analyze, build, manage, and work in teams; they synthesize ideas and put them to practical use. Someone has coined the term "imagineering," which means: "Let your imagination soar and then engineer it down to earth." This is what we will do—we will work with our brainstormed ideas to try and improve them through another round of creative thinking. In the engineer's mindset (Figure 7-1), we switch rapidly between quadrant D and quadrant A thinking, while keeping a positive, nonjudgmental attitude. We will look for ways to sort, group, combine, supplement, elaborate, develop, synthesize, and engineer ideas to make them into practical solutions to the original problem. Can you think of three or more reasons why green is a good color to represent the engineer's mindset?

Figure 7-1 A team of engineers.

Do you recall the three brainstorming rules? For creative idea evaluation, we also have three rules. Instead of quantity, we are now looking for quality. Instead of encouraging wild ideas, we are now looking for ways of making ideas better and more practical. We will continue to abstain from quick judgments and negative comments. A positive attitude will help us have additional creative ideas as we try to integrate imaginative, intuitive thinking with analytical, logical thinking.

THE THREE RULES OF CREATIVE IDEA EVALUATION

Look for quality, not quantity. Make ideas better.
Make "wild" ideas more practical.
Continue to defer judgment.

How soon after brainstorming should creative idea evaluation be done? It is a good idea to wait at least one day if the same team will be involved. Brainstorming and creative idea evaluation are both mentally exhausting activities and are thus more productive if done with fresh minds. Also, by leaving some time to let the conscious mind rest and the subconscious mind incubate the ideas that were generated, the thoughts that will come up during the evaluation phase will be more creative. This time lag will also give the facilitator a chance to do some preliminary organizing work with the ideas, if desired. In a typical brainstorming session lasting from 30 to 45 minutes, a team may come up with anywhere from 40 to over 200 ideas. Working with a large pool of ideas can be quite a task, unless we have a structured approach for making the job easier. Basically, our work is simplified if we can sort the ideas into groups or categories—thus this is the first step in the creative evaluation process.

Task 1—Sorting related ideas into categories: This organizing work can be done by the facilitator ahead of time or by the team at the start of the evaluation session. The ideas that were collected at the end of the brainstorming session are neatly transcribed onto 4x6 *Post-it* notes or plain index cards—one idea per card. Alternatively, if the idea pool is small, the facilitator can presort the ideas into categories and then have these typed up and duplicated for a handout to the team members. An example of this is given at the end of the chapter. If *Post-it* notes are used, a blackboard or smooth wall can be used (like the storyboard setup). Or the ideas can be grouped in the form of a mindmap on a large sheet of paper. If the team will do the idea sorting, especially if the idea pool is large, it is easiest if note cards are used. A large table is needed in this case for processing the ideas.

After the facilitator or the team members have written all the ideas on cards, they are randomly spread out over the table. The team gathers around the table to ponder the ideas in silence for a few minutes—to let the ideas sink into the subconscious mind. Then it is time to begin looking for similarities. Some ideas will appear to have a strong affinity for each other—they seem to naturally want to be together. For these similar ideas, a "title" card (perhaps in a different color) can be made up, and any ideas that seem to fit can be placed under this category. At this point, do not make these categories too broad and bunch ideas together that do not have much in common. It is quite all right to have many different categories. Team members can have brief discussions about where the ideas should go, but do not get bogged down with quibbling. If an idea seems to fit into more than one category, make up a duplicate card and enter the idea in both. If new ideas come to mind, they can be jotted down on new cards and added to the pool. In our experience, we have found that the sorting process is accomplished rather quickly. Ideas that do not fit into any category can be placed in the "odd ideas" category. With the title card on top, the idea cards in each category are bundled together with a rubber band. If more than seven categories are present, the process can be repeated by combining two or more subcategories into a new "umbrella" category. For some topics, it may be difficult to come up with category headings. In this case, ideas can be sorted according to well-known ideas, novel ideas, and wild ideas, or according to the degree of difficulty of implementation—simple (inexpensive) ideas, "meaty" (more challenging) ideas, and difficult ideas requiring major resources and change.

Task 2—Developing quality ideas within a category: After all ideas have been sorted into categories, the team needs to work with one category at a time. If the team is large, categories may be assigned to subteams of three to five members with different thinking preferences. Have a breakout room or widely separated tables ready for the subteams to work on their assigned categories. At the start of Task 2, conduct a brief creative thinking warm-up. The objective now is to engineer the many ideas or idea fragments within the category down to fewer, but more completely developed, practical, and better-quality ideas. The team members can consider and discuss the ideas in the category; they can add detail; they can elaborate; they can hitchhike other ideas; they can force-fit and combine ideas. Idea synthesis—combining several concepts or ideas in a new way—is a key mental process that should be especially encouraged and practiced. Synthesis and integration are illustrated in Table 7-1 with ideas from a brainstorming session with high school and college students. Wild ideas can be used as triggers for further creative thinking. For example, how could the idea of a "shooting range" be used to improve schools? Possible ideas are: (1) Students set targets that they want to achieve in independent learning. (2) Have more applied and hands-on science and mathematics labs—including the study of ballistics. (3) Have challenging and practical physical education activities during the day as well as evening.

Ten-Minute Team Acitivity 7-1: Idea Synthesis

Starting with the nine ideas given in Table 7-1, use the integrated problem-solving approach to synthesize a different solution from the one given in the example. Do you think that your solution will solve the original problem?

Use the wild idea, "Flood the school's hallways in the winter; keep the doors open at night and create ice tracks for sliding," for insight into underlying needs as well as a stepping-stone for further creative idea generation and synthesis. You may want to start the process by expanding the problem through asking a chain of "why" questions.

Table 7-1
Example of Idea Synthesis

Brainstorming topic: How can schools be made better?

Here is one category with nine ideas (from a total of 262 ideas):

> COUNTYWIDE SCHOOL SYSTEM CHANGES
> 1. *Specific academies at different schools.*
> 2. *Skill centers at different high schools.*
> 3. *Separate high schools for gifted students.*
> 4. *Saturday school taught by engineers.*
> 5. *Create many boarding schools (wild idea).*
> 6. *More business and trade schools.*
> 7. *Schools in factories.*
> 8. *Areas of excellence in all schools.*
> 9. *Students select the school they want to attend.*

Using the technique of integrated problem solving, can Ideas 1 and 2 be combined? Yes, and two alternatives come to mind easily—combine academies and skill centers at each school, either with the same area of emphasis (such as math and science, languages, or the arts and applied arts) or with a complementary emphasis. It seems that going with the idea of complementary emphasis would lead to a higher-quality school, so let's keep going with this idea. Can it be combined with Idea 3? One of these academies/skill centers could be designated strictly for gifted students. But with several excellent academies/skill centers in a larger community, the gifted would have a challenging environment and would by their very presence help improve the quality of the schools even more. Thus *all* the academics/skill centers should incorporate programs for the gifted.

What about Idea 4—Saturday school taught by engineers? Saturday school—that's an interesting concept; it could be used to enrich the academic and cultural programs at these academies/skill centers even further. Yes, let's go with this idea, but let's include other professional people from many walks of life, not just engineers. Wild Idea 5 looks impractical—but what about creating inviting, home-like areas in existing schools for neighborhood group study under the supervision of parents or older student mentors? When Ideas 6, 7, 8, and 9 are integrated to arrive at a comprehensive idea, one possible outcome is:

> NEW COUNTYWIDE SCHOOL SYSTEM CONCEPT
> *The school system will be restructured to have diverse, combined academy/skill centers with special programs for the gifted as well as for business and trades (with sponsors from the community); other schools will have special centers of excellence, and all will have Saturday enrichment programs and innovative curricula. Students select the school they want to attend. Schools will be open until 10 p.m. for group study with mentors and as community activity centers—with emphasis on community support for learning and culture by people of all ages.*

When two ideas are combined, this is considered to be a new idea. To save time, changes and additions to ideas can be made directly on the respective cards. Fasten cards together that have been combined into one idea, with the most developed, synthesized idea placed on top of the stack. During this process, try to make well-known ideas better. Examine each novel idea closely. The danger here is that the team may suddenly get carried away with one of the ideas. If this happens, do not stop evaluating all other ideas. Continue to look for ways to improve and synthesize ideas to come up with fewer, but higher-quality solutions. Use wild ideas as thought-starters to generate additional creative ideas. It is possible that the most useful and innovative solution to the original problem is hiding out among the wild ideas. Try to force-fit simple, meaty, and difficult ideas. Task 2 is shown symbolically in Figure 7-2—ideas are fewer but more complex.

Through this process of examining and discussing ideas and trying to synthesize ideas to engineer better, more practical solutions, the team gains an understanding of the logic, meaning, and purpose of the ideas. This is one of the most important benefits of this approach; it enables the team to find high-quality solutions. Thus this activity should not be rushed. Creative idea evaluation can easily take two or three times as long as the original brainstorming, even when the facilitator has done the grouping and even when Task 2 has been subdivided

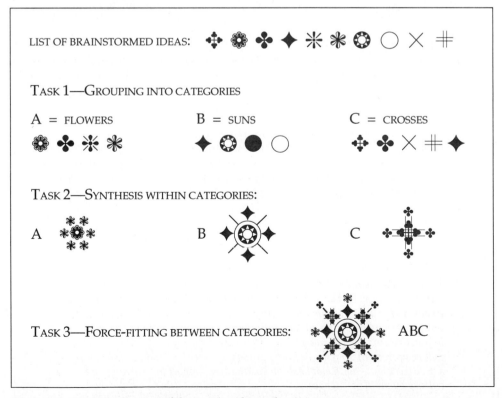

Figure 7-2 *Symbolic diagram of the creative idea evaluation process.*

among several teams. One benefit of this second round of brainstorming is that completely new ideas may pop up—it is not unusual that the best idea for solving the problem is generated at this time. When the teams have completed the task of getting fewer but better ideas for each category, the final, complete ideas for each team can ideally be written on large sheets posted on a blackboard or wall to facilitate the next step. Alternatively, depending on the types of categories and the original problem, we have seen teams who used tape or tacks to arrange the categories and the improved ideas on a wall board as elements of a storyboard.

Task 3—Force-fitting unrelated ideas between categories: In this step, the team or teams can try to take the most developed ideas from each of the final categories to come up with yet more creative solutions. This is truly a force-fitting activity because these ideas will most likely be quite different, with very little in common. Completely new and interesting ideas may be generated through this process. Again, post the improved, final ideas. However, for some types of problems, it is impossible to distill the large number of original ideas down to a few comprehensive solutions—creative idea evaluation results instead in lists of valuable ideas that, when implemented together, will solve the problem. For example, this could be a list of measures for improving the energy efficiency of a building. In this case, the entire list needs to be carried forward to the next step of idea judgment. In Figure 7-2, force-fitting the complex flower, sun, and cross symbols resulted in something new: an intricate, harmonious "sunburst" quilting design.

And now comes the last step—STOP! We have to realize that it is not possible to get a perfect solution. We do not want to become so exhausted or bored with the entire process that we are tempted to drop the project. At this point we probably still have quite a number of possible solutions; that's just fine. From this point on it will be the responsibility of the judge to determine which of the solutions will be best and should be implemented.

> *At every stage, synthesis involves the generation of alternative solutions,*
> *that is innovation, evaluation, and decision making.*
> *Commission on Engineering and Technical Systems, National Research Council, 1991.*

> *You can be wrong, you can commit errors in logic, even record inconsistencies,*
> *but I won't care if you can help me to useful new combinations.*
> *J. W. Haefele, Procter & Gamble, Cincinnati, Ohio*
> *in "The Relation of the Industrial Technical Literature to Creativity."*

⏱ *TEAM ACTIVITY 7-2: CREATIVE IDEA EVALUATION EXERCISE*

Conduct the creative idea evaluation with the brainstorming ideas from Team Activity 6-3. Start with a creative thinking warm-up. Summarize the improved ideas for each category on a large sheet of paper for posting on the wall to facilitate later idea judgment by the team.

After you have reduced the number of ideas to fewer but more practical solutions, take the time to analyze the process and the results. Was it easy to avoid negative criticism? Were you able to generate additional creative ideas? Are your synthesized solutions quite different from the original brainstorming ideas? How were the team interactions contributing to the results?

IDEA JUDGMENT—RANKING IDEAS

Our next step in creative problem solving is to find the ideas or solutions among the better ideas that will best solve the original problem. During the judgment phase, we establish judgment criteria and then sift and rank the ideas and solutions according to the criteria. We will look at the potential solutions to the problem and decide which ones are best and which ones have flaws that could cause future problems. The objective of the critical idea judgment phase is to identify the ideas that will best solve the problem.

Figure 7-3   The judge.

The Role of the Judge

For ranking ideas and deciding which should be implemented or which should be discarded, we need to adopt the judge's mindset (Figure 7-3). Here, the critical, conscious mind comes into full action. In some ways, the role of judge seems to be natural since it is easier to criticize than to explore or to transform ideas or to act. But if we spend all our time being a judge, we won't accomplish much. Also, it is important to remain impartial about the ideas we are judging. As a judge, it is our job to find the best idea and not wait for the perfect idea. Judges themselves are not perfect, but they need to make wise decisions based on "evidence" and principles. Why do you suppose the color purple has been selected to represent the judge's mindset?

*Good judgment comes from experience.
Experience comes from poor judgment.*
                              Ziggy.

Judges need a sense of timing to figure out which decisions should be made quickly and which decisions should be made only after a long, careful study. Judges have the responsibility to note flaws and then devise ways of overcoming them with creative thinking. Being a good judge takes practice because it is not easy to recognize the shortcomings of the ideas while still being able to keep an eye on the positive features. As judges we want to give the producer a solution worth defending—one that will be as trouble-free as possible. Thus a judge's responsibility is to consider the risks involved in the proposed solutions. All solutions have some possibility of failure—nothing is entirely foolproof. Even very carefully planned implementations carry an element of surprise, especially when dealing with an innovative idea, because we cannot predict all the reactions to such a solution. As we have seen earlier, Japanese companies view failures and mistakes as great opportunities for learning and improvement. Thus failures are not necessarily seen as bad; failures are starting points and motivation for growth. Also, judges have to decide if the timing is right for a new idea—a 1994 idea for the 1995 marketplace could ruin a company. Thus judges need to have future-oriented quadrant D thinking abilities to balance the quadrant A critical and analytical modes and the quadrant B risk-averse mindset.

 THREE-MINUTE ACTIVITY 7-3: FAILURE AND WISDOM

In groups of three, discuss the statement, "Experiencing failure makes us better judges." Then share your most interesting insight with a larger group. Especially focus on the aspect of wisdom—how can it be acquired?

Who should be the judge? Should a single person—for example, the facilitator or a middle-level manager—be the "idea judge" or should a team be involved in the judgment process? Sometimes this will be the same team that did the brainstorming or the creative idea evaluation. Sometimes this will be a different team. If much time is available, if a high-quality solution is needed, if there could be a problem with acceptance of the final solution, and if other people need this learning experience or training, then the team approach is best. If none of these factors are present, judgment by an individual will be more expedient. A combined approach is possible. An individual can do the preliminary selection, and a committee (the team) can make the final choice. Or a group can make the preliminary selection, with an individual (or a smaller group) making the final decision. An example of this last option is the procedure that was used in 1987 to select a new chancellor for the University of Michigan-Dearborn. A search committee made up of faculty members and staff narrowed a list of about 70 candidates down to three, and the Regents of the University of Michigan made the final choice.

What Is Good Judgment?

On a product's warranty statement, we recently saw this warning: "We cannot be responsible for the product used in situations which simply make no sense." Good sense—or good judgment—is difficult to teach, because it is best learned through experience with failure. However, we do not need to experience the failure personally; we can learn from studying the failures of others. This is particularly true in the field of engineering. A story by Galileo is retold in the November-December 1992 issue of the *American Scientist*, p. 525. A column was stored horizontally by supporting its ends on piles of timber. But since it was possible that the column could break in the middle under its own weight (as had been observed in various situations in the past), someone suggested that a third support be added at the center. Everyone consulted agreed that this would improve the safety of the column, and the idea was implemented. A few months later, the column broke in two anyway, at the center. The cause of the failure was the new support, which failed to settle at the same rate as the end supports. The column broke when too much of its weight was no longer supported at the ends.

How relevant is this story to today's technology? It serves as a reminder that solutions to problems can be the direct causes of failures. Recent notable examples are the collapse of the skywalks in the Kansas City Hyatt Regency Hotel (where changes in the support rods weakened the structure) or the space shuttle *Challenger* (where extra O-rings in the booster rocket were accepted uncritically).

> *Nine out of ten recent failures [of dams] occurred not because of inadequacies in the state of the art, but because of oversights that could and should have been avoided.*
> *Ralph Peck, foundation engineer, 1981.*

> *Bad design results from errors of engineering judgment, which is not reducible to science or mathematics.*
> *Eugene S. Ferguson,* Engineering and the Mind's Eye.

With the increased use of technology, good judgment becomes critically important. There has been a tendency to rely increasingly on computer models in place of hands-on experience, instead of integrating the two. One source of danger is that the user of a complex analytical computer program may not be able to discover all the simplifying assumptions made by the program designer. The "precision" of the computer's numerical output can give a false sense of security as to the validity of the calculations, even when critical factors unique to the particular problem are not included. Designers introduce another level of potential flaws when their high-tech designs do not consider the user interface. Eugene S. Ferguson, history professor emeritus at the University of Delaware, describes an example in "How Engineers Lose Touch," *Invention & Technology*, Winter 1993, p. 24:

> The designer of the Aegis [on the missile cruiser USS *Vincennes*], which is the proto-type system for the Strategic Defense Initiative, greatly underestimated the demands that their designs would place on the operators, who often lack the knowledge of the idiosyncracies and limitations built into the system. Disastrous errors of judgment are inevitable so long as operator error rather than designer error is routinely consid-ered the cause of disasters. Hubris and an absence of common sense in the design process set the conditions that produce the confusingly overcomplicated tasks that the equipment demands of operators.

The problem here was that the operators aboard the *Vincennes* were overwhelmed with more information than they could assimilate in the few seconds before a crucial judgment had to be made about shooting down a plane. Human abilities (and limitations) must be considered as part of the context in any design or solution.

A new engineering design (which is the solution to a problem or need) must combine analysis with intuitive knowledge gained from past experience. Even then, the judgment involved will contain a degree of uncertainty. Thus judges have the responsibility to detect errors that could have been made at any point during the design or problem-solving process, to eliminate flaws, and to evaluate the risks, consequences, and uncertainties of alternative solutions to the best of their abilities, before making the final decision on which solutions are to be sent on to implementation. Judges must be able to imagine all the things that could possibly go wrong with the solutions under consideration; yet judges must also have a flexible mindset that allows them to view uncertainty in a positive light. This idea will be discussed in more detail later in this chapter.

Idea judging is a two-step process: first, ideas and solutions are ranked using carefully developed judgment criteria (quadrant A thinking); then the final decision is made on which idea—considering needs, values, context, benefits, and risks (quadrant B thinking)—will best solve the problem and should be recommended for implementation. Judgment can only be as good as the information and values on which it is based; thus having a valid set of criteria is crucial to good judgment. With good judgment, decision making can become a relatively simple matter of choosing one or more of the highest-ranking ideas for implementation—if creative problem solving has been followed to involve all quadrants of the brain.

Developing a List of Judgment Criteria

A good list of criteria includes all factors that have an influence on the problem or decision. Let's say you have been looking for a new job and are very fortunate to get four different offers. Which one should you take, when the fifth option is to remain in your current position? How do you make the best decision? What would be some of the important factors and feelings that should be considered? It takes time to make up a valid list of criteria. The list can be developed through regular brainstorming—the more criteria, the better! Through creative evaluation, the criteria are further refined, and the most useful and important criteria are selected. Make sure the evaluation is balanced between analytical and intuitive criteria. Sometimes a weighting system is used. It can simply be based on rank, for instance from 1 to 5, with the highest number assigned to the most important criterion. Or the weighting factor for each criterion can be voted on by the team members (or some other qualified panel) and be assigned this averaged value.

Criteria can also be thought of as the boundaries or limits that the solution must fit to solve the problem. For example, government laws and regulations must be observed. A component of a larger system may be constrained by size, weight, and other physical limitations. However, if time permits, limits should be questioned. Are they merely arbitrary conventions? Why do they exist? Could the limits be overcome through creative thinking or the development of a new paradigm? Think of specs not as chains but as challenges! As a judge, you need to also pay attention to intuition—what attributes do you "feel" the ideal solution should have? Why does a certain criterion or solution seem "right" and another "wrong"? When we develop a list of criteria, we have an opportunity to put the solution into a larger context. We need to look to the future and consider the factors that will make implementation easier and more successful. A good list of criteria will help us understand all the important factors that are involved. In particular, we should look for criteria that will answer some of the following concerns:

1. **Motivation:** Why would people want to accept the solution—what are their motives? How can we motivate them to buy the product? Does the idea, service, or product meet the needs it was designed for?

2. **People:** How will people be affected by the solution? Will it be difficult to use the solution? Will they need to make changes in their lives? Is the product marketable? Who are the customers?

3. **Cost:** What will be the costs to you, to others? Will the solution be affordable? Will the product be easy to manufacture? Will it be easy to service or maintain? Can it be reused or recycled? Does the idea have other applications? Is the idea feasible? Does implementation require new technology?

4. **Support:** What support is available for implementation? What resources— such as materials, equipment, information, training, or people—will be needed to successfully implement the solution?

5. **Values:** What social values are involved? What will be the benefits to people? What are the safety issues? What are the dangers to the environment?

6. **Time:** Will the solution take a long time to implement? Will there be a short-term or long-term application for the solution?

7. **Effects:** What will be the consequences of the solution? What effects will it have on other activities in your organization, in your life, in your community?

Thinking about consequences is difficult, especially for young people, but it is also very important, because once we make a choice of one thing, other things will no longer be possible. We cannot have the cake and eat it, too. When people decide to get married, their new commitment separates them from the singles dating scene. If someone chooses to get into drugs, he or she may lose job, health, family, and reputation. Thinking about consequences can help us make decisions on the best timing for implementing a solution. One mother was asked by her son for a loan so he could buy new tires for his car—the old tires were in terrible shape, and he used the car daily to go to work. Since he still owed her money on a previous car maintenance loan, she was at first reluctant to advance still more. But when she thought about the consequences of postponing the tire replacement (and the increased danger of having a blowout while driving on a freeway during rush-hour traffic, especially with the approaching winter season), she decided the risk was not worth it. She loaned him the money immediately instead of waiting a few weeks.

> *Whatsoever a man soweth, that shall he also reap.*
> *Paul's Letter to the Galatians 8:7, New Testament (King James Version).*

> *To attain knowledge, add things every day. To attain wisdom, remove things every day.*
> *Lao-tse in the* Tao Te Ching.

Here is another example. During the 1989 TECHNORAMA at the University of Toledo, Molly Brennan, a young engineer from General Motors, was invited to the campus as the main speaker. She was one of the drivers of the *Sunraycer*, a solar car that won the race across Australia a year earlier. During a conversation after her speech, she shared an interesting story. She explained that her sister had come home from school one day with her planned schedule of classes. Their mother—an English teacher—noticed that her daughter had not signed up for physics. When asked for an explaination, the girl replied that she did not want to take physics because none of her girlfriends were taking the class. But the mother was insistent that her daughter not shortchange her future options. So the girl talked her friends into taking physics (even though the school counselor tried to dissuade them). Do you know what happened to the four girls? One is now a researcher in science, one is completing her residency as a medical doctor, and two are engineers!

> *Integration, or even the word "organic" itself, means that nothing is of value*
> *except as it is naturally related to the whole in the direction of some living purpose.*
> *Frank Lloyd Wright,* architect.

> *The critical power ... tends to make an intellectual situation of which*
> *the creative power can profitably avail itself ... to make the best ideas prevail.*
> *Matthew Arnold,* The Function of Criticism at the Present Time [1864].

> *I don't think that you can make change in an area this important unless you also know*
> *what has to be maintained, unless you have people of real seasoning and judgment.*
> *President-elect Bill Clinton, December 22, 1992, defending his choices for cabinet posts.*

It is important at this stage of the idea judgment process to maintain an enthusiastic spirit and to select a judgment technique that will avoid our natural tendency to automatically react to new ideas by focusing on what is wrong with them. Typically, if an idea is 90 percent right and 10 percent inadequate, people will jump on the flaw and imply that anyone who put this kind of faulty idea forward must be a fool. The result of such a negative outlook is that the whole idea is seen as worthless. As idea judges, we must guard against such an attitude, or valuable ideas and solutions will be discarded needlessly. As judges, we must remember that people are vulnerable—we must continue to maintain a safe environment for expressing ideas. This requires great sensitivity and wisdom since criticism of an idea is often taken as a personal attack and can destroy relationships as well as good solutions. Wisdom is illustrated with this little story we found in the December 1992 issue of *Reader's Digest* (contributed by Roderick McFarlane):

> On her golden wedding anniversary, my grandmother revealed the secret of her long
> and happy marriage. "On my wedding day, I decided to choose ten of my husband's
> faults which, for the sake of our marriage, I would overlook," she explained. A guest
> asked her to name some of the faults. "To tell the truth," she replied, "I never did
> get around to listing them. But whenever my husband did something that made me
> hopping mad, I would say to myself, 'Lucky for him that's one of the ten.'"

A list of criteria is very useful for analyzing the quality of different ideas and solutions and their capability for solving the original problem. Criteria can point out areas of weakness and can identify ideas that have too many shortcomings and should thus be dropped. But very rarely will an idea emerge as a clear "winner" that will satisfy all the criteria. Thus some additional evaluation techniques need to be employed if necessary in order to further sift and rank the ideas.

Techniques for Idea Judgment

To find the best ideas or solutions, a number of techniques are available. Most of these methods work best when they are supported with a good list of criteria.

JUDGMENT EVALUATION TECHNIQUES

Voting Procedures
Single Criterion

Advantage/DisadvantageTechniques
Separate listing for each option
Advantages, limitations, and potential (ALP)
Paired comparison analysis
Advantage/disadvantage matrix
Matrix with benchmark (Pugh method)
Matrix with weighting factors (QFD house of quality)

Advocacy Method
Reverse Brainstorming
Experimentation

When we do not have time to develop a good list of criteria that will allow us to rank ideas, we can use some type of judgment by **vote**, using a group. Or an individual judge can make a quick decision based on a **single criterion**. These two techniques are occasionally useful for cutting down a large number of ideas to a more manageable level. Voting can be done in a number of ways, openly or preferably by secret ballot. A major disadvantage of quick voting is a lack of explicit criteria. Each person makes decisions based on his or her own values or prejudices; there is no common, agreed-upon standard by which the judgment is made. Sometimes people make judgments without knowing why they make the choice, or they are swayed by peer pressure or "groupthink." Another disadvantage of a quick vote is that it tends to discourage the discussion of flaws in the ideas under consideration. If an idea is voted "the best," this seems to tacitly imply that it cannot have any faults because it has been accepted by the majority.

A large number of ideas can be reduced to a more manageable level through the use of a single criterion, such as cost. This technique is suitable for use by a group or by an individual. Caution is in order because the limits of a single criterion can often be overcome with additional brainstorming; thus a hasty decision here could eliminate some very creative solutions (or some very qualified people). For example, let us assume that in the search for hiring a research engineer the single criterion used to cut down the list of applicants is the requirement of having a Ph.D. degree. It could happen that the best candidate could be someone only two months away from getting the degree, or a very experienced person who made important discoveries working in a new field and then never found the time to complete the academic work for the doctorate. Thus the best judgment is rendered when a list of carefully thought-out criteria is used to measure the worth of the ideas that have made it to this step in the creative problem-solving process.

THREE-MINUTE TEAM ACTIVITY 7-4: CRITERIA AND VOTING
Select one of the following topics:

a. In groups of three people, share and discuss an experience with voting in which the lack of specific criteria caused problems. How were the problems resolved? What was learned from the experience?

b. Alternatively—especially in an election year—discuss some of the reasons mentioned in the media that people use to judge a candidate's suitability for political office.

c. In a group of six, brainstorm a list of ten important criteria for one of the following: Supreme Court justice; President of the United States, mayor or manager of your city, member of the local school board, U.S. ambassador to the United Nations. Was it easy or difficult to reach a consensus on the ten most important criteria?

First among the advantage/disadvantage techniques, we have the simplest approach—**a separate list for each idea**, with one column for all its advantages (positive marks or pros) and one column for all its disadvantages (negative marks or cons). The idea with the most advantages and least disadvantages "wins." This method has a major weakness because one negative can be so important that it could outweigh several or even all positives. Let's say we are developing an inexpensive consumer product. If one idea has all positive responses to the list of criteria (or a long list of advantages) except for high manufacturing costs, this one negative mark will be a serious barrier to our ultimate success. Thus we must not add up the positive and negative marks and take the arithmetic results without some critical thinking about what each negative mark implies. If sufficient time is available, the negatives can prompt another round of creative problem solving.

We can add a third column to this evaluation, in which we also take the long-range potential of each idea into account. This method is called **advantages, limitations, and potential** and makes it somewhat easier to give a fair evaluation to untried, creative ideas that depend on their potential benefits for acceptance. When we are interested in setting priorities or weighting factors among the criteria, we can conduct a **paired comparison analysis** in order to rank the criteria relative to each other. This method works best when the number of criteria and the number of solution alternatives are relatively small.

If we construct an **advantage/disadvantage matrix** with the list of criteria in a column to the left and the ideas to be evaluated across the top toward the right, we have a method that compares each idea with all the others for each criterion. To illustrate this technique with a simple example, let's take another look at the job selection problem. As shown in Table 7-2, each of the five job options has advantages (√) and disadvantages (0). So, which option should you choose? Let's say you brainstormed with your family and came up with the criteria listed in the left-hand column. The job options are arranged across the top of the matrix, and each job is evaluated against the criteria. For Job 1, the salary offer is very good, and this advantage receives a check mark. For Job 2, the pay is low (a disadvantage) and this is scored a "zero." This process is repeated for each criterion and each job.

Table 7-2
Example of an Advantage/Disadvantage Matrix: Evaluating Job Options

LIST OF CRITERIA	JOB OPTIONS				
	1	2	3	4	5
Pay	√	0	0	0	√
Other Benefits	√	√	0	0	√
Personal Growth	0	√	√	0	0
Good for the Family	√	√	0	√	0
Independence	0	√	√	√	0
Status	0	√	0	0	√
Excitement/Adventure	√	√	√	0	0
Quality Coworkers	√	0	√	√	√
Supportive Boss	√	√	0	√	0
Fits with Life Goals	√	√	0	0	√
TOTAL SCORE: √	7	8	4	4	5
0	3	2	6	6	5

When the matrix is completed, the scores are added separately for checks and zeroes. In this example, Options 1 and 2 are fairly close, with the next three (including 5, the present job) separated from the top two by a larger gap. Small differences in points are not important; thus the two top options must be considered further. Can any of the negatives be removed by additional negotiation? For example, could the salary offer be improved in Job 2? Or perhaps the lack of quality coworkers is only a temporary condition that can be expected to improve.

The criteria need to be reviewed carefully and perhaps supplemented with a **weighting system** for the entire list, not just for those items where the options differ. In the example, is pay more important, or should the potential for personal growth receive priority? When weighting factors are used, the final results will probably have a much larger spread, and it will be easier to select the best solution. What we want to illustrate with this example is the importance of selecting the right criteria. We must include all the important parameters if we want a true indication of the best options. The advantage/disadvantage matrix is a useful tool for ranking ideas and making decisions, because people working out the matrix will become aware of why ideas are ranked high, since they have an opportunity to extensively discuss and modify the evaluation criteria.

When the advantage/disadvantage matrix employs a benchmark product or an existing idea as the standard against which the new ideas are compared on a three-way scale, the technique is known as the **Pugh method.** (It is discussed in more detail in Chapter 10.) The Pugh method is a team approach to creative design concept evaluation, but it can be used for all kinds of ideas, not just design concepts. An existing product or idea is used as the datum. Each new idea or design concept is compared against the datum for each criterion and judged to be substantially better (+), essentially the same (S), or considerably worse (–). The Pugh method is an iterative technique; it goes through many cycles. The highest-scoring idea or concept of each round is chosen as the datum for the subsequent round, and ideas are continuously being improved with further creative thinking and synthesis among the ideas evaluated on the matrix. This process finally results in a consensus on an idea or concept that cannot be improved any further. Weighting factors may be added in the last round to confirm the best solution. The QFD House of Quality is an example of a matrix employing weighting factors (see Appendix A).

Other judgment techniques may be useful in special circumstances to help us rank ideas. When we have only a small number of ideas, the **advocacy method** can be used. Here the group members are assigned one or two ideas each and have the task of defending them to the group. Basically, they take turns emphasizing the positive aspects of their ideas (and how these ideas meet the criteria). This method has one disadvantage in that some serious weaknesses of the ideas may be over-looked, especially if the process is not accompanied by a good list of criteria. However, the procedure is valuable in that it gets excitement and intuition about innovative ideas back into the judgment process.

Reverse brainstorming is the opposite of the advocacy method. Here the group members criticize the weaknesses and flaws of each idea. This approach is an advantage for successful implementation because this knowledge will enable you to plan ahead to overcome the weaknesses. However, this technique must be used with other, more positive methods to overcome this negative thinking mode, and a strong effort must be made to develop "cures" for the weaknesses. Here, too, having a list of criteria will provide guidance to critical thinking. Reverse brain-storming tends to be used for evaluating ideas that are "not invented here," and care should be taken not to use a double standard and misjudge valuable ideas just because someone else thought of them.

Sometimes more data are required to make a judgment. **Experimentation** is the traditional approach to obtain such data. If only a few solutions have to be evaluated, we can choose **Edison's trial-and-error method.** For a larger number of solutions, this approach can be very expensive. The Japanese have invented a technique based on a statistical approach—**the Taguchi method** of design of experiments (see Appendix D)—that is very efficient for evaluating a large number of options and interdependent parameters. Finally, we can use another round of voting based on explicit criteria to determine the final ranking of the ideas.

Which one of the many available techniques should be chosen for a specific application? Use a technique that you are comfortable with and that matches the level of sophistication and complexity of your problem. Your choice will also depend on the time that you have available. During the judgment discussion, good communication and interpersonal skills are very important, as is an awareness of values and personal bias. We will discuss ethics and values in a moment; hints for good communication are presented in Chapter 11.

The methods we have just discussed result in ranked ideas, but they do not make the final decision for us. Criteria clarify priorities. Criteria may give a good indication of which may be the best solution for implementation. Decision making is a separate judgment activity.

⌚ *TEAM ACTIVITY 7-5: JUDGMENT EXERCISE*

If your team problem from Activity 7-2 is a "soft" problem, brainstorm a list of criteria with your team. Don't forget the creative thinking warm-up! Then engineer the ideas to obtain an improved list of valid criteria. Use your list with an appropriate judgment technique (such as the Pugh method) to find the top-ranking ideas. Discuss flaws and how they can be improved by combining the best features of different ideas or through additional creative thinking. Also consider the potential risks of implementation. Write the top-ranking ideas from each category on a new sheet of paper in preparation for team decision making.

If your team problem from Activity 7-2 is a "hard" (engineering design) problem, you may have to conduct idea evaluation and judgment in several iterations. In the first round, brainstorm a short list of important criteria to evaluate the broad concepts or approaches that are your "better" ideas. Or, depending on the type of problem you are working with and the results of the creative idea evaluation, you may be ready to brainstorm a list of design criteria for your conceptual designs. The teams then take time to develop and sketch several different concepts. The teams (or the facilitator) can work out a list of judgment criteria (and select the datum) against which the design concepts will be evaluated when using the Pugh method of creative design concept evaluation. If at all possible, do at least three rounds of evaluation.

Values, Presuppositions, and Bias

Judges are required to do much critical thinking. As we shall see later in this chapter, critical thinking is not taught well in schools, where only the analytical aspects of logical reasoning may be introduced, even though effective critical thinking and decision making are whole-brain processes. Experience also enters the picture. How do we know that we have made a good judgment? Judgments need the test of time, and we learn from the outcomes, from our failures, from our experiences with judgment. In addition to critical thinking skills and experience, a third factor influences our ability to render a good judgment, and that is our personal belief system of values, principles, and moral standards with its presuppositions and biases. Presuppositions are strongly-held, implicit paradigms that influence thinking and can either prevent or help a judge see or accept the merits (or flaws) of particular solutions. Thus one advantage of having a whole-brain team involved in the judgment process is to keep inappropriate paradigms from dominating the judgment. To be good judges, we not only have to be cognizant of our own personal biases, we must be aware of cultural bias, prejudice, and false assumptions that could influence the decision-making process. We all have heard of cases when a judge had to excuse himself or herself from a case because of personal involvement that would make it difficult to give an impartial judgment.

The only tyrant I accept in this world is the "still small voice" within me.
Mahatma Gandhi.

Principles and moral standards guide human behavior and thus are linked to survival. It is a concern to many educators that such values as honesty, loyalty, discipline, responsibility, and accountability—all aspects of personal integrity—are increasingly being lost in our society, even though we seem to demand them in our top political leaders. Why do we expect leaders, judges, and professional people to follow high ethical standards (ethics being the code of morals or standard of conduct of a particular person, religion, group, or profession)? Since morality involves the principles of right and wrong in conduct and character, who determines what is "right" or "wrong"? Where do personal and cultural values come from? How is conscience developed? Who has the responsibility for teaching moral values? These are difficult but very important questions that are often neglected, as is the related teaching about personal responsibility—knowing what is right and accepting the consequences for one's decisions and actions.

⏱ *TEAM ACTIVITY 7-6: CULTURAL VALUES*

In a group of six people made up of people from at least three different cultural backgrounds, discuss the following hypothetical situation without making a judgment as to which solution is "better"; instead, look at the different outcomes in terms of the underlying cultural values.

A man is in a building with his mother, wife, and child. Suddenly, there is an explosion, followed by a rapidly spreading fire. The man has to make a quick decision: which person in his family should he carry to safety first, since all three have suffered injuries that keep them from walking away on their own?

If you cannot find a multicultural group, discuss and develop reasons and explanations for saving the mother, the wife, or the child. What would be some of the values underlying each case? Look at the list of values that have been brought out in your discussion. Can you make a distinction between personal values and societal/cultural values? What happens when personal values conflict with societal values? Would the outcome of the discussion be different if a woman had to decide which family member to save: husband, father, or son?

Whistleblowers: Taking an ethical stand in today's materialistic world can be very costly. Two out of three whistle-blowers in the past lost their jobs in the organizations whose wrongdoing they exposed. The results are economic hardship, anger, depression, persecution, and isolation for the "ethical resisters" and their families. Blacklisting makes it almost impossible to find a job in a similar field. Why do these courageous men and women stand up for what they believe despite the high cost? Legislation was passed under the Bush administration which offers some protection to whistleblowers both in the private and public sector. Roger Boisjoly, the engineer at Morton Thiokol who tried to prevent the launch of the *Challenger* space shuttle in January 1986, now frequently speaks on university campuses. He is a champion for training professionals in ethical sensitivity. From painful personal experience, he knows that technical education alone is not enough to meet the challenges of the modern workplace.

> *If you have God, the law, the press, and the facts on your side,*
> *you have a fifty-fifty chance of defeating the bureaucracy.*
> *Hugh Kaufman, EPA whistleblower.*

> *The only thing necessary for the triumph of evil is for good men to do nothing.*
> *Attributed to Edmund Burke, 1729-1797.*

Attributes of critical thinking: When we have learned to think critically, we should possess the following characteristics: (a) We are aware of the potential for distortion in the way the world is presented by the media. (b) We are aware of physical limitations in the perception of reality and its interpretation by our mind (i.e., we know how the lack of sleep can affect judgment). (c) We are able to recognize mental blocks, overgeneralization, and false rationalization. (d) We are able to assess the "language" and thinking preferences involved in the verbal description of problems and ideas. (e) We are honest with ourselves. (f) We recognize and value evidence and feelings. (g) We resist manipulation, we overcome confusion, we ask the right questions, and we seek connections—we make a balanced judgment independent of peer pressure.

A poignant illustration of the lack of critical thinking and its consequences is provided by the following old legend retold by Iron Eyes Cody in *Guideposts* in July 1988—we leave it up to you to think of situations where the moral of this story would be relevant in today's world:

> Many years ago, Indian youths would go away in solitude to prepare for manhood. One such youth hiked into a beautiful valley, green with trees, bright with flowers. There he fasted. But on the third day, as he looked up at the surrounding mountains, he noticed one tall rugged peak, capped with dazzling snow.
>
> *I will test myself against that mountain,* he thought. He put on his buffalo-hide shirt, threw his blanket over his shoulders and set off to climb the peak.
>
> When he reached the top he stood on the rim of the world. He could see forever, and his heart swelled with pride. Then he heard a rustle at his feet, and looking down, he saw a snake. Before he could move, the snake spoke:
>
> "I am about to die," said the snake. "It is too cold for me up here and I am freezing. There is no food and I am starving. Put me under your shirt and take me down to the valley."
>
> "No," said the youth. "I am forewarned. I know your kind. You are a rattlesnake. If I pick you up, you will bite, and your bite will kill me."
>
> "Not so," said the snake. "I will treat you differently. If you do this for me, you will be special. I will not harm you."
>
> The youth resisted awhile, but this was a very persuasive snake with beautiful markings. At last the youth tucked it under his shirt and carried it down to the valley. There he laid it gently on the grass, when suddenly the snake coiled, rattled and leapt, biting him on the leg.
>
> "But you promised—" cried the youth.
>
> "You knew what I was when you picked me up," said the snake as it slithered away.

Ignorance is not bliss during the judgment phase. Bigotry and a "politically correct" view inhibit reasoning. As a judge, we must allow dialogue and explore beyond the limits of our own group or tribe (or our own mindset and preferred ways of thinking). Only when scientific (left-brain) and spiritual (right-brain) reasoning are integrated will society's problems be solvable and solved. This requires thinking along new paths. It has been said that the technological development and achievements of *homo sapiens* have far outstripped moral and ethical development. Can you support this opinion with concrete evidence? Can you cite evidence that would support an opposing view?

🕐 THINKING ACTIVITY 7-7: PERCEPTION

Alone, or with one other person, look at the shapes below. Can you see patterns in the individual shapes? Can you discern meaning from the sequence and repetition? If your paradigms and usual habits of perception are keeping you from recognizing the message in the shapes, continue with the next paragraph.

Instead of analyzing the individual symbols, look at the sequence of shapes differently—what if the meaning were hidden in the empty spaces between the shapes? Why do you think it is difficult for some people to immediately "see" the answer? What were your presuppositions about the nature of the assignment and how to solve it? Drawing this figure was at first a very frustrating task in the word processing program, since the approach was to draw the outline of the shapes. But when the problem was turned around and perceived as solids (not lines), it was a breeze to build the individual shapes with a series of rectangles.

IDEA JUDGMENT—DECISION MAKING

Decision making has been defined as selecting a course of action to achieve a desired purpose. As a judge, how can we be sure to make good decisions? Let us first look at some common ways of how people make decisions. Some of these techniques are used by teams, some by individuals. Some of these approaches are suitable for choosing ideas or solutions for implementation. The selection of the most appropriate method depends on the particular circumstances.

FORMS OF DECISION MAKING

Toss of the Coin
Easy Way Out
Checklist
Advantages Versus Disadvantages
Highest Rank
Voting
Common Consensus
Compromise
Compound Decision (Improved Team Solutions)
Delay
"No" Decision

When we have two options that are equally good, a **toss of the coin** can help us decide which one to pick, since either choice will give a good result. Sometimes it helps to focus on the consequences—make the decision based on the least troublesome or least risky alternative. If we have to choose from a number of equally good solutions, the **easy way out** will lead to the least painful, quickest resolution to a problem. Care should be taken not to ignore the long-term implications. We can also make up a **checklist** that needs to be satisfied by the best solution. The quality of the solution will depend on the quality of the checklist. A checklist is a list of minimum criteria and is useful if all important points that the solution must meet are included. Such a checklist helps to point out solutions that will really solve the original problem.

The **advantage/disadvantage matrix** has been discussed in some detail. For best results, it should come with a weighting system and include much thought about the nonremovable disadvantages. It is a useful tool for identifying the most promising options or solutions. We can then choose the **highest-ranking solution** that emerged from applying a list of criteria in a matrix or voting procedure. The more care that is spent on developing the criteria, the better the results. When **voting**, the team members should understand why they are casting their votes for particular ideas. If we feel that we want to vote for an idea even if it is ranked low in an evaluation matrix, we need to explore this prompting of our intuition—it is probable that we have used a very analytical approach when developing the criteria; this may have led us to leave out important values that need to be brought out into the open and considered.

There are three different levels of group decision making. A decision that is reached quickly by **common consensus** is usually a mediocre solution because only what the majority likes and agrees with is being incorporated in the solution and thus implemented. Creative solutions have a knack of stirring things up—thus they are not easily accepted. When a quick decision has to be made, people tend to throw out ideas they don't like, ideas that make them uncomfortable initially, or ideas that would require change. It takes time to make creative ideas understood and accepted. Common consensus may be expedient for a short-term solution to an urgent problem, but for the long term, a better-quality solution should be sought. People with widely differing views may choose solutions through **compromise**. Compromise is a trade-off, that is, some good parts are given up by both parties to gain acceptance of part of the solution. This concept is used frequently in government. Although a compromise may make the solution acceptable to a wider constituency, it may not be the best solution for the entire organization or community, because some good features have to be traded off to make the compromise acceptable. The **compound team decision** process can give a superior solution because the team concentrates on making the solution incorporate all the best features of several ideas, to the point where everyone agrees that no further improvement is possible. This is the approach used in the Pugh method because all objections and weak points have been overcome with additional creative thinking. This is defined as a compound decision, although some people use the word consensus here, too. When a group does a careful job during the entire evaluation process and when a champion has time to work for acceptance of new ideas, an excellent compound decision will be the result. In this process, what people don't like gets improved, not thrown out.

Delay is a decision alternative that may have its place. It may give you time to get more data and find a better solution, or perhaps the extra time will make the problem go away. Perhaps for political reasons, you want to avoid making a decision. By delaying the decision past a specified deadline, you can exercise what is known as the "pocket veto." Sometimes, a **"no" decision** is appropriate. Perhaps the situation has changed and the original problem no longer exists. Or the problem has changed so much that using any of the proposed solutions would make the problem worse. In this case, the decision to stop the problem-solving project before implementation is a good decision. A changed situation requires a new round of creative problem solving, starting with problem definition.

As a judge, you will need to appraise the situation and decide which form of decision making is the most appropriate for the problem and situation at hand. Important decisions with long-term effects and strong organizational impact require more thought, care, and time, whereas decisions on minor problems and issues can be made quickly, as a matter of routine. Established procedures, standards, and policies in an organization are useful since they form a framework for decision making that can reduce time and error. As a judge, you should also realize that it is practically impossible to please everyone. Being a judge requires wisdom—sometimes any kind of decision is better than waiting, and changes in the solution can be made later once the project is under way. For example, some universities lagged behind in acquiring computers for students because committees spent years in trying to decide what type of system would be best to avoid quick obsolescence, given a limited budget. But with the rapid developments in the field, it really did not matter much which system was purchased—they would all become obsolete. It was more important to get students into using computers—any computer; giving students proficiency in more than one system turned out to be a competitive advantage. Such decisions involve changing paradigms and calculated risk; paradigm pioneering requires good judgment balanced with flexibility.

 Three-Minute Team Activity 7-8: Consequences

With two other people, discuss examples of consequences that are reversible with hard work and some that are not. How are value systems related to dealing with consequences?

Our cultural heritage influences our decision making. Western civilization is in philosophy and practice a competitive system in which opposing views are argued and fought out in court, in politics, in business, and in everyday living. But through this process of attack and defense, positions become more rigid and thus creative solutions more difficult. Also, the losers are not in a mood to support the winning ideas, and all the participants lose credibility. Just look at a recent political campaign for an illustration. In contrast, other societies (especially in the Far East) have a long cultural tradition of teamwork and cooperation.

What do you do if you still have more than one best solution at this stage in the judgment process? One or more of the following steps may help you make the final selection—an example is worked out at the end of this chapter.

CHECKLIST FOR FINAL IDEA SELECTION

_____ *Can ideas be combined to obtain a higher-quality solution?*

_____ *Can different ideas of equal quality be implemented all at once or in sequence?*

_____ *How well do the ideas solve the problem? Use a 7-point rating system.*

_____ *Do the ideas meet all needs? If they pass this go/no go checkpoint, rank them according to any extra wants that they satisfy.*

_____ *Do a risk analysis on implementation with the top three ideas.*

_____ *Make a cost/value analysis.*

These final steps in the decision-making process can be done by the team or by management. Select the steps that are most appropriate to solve the problem. For example, a risk analysis (perhaps using Kepner-Tregoe) is only cost-effective for complicated, expensive solutions. The final decision should not be made strictly on the basis of return on investment (ROI) because instrinsic values or benefits to society or the environment can rarely be assigned a precise dollar figure. However, it may help to compare the long-term as well as the short-term costs and implications of implementation.

The checklist has two purposes: It provides a last opportunity for improving the final ideas, and the results of this analysis facilitate the final selection. This last judgment activity is immediately followed by decisions on what actions to take to implement the optimum solution or solutions and assigning this responsibility to specific teams or individuals. Implementation will be the topic of the next chapter.

⏱ *TEAM ACTIVITY 7-9: DECISION-MAKING EXERCISE*

Using the results of Team Activity 7-5, make your decision on which of the final ideas or solutions should be implemented. If you worked with a complicated problem that had many categories, you may have to apply some of the techniques included in the final checklist. If you had a design problem or otherwise used the Pugh method, explicitly make the decision to implement the top-ranking design or the highest-ranking solution (or combination of solutions). Include a brief explanation of how you made the final decision and why the selected solution is best.

*You need to develop a careful balance between making judgments
based on past experiences and keeping your mind open to new possibilities.
Mark Von Wodtke, Mind Over Media, 1993.*

Creative Decision Making

The decision-making approaches that we have discussed so far are primarily analytical procedures involving left-brain thinking. Some people make decisions intuitively, without consciously reasoning through the process or working out an explicit set of criteria. Then, in order to explain their decision to others, they may "invent" some rational reasons for their choice. This right-brain approach has been found to work quite well with people who have learned to trust their intuition and its reliability in making good judgment in particular situations. But because we live in changing times, where the future is very unpredictable, we need to use decision-making tactics that involve both left-brain and right-brain thinking processes. Dr. II. B. Gelatt, an educator and psychologist, career consultant, author, and trainer, has written the book *Creative Decision Making: Using Positive Uncertainty*, which uses both rational and intuitive techniques for making the best decisions. Because this small workbook provides some interesting insights into whole-brain decision making and is drawing parallels to the creative problem-solving process, we are summarizing it here.

Uncertainty is present in problem solving when we have too little information. It is equally present when we have too much information, especially when this data is irrelevant, conflicting, incomplete, unconnected, or even wrong. Also, what we know is not the only basis for decision making—both what we want and what we believe strongly influence what we decide to do. This viewpoint is expressed in a four-step framework, and it is based on an attitude that sees uncertainty as positive!

1. **Goal:** Be focused (left-brain) and flexible (right-brain) about what you want. Goals are not fixed in concrete; they are just guides. Thus be open to change in response to changing conditions and changing expectations. As you achieve your goals, be open to unexpected discoveries.

2. **Knowledge:** Be wary (left-brain) and aware (right-brain) about what you know. Knowledge is power. But ignorance can be bliss: you haven't learned yet what doesn't work; imagination is valuable, and memory can't always be trusted.

3. **Belief:** Be objective (left-brain) and optimistic (right-brain) about what you believe. Wishful thinking has a rightful place. What is reality? Positive thinking can be a self-fulfilling prophesy.

4. **Action:** Be practical (left-brain) and magical (right-brain) about what you do. Trust intuition. Respond to change, but also cause change. Planning leads to learning and vice versa. What are your personal paradigm shifts? Be playful, not fearful, when making decisions. Play with the limitations of your logic and the bounties of your intuition.

The ideal situation is to develop a balanced approach. Evaluate the actions you could take and the possible results and uncertainties involved—the options, consequences, and probability of success. Have an attitude that asks: "What else?"

Use different mindsets: be objective —> detective, positive —> explorer, emotional —> artist, creative/integrating —> engineer, negative —> judge, and controlling —> producer/implementer. Thus decision making is at its best when it employs all six mindsets of the creative problem-solving process. Decisions here are seen as having four outcomes: Either they result in a plus or a negative for the self, and/or a plus or negative for others. When options are being evaluated, they can be passed through this **outcome matrix**:

	PLUS	NEGATIVE
SELF		
OTHERS		

Practice imagining or "inventing" the future. Does getting the facts equal knowledge? How can old knowledge block new thinking? Life is a journey, not a destination. Get advice; collect different opinions. Use the process of internal debate—have your left brain supply rational arguments; then have your right brain comment on how you feel about each argument. Consider other people—who is on your "left" and who is on your "right"? What do their positions tell you? Who do you want to be like? Why? Whose opinion do you trust? Be optimistic about what might happen—you can change and influence what will happen. Your beliefs can determine what you do, what you want, and what you know. Be versatile; adapt to change. As an idea judge and decision maker, be sure to review the available information, options, beliefs, and goals.

This approach was developed to give "decision advice that is more closely related to what people actually do than to what experts say they should do." Positive uncertainty paradoxically combines intellectual/objective techniques and imaginative/subjective techniques into an unconventional wisdom for future planning and decision making. Does using this process make decision making easier? To answer this, we must first ask: "When is decision making easy?" It is easy when we have developed good solutions through the creative problem-solving process; it is difficult when none of the available options really solve the problem. The problem is that creating an effective solution usually takes longer than the time allowed for making the decision. But to create an effective solution to an unstructured problem requires that we employ the capabilities of the whole brain. We can use a decision "circle": Quadrant A—What is the evidence? Can it be trusted? Quadrant B—What must we do with what we know? How do we take action? Quadrant C—Do we value the outcome? How are others affected? Quadrant D— What are the possible outcomes? Can we imagine what might happen?

> *There seems to be no invention, no matter how sophisticated, that can equal the power,*
> *flexibility, and user-friendliness of the whole human mind.*
> *We all possess the world's finest multisensory decision-making machine right in our heads.*
> *All we have to do is to learn how to use it.*
> H. B. Gelatt, Creative Decision Making, *p. 54.*

Idea Judgment in a Nutshell

This whole discussion on judgment may seem complicated to you, especially where it involves teams. It helps to keep things in perspective if you remember the objective of idea judgment—finding the best solution. Unknowingly, you are probably doing many of these steps already. But just to summarize, here is a simple checksheet for a judge that you can apply as an individual when judging ideas.

CHECKSHEET FOR A JUDGE

1. OBJECTIVE   ____ *What is the current problem situation?*
                     ____ *What is the idea trying to do?*

2. POSITIVES   ____ *What is worth building on?*

3. NEGATIVES   ____ *What are the drawbacks?*
                     ____ *What is the worst thing that could happen?*

4. PROBABILITY ____ *What are the chances of success?*
                     ____ *If the idea fails, what can be learned?*

5. TIMING   ____ *Is the timing right for this idea?*
                     ____ *How long do you have to make your decision?*

6. BIAS   ____ *What assumptions are you making?*
                     ____ *Are these assumptions still valid?*
                     ____ *Do you have some blind spots?*

7. THE VERDICT ____ *What is your decision?*
                     ____ *How will it affect people?*
                     ____ *What is to be done next?*

CRITICAL THINKING AND CREATIVE PROBLEM SOLVING

Critical thinking, as taught at the secondary school and even at the undergraduate college level, may include the following features:
• Critical thinkers decide on what they think and why they think it.
• Critical thinkers seek other views and evidence beyond their own knowledge.
• Critical thinkers decide which view is the most reasonable, based on all the evidence.
• Critical thinkers make sure that they use reliable facts and sources of information; when they state a fact that is not common knowledge, they will briefly say where they have obtained the information.
• When critical thinkers state an opinion, they anticipate questions others might ask and thus have thoughtful answers ready to support their opinion.
The problem is that this teaching is often narrowly focused, mostly in the area of literary (or artistic) criticism, and the connections to everyday life (from the workplace, human relationships, the media, and responsible citizenship) are missing.

> *To criticize is to appreciate, to appropriate, to take intellectual possession,*
> *to establish in fine a relation with the criticized thing and to make it one's own.*
> Henry James.

For example, a piece of writing can be evaluated on how well it transmits the author's thoughts, intent, and message on several levels, where these levels correspond to the four quadrants of thinking in the Herrmann model. We can begin with quadrant B and examine if the work follows accepted rules of grammar, spelling, and form appropriate to the work (i.e., essay, short story, poem, technical report). Then in quadrant A we can analyze whether thoughts, arguments, and conclusions are connected logically, based on the available facts, situation, or data. In quadrant C, we can evaluate how effectively sensory images are created and emotions evoked through the chosen words, visualizations, and settings—and how we personally are responding to the work. Finally in quadrant D, we can explore the underlying metaphorical, mythical, contextual, visionary aspects and meanings.

Three views are prominent in current literature on teaching critical thinking at the college level: argument skills, cognitive processes, and intellectual development. **Argument skills**—Students are taught the skills of analyzing and constructing arguments based on informal logic. This emphasis on analytical skills may improve the students' ability to justify beliefs they already hold. But it has been found that students are unable to translate this learning to everyday issues. **Cognitive processes**—Here students interpret problems or phenomena based on what they already know or believe. They construct a mental model of the problem or situation around a claim or hypothesis which is supported by reasoning and evidence. Three kinds of knowledge contribute to the model: the facts involved in the particular discipline, knowing the procedures on how to reason in the discipline, and meta-cognition, which means evaluating the goals, the context, the cause-and-effect relationships, and the progress of inquiry or problem solving. However, new learning is not stored as a collection of isolated facts, but as meaning constructed into patterns or scripts as understood by the student. Professors rarely teach the strategies, procedures, and metacognition explicitly—thus students rarely learn how to apply knowledge and critical thinking in unfamiliar situations. **Intellectual development**—this approach examines students' relationship to belief and knowledge. The best-known model has been developed by William Perry and expanded by Mary Belenky and associates. We will discuss this model in some detail since it forms the basis of much of the efforts and research in the area of critical thinking done today at the college level.

Four-Stage Development Model of Critical Thinking

Each stage is described with a metaphor. After identifying the students' thinking characteristics at each stage, we will draw connections to corresponding steps in the creative problem-solving process and quadrants of the Herrmann model of thinking preferences. In the traditional college curriculum, the majority of students do not progress beyond the second stage, and some find the process in which they are required to question authority and their own beliefs very stressful.

We believe that students who are taught the creative problem-solving process will learn to do good critical thinking. Perry has found that students do not progress through the four stages uniformly. They may be at one level in one discipline and at a lower level in another subject or situation. Research is needed to quantitatively assess how creative problem solving improves the acquisition of critical thinking skills, and how this progress is related to the students' brain dominance profiles. Also, as more information becomes available on how women and minority students learn in cooperative environments, the four-stage model may undergo significant changes. Here is a summary of the four stages of the current model:

STAGE 1—AUTHORITIES HAVE "THE ANSWER":
Metaphor: Police Officer—because answers are seen as either right or wrong.

Answers are found by rules, formulas, procedures. Sometimes students will determine what authorities want (i.e., parents), then will do the opposite. Differing peer views can be distracting or confusing, and some (especially women) may be intimidated by male authority figures. Students are taught WHAT to think.

Students have no empathy or tolerance for differing views or ambiguity. They find interpretation very difficult in a world they see is much more complex than they thought. They do not recognize that the knowledge they receive is preselected and biased.

Mindset: The primary associated thinking preference is the quadrant B mode; the primary associated creative problem-solving mindset is the detective.

How can we prepare students to move on to the next level?
✦ Show that problems can have more than one right answer—things are not always black or white. Diverse views are legitimate. This uncertainty makes most students uncomfortable since throughout their early school years, they may have been taught that problems do only have one right answer (which they should be able to find within a few minutes).
✦ Students need facts and details, but they must become aware that facts may not be complete or reliable; theories or conclusions drawn may be false or incomplete.
✦ Demonstrate that imagination is needed; logic is no guarantee of finding the best answer. Encourage openness to new ideas; let students begin to explore their own inner resources.
✦ Discuss how the scientific method—when presented as a straightforward process of finding the answer to a problem—can be misleading, since in actual applications, scientists encounter many detours, blind alleys, dead ends, interactions with other minds (including serious disagreement), idea synthesis in their own subconscious mind, none of which are usually reported when they present the results of their research and how they arrived at the solutions.
✦ Provide a problem-solving climate that is cooperative, not authoritarian.
✦ Teach and apply creative problem solving, so students learn to recognize that many problems are open-ended and that finding a best solution takes much time, discussion, and reflection. Students thus learn to be comfortable with quadrant D thinking; they are taught to overcome the mental blocks of expecting only one right answer, following the rules, and being uncomfortable with ambiguity. They also learn to define the problem, ask questions, and look at the context to obtain a clearer

understanding of the situation—in the mindset of the detective and explorer. The process of doing problem solving in teams with people who think differently gives them an appreciation for other viewpoints.

STAGE 2—ONE'S OWN OPINION IS VALUABLE:

Metaphor: Supermarket—because diverse opinions are acknowledged.

Answers: When facts are not known, subjective opinions are valid and an acceptable source of knowledge. Students evaluate opinions based on feelings, intuition, and common sense. Students are encouraged to inquire: WHAT ELSE?

Students recognize that authority can't always be trusted. Also, logic is no guarantee to finding the right approach—theory comes from imagination. Students come to accept that conflicting opinions are a fact of life and legitimate. They grow to have an attitude of infinite tolerance for other views, but they do not yet have a way for making critical judgments.

Mindset: The primary associated thinking preferences are quadrant C and D modes; the creative problem solving mindsets are the explorer and artist.

How can we prepare students to move on to the next level?

✦ Instructors need to honor students' opinions without judging—let students express and explain ideas in a peer-oriented, supportive environment. Respect that students have not yet thought about the sources of their personal value systems.

✦ Through open, empathetic discussion, let students discover that opinion alone is insufficient. Just because something is possible does not mean that it is acceptable and "good enough" for carrying out; ideas need to be evaluated in a more objective, rational manner.

✦ Students need to be taught explicitly the criteria and argumentation for each field of knowledge. Evaluation in literature is different from evaluation in physics, not so much in the thinking skills required, but in vocabulary and format.

✦ Students must then be taught to compare different ideas based on these criteria.

✦ Show how imaginative thinking is used to develop theories and improve criteria.

✦ Finally, demonstrate how people's paradigms are influenced by their social context and past experience.

✦ When you teach creative problem solving, the explorer's mindset helps students investigate the context of the situation, and the artist's mindset encourages them to value their own ideas and opinions, as well as the ideas of others. Then, in the engineer's mindset, they are encouraged and taught to positively evaluate ideas and creatively synthesize them to come up with better answers.

STAGE 3—CONNECTING KNOWLEDGE AND PEOPLE:

Metaphor: Games—because "truth" and "rules" depend on their frame of reference.

Answers are found by seeking to understand—in an empathetic, nonjudgmental way—why other people think and do as they do. The context of a situation is important. Opinions differ in quality—some reasons and arguments are better than others. It is O.K. to be skeptical. Students are guided toward asking: WHY?

Students understand the "teacher's game" but are still unable to make impartial, analytical judgments. Students also come to understand the game or context in their particular fields of study, but they do not yet look for larger connections. The emphasis is on learning discipline-specific methods of reasoning.

Mindset: The primary associated thinking preferences are quadrants D and A modes; the primary creative problem-solving mindset is the engineer.

How can we prepare students to move on to the next level?

✦ Demonstrate how the disciplined approach can enhance "the inner voice."

✦ Discuss the relationship between analytical thinking and underlying presuppositions. Investigate the origin and importance of values and ethics.

✦ Discuss how choices among different approaches and applications reflect values and different paradigms.

✦ The implicit value content in each field should be made explicit. Also include discussions of personal and cultural values and how these can be strengthened to achieve desired objectives and outcomes.

✦ Show how empathetic imagination together with logical reasoning and argument can be used to support a position and illuminate a situation.

✦ Affirm that holding a position in the face of uncertainty requires courage.

✦ Through creative problem solving, the students learn to synthesize creative ideas and positive analytical approaches in the engineer's mindset. In the judge's mindset, they freely iterate (move around) between analytical quadrant A thinking, interpersonal, value-based quadrant C thinking, and contextual quadrant D thinking during the process of ranking and judging ideas. The application of creative problem solving in many different fields should be demonstrated and encouraged.

STAGE 4—CONTEXTUALLY APPROPRIATE DECISIONS AND COMMITMENT:

Metaphor: Life—because knowledge and rational thinking are integrated with "inner truth," leading to commitment, caring, and action—the ultimate goal of education.

Answers: Creative thinking and exploration go beyond the students' own feelings and fixed paradigms. Students have learned HOW to think and nurture ideas (not just criticize). Judgment is appropriate to the problem and its context; answers are complex and include trade-offs and compromise.

Students understand the values inherent in different fields and disciplines and can connect them in appropriate ways to their own values. They are empathetic but have developed reasonable limits to their tolerance for erroneous, illogical ideas. Students are involved in issues and know that they make a difference. They are truly educated and can connect learning and insight from liberal arts with professional training. They appreciate and know the rational (and spiritual) basis of their own value system and beliefs. Thus students are able to make good decisions; they develop self-confidence and commitment to their decisions and can apply this thinking skill to real-life situations, since they have an understanding and willingness to accept the attendant risks and uncertainties. Students at this stage know why they have come to the answers and conclusions they have worked out; they can defend the positives, are aware of the shortcomings, and are open to discuss additional or alternative ideas. They have the communication skills to explain their thinking and solutions to others. The students are competent creative problem solvers and producers.

Mindset: The associated thinking preferences are all four quadrants (A, B, C, D), since these critical thinkers can apply whatever mode is appropriate to a particular situation. The primary associated creative problem-solving mindsets are the judge and the producer; these quadrant B thinkers can fluently use all the other mindsets—they are whole-brain thinkers.

Both the teaching of creative problem solving and critical thinking require changes in the instructors. Students must be given opportunities to express their ideas in a "safe" environment free from negative criticism, sarcasm, and ridicule. Instructors need to share information about their own growth, mistakes, and tentative views; they need to acknowledge the joys and frustrations of change, and they need to explain their own value commitment. Their thinking must be more transparent: they should not exclusively present the final, neatly packaged results of their own thinking and problem solving. Instead, they need to demonstrate the process of "thinking through a problem with others" (which is part of cooperative learning). As teachers, we cannot shirk this responsibility—critical thinking is needed to deal with the crucial issues and responsibilities of our times.

 ⏱ *Discussion Activity 7:10: Critical Thinking*
Form a discussion group of five people.

Based on the four-stage model of critical thinking development by Perry and modified by Belenky and associate researchers, briefly analyze (with supporting evidence) how far along you are in the process of learning critical thinking. How did you learn the skills and attitudes of each stage—or what do you think would help you progress to the next stage?

Now select one of the discussion topics below. How important are critical thinking skills for finding solutions to these problems?

✦ *What new mores (moral values in society) are required to allow for what is called "fractional responsibility," for example, in energy conservation, ecology, caring for the homeless, etc., where no one person can solve the problem, but where the participation of many is required to make a difference?*

✦ *What is the relationship between critical thinking and reducing bias in society (as related to sex, race, religion, nationality, economic status, and level of education)?*

✦ *How is critical thinking connected to communications in a high-tech society?*

✦ *How important is an education in liberal arts to functioning in a high-tech society? How important is technological literacy?*

HINT: If you are uncomfortable about sharing personal views on controversial subjects, you can still participate in the discussion by contributing answers to the question, "What have you heard about . . . ?" Or ask the group to collectively switch roles, playing "devil's advocate" for several differing viewpoints. You do not have to personally believe something is true in order to understand why others may find a theory or model useful.

Optionally, select a principle that you strongly believe in and explain the underlying rational framework as well as your personal reasons why you support this belief.

Becoming Critical Thinkers (Stephen Brookfield Model)

In this model, people break out of the analytical pattern of critical thinking that they have been taught in school through some trigger event (which can be either a positive happening or a tragedy). They go through a period of self-appraisal and examination; they search for, explore, and test alternatives and new paradigms. Through this process, they develop new perspectives; they then try to choose the "best" and integrate it into their life. This results in changed attitudes, confirmed beliefs, and altered subconscious feelings. Others (including instructors) can assist in the process by affirming self-worth, listening attentively, and showing support of the effort. They can provide motivation and encouragement for risk taking, evaluate progress, and supply a contextual network and resources.

Critical thinking that includes teamwork is needed to maintain a healthy democracy. Pressure against critical thinking comes from people in power who want to preserve the status quo, especially if it is inequitable. Examples are political dictators, labor leaders, employers, teachers, family members, professional groups, anyone with a vested interest in continuing an existing paradigm and hierarchy. Critical thinking is also needed to counter the bias and influence of the media, both in the entertainment sector and in the reporting of news.

How do we recognize critical thinking—what are its characteristics?
- Critical thinking is a process, not a result; it includes the continuous questioning of assumptions. It is important to understand the context of problems (and the underlying assumptions and social value system).
- Critical thinking is a productive and positive activity: it includes creativity and innovation. Imagination is practiced; possibilities and alternatives are explored. This leads to reflective skepticism—change is not simply accepted because it is new. Consequences of actions are anticipated.
- Critical thinking is emotional as well as rational—it is whole-brain thinking where we recognize our assumptions within the framework of our personal beliefs and commitments as well as within the context of the world around us. Criteria are not strictly objective but subjective. Role playing, decision simulation, and preferred scenarios and futures are valid creative thinking strategies. Poetry, fantasy, drawing and painting, songs, and drama are means to release creative imagination and thus help in developing critical thinking.
- Critical thinkers are curious, flexible, honest, and skeptical—they can distinguish bias from reason, facts from opinion. They can use thought rationally and purposefully together with feelings and intuition to move toward a future goal.

Critical Thinking Versus Problem Solving

The current view appears to consider critical thinking as a form of problem solving, with critical thinking involving open-ended or unstructured problems, in contrast to specific problem solving where the focus is much more narrow. But problem solving as used by these researchers is analytical problem solving (a left-brain activity), not creative problem solving. A model of the problem situation is constructed, the current state is analyzed, the constraints are identified, data are collected, one or more hypotheses are postulated and then tested, until the goal or resolution of the problem is achieved. The problems used in laboratory studies can be complex but usually have only one or at most a very limited number of solutions that will work. In contrast, critical thinking uses inductive reasoning about problems that have a multitude of possible answers. In this view, critical thinking closely resembles the thinking processes used in creative problem solving, as we have seen in the discussion of the four-stage model.

But the goal of critical thinking is not necessarily to find a solution, but to construct a logical representation of a situation or position based on plausible arguments and evidence, where the "truth"of the model cannot be tested. This is a major difference from the creative problem-solving approach. Let's illustrate this difference in an example. In a court of law, a couple is involved in a custody case, in which the opposing lawyers are trying to build the strongest case (by argument and supporting evidence) for their client's position. If creative problem solving were used, the estranged parents would try to define the real problem and work together to develop a solution that would be acceptable to all, but would above all consider the needs of their child. What if creative problem solving had been used at an earlier time in the marriage—would it have helped to build a strong family instead of an adversarial relationship?

In professional practice such as business, engineering, teaching, or architecture, reasoning combines aspects of critical thinking and problem solving. Problems are ill-defined—thus they call for creative problem solving. In the current (traditional) view of critical thinking, the central element is the ability to raise relevant questions and critique solutions without necessarily posing alternatives. We believe, as do Stephen Brookfield and H. B. Gelatt, that critical thinking needs a more broadly defined concept that includes playing with ideas and creatively developing analogies and metaphors, not just logical reasoning. In creative problem solving, solutions, ideas, materials, presentations, writings, etc., are critiqued, but the process is taken an additional step in that ideas for making improvements must be proposed or considered. Creating these ideas requires right-brain thinking and a positive attitude to balance a critical mindset that looks only for flaws.

How effective is training in creative problem solving to developing critical thinkers? We do not yet have quantitative data to confirm a direct beneficial influence. We would like to evaluate the critical thinking skills of some of the secondary school students in Toledo, Ohio, who have completed Level 3 of our Math/Science Academy program based on creative problem solving and compare them with students of similar academic abilities and grade level who have not had this training. We do have the feedback of the teachers who have worked with these students and who have commented on the surprising quality of thinking—often equivalent or superior to that of the adults on their teams (to whom creative problem solving was a new skill). As you read through the case studies in the remainder of this book, make a judgment on the quality of critical thinking that you can discern and how it might have been influenced by the problem-solving process.

Case Study in Idea Engineering and Judgment

PROBLEM DEFINITION STATEMENT

> How can stress be reduced for employees faced with major changes
> in job status (dismissal, transfer, or plant closing)?

This problem was brainstormed with a group of managers and engineers. The first brainstorming session was short and resulted in 29 different ideas which were presorted into five categories by the facilitator, typed up, and handed back to the same group the following day. During creative idea evaluation, the group decided to focus on *things that managers can do.* With this viewpoint, ideas were then improved and added in each category. These additions are shown in boldface lettering below. Later, during idea judgment, the group decided to concentrate on novel ideas. This illustrates the use of a single criterion.

Next, they ranked those novel ideas in order of difficulty, with the easy ones to be implemented first. This ranking order is indicated in the example in brackets. Ideas 1 and 2 were judged to be easy, ideas 3, 4, and 5 more difficult, and ideas 6, 7, and 8 the most difficult to implement. Note that most of the final ideas did not appear in the first brainstorming session but surfaced or were engineered during creative idea evaluation. This is why this second round of creative, but more focused, thinking in the engineer's mindset is very important and should not be skipped. In this workshop problem, the teams worked from typed-up idea summaries, not idea cards, since the idea pool for each category was small. The "better" and "best" ideas were then developed on larger sheets of paper.

CATEGORY A—EDUCATION AND TRAINING TO GIVE PEOPLE OPTIONS:

1. Set up job rotation, so workers will be more versatile and can move to other positions within the company if their position is abolished.
2. Train workers **continuously** in new technology **and languages** so they are qualified for new jobs (either in the old company or elsewhere).
3. People should be educated in the schools **and in the media** to expect change. With this mindset, it will be accepted practice to always have an alternative option or two to fall back on if necessary.
4. Managers need to be continuously informed about the changes technology brings to their companies.
5. Workers have to be given the time by management to become competent in their new jobs in high tech (this may take as long as a year or more for complicated computers).
6. New employees should be trained for the job that they are expected to do. Also, they need to have a clear job description.
7. **Managers need to talk to school boards, influence media. [Idea 6]**
8. **Unions and management need to brainstorm together. [Idea 7]**

CATEGORY B—MEASURES FOR HELPING DISMISSED WORKERS:

1. Unemployment insurance.
2. Job placement programs by the company or the government (paid by the company).
3. Counseling to help locate another job (paid by the company).
4. Counseling for employee and family to cope with this change (paid by the company).
5. Assistance with relocation (real estate, finding a job for spouse, etc.).
6. Set up an organization like a personalized chamber of commerce to assist relocating workers.
7. The company should offer comprehensive retraining (**including languages**) or support/sponsor the employee for further education at a college or other school. **[Idea 4]**
8. National **or global** data bank to match workers with job openings in the whole country **or overseas**. **[Idea 5a]**
9. **Managers should be on the lookout for networking with other companies that may be able to use workers (make pensions portable). [Idea 5b]**

CATEGORY C—THINGS MANAGEMENT CAN DO:

1. Management needs to organize company to allow for horizontal interaction (information flow and movement of workers).
2. Management needs to be aware of trends in society and the marketplace, watch for new opportunities, prepare new products to meet the new needs. This requires creative thinking. **[Idea 3a]**
3. Management needs to understand change and technology, as well as the impact these factors have on the company and the employees. **Manage innovation to provide jobs. [Idea 3b]**
4. Training for workers must be continuous.
5. Training for managers must be continuous; they must be prepared to deal with change creatively.
6. **Reduce bureaucracy.**

CATEGORY D—MEASURES TO AVOID OR PREVENT DISMISSAL OF WORKERS:

1. All employees agree to voluntary pay cuts in order to keep workers from being dismissed. **Managers support this plan.**
2. All employees agree to reduced work hours (especially when the company's difficulties are expected to be only temporary); this will avoid laying off people. **Managers support this plan.**
3. Personnel surveys should be taken to match people to jobs for best productivity and morale.
4. **Have bonus or profit sharing.**

CATEGORY E—MEASURES THAT REDUCE STRESS IN THE COMPANY:

1. Assess quota levels fairly and adjust them when changes have occurred.
2. Have a mediator to minimize/remedy interpersonal conflicts.
3. Reward company loyalty; give merit recognition.
4. Pair each new person with a mentor. **[Idea 1]**
5. Allow for mistakes; look at mistakes constructively (like the Japanese). Mistakes are a learning opportunity. This will avoid a cover-up of mistakes which can be damaging to the company. **[Idea 2]**
6. Arrange for social activities for employees and management together to make people more comfortable with each other.
7. Foster a spirit of cooperation, not competition. Emphasize the benefits of teamwork.
8. **Influence government policies to avoid those that are counterproductive. [Idea 8]**

> *No idea is a "bad" idea—look for the good and build on it for a successful life.*
> *Jason Tansel, engineering student.*

> *Think smart—find and use the "best" in even the "dumbest" idea!*
> *Adam Macklin, engineering student.*

> *Optimization only happens if you want it that way. It takes extra effort to get that extra plus.*
> *Sidney F. Love,* Planning and Creating Successful Engineered Designs, *1986.*

Case Study of Final Ideal Selection

The problem of "How can high schools make learning more relevant?" was brainstormed by a class of twenty-two first-year honor students in engineering. For creative idea evaluation and idea judgment, the class was divided into four teams. Each team worked with the brainstorming ideas in assigned categories to come up with fewer but more complete ideas. Then these ideas were posted on the wall; each team chose a small list of criteria and then ranked the final ideas based on these criteria. The teams and their top-ranked ideas were as follows:

Team 1—Curriculum: Strengthen the curriculum with special academic programs, including a new creative thinking class to increase practical applications and use of problem-solving methods.

Team 2—Teaching: Teachers should be tested before they are hired, not only to determine their amount of knowledge in their field but also to judge their ability to teach and convey this knowledge to others.

Team 3—Environment: Set up "career visits" to businesses and industry.

Team 4—Structure: Restructure high schools to follow a flexible, college-type class schedule and atmosphere, *coupled with a positive grading system.* (Note that the concept in italics was added after the first round of evaluation.)

These final ideas were evaluated with an advantage/disadvantage matrix to determine their weak points with respect to implementation (indicated by o in the matrix). To improve the overall acceptability and chances for success for a wide variety of students, the final idea of Team 4 was slightly modified to include the concept of a positive grading system (Idea 4b).

EVALUATION CRITERIA		IDEAS: 1	2	3	4a	4b
1. Will the solution improve student learning?	yes—no	√	√	√	o	√
2. What will the implementation process be like?	easy—difficult	√	o	o	o	√
3. Will implementation lead to change/innovation?	yes—no	√	√	√	√	√
4. What will be the costs of implementation?	low—high	√	√	√	√	√
5. How many students will be served?	many—few	√	√	√	√	√
6. Will the solution decrease the dropout rate?	yes—no	√	√	√	o	√
7. Will the solution increase college-bound students?	yes—no	√	√	√	√	√
8. Will the solution impact the disadvantaged?	much—little	√	√	√	o	√
9. Can the solution get community support?	yes—no	√	√	√	o	√
10. What is the degree of risk?	low—high	√	√	√	o	√
11. What is its effect on school morale?	up—down	√	√	√	√	√
12. Will teachers accept the solution easily?	yes—no	√	o	√	o	√

Next, these ideas were examined in light of the final idea selection questions:

Can these four ideas be combined to obtain a higher-quality solution? Career visits and a college-type schedule address different students—thus both of these ideas are ranked of equal importance. Quality teachers are also essential, as are the stated curriculum improvements (stronger academics together with a creative thinking/problem-solving class) if high schools are to make learning more relevant. These different ideas cannot be combined; they must all be implemented.

Can these ideas be implemented all at once or in sequence? The career visits will be easiest to implement. The creative thinking/problem-solving class will require teacher training, different room layouts, and some adjustments in scheduling; after-school, Saturday, or summer seminars are also options. These changes will take some time to implement. Implementation of the other two ideas will be more difficult. The college-type schedule could be tried in smaller schools first. However, encouraging teachers to use a more positive grading system should be fairly easy to do. Implementing teacher testing is a rather thorny issue—but it is beginning to be done by several states as one of many approaches to increasing the competence and quality of teachers.

How well do the ideas solve the problem? The creative thinking/problem-solving class will benefit all students. Quality teachers are also an essential prerequisite. The career visits will serve noncollege-bound students especially, whereas the college-type schedule of course will benefit the college-bound students the most. However, all four ideas in combination provide a good climate and a more complete solution to the problem.

Do the ideas meet all needs? Each idea does not meet all needs—the list of required improvements in public education is simply too long. (Many of the original brainstorming ideas addressed specific needs and should be implemented later in stages.) Implementing the four top-ranked ideas will form a foundation to build on, with further improvements possible as needed in both the academic and vocational areas.

What are the costs and risks involved? The schedule change would place more responsibility on the students; thus it comes with some increased risk. Creative thinking requires more flexibility from the teachers. But the overall risk for the country as a whole is much larger if nothing is done to improve schools— thus we cannot afford to wait. A firm determination is required to do whatever is necessary to make the schools better—there is no other priority that is more urgent. The costs of implementing these four ideas will actually be quite reasonable, and it should be possible to develop strong community support from taxpayers and parents, as well as from the business sector.

> *There is no evidence that any of the skills of critical thinking learned in schools*
> *and colleges have much transferability to the contexts of adult life.*
> Stephen D. Brookfield, Developing Critical Thinkers

Case Studies in Engineering Ethics

Engineering Times, the monthly paper published by the National Society of Professional Engineers (NSPE), contains a regular feature on engineering ethics. In "You Be the Judge" the following statement is given preceding each case:

> Although engineering is a profession of precise answers based on scientific prin-
> ciples, engineers work in the real world where business, ethical, or even human-
> relations questions have no easy answers. Many engineers have turned through the
> years to NSPE's Board of Ethical Review for impartial help in making ethics judg-
> ment calls. Do you want to try your hand at deciding a case? Below are some
> situations posed to the board. Note. It should be understood that each ethics case
> has its own answer; each case is unique. The general response in one case may not
> fit what appears to be a similar problem.

Case 1—Utility Cost Consultant

The situation: N. R. Gee, P.E., a specialist in utility systems, offers industrial clients the following service package: a technical evaluation of the client's use of utility services (electricity, gas, telephone, etc.); recommendations, where appropriate, for changes in the utility facilities and systems; methods for how to pay for such utilities; a study of pertinent rating schedules; discussions with utility suppliers on rate charges; and renegotiation of rate schedules. Gee is compensated for those services solely on the basis of how much money the client saves on utility costs.

What do you think? Is it ethical for Gee to be compensated in such a manner?

What the board said: Gee is acting ethically in accepting such a contingent contract arrangement. (For an explanation, see the February 1989 issue, p. 3.)

CASE 2—EXPERT WITNESS

The situation: X. Burt, P.E., was retained by the federal government to study the causes of a dam failure. Later, Burt was retained as an expert witness by a contractor who filed a claim against the federal government demanding additional compensation for work performed on the dam project.

What do you think? Was it ethical for Burt to be retained as an expert witness under these circumstances?

What the board said: The Board of Ethical Review found that Burt's actions were unethical. (For an explanation, see the April 1989 issue, p. 3.)

CASE 3—RECRUITING

The situation: Hook & Cook, an engineering firm, sent a form letter reciting its history and policies to all engineers at the engineering firm of Ready, Willing & Able. The letter concluded with the statement:

> We enclose for your consideration a summary of the aims and objectives of our firm, as well as the various advantages offered to those who join us. We hope that you will read it and perhaps refer to us those individuals whose professional philosophy matches our own.

The enclosure reference was to a twenty-page booklet covering the history, aims, benefits, and rules of Hook & Cook.

What do you think? Was the recruitment of engineering personnel through the method described consistent with the NSPE Code of Ethics?

What the board said: The board found that the described recruitment was ethical under the Code. (For an explanation, see the June 1989 issue, p. 3.)

✋ *CRITICAL THINKING ACTIVITY 7-11: ETHICS AND PEER PRESSURE*

Select one of the engineering ethics cases above and present a list of arguments why you think that the person involved acted ethically or not. You may come up with a different conclusion from the board's. Then check through the referenced issues of Engineering Times *and compare your reasoning with that of the board. Discuss the differences. This exercise can be done by an individual or a group.*

If you have a group of fifteen or more young people, you can conduct an interesting demonstration of peer pressure using the results of Team Activity 6-2. Have the group brainstorm a creative name for itself (get a list of at least ten ideas). Each person is then given five sticky dots to vote for the preferred names—they can distribute their "votes" in any way they wish, including giving multiple votes to one or more ideas. But before the voting starts, ask one person to vote a secret ballot instead of using the sticky dots (also with five votes). After the group has finished voting with the dots, use a differently colored marker (or dots) to show the vote of the secret ballot. Then have a discussion of the results—how did the group's vote show peer pressure compared to the secret ballot vote?

RESOURCES FOR FURTHER LEARNING

7-1   Mary F. Belenky, Blythe M. Clinchy, Nancy R. Goldberger, and Jill M. Tarule, *Women's Ways of Knowing: The Development of Self, Voice, and Mind,* BasicBooks, New York, 1986. This work is now available in paperback. It has been highly recommended to us in a creative thinking workshop at the University of Toledo for raising sensitivity to social questions and women's point of view. These researchers have found that Perry's model of critical thinking may not be valid for female students.

7-2   Richard N. Bolles, *What Color Is Your Parachute?* Ten-Speed Press, Berkeley, California, 1989. This book is updated every year and helps people evaluate their own aptitudes and interests on the way to choosing a career.

7-3   Stephen D. Brookfield, *Developing Critical Thinkers—Challenging Adults to Explore Alternative Ways of Thinking and Acting,* Jossey-Bass, San Francisco, 1988. This book shows that critical thinking is not simply an abstract, academic exercise but a productive process enabling people to be more effective and innovative in many aspects of life and work. Available by the same author is an audiotape, *Becoming Critical Thinkers: Learning to Recognize Assumptions That Shape Ideas and Actions,* Jossey-Bass, San Francisco, 1991. This tape demonstrates through interviews and examples that critical thinking is a productive process that can guide people to deal with problems more innovatively and effectively.

7-4   Eugene S. Ferguson, *Engineering and the Mind's Eye,* MIT Press, Cambridge, Massachusetts, 1992. This book examines how engineers lose touch with the real world through too much reliance on computer models and a lack of hands-on experience in their education and workplace.

7-5   Michael J. French, *Invention and Evolution: Design in Nature and Engineering,* Cambridge University Press, New York, 1988. This paperback book contains many examples of designs and products. Judgment and design decision making are taught implicitly by example; there is little mention of explicit procedures.

7-6   H. B. Gelatt, *Creative Decision Making: Using Positive Uncertainty,* Crisp Publications, Los Altos, California, 1991. This small workbook leads the reader to explore both rational and intuitive techniques to make the best decisions.

7-7   Myron Peretz Glazer and Penina Migdal Glazer, *The Whistleblowers: Exposing Corruption in Government and Industry,* BasicBooks, New York, 1989. This paperback summarizes the experiences (and the price paid) of sixty-four courageous ethical resisters and their spouses; it is a study of the values that lead people to put successful careers and families at risk to warn the public of dangerous and illegal situations.

7-8   Spencer Johnson, *"Yes" or "No": The Guide to Better Decisions,* Harper Business, New York, 1992. The fictional story of a businessman's hike up a mountain teaches important decision-making concepts: focusing on what you really need (not just want); recognizing and examining all options; thinking through decisions completely; anticipating the likely consequences of actions; and drawing on past experiences and abilities to make better decisions. It comes with a wallet-sized tool—the "decision map."

7-9   Joanne G. Kurfiss, *Critical Thinking: Theory, Research, Practice, and Possibilities,* ASHE-ERIC Higher Education Report 2, Clearing House on Higher Education, George Washington University, 1988. This report surveys theories and research into current college practices of teaching critical thinking in the area of argument skills, cognitive processes, and intellectual development.

7-10   Chet Meyers, *Teaching Students to Think Critically: A Guide for Faculty in All Disciplines,* Jossey-Bass, San Francisco, 1986. The author demonstrates why critical thinking should be taught as a part of every course; included are guidelines for class discussions and assignments that can motivate students and teach rigorous critical thought. The book also shows how faculty and administrators can create a campus climate for teaching critical thinking.

7-11   William Perry, *Forms of Intellectual and Ethical Development in the College Years: A Scheme,* Holt, Austin, Texas, 1970. This book contains the details of the Perry model, which is the basis of much current research in the area of critical thinking.

7-12   Henry Petroski, *To Engineer Is Human,* St. Martin's Press, New York, 1985. This book presents studies of famous engineering failures.

7-13   Win Wenger, *A Method for Personal Growth and Development,* United Educational Services, 1991. This sourcebook on image streaming gives step-by-step instructions on how to learn this technique as an individual and how to teach it to groups.

ACTIVITIES AND EXERCISES FOR ENGINEERS

7-1   SENSORY EXPERIENCES AND SALES AD

First, buy a fruit or a vegetable that you have never eaten before. Examine it, taste it, eat it. Use all five senses (sight, touch, smell, taste, and hearing) to describe and appreciate this new experience. Write each statement on a separate card. Note the shape, color, flaws, textures, flavor, sound-producing aspects, odor, temperature, possible uses. Draw many analogies as you go along, finding image-filled ways to describe the event. Be wildly poetic!

Next, sort the statements with the method of creative idea evaluation. Make up several categories; combine ideas within the categories and then between the categories. Use one of these improved ideas and write a sales ad for this fruit or vegetable. Would you buy this fruit or veggie based on your experience? Would you buy it based on your ad? Test this last question on several of your friends.

7-2   DISASTER—SO WHAT?

a. Suppose that while you are out of town for a relaxing weekend with your family or friends, your car with all your money, luggage, and everything is stolen. Find at least ten ways to turn this apparent disaster into an interesting, positive, or enjoyable experience.
b. Do a creative evaluation—can you engineer and integrate these ideas into one or two practical solutions?
c. Discuss the results and application of this exercise with two or three friends—will the results affect the way you will plan your trips in the future?

7-3   OBSERVATIONS AND CREATIVE WRITING

Take a walk around the block. Bring a notepad. Pause to listen—what do you hear? Analyze all the sounds, make notes. Walk a bit further. Pause to observe—what do you see? Big things, little things, near, far, up, down, back, front, changes, trends? Walk a bit further. What do you feel? The sun on your skin (what does it feel like?), the rain in your face (what does it feel like?), the wind in your hair and in your clothes? Can you touch a tree, a wooden fence, an animal? Walk a bit further. Pause to notice—what can you smell and taste? Did you find something that you had never noticed before? Now write down all the thoughts and ideas on cards and try sorting them into different categories. Combine ideas—does this help you come up with especially creative phrases? Then mix up the cards again and try a different set of categories and different combinations of thoughts. Or invite a friend to do this second grouping activity. Are the results different this time?

7-4   BRAINSTORMING AND ENGINEERING IDEAS

Brainstorm a problem from the following list, either alone or with a group. Then do a creative evaluation with the brainstormed ideas a day or two later to come up with improved ideas.
a. How can I make more time in my daily schedule for regular exercise?
b. In what way can college classrooms be improved (at low cost) to encourage learning?

c. How can we develop a habit of lifelong continuing education?
d. In what way can teachers improve the creative climate in their classes?
e. In what way can paperwork (and bureaucracy) in our organization be reduced?
f. In what way can a particular procedure (specify) be improved?
g. Design a child's playground toy that is sturdy, safe, yet completely recyclable.
h. How would you improve communications between parents and teenagers?
i. How would you explain the concept of a bicycle to a blind person?
j. What can an individual do to help the homeless?
k. How can parents teach their children goal-setting and time-management skills?
l. Come up with your own brainstorming topic—select a small but difficult task.

7-5   DIFFERENT CULTURES

It has been said that the hardest part of being accepted in a new culture is learning what not to say. In groups of four or five (preferably with a similar cultural background), brainstorm a list of things that a person coming to your community in the United States should not say. Then do a creative idea evaluation—can you simplify or condense the list to a few key concepts? Finally, compare your list with that of groups who have a different cultural background from your own.

Interview some foreign students or foreign-born people among your acquaintances and ask them for ideas about what not to say if you were to visit their native country.

7-6   *IMAGE STREAMING*

Image streaming is a technique developed by Dr. Win Wenger, president of the Institute for Visual Thinking in Gaithersburg, Maryland. By integrating right-brain and left-brain thinking, it can help improve your mental abilities. To do this exercise, you need another person or a tape recorder. For this exercise to be effective, you must talk out loud, not just think to yourself. You may also want to use a timer set at 20 minutes. Here are the steps:

1. *Close your eyes and turn on the tape recorder (or ask your friend to listen attentively).*

2. *Start describing what you "see" (blotches, patterns, images from your memory, a person, object, or scene from your past). Describe all aspects of the image: smells, sounds, colors, feelings of texture, temperature, whatever sensory information is attached to the image. Since visualizing, intellectually analyzing, and speaking aloud all use different parts of the brain, this exercise does what is called "pole bridging" in your brain.*

3. *Continue to follow your image streams with rapid talk until the time is up. Remember to report everything that comes up, even if you think a particular impression is not important. Look for as much detail as possible.*

4. *During or at the end of a session, develop a humorous interpretation for the "messages" that have come to mind, if you can.*

Practice this technique on a regular basis, in sessions from 10 to 30 minutes; this will make it easier for you to access your right brain during problem solving for new and useful creative ideas. Image streaming is closely related to an ancient method of learning. Socrates, through asking questions, would cause his students to examine their inner and external perceptions; they had to describe what they found. Through this technique, they gained understanding and personal growth, the mark of true education.

> *It is impossible to go through life without making judgments about people.*
> *How well you make those judgments is critical to the quality of your life.*
> *Before you judge someone else, you should judge yourself.*
> *M. Scott Peck, MD, psychiatrist and author.*

---

*Asterisks denote exercises for more advanced learning.

ACTIVITIES AND EXERCISES FOR JUDGES

7-7 CHECK YOUR ASSUMPTIONS
Ann and Barnaby are found dead on the living room floor in the middle of a pool of water and broken glass. Write a story of what happened. (Hint: Examine your assumptions about the facts given.)

7-8 FABLE
First, examine one of Aesop's fables and analyze the moral value that is being taught. Next, write your own fable.

7-9 VALUES
Think about a decision that you have made during the past week. Then write a letter to your best friend, explaining the reasoning and values you used to make the decision.

7-10 FAILURE
Imagine that you are a senior citizen giving a talk to a group of high school students. What would you tell them about the value of failure? Include a funny story or two (true or invented) about your personal failures in school (or in life) and what you learned from them. Especially relate how the failures helped you develop good judgment.

7-11 *THE POWER OF THE TELEPHONE*
Do not answer the phone the next two times it rings. How does this make you feel? Was it difficult to do? How much power does the phone have over you? How would you reason with someone who thinks that an emergency call might be missed if the phone is not answered? What strategies could be used to let you decide when you want to talk or when you do not want to drop what you are doing to answer the phone?

7-12 NEGATIVE AND CONSTRUCTIVE CRITICISM
Here are some questions that require reflection before answering. What is constructive criticism? Why is negative criticism the opposite of good thinking, according to Edward de Bono? What technique should you use when you must criticize someone? How can you overcome a habit of criticism—let's face it, it does give a nice (but only temporary) feeling of power. Can you think of other ways of dealing with people that will make you feel good and at the same time will have much better long-term results in maintaining good relationships and in getting other people to do their best to please you? In what way do critical thinkers have open minds? Explain the difference between critical thinking and criticizing ideas and people in a negative way. Compare your answers with those of two or more colleagues or friends. How was your interaction—supportive or negatively critical?

7-13 HOW TO CRITICIZE
Make up a scenario in which you have to criticize someone. Write it in such a way that you start out with two positive statements. Then make a wishful statement about the item you want to change, followed by another positive statement about the other person. Then conclude with a hopeful, cooperative, positive statement.

EXAMPLE:
Critical statement: Ugh, you smell awful; why can't you quit smoking!
Better way: I appreciate your visits—you have a way of cheering me up. And it is so thoughtful of you to take off your sneakers when walking across my nicely buffed floor. I wish you could take the same care with your health and quit smoking. I bet this could even increase your endurance—you might win the marathon next time! Let's make a pact for mutual support and encouragement—I'm willing to give up snacking on junk food; this way we'll both be winners.

7-14   How to Accept Criticism

It is easy to just feel dejected. It is normal to be put on the defensive when receiving criticism. It is abnormal *to look at the criticism as an opportunity for self-improvement. Be abnormal! Think of a situation recently when you were criticized. But instead of thinking of defenses or feeling hurt, place yourself "outside" the situation. Analyze the criticism. Was there a basis for it? What situation brought it about? Did it hit home? What should you change to avoid this situation in the future? If the criticism is unjustified, mentally write it on a piece of paper, then imagine throwing it in the trash (or down the toilet). Then let the matter rest.*

7-15   *Judging a Television Program*

a. Alone or with a group, watch a television program. Record it on a VCR for a later rerun. After the program ends, judge it quickly on the basis of positives and negatives. Write down these judgments.

b. Develop a set of thoughtful criteria for judging TV programs (including the news). Also consider your values — on what are you basing your criticism or choice of criteria? If you can, involve people of different age groups when you make up the list of criteria.

c. Now run the taped program again and judge it using the list of criteria. Is your judgment different this time? From working out the list of criteria, did you gain some insight into what makes a "good" program and which programs are just a waste of time (or, even worse, garbage for your subconscious mind)? Are there differences in the criteria based on age, or are there universal criteria?

d. Why do you like your favorite programs? Do you feel guilty when your viewing does not include many educational programs?

7-16   The Power of Visual Images

a. Compare the power of the visual image on television versus the spoken message. For example, radio listeners thought that Richard Nixon had won the presidential debate based on his grasp of ideas and issues, whereas TV viewers gave Kennedy the win, based on his projection of a calm demeanor that they thought would make a better president. How much of the effect is due to the substance of the ideas discussed; how much is due to staging (angles, color, surroundings, makeup) that can make people look better, stronger, more trustworthy?

b. Watch a current television debate on some issue. How much control is exerted by asking leading questions?

c. Watch a news program. In what way could the presentations be slanted because the message is mostly visual? Who decides what is shown in a 30-minute newscast, and how much time is allotted to each subject? What are some biases that influence television programming in general and the contents of an individual broadcast? How much time do you think is given in a typical news report to the context? Use a stopwatch and take some data to answer this question.

7-17   Personal Values

a. How would you explain to a 5-year-old child that taking a candy bar in a store without paying for it is wrong? Would your explanation be different if the child were your brother, a friend of your brother's, or a stranger? Why or why not? What if the child were a teenager?

b. How valuable is a good name? Brainstorm this question with a group of your peers and with a multigenerational group. Do the answers come out differently, or is there a common ground? What values are being expressed by the participants in the discussion?

c. Discuss who gets hurt when students cheat on an exam because they did not make the effort to thoroughly learn the material. What are the consequences if you fail to practice new skills?

d. Find examples of people who have overcome handicaps or personal tragedies. How did they do it? What inner resources do they have? How did their beliefs change because of these experiences?

7-18 *Cultural Values*

Surveys have found a conflict between personal and cultural values. Trial by a jury of one's peers is considered to be an important value in our democratic culture, yet young people are increasingly unwilling to serve on a jury. What personal values do you think these young persons have that conflict with the cultural value? How are freedom and personal commitment related? What values must a democratic society have to survive? How important are hard work, discipline, respect for law and order, service, tolerance, and honesty to the survival of democracy? How prevalent and respected are these values in our society today? What values undergird a caring community? What are some important values in a society dominated by scarcity? What are important values in an affluent society?

7-19 Pluses and Minuses of Computers

How would a broad shift to computer-based education influence societal values?

7-20 Five-Minute Decision-Making Exercise

With a group, do a 5-minute decision-making exercise, using a timer.
Use 1 minute to choose and define a target (problem definition).
Use 2 minutes to expand and explore the problem (idea generation).
Use 2 minutes to contract and conclude the problem (idea evaluation and judgment).

7-21 Whom Do You Admire Most?

Name a person whom you admire very much. Why? Do you want to be like this person? What changes would you have to make in your life-style to become more like this person? What patterns of behavior would you need to establish that you don't already have? What bad habits would you need to break? What would you need to accomplish in the next year?

7-22 *Evaluations of Conflicting Opinions*

Find newspaper or journal articles that give two opposing points of view on a certain subject. Give a brief summary of each; then indicate your agreement or disagreement with the expressed views. Support your viewpoint with additional facts or point out where the writers should have supplied more information for their opinions.

7-23 Applications of Critical Thinking

Practice critical thinking when evaluating television and other commercials, printed ads, music, TV programs, magazines, newspaper articles, and political speeches and "buzz" words. For example, do "pro-life" and "anti-abortion" have the same meaning? Is "pro-choice" a good word to use for someone talking about abortion rights, or does it disclose a fallacy in thinking—should a "choice" about responsible behavior be made at an earlier point in time (choosing abstinence, or deciding to use protection before sex or being mature enough to take on the consequences)? Is freedom always coupled with responsibility in a democratic society (and in what way)? In a group, discuss and answer questions such as: Does everyone have the right to drive a car, or is it a privilege? Who decides what is right—is there such a thing as absolute truth? Do the ways the questions in this problem were posed reveal the underlying values and bias of the authors? Is it possible to do value-neutral teaching?

7-24 *Fraud in Science*

In a November 1991 survey of 1500 scientists conducted by the American Association for the Advancement of Science, more than a quarter of the people responding indicated that they had personally encountered dishonesty in scientific work, such as falsified data. What is the difference between sloppy science and fraud? Do a library search and report on one of the following cases: the recent cold-fusion controversy; John Darsee, heart researcher at Emory University and Harvard; William T. Summerlin, cancer researcher at the Sloan-Kettering Institute; Cyril Burt, British psychologist; Robert A. Millikan, 1923 Nobel prize winner; John Dalton, chemist; Gregor Mendel, geneticist; Johann Kepler, astronomer.

CHAPTER 7 — SUMMARY

Your Role as Engineer: Group, sort, organize, build on, develop, integrate, engineer, and synthesize ideas to create good solutions to the original problem.

Creative Idea Evaluation: It has three rules: (1) Quality, not quantity, counts. (2) Make the ideas more practical. (3) Judgment is still deferred until later. The process is done in three steps: (1) Grouping the ideas into categories. (2) Developing and synthesizing ideas within the categories. (3) Force-fitting ideas between categories. Wild ideas are stepping-stones to further creative thinking.

Your Role as Judge: Use a critical, conscious mindset together with positive, creative thinking to decide which ideas are best. Look ahead and consider the impact of the solution; assess values and bias. It takes experience with failure to develop good judgment. Judges also have to deal with uncertainty, risk, and ethics.

Ranking Ideas: Develop a list of criteria. Criteria are standards used for judging; they are best developed through brainstorming with a team and should consider such factors as motivation, people, cost, support, values, time, and consequences. A number of judgment techniques are available to help judge and rank ideas.

Decision Making: Traditional forms are mostly analytical. Cultural values influence attitudes in decision making (i.e., cooperative or adversarial). Creative decision making sees uncertainty as positive. Features are: (1) Be focused and flexible about goals. (2) Be wary and aware about knowledge. (3) Be objective and optimistic about beliefs. (4) Be practical and imaginative about actions. Use all six creative problem-solving mindsets to make the best decisions!

Critical Thinking and Creative Problem Solving: Traditional teaching focuses on teaching students argument skills. It is concerned with the cognitive processes; students construct meaning out of the knowledge and procedures learned in particular disciplines. Critical thinking can also be modeled as a process of intellectual development. The model developed by Perry and modified by Belenky and associate researchers consists of four stages: **1. Authority**—Answers are either right or wrong; and there is only one right answer. Students are intolerant of differing views and cannot recognize bias in knowledge. **2. Own Opinion**—Many viewpoints can be valid; knowledge can be gained through listening to the "inner voice" and imagination. **3. Context**—"Truth" depends on its frame of reference. Arguments can be supported through discipline-specific reasoning. **4. Mature Decision**—Rational thinking, caring, intuition, and values are integrated and result in commitment—the ultimate goal of education.

In the traditional college curriculum, less than half the students progress beyond the second stage. Each stage can be associated with steps or specific mindsets in the creative problem-solving process; thus practicing creative problem solving should help students develop critical thinking skills. Critical thinking as applied in real-life situations is a whole-brain thinking process. Instructors can provide supportive climates to enable students to progress to higher stages in critical thinking; they can share information about their own growth, tentative views, change, and values.

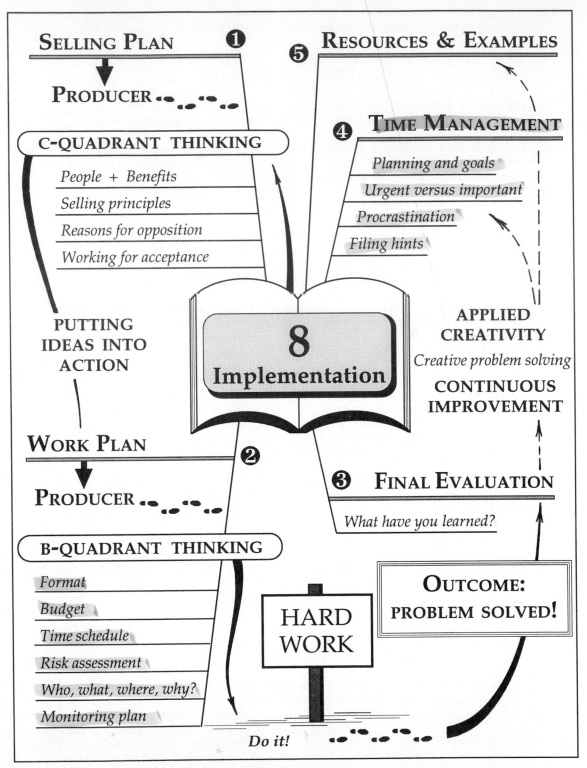

SELLING PLAN ❶

PRODUCER

C-QUADRANT THINKING

People + Benefits

Selling principles

Reasons for opposition

Working for acceptance

PUTTING IDEAS INTO ACTION

WORK PLAN ❷

PRODUCER

B-QUADRANT THINKING

Format

Budget

Time schedule

Risk assessment

Who, what, where, why?

Monitoring plan

8 Implementation

HARD WORK

Do it!

❺ **RESOURCES & EXAMPLES**

❹ **TIME MANAGEMENT**

Planning and goals

Urgent versus important

Procrastination

Filing hints

APPLIED CREATIVITY

Creative problem solving

CONTINUOUS IMPROVEMENT

❸ **FINAL EVALUATION**

What have you learned?

OUTCOME: PROBLEM SOLVED!

Mindmap of Chapter 8. Refer to Chapter 2 (pp. 55-56) to learn more about mindmapping.

8

SOLUTION IMPLEMENTATION

WHAT YOU CAN LEARN FROM THIS CHAPTER:
• Putting your idea into action— the role of producer.
• Step 1—selling your idea: strategic planning and working for acceptance.
• Step 2—the work plan: who does what, when, and why; schedules and budgets; risk assessment.
• Step 3—implementation monitoring and follow-up; final evaluation.
• Time management—a cure for procrastination and other time wasters.
• Resources for further learning: case studies, references, exercises and activities for producers, and summary.
   Review of Part 2—critical judgment and personal application.

> The most difficult thing in the world is to put your ideas into action.
> Johann Wolfgang von Goethe, German philosopher and poet.

PUTTING YOUR IDEAS INTO ACTION—
THE PRODUCER'S MINDSET

We have now reached the final step in the creative problem-solving process. Earlier, we defined a problem in the detective's and explorer's mindsets; we generated interesting ideas on how to solve the problem in the artist's mindset; then we worked to make these ideas more practical in the engineer's mindset, followed by the judge's mindset when we evaluated and ranked ideas and then made decisions on which solution to implement to solve the original problem best. Do you think what we have done so far was difficult? Perhaps it has been difficult since creative problem solving with its advanced levels of critical thinking may have been a different process from what you are used to doing when solving problems. When you have more practice, the first four steps are usually quite a bit of fun. It is in the last step, implementation, where most of the hard work comes in. This is where you become the producer; this is where you need your energy, your persistence, your careful planning and self-discipline as well as your interpersonal skills—your quadrant B and C thinking. The producer carries the idea from the what-if stage to action—ideally to a successfully solved problem.

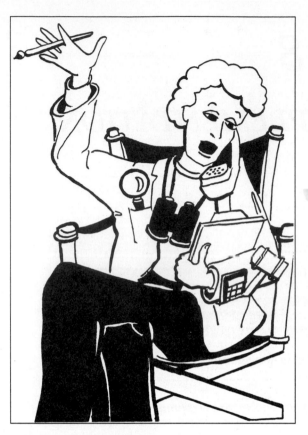

Figure 8-1   The producer.

The producer receives the pay-off for the creative problem-solving process. As detectives, explorers, artists, engineers, and judges, we did much thinking and "playing" with ideas. In the producer's mind-set (Figure 8-1) we take action—well-planned action. What are some characteristics of producers? Producers are managers; they have something to fight for. They follow a strategy, they are persistent, and they are good communicators. As paradigm pioneers, they have a great deal of courage. This word is related to the word "heart." Producers "take heart," are optimistic, do not give up. We will see how this attitude will help us implement our ideas—implementation is time-consuming. Although quadrants B and C thinking are emphasized, producers use the entire creative problem-solving process—the whole brain—since implementation is a new, unstructured problem. Why would red be an appropriate color for the producer's mindset?

Dr. Robert Warner, Dean of the School of Information and Library Studies at the University of Michigan and formerly in charge of the National Archives in Washington, D.C., said this in a speech to honor students on the Dearborn campus:

> Dream the impossible dream; if something is important enough to you, take risks and get involved, even if the odds are against you. Pick a good cause, work hard with the right people, do the things that will get you growing support, and with a bit of luck, your impossible dream will become reality.

Idea implementation is not easy; it takes careful preparation and hard work. People come up with ideas all the time, but most ideas never get anywhere. How often have you had a really good idea—yet nothing was done? This probably happens more than once a day. It is quite easy to have ideas; it is much harder to get the ideas implemented because implementation plainly takes much effort. For many people with strong quadrant D thinking (and thus many innovative ideas), quadrant B is the least preferred thinking preference; thus the organization, structure, and attention to detail required for implementation are especially difficult to do. Idea implementation has to be carefully planned; the planning cannot be skipped, and you cannot rush through it, or you will not get good implementation.

People seldom hit what they do not aim at.
Henry David Thoreau, writer, engineer, naturalist.

But even with the most careful planning, do not expect everything to go perfectly and smoothly. You will be dealing with people and change, risk and uncertainty, and thus you can expect opposition. We will look at some strategies we can use to overcome opposition and work for acceptance. Opposition is natural, because when you implement an idea—especially if it is an innovative idea—you are asking people to change. People often see change as threatening—or expressed more humorously, "The only person who likes a change is a wet baby." Idea implementation is difficult also because by this time, you (and any other people closely involved in the process) may be sick and tired of the whole problem; you are exhausted, you have run out of physical and mental energy and just want to get the whole thing over with. But it is also possible that you are so excited and eager about implementing your solution that you may skip the required care in tactical planning and support building. Implementation is *the* place for quadrant B people—since they are concerned with safekeeping, they carefully plan to minimize risk.

The first task that you need to do to implement your idea or solution is to analyze which portions of the process will be your responsibility and which parts will be done by other people. In other words, you have to plan your strategy. Will you have to convince others of the benefits of your ideas, so they will accept, fund, and implement your solution? Or will you and your team be in charge of the entire implementation process? You will have to check that the solution is working as designed and also evaluate the creative problem-solving process and what you have learned from it. We will discuss each of the three responsibilities in turn.

SELLING YOUR IDEA

Gaining acceptance for your idea involves careful planning. Analyze your targeted audience; prepare a list of benefits; develop a strategy on how to make an effective sales presentation. Use the whole brain, but concentrate on quadrant C as you develop your selling strategy and apply good communication skills.

Analyze the context—develop your selling plan: What will be your plan of attack for implementing your idea or solution? This depends on your situation. Do you have an innovative idea that could help your organization solve a problem? In this case, your primary objective is to sell your idea to management. Or did you and your team work out a solution to an assigned problem by going through all the stages of creative problem solving? In this case, you and your team may have the responsibility for completing the project and organizing the entire implementation yourselves, and your biggest problem may be getting the affected people to accept the solution. Are you an inventor of a creative idea or product who has to convince someone to bankroll the new venture? Perhaps you have to enlist the help of a spouse or parent to get your idea implemented. In all these cases, your primary goal is convincing your audience of the benefits of your idea. You must determine who your audience is and what these people may want from you. What would be the best approach to use? Chapter 11 on communications in Part 3 of this book gives helpful hints on how to plan an effective approach.

Why do you need a selling plan? The selling plan will prepare you to make a sale. You can have the best idea in the world, but if it is poorly presented, the result could be "no sale." Conversely, a good presentation often sells an idea that is only fair. We have all seen cases of mediocre products that are well presented and are thus finding ready acceptance in the marketplace. We can also find some examples in recent political races. That is why a good selling plan is very, very important. You need to know what you are selling, you need to know some selling strategies, and you need to employ some techniques that make for an effective presentation.

The list of benefits: The central part of your selling presentation is the list of benefits. What benefits can be reaped by implementing your idea or solution? What unsatisfactory situation will be remedied? And even more importantly—how will these advantages directly benefit your audience? Brainstorm as many direct (and indirect) benefits as you can think of. Look back to the list of judgment criteria and the results of the final evaluation matrix (if you used one); these are good starting points for bringing out the strengths of your idea. If your team went through several rounds of creative idea evaluation with the Pugh method, you will all know why your idea is the best possible solution. Cast these benefits into words that will be easily understood by your audience—take their thinking preferences into account. Having a list of benefits can also serve for self-motivation, to encourage you not to give up in case implementation runs into resistance or other snags. Also try to imagine and address possible arguments that may be brought up by people opposing your idea. Write down your objectives and those of your audience. What is your idea supposed to accomplish? What good will it do? What will motivate people to accept the idea? Ask yourself critically: What are these people's perceptions about your idea—will you need to correct misunderstandings or can you build on existing expectations?

Principles of selling: Let's look at some general principles of selling. They are applicable whether you are selling to a large group or an audience of just one person. But keep in mind that selling is not a one-shot deal—consider it in the whole context of gaining acceptance and overcoming opposition. A good part of your success will also depend on the kind of person you are—your character, your reputation, your integrity, your good name. Never jeopardize these to get a quick sale. Do not look at "making the sale" as winning a battle of wits; think of it in terms of building long-term interpersonal relationships. Contextual thinking is important—whether you are selling an idea or a product, the "buyers" are very much interested in the associated service that they expect you to provide. Table 8-1 summarizes some selling principles you may want to keep in mind.

When selling something, present both sides and act modestly and a bit doubtful.
Benjamin Franklin.

Table 8-1
Principles of Idea Selling

1. DON'T OVERSELL.

   Be moderate. Don't be arrogant; don't exaggerate. Listen to what your audience has to say. As soon as you have made the sale, stop. If you keep talking, the "buyer" may change his or her mind!

2. DON'T GIVE UP TOO SOON.

   Selling may take longer than expected—thus don't give up too soon. It took Columbus two years to persuade Ferdinand and Isabella to let him have his ships. It took Chester Carlson seven years to sell his idea of the Xerox process. If you have a worthwhile idea, persist! Try a different approach—it just may succeed.

3. WATCH YOUR TIMING.

   Don't be impatient; watch for the right moment. Don't try to sell an idea to people when they are tired, annoyed, or unwell. Don't approach the boss just after he or she has had some bad financial reports (unless of course the idea happens to be the perfect solution to that problem—then that would indeed be auspicious timing). Also be sure *you* are feeling well. At work, midweek is usually better than Monday or Friday; mid-morning or midafternoon is better than just before lunch or quitting time.

4. BE BRIEF AND TO THE POINT.

   Chapter 11 gives an outline on how to prepare an effective 30-second message. People are busy—30 seconds may be all you have to capture their interest and sell your idea. Boring your audience may be fatal to your sale.

5. PLAN YOUR PRESENTATION CAREFULLY—USE VISUAL AIDS.

   This way, people will be able to visualize and remember your ideas better. Check that the visual aids are of good quality and easy to understand; also check that all your equipment is in working order.

6. MAKE YOUR IDEAS EASY TO ACCEPT.

   Emphasize the benefits that your idea and the accompanying changes will bring to the people in your audience. If appropriate, describe in some detail *how* the implementation could be done. Have someone who is supportive of your idea in the audience or in your organization.

7. AVOID CONFRONTATION AND CONTROVERSY.

   Expect opposition. Don't argue. It is better to stop the discussion on good terms and come back again another day. Also consider whether you would be willing to compromise in order to get at least some part of the idea accepted.

There is only one way under high Heaven to get anybody to do anything.
And that is by making the other person want to do it.
Dale Carnegie.

Dealing with opposition—working for acceptance: As you develop your selling plan, be prepared to deal with opposition diplomatically. You must not be surprised if you run into opposition to your ideas at some point in the implementation process—you can expect it, especially if you are working to implement a very creative idea! A person opposing your idea can have any number of reasons. Table 8-2 summarizes possible reasons for opposition and presents ideas on how to develop acceptance.

The following points have also been found to be helpful for building easier acceptance of new ideas:

1. IDEAS ARE "TRYABLE"—The implementation and benefits can be demonstrated with a small pilot program or in a test market.

2. THE INNOVATON IS REVERSIBLE—If the idea doesn't work, it is possible to go back to the way things were done before, without a lot of hassle and expense.

3. THE IDEA IS DIVISIBLE—The idea can be implemented in easy steps or stages that require only small changes each time.

4. YOU CAN BUILD ON WHAT IS ALREADY THERE—The idea represents an improvement, not a radical change; it maximizes previous investment and relates to things people are already familiar with. It fits in with the culture and explicit long-term goals of the organization.

5. THE IDEA IS CONCRETE—The idea is something that can be visualized, that is real, that can be seen, felt, tasted, and smelled. The idea will have tangible results, not vague benefits, and it will have strong publicity value and good potential for increased status for the people who will be involved.

If none of these favorable conditions and approaches apply, investigate whether you can do the project on your own. Do it on the side, unobtrusively. Don't seek the limelight until you have your project running well. Many successful small companies have taken this approach; they were able to develop a sizable market share before they were noticed by their large, powerful competitors.

 TEAM ACTIVITY 8-1: OVERCOMING OPPOSITION

In a team of four, select one of the reasons for opposition listed in Table 8-2. Pick a good idea (preferably from an earlier brainstorming and creative idea evaluation exercise). Then prepare a small skit in which two team members are trying to sell the idea to the other two who adopt the selected opposing mindset. Then switch roles and repeat the exercise, with the same or a different reason for opposition. Finally, evaluate the results of the exercise—how easy was it to put yourself into an opposing point of view and come up with positive arguments and responses to sell the idea?

Sometimes the most important person who has to accept change is you! Or you may be in a position where you are the lone champion of a novel idea. Use the list in Table 8-3 for encouragement and self-motivation. Then give it your best shot. Get on your feet—you are the producer who gets the action rolling! If you have a habit of procrastination, study the section on time management in this chapter.

Table 8-2
Reasons for Opposition and How to Gain Acceptance of Ideas

---

1. LACK OF UNDERSTANDING; NO DIRECT BENEFITS.

   Criticism may merely be someone saying "I don't understand what you are doing and why you are doing it." *Make sure you explain your ideas in terms your audience can understand. Be empathetic.* Your idea might not have much in it that directly benefits your opposition. Many people first ask: "What's in it for me?" *Consider your audience's priorities.*

2. YOU ARE NOT FOLLOWING THE RULES.

   People—especially quadrant B thinkers and organizations—like things to go on in their customary ways. They have not yet learned to use imagination. *Have an attitude of sympathetically agreeing with your audience and their legitimate right to voice their concerns. Listen carefully to their arguments; they may bring out ideas that you can combine with yours to make it "their" project. Present your implementation plan and budget in a structured format. Get others involved in the problem-solving process right from the start; do not try to do everything yourself. Emphasize the mutual benefits. By all means avoid being patronizing. Positive thinking and enthusiasm can be contagious!*

3. PREJUDICE AND OTHER LOYALTIES.

   People may not have anything against your ideas, they just don't care about you perhaps because of subconscious prejudices. It's simply a fact of life—an extra barrier to be overcome. If you are young or otherwise an outsider, you can count on this type of opposition. *Can you find some common ground? Or can you find champions in the "in" group, one at a time?* Also, people may have other loyalties or commitments. You may have a better idea than your colleague; but if the boss is his girlfriend or his grandmother, she may support his idea in preference to yours. *People skills are important; sometimes it takes years to build a foundation of trust, as American companies that want to do business with organizations in the Far East have discovered, for example. Do not take the opposition personally; answer objectively and do not argue back; be calm and friendly. Above all, do not bully, counterattack, or ridicule the opposition. Good-natured humor (at your own expense) can defuse hostile moments.*

4. IMPLIED CRITICISM AND FEAR OF CHANGE.

   People take your idea as criticism that their ways are no good. *This is an important stumbling block in an organization and requires sensitivity and diplomacy. Here, an established policy that encourages an attitude of continuous improvement helps.* People may fear your idea will bring change and a loss of status, influence, or position. *Do not underestimate the force of this factor. Try to anticipate these concerns in your planning and presentation—acknowledge that these points are valid and deserve to be taken seriously. Concentrate on the positive aspects of your idea; emphasize how the idea will benefit the person as well as the organization, even though there will be a short time span when things may be more difficult (until everyone has become accustomed to the improved conditions).*

Table 8-3
Checklist for Self-Motivation and Lone Champions

___ *I have a good plan of attack.*
___ *I can be proud of my past successes.*
___ *There is a big potential payoff.*
___ *I'm getting encouragement from one or more supporters.*
___ *I believe in myself.*
___ *I have faith in my idea—it's a great idea!*
___ *I have an alternate plan B to fall back on.*
___ *I have to succeed—there is simply no acceptable alternative.*

Whether you think you can or can't, you're right.
Henry Ford.

As a producer, you must overcome pessimism. Many people—even young people—get discouraged when they look at the problems in the world. They feel helpless and powerless to do anything about these conditions. They ask: "What can one person do anyway?" You need to look for positive examples. Read about people who are champions—they all started alone, with one idea. Look around and notice what is good in your community. How did those things get established? Many times the answer will be "because someone had a good idea and sold the idea to others who had the means to get it implemented." You can find examples all around you. Here is an illustration from our personal experience.

Case Study: Exercise Trail

As a family with four young children we were vacationing in Switzerland and really enjoyed the exercise trails that were the rage over there twenty-five years ago. Each community had its own trail, and people were out at all hours of the day and night exercising. When we returned home to Brookings, South Dakota, one of our children asked: "What if our town had one of these trails?" Since these trails were usually laid out in a forested area, shade would be a problem—summers in South Dakota are much warmer than in Switzerland. Also, a major life insurance company was sponsoring these trails by furnishing signs and some equipment. We assembled the snapshots we had taken of our family exercising at various "exercise stations" and took them to the local parks and recreation department. To our considerable surprise, the idea was very well received. We then obtained the materials about the trails from the sponsoring insurance company in Switzerland. Because the instructions for each station's exercises were copyrighted (and in German), we translated the material and made new sketches to go with the instructions. A local artist improved these sketches, the 3M company donated a strip of land which included a wide windbreak for shading, and the National Guard did the actual construction of the trail and exercise stations. Thus the community had its exercise trail—one of the first in the United States—at little cost.

There is a sequel to this story. A few months later, just two weeks after we moved to Tennessee, we read an article in the daily paper by a columnist who was reporting a conversation he had with a lady, a native of Switzerland, about the exercise trails there. We contacted the columnist and asked to be put in contact with this lady. We met a delightful middle-aged Swiss couple. Since the men were rather busy, the women decided to work together to try and sell the idea to the city of Knoxville. Well, there it got hung up in red tape, but people in a community nearby heard about it. So one evening, the two housewives (with shaking knees and pounding hearts) gave a presentation about these trails to the Oak Ridge City Council. Thus it came about that a year later that community had its own safe trail located in a lovely park adjacent to the police department. And the two families became lifelong friends in the process! So don't be too timid in trying to sell your ideas—the results can surprise you, even if the outcome may not be exactly what you set out to do!

> ⌚ *TEAM EXERCISE 8-2: SELLING PLAN*
> *For the final "best" solution from the judgment phase, prepare a list of benefits and a selling strategy. Think about your targeted audience and the presentation format that will be required. What potential opposition will you have to address?*

In a well-organized system, all of the components work together to support each other.
W. Edwards Deming, quality expert.

THE WORK PLAN AND IMPLEMENTATION

As we have seen in the preceding section, selling the idea may be the only task that you need to do as a producer. In this case, the job of setting up a work plan and doing the actual implementation will be someone else's responsibility. But what if you are that person who is given the task of implementing the idea or solution? Or you may have been the person in charge right from the start of creative problem solving and will have to see the problem through to the end. Or else, you may have the opportunity or the job of implementing an idea someone else has sold to you. Thus, how do we go about the task of preparing for an actual implementation? The predominant thinking preference here will be quadrant B, because the work plan maps out the exact steps needed for implementation—who does what, when, and why. You also address the prevention of possible failure; you prepare time schedules and cost budgets. But since implementation is an unstructured problem, you must be prepared to use the other five mindsets to consider alternative ideas, the context, and the people interface.

We will discuss a variety of work-plan models in a moment, but first we want to examine a simple example given in Table 8-4. Let's say you seriously want to change a bad habit. You will not accomplish your goal unless you adopt a "battle plan." Here is an illustration that we have adapted from a plan to stop smoking recommended by the American Cancer Society.

Table 8-4
Example of a Work Plan: "How to Stop Smoking"

1. Why do you want to quit smoking? List as many reasons as you can think of in support of a smoke-free "you." What are the obstacles—what are some reasons for smoking that you will need to overcome? *To sell the plan to yourself, you need a list of benefits and an analysis of the opposition you will have to overcome.*

2. Change to a low-tar, low-nicotine cigarette and select a Quit Day two or three weeks in the future. Mark the day on your calendar! *You are preparing your mind for action.*

3. Chart your smoking habits for at least two weeks. Write down how many cigarettes you smoke a day and when you smoke them. Go over this list and rank which cigarette you think is the most important or desirable to you, such as the one with morning coffee; the next most important one; and so on, down to the least important. *Habits are easier to change when you bring them from the subconscious mind to undergo critical judgment.*

4. Eliminate one of the cigarettes you routinely smoke. It may be the most important one, or the one in the middle of your list, or the least important one. Secure a supply of "oral substitutes"—mints, gum, ginger root, dried fruits and nuts, raw vegetable sticks, popcorn, or even mouthwash—and use them instead of reaching for a cigarette. *You are beginning to put your plan into action with an easy first step.*

5. Repeat each night, at least ten times, one of your reasons for not smoking, from your "list of benefits." Visualize a healthier, more attractive "you." *You are maintaining your motivation and programming your subconscious mind to support your conscious efforts.*

6. Quit on Quit Day. Try different substitutes as the urge to smoke recurs. Enlist a friend or family member in a series of busy events, such as going to the movies or theater, playing tennis, or taking several long walks. Keep reminding yourself, again and again, of the shocking risks of cigarette smoking to yourself and to those you love. *Strengthen your motivation by looking at the long-term benefits and consequences.*

7. Set up a daily or weekly reward structure for success—treat yourself to something nice that you can now afford to buy with the money you saved by not smoking, or make a contribution to a worthy cause. If you give in to temptation, that is no reason to give up your plan. Analyze the cause of the failure; learn from it; eliminate the contributing factors; forgive yourself, and start again. Can you strengthen your support system? Can you recruit others (perhaps including medical help) to encourage you? Can you make this a true team effort with family members and friends? *Believe in yourself—you have a good plan and you will succeed!*

Implementation Strategies

The purpose of a work plan is to make sure that the idea or solution will be put to work—that it will work right and be on time as well as within budget. Several different procedures are available for the work plan; the complexity of the problem will determine which approach should be taken. For example, a very complicated implementation involving many people and tasks should employ a PERT chart for planning and progress monitoring; moderately complex problems can be handled with a flowchart, and simple projects will need nothing more than a time/task analysis or answering the 5-W questions. Depending on the time available, you may want to develop or adapt your own method, or you may well decide to copy an approach that worked successfully for you in the past for a similar problem. When you are dealing with a complicated, risky, expensive implementation, you may want to include a risk analysis as you prepare your work plan. Consider all the resources that are available to help make a successful implementation.

IMPLEMENTATION STRATEGIES

Copycat
5-W Method
Time/Task Analysis
Flowcharts
PERT Charts
Budgets and Time Schedules
Risk Analysis

If your idea is similar to one that has been successfully implemented before, you will save a lot of time and trouble if you can just copy the procedure, maybe with some minor adjustments. Be a **copycat**!

The **5-W method** asks the questions "who, what, where, when, and why." It is not only useful in problem definition and during brainstorming, it can also be quite efficient for identifying the tasks of a simple implementation. Begin by listing the required tasks, then ask a series of questions about each task. For example, ask: Who will do this task? What will they do? Where will they do it? When will they do it? Answer each of the questions as specifically as possible. Next, ask "why" for each of the questions. Why should these people do this particular task? By asking "why," you provide a reason or justification for each action to ensure that no major activity is overlooked. Finally, give or obtain the go-ahead for the implementation and set the plan in motion. The storyboard format can be employed very nicely with this method. Note that the "how" is not being specified. If you feel uncomfortable about letting people make their own decisions on how they will do their assigned tasks, try to restrict your directions to broad outlines only. Most likely you will get much better performance and cooperation when you let people do their own thinking and decision making about their jobs—avoid "micromanaging" the project. However, when it is critical to the project that certain procedures be followed, be sure to specify these and explain why.

Time/task analysis is one of the simplest work plan formats and visually presents the time requirements of each implementation task. Typically, every task that must be completed to implement the idea is listed in the left-hand column of a lined chart, with the time scale across the top. Then the time required to complete each task is estimated as accurately as possible, as well as the target date for completion of the tasks. These estimates must be realistic; it is a common mistake to underestimate the time required to complete each task. For each task, a time line is drawn from starting date to the projected completion date. The chart clearly shows simultaneous or overlapping activities. An example that was used in a research project proposed to the Department of Energy is shown in Table 8-5.

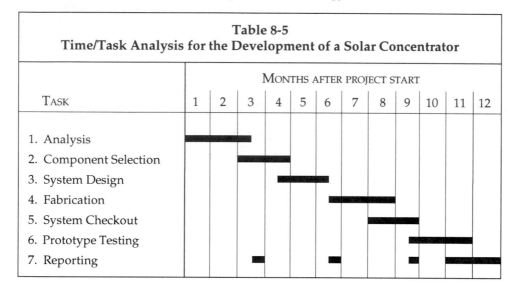

Table 8-5
Time/Task Analysis for the Development of a Solar Concentrator

	Months after project start											
Task	1	2	3	4	5	6	7	8	9	10	11	12
1. Analysis												
2. Component Selection												
3. System Design												
4. Fabrication												
5. System Checkout												
6. Prototype Testing												
7. Reporting												

Flowcharts visually present all the activities that must be performed for idea implementation in a sequential arrangement. The basic elements of the flowchart are activities (designated by rectangles), decision points (indicated by diamonds), and arrows. The flowchart is then used as a guide for implementing the idea. Time estimates for each activity may be added to the flowchart. Flowcharts are useful for showing simultaneous activities (through different branches) and prerequisites.

The **Program Evaluation and Review Technique (PERT)** method is a plannning tool as well as a progress monitoring tool for large, complicated engineering projects. It consists of events (start or completion of a task) connected by activities (actual performance of the task). Time estimates are given for each activity. The completed chart represents a network of interconnected activities and can identify bottlenecks and critical paths to the successful completion of the project. Complex networks are handled with computer programs that can track thousands of activities and events. The most difficult part of setting up a PERT network is identifying all the activities that precede the completion of specified events. Although this type of detailed planning takes a considerable effort, it leads to implementation that is quite routine.

The work plan can help in the preparation of the **cost budget** for implementation. When you have determined who does what and for how long, you will have a good grasp of the labor costs. Usually, it is easiest if the costs for each task are estimated first for various cost categories such as salaries and wages (including fringe benefits), equipment, materials and supplies, travel, and communications (computers, telephone, etc.). Then the totals for each category can be calculated. Are the required funds available? For projects that are funded from an outside source, you may be able to include a certain percentage for overhead costs. If a ceiling for the total budget exists, various adjustments and trade-offs may need to be made between the different tasks and their estimated costs. If the budget cannot be met without seriously jeopardizing the implementation, fund-raising may have to be added as one of the primary activities in the work plan. Fund-raising is in itself a new problem that will require creative problem solving. The task of making up a budget is much easier if you can follow a specified format and examples of budgets for similar projects.

Even if you are the only person scheduled to do the work, you still need to make a work plan or list of what needs to be done and when. Estimate how much time it will take and what support items you need to get each task done. Also find out where you need more information or training or where you could or should get other people involved. Do any of these people need special training? Be sure to include these activities in the **time schedule** for a successful implementation. If you are a procrastinator, you have to make up your mind to do each day what you have scheduled to do. Set a starting date and a reasonable deadline—add intermediate dates for milestones for longer projects, so you can gauge if the implementation is proceeding on track.

A number of techniques are available for **risk analysis**. The Kepner-Tregoe problem solving method uses a special method during the implementation process to anticipate possible difficulties and develop countermeasures. This technique, known as potential problem analysis, should be applied to very important projects where major obstacles to implementation are anticipated. This procedure requires a considerable investment in time and effort and thus would not be cost effective on a routine basis. Alternatively, an FMEA can be conducted on the implementation process in order to identify areas that have high probability scores for possible failures. This will allow you to develop contingency plans for the most critical areas. Other risks may have been identified during reverse brainstorming and other judgment activities; appropriate measures to deal with these risks should be incorporated into the work plan.

Producers are people who are committed to seeing their ideas implemented; they make great personal sacrifices and shoulder the risks. When we studied the mental blocks and the fear of failure in Chapter 4, we learned about Art Fry and how he invented the *Post-it* notepads. How did he get his giant organization, the 3M company, to adopt and implement his idea? It makes quite a story:

Post-it notes were by no means an instant success. First, Art Fry had to sell the idea to his boss who initially was very skeptical because he foresaw that these notes would be rather expensive. But they agreed that the idea was worth testing. The two of them distributed samples throughout the company. When secretaries began using the notes, their bosses wanted to get in on this product, too. Soon, the 3M people were sold on the idea. Also, they suggested various improvements to make the notes practical in many different ways.

The next problem was—how to produce these pads, since 3M was geared to manufacture many different types of tapes and other products that came in rolls only. Here, the personal commitment really came in. Art Fry finally had to invent and construct an assembly machine in the basement of his home. When it was complete, a part of his basement wall had to be torn down to get the machine out.

In 1977, three years after Fry first thought of the idea, 3M was ready to test-market the product in four cities with eye-catching, fancy displays and large newspaper ads. The tests were an absolute failure—so much so that the company decided to kill the product.

But Art Fry persuaded the commercial office supply people to try another approach—they needed to talk to the customers to find out what the problems were. They found out that people who used the product loved it and wanted more, but the others had no idea what the notes were. This was one product that had to be "experienced" before it would be bought. Thus for the next marketing test, thousands of samples were given away. The company figured that a 50-percent repurchase would indicate a wild success. However, to their astonishment, the test resulted in a 94-percent repurchase, and sales took off. In two years, distribution was nationwide and across Canada; the following year the product was marketed throughout Europe.

The moral of the story is—don't give up too soon. Maintain your personal commitment and get the support of others. Even good or especially creative ideas can and do fail in the beginning. Just keep working to overcome objections and barriers with creative problem solving; persist, and you will succeed.

⌚ *TEAM ACTIVITY 8-3: WORK PLAN*

For the implementation problem in Activity 8-2, prepare a work plan. This may involve an actual implementation, or it may be a final project presentation. Determine who will need to do what, when, why, and where. Prepare relevant charts, schedules, and budgets.

IMPLEMENTATION MONITORING AND FINAL PROJECT EVALUATION

Your idea has found acceptance; work plans have been prepared and approved; your project has been funded; and the work has been executed by those assigned to carry it out. Thus you now have only one remaining responsibility—to make sure that the solution actually works. In a small project, it may be possible to make arrangements to personally check up on the success of the implementation. Depending on the type of problem and organization, you may need to work out an implementation monitoring plan as part of the work plan. There is nothing complicated about a monitoring plan; you are probably using one for keeping track of your car's maintenance or for doing seasonal jobs around your home. A monitoring plan provides for periodic checks to make sure things keep on working right as part of preventative maintenance and follow-up. Finally, upon completion of the implementation, reflect on the results as well as the entire creative problem-solving process as a learning experience.

Implementation Monitoring and Follow-Up

Long-term monitoring of the implementation is not usually the responsibility of the creative problem-solving team, although the team may develop procedures able to confirm that the solution has been implemented correctly and that the problem has been solved without causing other problems. Has any resistance to the change appeared, and has this been handled properly? Have unexpected effects appeared in related areas? Have modifications been made to the solution to make it "fit" better? A first review should be scheduled within two weeks after implementation has been completed, with a follow-up in six to twelve months.

The team involved in problem definition is probably best suited for monitoring the implementation of the solution and should take on the responsibility for follow-up. In a manufacturing plant, statistical process control methods (SPC) can yield quantitative data to verify the improvements that have been achieved with the solution. These diagnostic tools should be employed on a specified schedule, not only to provide continuous monitoring over the long term but also to pinpoint new problems that may arise due to the implementation or to identify modifications that have been made (purposely or inadvertently) during implementation or later. Specific testing may also be required to confirm the expected results. This is particularly the case when Taguchi methods are employed to optimize products and processes; the analytical results should be verified through testing before full-scale implementation of the required changes is authorized.

The following tasks are usually included in the confirmation process:
- Make all required changes in the appropriate drawings.
- Make all required changes in the affected processes and procedures.
- Dispose of all "old" forms, drawings, equipment, and materials.
- Notify all individuals impacted by the change (at all levels in the organization).
- Set up audit functions for easily-reversible situations.

Additional responsibilities may be assigned to the monitoring team:
- Conduct a midcourse review and make appropriate corrections if needed to stay on target.
- Obtain customer feedback on the improvement.
- Investigate wider applications of the successful solution or change.
- Obtain employee feedback (acceptance or resistance) to the change.
- Monitor if longer-term associated requirements have been instituted, such as training for employees and the development of new instructions manuals and management procedures.
- Evaluate and improve the creative climate in the organization.

Do you have a favorable climate, a mechanism, and a team responsible for receiving and evaluating suggestions for continuous improvement and innovation within the organization? Do any of the procedures used during the problem-solving process need refinement? If you had a team (or several) working with you in the creative problem-solving process, inform them of the success of the implementation and thank them for their participation, cooperation, and contributions.

Final Evaluation

It is very helpful if you keep a journal or take notes during the entire creative problem-solving process. When the project has been completed, write a brief summary of the results and your achievements. Then you can sit back and review what the process has done for you. Did it help you grow? What have you learned? Can you use this same idea somewhere else? Did you achieve all the goals or some of the goals? How did the process help you relate to and communicate with people? Did the process open new future opportunities for you? If the solution did not work out, you can still ask what you have learned from it. Sometimes we learn more from our so-called failures than from things that are too easy. As a producer, reflect on where you could have avoided needless battles. If you were involved in the project as a supervisor, facilitator, or instructor, identify areas where you can give your people positive feedback. Continue to help them focus on accomplishments as well as learning; encourage them to be supportive of each other by recognizing the team's collective contributions.

Keep your summary and conclusions in a file. As you become more practiced in creative problem solving, this projects file will grow into a valuable database. It will help you become more efficient, since you can reuse "ideas that worked" and avoid (or improve) ideas that did not work. As part of honing your producer's mindset, review the time management skills discussed in the next section.

⌚ *TEAM ACTIVITY 8-4: PRESENTATION OF PROJECT RESULTS*

To conclude your team problem-solving project as well as to give you another opportunity to practice your skills, prepare a final team presentation. Use the thinking preferences of your team members to advantage; different people can take on different roles and leadership at various stages in the process; all must be involved and make a substantial contribution to the final presentation. Depending on the subject of your team project and the setting (formal class, workshop, self-study, assigned problem by a client, etc.), you can choose the appropriate format (oral presentation, skit, written report—possibly including visual aids and a "selling" video).

The presentation should include: (1) a summary of the problem and goals, (2) a description of the creative problem-solving process, (3) the solution ideas, (4) the list of benefits, (5) a work plan (if appropriate), (6) risk prevention and how to motivate people to accept the solution, (7) a description of implementation results (if available from a pilot project), and (8) an evaluation of what was learned (or how the success of the solution could be measured). In the presentation, be enthusiastic; be positive! Show a willingness to work hard to see your idea implemented.

Finally, critique the presentation—what ideas were especially effective and creative? Jot down ideas used by others that you may want to try yourself sometime.

TIME MANAGEMENT—A CURE FOR PROCRASTINATION

One of the most serious hindrances to being a good producer and getting things accomplished is poor time management. If you want to improve your skills in this area and especially if you want to overcome a habit of procrastination, the following discussion is for you. Feedback we have received from students (including those in honors classes) indicates that procrastination is one of their most serious stumbling blocks to being successful in college and achieving goals.

> **If you don't take time to plan,**
> **you are planning to waste time.**

Planning

Collecting data: Planning must begin with an analysis of your goals and priorities in life. People usually take the time to do the things they really want to do. The question is—are the things that you are doing in tune with your priorities and goals? You have to take a careful look at your long-range goals, not just your short-term goals. Make sure that your short-term goals do not prevent you from reaching your long-term goals. What are your priorities? How are you actually spending your time now?

 ASSIGNMENT 8-5: GOAL SETTING
1. *Make a list of your long-term goals (beyond one year to about ten years).*
2. *Make a list of your short-term goals (things you want to accomplish within a year).*
3. *Make a list of priorities.*
4. *Complete a daily time log for two consecutive days, in 15-minute increments.*
5. *Complete a weekly work schedule (or study schedule if you are a student).*

As you are preparing your weekly schedule, take your daily energy cycle into account. Do you do your best work when you get up at 5 a.m. or do you slowly come alive in mid-afternoon, reaching peak performance when most people have gone to bed? If you have a choice, schedule your most important work or classes during your peak times for better alertness and learning, and do your routine tasks in your "low" periods. Generally, one hour of mental activity during daylight hours is worth one-and-a-half hours of night-time learning and studying. The information that you are asked to collect in the assignment above provides a database for planning.

Priorities: What are some of the things that are important in life? Starting from a young age, education takes a considerable time commitment. Later, the emphasis shifts to work and career, although learning activities of one sort or another should never be discontinued to keep our minds functioning well into old age. To have a balanced life, we must make time for all of the following aspects:

> *Spiritual, Family, Career, Social, Self, Health, Leisure, Finance*

Your weekly schedule: Include all the routine tasks you normally do in a week: work, commuting, shopping, meetings, learning, social time, recreation and exercise, hobbies, meal preparation, household maintenance, bookkeeping, correspondence, eating, sleeping, health and personal care, etc. If you are a student, each college course on the average requires two hours of study time for each hour in class (or each credit hour for lab courses). It is important that new college students set up this type of discipline at the very beginning of the first term. If you slack off for three weeks, you have set up a habit that will take an extra effort to correct. If you got "A" grades in high school without doing much studying, you may be in for a shock if you do not establish good study habits when starting college.

Now look at your list of priorities. What other things besides working or learning are important aspects in your life right now? Do you have family responsibilities? Mark time for these on your schedule. What about your spiritual life? Mark time for your religious commitments. What about your health? Fill in time on your schedule for regular meals, sleep, and exercise—an aerobic workout at least a half hour three times a week. What about social activities (including helping others) and leisure? Are you running out of uncommitted time slots on your schedule? Can your social activities be combined with other items? Could you exercise with family and friends? Could mealtimes be used for maintaining your social relationships? Can you and your "significant other" be together during religious and hobby activities? Some creative thinking is required here to come up with solutions that will help you get all your priorities into your weekly schedule.

What about finances? If you are working and studying, you must schedule both, with adequate commuting time. If you must work while in college, you will have to reduce your class load. Here is a formula that can give you a starting point:

Credit-hour load   =   1/3 [48 – (number of hours of work per week)]

You can adjust your course load as you go along, depending on how well you are managing in your course of study. If you are continuing your education while working, some other areas may have to be temporarily curtailed, such as recreation. Television watching is often an area where substantial cuts can be made. Are you merely working to finance a car? Calculate the financial gain possible by giving up the car and graduating a year earlier. Are loans (instead of work) a good option?

Last, but not least, you must schedule time for yourself— for "personal care and maintenance." This includes personal grooming, household chores such as doing the laundry and keeping your abode in reasonable order, taking care of correspondence (including paying bills), as well as planning and evaluation. Also schedule one hour a week just for yourself—to daydream, think, reflect, give yourself a special treat, indulge in a hobby, or soak in the bathtub. To prevent running out of energy and affecting your creative thinking ability, leave yourself some breathing space between various activities as you experiment with different arrangements in scheduling your week. You will encounter Murphy's law ("If anything con go wrong, it will"), and life will be less stressful if you have some built-in flexibility. Try the "best" version of your schedule for a week or two. Basic flaws in the schedule will quickly become apparent—make the changes needed. You may also have to learn some new attitudes about keeping to a schedule: Do you tend to be rigid—could you benefit from loosening up? Or are you easily distracted and casual about time management—do you need to develop self-discipline?

Progress toward your goals: Compare your list of long-term goals with the completed weekly schedule. Circle those items on the schedule that have anything at all to do with achieving your top three goals. If you have many circles—good for you. If you don't have any circles, you need to evaluate your goals and your daily activities. If you are not incorporating actions into your daily schedule that move you toward your goals, you will not reach those goals; you will be under stress and dissatisfied with your life. If you want to be an engineer but your class schedule is made up entirely of art, music, and social studies, your activities will not move you toward your goal. Why did you pick this particular selection of courses? Why do you want to be an engineer? Is this your personal goal, or is someone pressuring you? Perhaps you need to change your goal! If you want to be a writer but your present responsibilities do not leave you with a free minute during the week, you must restructure your life to make time for working toward your goal. Thinking, planning, and managing your time may not be as exciting as other things in your life, but they are the way to fulfill your ambitions. As you get practice in planning and time management, it becomes easier, and you will see results! Do this type of goal evaluation at least once a year (perhaps at year's end, or on your birthday).

Personality and Modes of Thinking

If you have trouble with time management, this may be part of your personality style and preferred modes of thinking. Here is an overview of four different personality types—see which one fits you most closely. This will give you an idea of the problems that you may have with time management and the best remedies that you can use. Each of these personality types corresponds to some degree with the modes of thinking of the Herrmann four-quadrant brain model.

Personality A is a perfectionist who does everything right the first time. These people organize everything logically, can concentrate well, and work unusual hours to get the job done. The difficulties come when they do not delegate work; they want everything perfect, and they underestimate the time needs for doing their tasks. If you are such a person, you can manage time better if you set priorities and review these often, if you don't get lost in details, and if you learn to delegate tasks (without being too critical). Be realistic in your time estimates.

Personality B is steadfast, dependable, and punctual, yet can tolerate the lack of time sense in others. Potential difficulties arise because these people get upset with rescheduling; they won't deviate from schedules, which can result in unfinished work. Basically, they are inflexible. The remedy is learning to be more flexible. If you recognize yourself in this category, analyze your time use. Work unusual hours when needed. Adopt new ways of doing things; avoid ruts.

Personality C is a "people person" who is aware of time problems and accepts suggestions. These individuals make excellent team workers. Difficulties arise when others impose on their time. Also, they are prone to procrastination, and they waste time by being very talkative. If you are such a person, you need to learn to use calendars and schedules to plan and identify priorities. You must write out "to do" lists and obey them; you must follow written instructions, even though this is boring. You need to develop listening skills so you will not become easily distracted. The daily time log analysis can be very helpful in identifying where time is wasted.

Personality D is a go-getter and a person who sees the big picture, who works long hours, who has abundant energy—one who barges right in when a job needs doing. Conversely, such a person is too impatient to do tactical planning and to analyze the time used. The results are schedule conflicts and pressure because time requirements are frequently underestimated. These difficulties can be overcome by learning to plan in more detail, to be aware of the shortcomings of not being analytically minded. If you have these traits, try to establish a follow-up evaluation routine. Learn from the analysis, and respect the input of others.

It is possible to have the traits of more than one of these thinking and behavior modes. These personality types can each make valuable contributions with their talents and special ways of thinking, but all will function better and interact better with other people if they learn good time management. Keep a positive attitude as you try your schedule and establish new habits. Don't get bogged down in minor details and stressed out if unexpected things throw your schedule off. Most importantly, the goal is to find those time management strategies that work for you!

Daily Priorities: Separating the Important from the Urgent

As your last activity each evening (or as the first thing in the morning), get into the habit of spending a few minutes making a daily LIST OF THINGS TO DO. This list starts with the information from your weekly schedule. But many other things will be added to the list, such as appointments, things that did not get completed from the day before, or unexpected (and thus unscheduled) items. These items can be identified according to three different priority categories:

A	**=**	**First priority: things that absolutely have to be done.**
B	**=**	**Second priority: things that should be done.**
C	**=**	**Third priority: things that can be put off if necessary.**

This discussion assumes that you will not have enough hours in the day to do everything that you think you have to do, especially since other items usually crop up and have to be added to the daily list, thus interrupting your careful plans. How do you determine which things should be done and which can be neglected? Has it ever happened to you that you had so many urgent things to do in a day that you never got around to doing those items that were really important to you? Do you have a strategy that allows you to do more of the important things and less of the urgent? The approach described on the next page is a technique that can help you decide on priorities for your daily list.

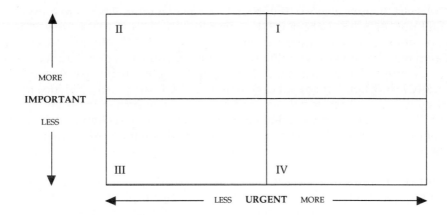

On a piece of scratch paper, first list all the things you think you have to do that day. Then draw a large rectangle (as indicated above) on a blank piece of paper and divide the box into four quadrants. Label the vertical axis on the left IMPORTANT and the horizontal axis URGENT. Add arrows to the axes to indicate increasing importance and increasing urgency, the further away you move from the bottom left corner. Then label each quadrant for identification. Let's discuss each quadrant in turn, so you can decide into which category (or quadrant) a particular task or item on your scratch list should be placed.

Let's begin with Category IV—the **urgent but unimportant** items. These can be real time wasters and time killers. And if you spend time doing these, you will not have anything of substance accomplished. Why? Because these truly are unimportant! People will try to make you believe that these are important, but you will have to use your own judgment and wisdom in making this determination—don't let others rob you of time and put you under pressure. Watch advertisers. They are masters at this game: You must buy this article today or it will no longer be on sale (or no longer available, or you'll lose a golden, once-in-a-lifetime opportunity, etc.). If you think critically about it, you will usually come to the conclusion that (1) you don't really need this article, (2) it is not such a bargain—you saw it for less somewhere else, or (3) you could use it, but you will go out and buy it when it is most convenient for you. What should you do with the items that fit into this category? Most can be tossed into the wastepaper basket; that's exactly where they belong. Cross them off your list. If there are no serious consequences for not doing these, don't waste time doing them. Those that do have serious consequences belong in Category I.

There is no question of what you need to do with Category I—**important and urgent.** These are items that you must do, the sooner the better. These will be your priority A activities; mark them as such on your list. People usually find time to do these—so this category is not a problem in time management. Especially if you have developed a habit of good planning, this category should not contain an overload of items.

Category II and Category III items take some planning. These are things that easily get put off precisely because they are not urgent. But if they are not done, they can either grow to be urgent, or they can grow to be important. "A stitch in time saves nine"—you've heard that saying. Or, "an ounce of prevention is worth a pound of cure." These expressions fit items that people consign to Category III—**potentially important but not yet urgent**; it would probably be a good idea to take care of as many of these as you can in your daily activities, but assign them to priority C. If you can't get them done during the week, the weekend might be less busy. Delegate—see if you can find someone else to do them. Assess each item and toss out things that are of little consequence—they are merely clutter in a busy life.

Now comes the most critical time management category: Category II—**important but not yet urgent.** This is where you really use your planning and time management skills. Into this category you need to put your larger projects, your book reports, term papers, long reading assignments, your home remodeling, your party planning, your Christmas shopping, whatever—all those things that you can't do in one day. These are important things, but when you first get the assignment or recognize the need, they are not an emergency, and thus for the time being, you are tempted to put them off. If you neglect to schedule them into your daily activities, they will eventually become urgent (or even critical), and then you will not have enough time to do them carefully. If you begin working on these activities early, in smaller "bites," you will be in control and you will reduce stress in your life tremendously. Therefore, make sure that you schedule items from this category into your daily A and B priorities. Remember that many Category II items may be connected to your achieving your long-term goals; you must schedule them into your weekly schedule! When you begin work early, your subconscious mind can incubate ideas and thus help you with good solutions as you progress through the project. When you wait until you are rushed and stressed, you are working against the best capabilities of your brain's functioning and cannot expect to do your best thinking and get good results.

> *My father can climb the highest mountain,*
> *My father can swim the deepest ocean,*
> *My father can fly the fastest plane,*
> *My father can fight the toughest tiger,*
> *But most of the time he just takes out the garbage.*
> *Author unknown.*

Time Wasters and Dealing with Procrastination

We have just seen that lack of planning (being ruled by the urgent but trivial) can be a big time waster. What are some other time wasters that we need to look out for? And how do we deal with the biggest time waster of all—procrastination?

Time wasters: Here is a list—check to see if any of these items apply to you.
Disorganization: This can include a cluttered work space, misplaced items, poor memorization skills, and poor scheduling habits—all symptoms of a general lack of attention to quadrant B thinking.

Attitude—lack of interest, poor listening habits, unnecessary perfectionism. This is largely a sign of imbalance, with too much focus on some all-consuming activity and too much attention to detail in quadrant B, while neglecting interpersonal relationships and consideration for others in quadrant C.

Lack of self-discipline—temper, impatience; inability to say no; carelessness. This may involve avoidance of quadrant B as well as quadrant C thinking. Some of these items have cause-and-effect relationships: a person who can't say no can get overcommitted with various responsibilities, which can lead to stress which in turn is expressed in temper, when the lack of planning is the root of the problem.

Interruptions—gossip, visitors, phone, junk mail; music, TV, parties, friends. These items have to do with priorities as well as with wisdom and judgment in making choices. Fun, relaxation, social interactions are desirable—within limits. The focus on quadrant C must be balanced with other activities that will lead to the attainment of short-term and long-term goals.

Lack of planning—waiting, unproductive meetings, unnecessary errands. This combined deficiency in quadrant B and quadrant D thinking leads to inefficiency in work and other areas. A few minutes of thinking ahead can prevent wasted hours. Get into the habit of making shopping lists and agendas.

Unused information and lack of communications: Good communication is needed when living and working with other people when we depend on others to do their part of the job. The group functions well only when each person knows what is expected. We can also waste time when we do not have enough information or when we have so much clutter that we do not even know where to start when we are looking for a specific item. Having at least a rudimentary system for organizing one's life and maintaining communications (quadrant B and quadrant C skills) will go a long way to prevent the loss of time here.

Procrastination: This is a major reason for wasting time and can have several causes which are discussed in more detail below.

Procrastination: Most of the items in the list above are rather self-explanatory. But what is procrastination? Procrastination is an attitude that says: "Do it later." It can cause all kinds of trouble; it is the major reason why the urgent and important category gets so crowded! Why are people doing it? Primarily they have developed a habit of putting off doing things that they see as being unpleasant or difficult. Procrastinators may have indecisive personalities; they may have a fear of failure—a fear of making mistakes. Thus these people welcome all kinds of interruptions in order to get out of having to do their tasks. How can you overcome such a habit? When faced with an unpleasant task, do a time study. How long does it really take to do the job (if no interruptions occur)? It usually comes as a big surprise to find that doing the job takes less than half the time estimated; more time is wasted in trying to get out of doing the task than in actually doing it. Alternate doing pleasant and unpleasant tasks; give yourself a reward for completing an unpleasant task quickly. As you learn to be more creative, it will become easier to overcome the fear of failure or the dread of making a mistake. If you are discouraged because you always seem to face jobs that are too big and too difficult for you, try breaking these jobs down into smaller tasks—then do these smaller things one at a time. Pat yourself on the back as you see your accomplishments grow.

Set up a filing system: Have a filing system to keep track of information and of things to do so you do not waste time looking for data you know is "somewhere around here." For example, students can keep a file of important and supplementary information for each of their classes. At the end of the semester, the file is cleared of unneeded information and moved to inactive status in accessible storage. Keeping annotated books (instead of selling them after the final exam) can save time in the future, because you will remember exactly where to find a specific piece of information—these books are a valuable databank. Also keep files for financial and tax records, insurance information, job information (for updating your résumé), and health information. When a medical test is done for you, ask for a copy of the results and file it (medical records do not always follow you when you move—this information could be vital in the future). Set up a "tickler" file for each month of the year. Then at the beginning of each month, make weekly schedules with those items. This system is great for keeping track of deadlines and cyclic activities—like visiting the dentist twice a year). If you have trouble keeping track of computer files and hard copy of important paperwork, keep a file in your filing cabinet that includes the paper, together with its computer file on a floppy disk.

Take the time to set up an organizational system that is right for you. One student designed his own system with note cards (each card had the task and time estimate, and the cards were stacked in order of priority). Be realistic, be flexible; create an environment that is functional as well as attractive, friendly, relaxing, stimulating—one that will help you do your best. Your needs change. Perhaps this is a good time to update your current system and learn some new skills. Do you need to get out of a rut and think up some creative solutions—or do you need to put a rein on chaos and implement routines for cleaning and organizing your things?

Other scheduling tips:

1. *For larger projects, make a PERT diagram and then follow it.*
2. *In your daily, weekly, and monthly schedules, include time for planning.*
3. *Do not overschedule; leave flex time—as much as 25 percent.*
4. *If you have too much to do, get help. Learn to delegate; don't volunteer for too many tasks. If you accept a new task, drop one of your other responsibilities.*
5. *Handle each piece of junk mail only once—make instant decisions.*
6. *Allow for breaks; do not work at a computer or study for more than two hours without an interruption. Allow incubation time for ideas.*
7. *Organize your work space efficiently.*
8. *Learn to take shortcuts; eliminate the trivial.*
9. **Frequently ask: what is the best use of my time right now?**

🕐 ACTIVITY 8-6: IMPLEMENTATION PLANNING

From the discussion on time management, select the one tip that you think would benefit you the most in terms of immediate benefits. Then make up an implementation plan and carry it out over the span of three weeks. Write a brief evaluation of your results and what you have learned about yourself.

CASE STUDIES OF FINAL PROJECT PRESENTATIONS

Below are summaries of four verbal team presentations and a written report on an implementation of creative problem solving in the workplace by a student.

Case 1—Coping with Peer Pressure (High School Students)

A team of ten students from several Detroit schools, *The Creative Thinkers*, selected and brainstormed the problem of "How can students cope with peer pressure so they won't drop out of school?" During creative idea evaluation, they came up with five categories of ideas (listed in the table). During idea judgment, they had to select one "best" idea. Since the team was split on two winning ideas (shown in italics), the students prepared work plans and a presentation for each.

Coping with Peer Pressure: List of Improved Ideas for Judgment

Build students' self-esteem and ability to make good judgments:
- Let students paint their lockers.
- Let students pick their counselors; have more and younger counselors.
- A course in etiquette can teach students manners, self-confidence, and how to get along with all kinds of people.
- Have seminars for students to develop insight about making good life choices.

Help students be stronger emotionally to resist pressure:
- Teach communications skills, especially on how to disagree without rejection or anger.
- Teach young adults about religious views, so they can develop a strong faith.
- Teach a course in middle school on "coping with real life."
- Help students explore psychology, sex, values, and how these fit together for good relationships.

Improve the school environment to reduce peer pressure:
- Make the academic curriculum challenging, interesting, goal-oriented.
- Have students and teachers participate together in sports, other activities.
- All students should train and participate in some sports, with facilities open 24 hours/day.
- Have more art, drama, music activities, with talent shows.

Students can work with their schools to make them a great place:
- *Students can produce their own videos about coping with peer pressure.*
- *"Voc Tech Fashions:" students could design, manufacture, and market attractive school clothes for all students.*
- A clothing allowance for schools could be discussed.
- Parent/teacher conferences should be eliminated; increase student involvement instead.
- Give peer group leaders training to make them good leaders.

Get support from other adults:
- Find adults who care, who listen, and who are trustworthy.
- Adults need to understand street language.
- There should be more "real life" TV programs.
- Have more recreational opportunities for young people in the community.

✧   **Class presentation on the video idea**: A work plan was developed on how to make the student-produced videos about coping with peer pressure. The following tasks were cited:
1. The principal of the school will have to be asked for permission.
2. An English teacher will need to be "drafted" as mentor for after-school supervision of the script writing. Or perhaps this activity could be made part of a composition class.
3. A drama teacher will have to be asked to guide the production, prop preparation, and acting.
4. Team members will need to develop plans, budgets, and do fund-raising for this project in the community (among businesses, parents, other supporters).
5. A community cable TV station will have to be contacted to help produce the videos and perhaps schedule airtime.

The students concluded that the project was quite feasible and could easily be done within one school year; they anticipated no major difficulties.

◆   **Class presentation on the Voc-Tech Fashions idea:** At first, the group had a rather difficult time selling the idea for "Voc Tech Fashions" to their skeptical peers. The objective was to have uniforms at each school, in order to cut down on student crime and the high expense connected with peer pressure for wearing designer clothing. However, to get around the traditional, drab uniforms, these modern, stylish voc-tech fashions (different for each school) would be designed by the students themselves. By concentrating on the list of benefits, the team managed to be quite convincing on how such a project would increase school spirit and pride as well as community support and identification with the school, thus attracting better teachers. There would be less crime, and the students associated with the entire production (design, manufacture, selling) would learn valuable business skills and make money for the schools—in short, Voc-Tech Fashions would be very good for Detroit! However, implementation would be difficult because students graduate and different groups have to take over each year.

Case 2—How to Make Schools Better (College Freshmen)

A class of freshmen engineering students—*The Guinea Pigs*— generated 262 ideas by the panel method. These ideas were sorted into five main categories. One of these categories was used by the instructors as an example; the remaining four categories were assigned to groups of seven students each. Team 2 worked with the "Outreach" category. During creative idea evaluation, the team came up with five topics and a total of twelve complete ideas as given in the following list. This was a difficult category because the ideas in the five topics were very different and could not easily be combined. During idea judgment, the team selected the first idea in the "business" category as its top solution because this important concept would have a long-term impact in making high schools better. The students met at night to prepare their presentation. They made up a script that nicely included all team members (who covered a wide range of verbal fluency). They focused on only one concept from their top solution for their presentation: "Businesses provide career help through on-the-job visits set up like a mini co-op program."

Making Schools Better: List of Final Ideas for Judgment

PARENTS

1. Schools need to keep in closer touch with parents to help parents monitor their children's progress.
2. Parents need to learn to provide the right kind of support to encourage kids to do well and study more, yet not apply too much pressure (i.e., have classes for parents).
3. Parents can become more involved by providing resources such as science projects or speakers (on the danger of cults and satanic games, or drug and alcohol use).

COLLEGES

1. Have more high school/college cooperative programs (with company sponsorships), such as research assistantships for high school students at the university; college students assisting in high school (science) classes; college students explaining the college majors and academic requirements.
2. Make college admissions requirements harder.
3. Make college cheaper (first two years free?) or provide more financial aid. Also, make financial aid application forms easier.

BUSINESSES

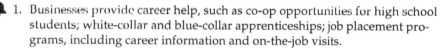

1. Businesses provide career help, such as co-op opportunities for high school students; white-collar and blue-collar apprenticeships; job placement programs, including career information and on-the-job visits.
2. Businesses provide sponsorships to schools which can include the following: speakers from industry; senior citizens (as speakers or tutors); peer tutoring programs; facilities to help students with learning disabilities; educational field trips; helping students volunteer in charitable or environmental organizations; funding for special, identified needs of individual teachers and subjects (especially in science and math) and for enrichment programs.

GOVERNMENT

1. Lobby federal and state governments for more funds for programs to help mentally and physically handicapped students.
2. Petition lawmakers so drug seizure money can be used to improve schools.

SCHOOL BOARDS

1. Countywide school boards need to brainstorm with teachers, administrators, and leaders in the community on alternative "structures" to develop areas of excellence for each high school.
2. Ideas to be considered are: letting students select the school they want to attend; a high school for the gifted; boarding schools; more business, trade, and "industry" schools; specific academies and complementary skill centers at different high schools; Saturday school (for brainstorming, enrichment, or remedial programs).

🔔 **Class presentation—businesses provide "A Day on the Job" program:** The speaking parts (with corresponding posters) had the following content:

🔔 PURPOSE

The purpose of bringing this co-op program into the schools is to give students a chance to see what it's like in the real business world. This new program will also help the students make up their minds on a career choice so they can be better prepared and motivated to take the proper courses in school.

🔔 GOALS

This program sets many goals which it will achieve in the schools for students. Some of the goals are as follows:

1. To give students an idea of what the real world has in store for them.
2. To help students make up their minds about different careers.
3. To help students select further high school courses that are required for their career choice (or for college preparation).
4. This co-op could lead to future employment with the co-op company. This could help students earn money to advance their education in college.

The team leader then handed out an information packet consisting of a nicely designed cover sheet/advertisement (using computer graphics) and an application form for the "A Day on the Job" program. On this form, in addition to name, address, school, age, sex, grade point average, and proposed occupation, the student had to answer questions about extracurricular activities, and in one paragraph describe why he or she was interested in the "A Day on the Job" program, why the particular profession was chosen, and what he or she hoped to gain from the program. A supervisor from the public schools was invited to the final presentation. This pilot class was unusual in that it included selected high school students on an experimental basis—they interacted well with the college students.

Case 3—Encouraging Students to Think Creatively (Teachers)

Twenty-four public school teachers for grades 3 through 12 from a suburban/ rural school district were divided into two teams to brainstorm the topic "How can teachers encourage creative thinking in their students?" Together, the teachers generated approximately 200 different ideas. Because of time constraints, they had to move through creative evaluation, judgment, and implementation quickly, before the ideas were completely synthesized or the "odd" ideas explored and integrated. Even so, the two teams came up with interesting solutions in their workshop class problem. Typically, in a two-day workshop, creative thinking, problem definition, and brainstorming are covered on Day 1 and creative evaluation, judgment, implementation, and various applications on Day 2.

Team A used the following criteria in the judgment phase to select the best ideas: (1) cost, (2) classroom time, (3) special training, (4) teacher motivation, (5) student benefits, (6) practicality, and (7) correlation with other material. Two main concepts surfaced as the most important, final ideas:

ENCOURAGING STUDENT CREATIVITY: FINAL IDEAS

☆ Offer a workshop to promote the philosophy and techniques of creativity in the classroom.

★ Design a list of creativity stimulators that have been implemented by teachers in their classrooms.

Since the teachers in Team A wanted to continue working with both of these ideas, they were split into two groups (Team A-1 and Team A-2) to work out separate implementation plans and class presentations.

☆ **Class Presentation A-1—Educator Workshop on Creative Thinking:** Team A-1 prepared posters listing the following goals, benefits, tasks, and possible failures involved in setting up a workshop on creative thinking for educators.

☆GOALS

• Present philosophy for promoting creative thinking.
• Provide specific techniques which teachers can use in the classroom.
• Establish a system for peer support in sharing experiences.

☆BENEFITS

• More stimulation for students.
• Students will develop thinking skills.
• Active participation by students in the learning process.
• Problem-solving skills carry over to life situations beyond "school."
• Motivation and stimulation for teachers.

☆TASKS

• Who?   County consultants (with local input).
• What?   Plan workshop. Obtain resource people. Plan location.
          Set agenda. Plan refreshments. Establish evaluation procedures.
• When?   In-service 1990.
• Why?   Maximum exposure to accomplish goals.

☆POSSIBLE FAILURES

• Funding.   • Teacher opposition.
• In-service already set.   • Lack of application of techniques.

Group 1 presented its results to the class in the form of a skit, with one group member impersonating a senior teacher and a second person acting as a member of the board of education. The senior teacher's job was to sell the workshop idea (with the prepared posters) to the board of education member. The two teachers did a marvelous job of spontaneous play-acting.

★ **Class Presentation A-2—The Creativity Resource Box:** Team A-2 prepared a skit to present the idea of creativity stimulators that can be used by teachers in the classroom. This skit involved all six group members. The group came up with the idea of a product, the creativity resource box.

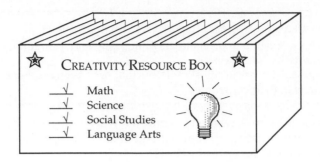

The resource box contains a filing system of creative ideas (color-coded by subject area for math, social studies, science, and language arts). In the skit, two enthusiastic teachers were visiting colleagues at other schools. In an exuberant, positive manner, they proceeded to "sell" the box with these neat, creative ideas to a teacher in each subject area (and at a different grade level). They were able to convince each teacher to try the box—free. The teacher's feedback and ideas were solicited, and further help and support were offered. This class presentation is an example of the creative thinking involved in coming up with an effective presentation resulting in a marketable product—an innovative business idea!

> *Truth persuades by teaching, but does not teach by persuading.*
> Tertullian, ca. A.D. 200.

★ **Class Presentation B—Brainstorming:**

Team B—when looking over the different categories during the idea judgment phase—decided to select "creative classroom activities" as a single criterion and voted to focus on brainstorming as their number one priority. Teachers should do more brainstorming, in many different ways, classes, and situations—for short periods (minutes) as well as in longer projects. The team prepared a benefits list, a list of goals, and a work plan. For their class presentation skit, they decided to ask their principal to give them about 20 minutes during the first teachers' meeting of the new school year to introduce to their fellow teachers the benefits of training students in creative thinking and brainstorming.

Team B began its presentation with a jingle over the school's intercom: "All aboard the *Be a Success Express,*" to introduce the title and theme of their presentation: **The Great Brainstorm Express!** Then a committee of three teachers invited their principal to a brief meeting where they presented him with their plan, using posters and illustrated overhead transparencies showing train and bee cartoons. Involving their own principal in this skit was a creative solution for these teachers to introduce their administrator in a nonthreatening, fun way to their future plans—certainly an interesting approach of working toward acceptance!

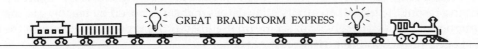

Brainstorming as a Creative Class Activity

The List of Benefits

- Improves student creative thinking skills.
- Better teacher-student relationships.
- Happier kids!!!
- Better learning ambience.
- Teaches cooperative learning and teamwork.
- Encourages originality, responsibility, and student self-determination.
- Kids feel better about themselves and school.
- Active learning; improved curriculum.
- No cost!!!
- Good public relations and community support for education and the school.

Implementation Plan

1. Present the plan to the principal; get his or her support.

2. Present a program to the school's faculty at the general faculty meeting.
   - Place "mystery" object at each table.
   - Creative thinking exercise; list five possible uses for the object.
   - Present benefits of brainstorming skills.
   - Explain procedure (rules of brainstorming).
   - Give list of results of the workshop (or get one idea each from the teachers for something else to brainstorm).
   - Will form teams of two experienced teachers to train the other teachers in the same grade.
   - Hand out committee-prepared packet to each teacher. Use audio-visual aids (or overhead transparencies) to describe.
   - Encourage each teacher to extend the creative brainstorming procedure to his or her classroom and share ideas with peers.

Goals

Brainstorming activities should be used frequently in all classrooms. These ideas have been judged as the top five applications from a long list of ideas:
- Use a brainstorming bulletin board.
- Do classification and evaluation with brainstorming ideas.
- Do brainstorming design exercises.
- Brainstorm methods to help students develop people skills (giving compliments, resolving conflicts, etc.).
- Brainstorm activities that students can do on the bus.

Final Comment: All three groups gave an excellent presentation. Just think how schools would be improved if all three ideas were adopted, since they complement each other: teacher's workshops on creative thinking, the resource box, and the brainstorming applications in the classroom.

Case 4—Companies Need Better Employees (People in Industry)

A group of twelve engineers and managers—*Marge's Engineers and Managers*— from a large corporation in the Midwest generated 50 ideas by a four-member panel; an additional 90 ideas were generated through written brainstorming by the group members in the "audience." This group redefined the original problem of how companies can help improve the quality of schools as: "Companies need better employees." The group changed the direction of the discussion away from elementary and secondary education toward higher education and training.

During creative idea evaluation, the ideas were divided into three categories. The brainstorming team was divided into three teams of four members each, and each team worked on one of the categories with the objective of dividing the ideas into subtopics and then combining the idea fragments into more complete, superior, practical ideas. During idea judgment, each team established criteria for selecting the most important idea(s) from their topics for submission to the class for judgment. These best ideas were posted on large sheets for a final round of idea synthesis. The concept of the industry/university coordinator did not surface until this last effort of idea synthesis at the very end of the idea judgment process.

TEAM 1—COMPANY AND SCHOOL INTERACTION

Criterion: Introducing the "voice of the customer" into schools. Also, this was the topic with the most interesting and the largest numbe of ideas.

Ideas:
1. Academic "sabbaticals" to industry.
2. Faculty/industry discussion groups.
3. Engineering (and other types of student) exchanges.
 4. Industry/university coordinators.

TEAM 2—COMPANY CLIMATE AND TRAINING

Criteria: Communications and getting the right people.

Ideas:
1. Establish hiring criteria for employees:
   a. Select facility locations that include concern for educational levels.
   b. Properly train the people who will do the hiring.

2. Set up an education program to assure that:
   a. People are trained to do the job they are hired to do.
   b. Cross-training is set up for all jobs.
   c. Clear lines of communications exist.
   d. Mandatory training time is instituted.

TEAM 3—RECRUITING AND HIRING

Criterion: The most important factors to get and keep good employees.

Ideas:
1. Create a hiring system that will focus on the top 3 percent of those who meet established hiring criteria.
2. Develop an incentive system to attract and maintain high-quality applicants and employees.

Two ideas were selected as most important for implementation—one long-term, one short-term. These engineers and managers were experienced brain-stormers (especially in the storyboard technique). However, they had not previously been taught to focus on creative idea evaluation and thus followed their habitual pattern of quickly moving to judgment and implementation without special attention to idea integration.

For solution implementation, the workshop participants were divided into two new teams—with one solution assigned to each team—in order to work out a brief implementation plan. Each team appointed a spokesperson who presented the plan. These presentations were the least interesting part of the entire problem-solving process because the teams (and especially the presenters) were not looking for an innovative approach but followed their usual procedures and jargon. Here are the summaries of the content of the two presentations:

 ESTABLISH A POLICY FOR HIRING PEOPLE
AND FOR TRAINING EMPLOYEES INTERNALLY (SHORT-TERM SOLUTION)

Goals and benefits: The company becomes more competitive.

Task assignment:
1. Establish a corporate committee to create the corporate hiring and incentive system, beginning immediately, since the corporation is losing performance strengths at all its facilities.
2. Failure prevention: Begin information and data exchanges with members of each plant to gain support and avoid confrontation.

 WORK WITH THE SCHOOL PIPELINE:
INDUSTRY/UNIVERSITY RELATIONSHIP (LONG-TERM SOLUTION)

Goal: Get better employees through industry/academic interaction.

Tasks:
1. Identify coordinators. Work out budget considerations.
2. Understand the needs (both from the university's and the corporation's viewpoint).
3. Set priorities and develop plans to meet the most urgent needs.
4. Meet the needs through mutual assistance and cooperation.

Final Comments: The main objective of a class exercise is to illustrate the different steps and mindsets of the creative problem-solving process—the quality of the results is incidental. In an actual problem-solving situation, creative idea evaluation is crucial for obtaining a high-quality solution. During an exercise, the team members should be encouraged to try for novel and innovative solutions. Otherwise, people who have little practice in right-brain thinking—experienced as well as inexperienced problem solvers—tend to pick very conservative ideas for their final solution, which of course defeats the purpose of the course or workshop.

Case 5—Creative Problem Solving Applied in the Workplace

This case study was written by a student in an extension class we taught at Delco-Chassis in Sandusky, Ohio, during the fall of 1991 as part of a pilot program in engineering in which nontraditional students in technical fields can earn a B.S. degree while continuing to work. Ellen chose this case study as the subject of her "thinking report" (which was a class requirement with the objective of making connections between classroom learning and something outside of class).

"THE ROOM TEMPERATURE COMPROMISE"
ELLEN STANTON

Problem Background: With a goal of increasing the productivity of the worker, we are concerned with the office environment. My job title is designer and I work with computer-aided design drafting software (CADD). Most of my time is spent at the CADD workstation or at my desk next to the workstation. My work area is located in a room 16 ft x 16 ft which contains two additional workstations. Their operators are the other members of our problem-solving team. Our problem is the air conditioner. It was installed in the ceiling to cool the space occupied by the computer terminals. The temperature is set by the thermostat inside the room, so that the room environment can be controlled separately from the larger office area outside.

Problem Definition: First, we surveyed the group members. The unit blows too much cold air for me. It's not cold enough for Mr. Polar Bear during the "off" cycle. It causes sinus aggravation for the third person. As detectives we attempted to describe the parameters of the problem. By inviting a cigar-smoking co-worker in, we observed that during the "on" cycle the air blew out from a directional finned vent; the adjacent vent is an intake.

By experiment, using a thermometer and a stopwatch, we found that at a specified temperature setting there is a consistent "on" and "off" cycle period of time. This varies day to day as the outer office temperature varies and mixes with our room temperature through the open door. On a specific day, readings were recorded for a thermostat setting of 74 degrees. Cooling started at 75 degrees and ran for 12 minutes; it stopped at 73 degrees and stayed off for 12 minutes. This cycle repeats predictably. For a setting of 76 degrees, cooling started at 77 degrees and ran for 10 minutes; it stopped at 75 degrees and stayed off for 28 minutes.

Idea Generation: After we gave up on changing the thermostat settings when no one was looking, our group began verbal brainstorming. One condition we accepted was that we would not replace the unit, but we would attempt to modify its effects. Our ideas are listed as follows:
- Remove the walls.
- Shut the system off.
- Redirect the flow with cardboard.
- Diffuse the flow with a box cover with holes in the box.
- Put pinwheels up, covering the vent area, to diffuse the flow.
- Turn the temperature up so the "on" cycle is short and "off" cycle long.
- Change the direction of the vent grate fins to point in the opposite direction.

Creative Idea Evaluation: We began to engineer the ideas by grouping them:

❄ SIMPLE:
- Shut system off.
- Turn temperature up.
- Change direction of grate.

❄ MEATY:
- Redirect flow with cardboard.
- Diffuse flow with cardboard box full of holes.
- Diffuse flow with pinwheels.

❄ DIFFICULT:
- Remove the walls.

We decided not to combine or modify the different ideas within the categories.

Idea Judgment: As judges we defined and selected our evaluation criteria to be: optimal people comfort, low cost, quick implementation, easy operation, simplicity of solution, and effectiveness of solution. We ranked the ideas as follows:
1. Change direction of grate (simple).
2. Redirect flow with cardboard (tricky to implement but reduces wind chill).
3. Turn temperature up (this reduces cycle "on" time but makes Mr. Polar Bear uncomfortable).
4. Cardboard box with holes (bulky and unaesthetic).
5. Pinwheels (aesthetic but distracting).
6. Remove walls (involves other workers' time and cost).
7. Turn system off (impractical because computer systems generate enough heat to make it uncomfortable for the workers in the room).

❄ **Solution Implementation:** We selected and implemented items 1 through 3. We reviewed the results at each step, before proceeding with the next one.
1. Change directional vent. —— Results are limited; air deflects off the opposite wall and results are unsatisfactory; the workers still have wind chill.
2. Redirect flow by taping placemat onto the grate. —— This makes the air flow toward the open door. The room conditions are improved, but now the worker outside the door is complaining of wind chill.
3. Turn the temperature up, leaving the placemat in position. —— This reduces the "on" cycle time and the wind chill, but Mr. Polar Bear is melting during the "off" cycle.

Final Evaluation: We had to go back and do some additional judgment. By Edison's trial and error technique we arrived at an improved solution. Mr. Polar Bear introduced a personal-sized fan that keeps him cooler during the "off" cycle and allows a higher temperature setting to decrease the "on" cycle time. This solution is not perfect, but it is a noteworthy exercise in compromise for everyone concerned. The solution was accepted since the three persons affected the most were part of the problem-solving team. The solution was "no cost" to the organization and thus did not require the approval of management.

Comment: This was a fine, practical application by a student who was learning the creative problem-solving process. The co-workers had no previous creative problem-solving instruction or experience. The engineer's mindset could possibly have benefited from additional creative thinking and synthesis.

Ideas are not just born—they need work to grow into useful applications.
Dave Stouffer, engineering student

There is always room for **kaizen** *(continuous improvement)—and creativity is the key!*
John Campbell, engineering student

Just because you fall upon a stumbling block doesn't mean you have to remain on the ground forever.
Use creative thinking to help you get up and get going again.
Deborah Header, engineering student

RESOURCES FOR FURTHER LEARNING

8-1   Mortimer V. Adler, *How to Speak, How to Listen,* Macmillan, New York, 1983. This book talks about two essential business skills—sales talk (including establishing your credibility, the use of persuasion, and giving effective lectures) and active listening.

8-2   Dale Carnegie, *How to Win Friends and Influence People,* Simon & Schuster, New York, 1937. This book (also available in paperback) provides perhaps the most widely used advice on how to get along with people and get them to accept your ideas. Dale Carnegie courses have been very successful in teaching people public speaking skills.

8-3   Ernst Dichter, *How Hot a Manager Are You?* McGraw-Hill, New York, 1986. Some of the topics discussed are decision-making styles, working to get ideas accepted, motivating people and teamwork, and open communication.

8-4   G. Ray Funkhauser, *The Power of Persuasion,* Times Books, New York, 1986. An understanding of the interpersonal dynamics and environment in the business world will help you develop sales strategies for your ideas. This book gives advice on wielding as well as resisting influence and power.

8-5   Donald C. Gause and Gerald M. Weinbert, *Are Your Lights On? How to Figure Out What the Problem Really Is,* Winthrop Publishers, Cambridge, Massachusetts, 1982. As preparation for planning to solve the problem of implementation, this little book provides a good review of problem definition and problem solving with its humorous and entertaining approach.

8-6   Marion E. Haynes, *Personal Time Management,* Crisp Publications, Los Altos, California, 1987. This book is especially recommended for college students.

8-7   Ernst Jacobi, *Writing at Work,* Hayden, Rochelle Park, New Jersey, 1976. This book shows how to make written communications forceful and interesting to get results. It includes a section on proposal writing. An updated version was published in 1986.

8-8   Victor Kiam, *Going for It! How to Succeed as an Entrepreneur,* Morrow, New York, 1986. Here you can find very good insight into the art of selling—which is not seen as a one-shot activity but something that needs to be done continuously.

8-9   Alfie Kohn, *No Contest—The Case against Competition: Why We Lose in Our Race to Win,* Houghton Mifflin, Boston, 1986. The author shows that gaining success by making others fail is an unproductive way to work or learn, whereas cooperation makes people happier, better communicators, more secure, and more productive.

8-10   John T. Molloy, *Dress for Success,* Warner Books, New York, 1976. When selling ideas (or products), appearance is important in creating a favorable climate and attracting the buyer's attention without detracting from the objectives.

8-11 Sunny Schlengler and Roberta Roesch, *How to Be Organized in Spite of Yourself: Time and Space Management That Works with Your Personal Style,* New American Library, New York, 1989. This book offers practical time and space management hints for ten types of people: perfectionist, hopper, allergic to detail, fence-sitter, cliff-hanger, everything out, nothing out, right angler, pack rat, or total slob.

Two previously cited references also provide information pertinent to the topic of this chapter. *The Rational Manager* by Charles H. Kepner and Benjamin B. Tregoe (Ref. 5-10) describes useful methods for risk analysis and implementation. *How to Model It: Problem Solving for the Computer Age* by Anthony M. Starfield, Karl A. Smith, and Andrew L. Bleloch (Ref. 2-15) contains a nice chapter on how to build a PERT chart and identify critical paths. This small softcover book also exemplifies thinking processes and decision points in problem solving.

IMPLEMENTATION EXERCISE AND EXAMPLE

A summary of a creative problem-solving process is given below from problem definition through idea judgment. The assignment is to study this example and then prepare an implementation plan (including solution monitoring). This is a team project and should involve all six mindsets to develop the best plans. This problem was brainstormed, evaluated, and judged by *The Orange Crush*—a team of managers and engineers from industry.

Problem briefing:
Some barriers to creative thinking and thus to creative problem solving:

---

1. "Creative" groups are seen as more difficult to manage.
2. Managers don't know how to encourage creativity.
3. Employees don't know how to think creatively.
4. Managers and employees fear risk or changing the status quo.
5. The old problem-solving habits are seen as adequate.
6. Managers give up too soon (no quick return on investment).
7. Creative thinking makes analytical thinkers uncomfortable.
8. The corporate structure discourages interdepartmental teamwork.

Problem definition:
The group played with various definitions and then chose this statement:

> How can an organization encourage creativity
> in individual employees?

Idea generation (brainstorming):
Although conditions were not optimum due to midafternoon scheduling, limited time for the exercise, and group members inexperienced in brainstorming together, 70 "raw" ideas were generated:

1. Participation.
2. Training.
3. Money.
4. Budget.
5. Time.
6. Fun.
7. Rewards.
8. Hawaii.
9. Contest.
10. Signs, posters.
11. Implement results.
12. Creativity newsletter.
13. Change organization.
14. Change management.
15. Change the environment (go elsewhere).
16. Change office.
17. Background music.
18. Correct tools.
19. Provide objective.
20. Provide incentive.
21. Increased (focused) responsibility.
22. Mix different groups.
23. Experts.
24. Switch jobs horizontally.
25. Switch jobs vertically.
26. Competition.
27. Add deadlines (some stress).
28. Importance.
29. Beauty of solution (study examples).
30. Listening.
31. Training in communicating.
32. Seminar.
33. Visual aids.
34. Art classes.
35. Promotions.
36. Reading.
37. Customers and designers meet.
38. Totally out of element.
39. Expand horizons (work in other areas).
40. Working lunch.
41. Field trips.
42. Films.
43. Employees develop training materials.
44. "Failure" is acceptable.
45. Try it out.
46. Everybody's input counts!
47. Give challenge.
48. Give opportunity.
49. Train/teach how to use resources.
50. Make resources available.
51. Less supervision.
52. More responsibility.
53. "Crazy Idea of the Month" Award.
54. Pay percentage of $ savings.
55. Recognition/publicity.
56. Entrepreneur training.
57. Independent committee (not management) for idea review.
58. Give opportunity (funding, time, equipment, labs, support staff) for implementation.
59. Promote group participation and decision making.
60. More supervision and interaction between management and employees.
61. (a) Give bottom-line objective.
    (b) Give the people at the bottom an objective—challenge them to do it.
62. In bad times especially (and all the time), give budget support for creativity.
63. Budget priorities.
64. Pride.
65. Prize.
66. Role models.
67. Suggestion award systems.
68. Get rid of "negative" people.
69. Freedom for decisions.
70. Freedom for change.

Creative idea evaluation:

The team leader sorted the ideas into preliminary categories. The team members reviewed these categories and ideas; they shifted and rearranged ideas; they assigned some ideas to multiple categories; they renamed some categories. Then subteams of two or three people worked with individual categories in an effort to make ideas more complete, to combine ideas, to synthesize ideas, to hitchhike ideas. During this process, the meaning of ideas was further discussed; some ideas were assigned or combined with ideas in other categories. The "wild" ideas—ideas that did not fit into specific categories or ideas that were ambiguous—received special attention. These developed ideas are listed below by categories. The numbers in parentheses refer to the ideas as identified on the brainstorming list.

---

1. MANAGEMENT: EMPHASIZE THE IMPORTANCE OF CREATIVITY (28).
a. Do more careful listening to employees (30); exchange positions (cross-training with employees and managers); be more personally involved and encourage employees (60).
b. Provide correct tools, resources—books, reading/fun time, posters and signs (6, 18, 36, 50).
c. New procedure: independent committee evaluates ideas, not management (57).
d. Negative people are fired or counseled/trained to make them positive contributors (13, 68).
e. Establish budget priorities (consistent support) for idea generation, development, and implementation (3, 4, 50, 58, 62, 63).

---

2. MAKE CREATIVE THINKING REWARDING (7, 20)
a. Recognition/publicity (55). Prizes; contests, competitions; "Crazy Idea of the Month" award (9, 26, 53, 65). Promotion of creative individuals (35).
b. Profit sharing; percent of cost savings (54).
c. Opportunity for experiments—try out the ideas (45); give time for creative thinking (5).
d. Frequently implement "small" as well as "big" creative ideas (11).

---

3. EMPLOYEE TRAINING AND EDUCATION (2)
a. Provide training on how to use resources (49), how to communicate better—left-brain and right-brain people (31), and how to be entrepreneurs (56).
b. Have art classes to develop visual thinking and sketching skills (34), have seminars on many different (mind-expanding) topics (32).
c. Provide role models, leaders, experts (23, 66).

---

4. EMPLOYEE DEVELOPMENT
a. Develop pride and self-esteem by providing opportunities to be successfully creative (64).
b. Employees develop training materials, visual aids, signs and posters; they use or make films. They publish a "Creativity Newsletter" (10, 12, 33, 42, 43).
c. Employees devise award system and brainstorm procedures (67).

---

5. MAKE THE ENVIRONMENT MORE CREATIVE
a. Place employees in a completely unfamiliar element (stimulating environment) (15, 38).
b. Change office (nice furniture, nice views, plants, artwork, background music) (16, 17).
c. Go on field trips (41).
d. Have working vacations in nice locations, i.e., Hawaii (8).

6. MANAGEMENT: ESTABLISH A CREATIVE PSYCHOLOGICAL CLIMATE

a. Everybody's input counts (46) and failure is acceptable (44); play is O.K. and ideas are "fun" as well as interesting (6).
b. More freedom for change (70); more opportunities to use creative thinking (48); more focused responsibility with objectives (19, 21, 27, 61a, 61b); more freedom for decisions (52, 69); less supervision (51). Challenge all employees to participate (1, 47).
c. Give option of working during lunch. Have working lunches when useful. (40).

7. TEAM BUILDING AND GROUP ACTIVITIES

a. Switch jobs horizontally and vertically (24, 25); have working lunches (40).
b. Mix different groups—from different departments: customers and designers. Visit other departments (22, 37, 39).
c. Promote many different group activities and also group decision making (59).

8. WILD, ODD, AMBIGUOUS IDEAS (NEED ADDITIONAL BRAINSTORMING)

a. Beauty of solution—how can it be taught, encouraged—what does it mean? (29).
b. Change the organization—people can move to other organizations (misfits?) or change the organization's culture? (13).
c. Change management; this may be very difficult to do—replace people or change their thinking habits (14).
d. Pride—explore the implications of this concept as related to creativity (64).

Idea judgment:

The following criterion was chosen as being the most relevant (most important) to the problem as defined: ENCOURAGE CREATIVE THINKING. At this point, the team made an assumption that the organization already had a budget to support creative activities and training. Then the teams ranked the categories and selected "rewards" and "things management must do" as most important:

REWARDS:

1. Recognition/publicity. Prizes; contests, competitions; "Crazy Idea of the Month" award. Promotion of creative individuals.
2. Profit sharing—a percentage of cost savings due to employee's creative idea.
3. Give opportunity for experiments (trying out the ideas). Provide time for creative thinking. Frequently implement "small" as well as "big" creative ideas.

THE MOST IMPORTANT THINGS MANAGEMENT MUST DO:

1. Personal involvement of management. Listen to employees, then act on ideas.
2. Cross-training for managers and employees.
3. Provide tools and resources for creative thinking and developing ideas.
4. Promote changes in psychological climate: failure is acceptable, everybody's input counts!
5. Provide opportunities, freedom, and responsibility for change and creativity.
6. Provide a creative physical environment.

⏱ TEAM ACTIVITY 8-7: IMPLEMENTATION FOR EXAMPLE PROBLEM

Due to severe time constraints within their workshop format, The Orange Crush *members were unable to complete problem solving. Thus, no plans were developed for selling or implementing the "winning" idea or best solution. This is now the job for you and your team.*

a. Because many potentially useful and interesting ideas were generated and developed in the example exercise, your assignment is to supply the missing step in creative problem solving by working out an implementation plan for some or all of the top-ranked ideas. If you do not agree with the selection of best ideas (that is, if you think you can come up with even better ideas), you and your team may backtrack to brainstorm and engineer other solutions.

b. Analyze your own working environment. Then choose one of the ideas in the problem above that you consider crucial for improving your own creative work environment (and that of your co-workers). Develop a selling plan (including a list of benefits). Then "sell" the idea to your superior or upper management. Report on your experience (degree of success)—what have you learned? How can you use the results in the future?

EXERCISES AND ACTIVITIES FOR PRODUCERS

8-1   *READ THE BIOGRAPHY OF A FAMOUS PERSON*
As you read the biography of a famous person (either a historical or a present-day personality), be on the lookout for producer traits. Has this person stood up for ideas? How has this person overcome opposition? Write a short report and focus on those areas of the person's life where ideas were put into action. What can you learn from these examples?

8-2   SELLING PLAN
Take one of the top ideas from an earlier brainstorming, creative evaluation, or judgment exercise and develop a selling plan for this idea. Indicate the benefits and especially what you would do to build acceptance and overcome opposition.

8-3   ROLE MODEL
History as well as society today is filled with people who are producers—people who are willing to stand up (and even die) for their beliefs and ideas. Find a role model and describe what qualities you admire in this person. What are your criteria—on what are you basing your judgment? Now repeat this activity with another person—someone who is very different from the first model in ethnic background, beliefs, education, sex, age, etc. Despite the differences, do they have some producer traits in common? What are the most striking aspects of their battles? Select one attribute of these producers that you think is especially important and describe how you could possibly make it a part of your life.

8-4   *SELLING VIDEO*
As a class project, make a video on how to sell "ideas" suitable for teenagers.

8-5   SELLING TECHNIQUE ANALYSIS
From the marketplace or business world, identify some selling techniques. What makes them effective (or not effective)? What are the objectives? How well do they address customer needs? How could you resist this type of approach?

---

*Asterisks denote projects for more advanced learning.

8-6   *CHARITABLE ORGANIZATIONS*
Scan your junk mail for a couple of weeks and collect as many letters from charitable organizations asking for support as you can. Then analyze the content. What approaches do you find appealing and persuasive—what approaches do you find distasteful and negative? Select the best and the worst and prepare a 3-minute presentation; include an analysis of how well the appeal addressed all four brain quadrants.

8-7   *WISE CONSUMER*
How do you educate children to become discerning consumers who will be able to resist hype in advertising? Research the topic and add your own conclusions.

8-8   *SELLING YOURSELF*
How do you come across to other people? Are you projecting a positive self-image? Observe how you interact with people. Do you make derogatory remarks about yourself (especially in response to a compliment)? Make a pact with a friend to monitor and encourage each other.

8-9   BODY MAINTENANCE
Are you in good physical condition? Are you eating a healthy diet? Do you exercise on a regular basis? This is the time in your life to develop good health habits or break a bad habit. If there is something you would like to change, write out an implementation plan. Make a schedule, a list of benefits, set goals, and then do it. As an example, review the "stop smoking" plan.

8-10   WRITE A LETTER
Write yourself a letter describing one thing that you hope to accomplish by the end of next year. Seal the letter and give it to a friend or relative to mail to you a year from now. Make out a work plan to schedule all the steps you will need to take to accomplish the goal; then start doing these things one step at a time to make your dream come true.

8-11   EVALUATING AN EXPERIENCE
Can you think of a happening in your life that was terrible at the time but turned out to be a blessing in disguise (mostly because you learned an important lesson that helped you mature)? Write this up and then send it to someone that you think (a) would get a good laugh out of it, or (b) would be encouraged by it.

8-12   HUGS
Hugs are good for you emotionally; hugs also help your mind think more creatively. Have you hugged someone today? Three hugs a day for a child is not too much, and we never outgrow this need.

8-13   *IMPROVING YOUR MIND*
Do something for your mind. This month, read an inspirational book, a good bibliography, a book about inventions, a book about other cultures, other people, other countries. Meet a person from another country and see how well you can understand each other. Find out which daily routines are done differently in that part of the world from what you have always assumed was the only way of doing them. Make a "work plan" for your reading. Plan an activity (a party?) to meet a person from another country. Would you consider hosting an exchange student?

8-14   *EXODUS CONTINUED*
This is a follow-on to Problem 3-3. In the Bible (Old Testament), analyze Exodus Chapter 4, Verses 1-17. This is a fascinating example of God teaching Moses creative problem solving. Can you identify problem definition, idea generation, creative idea evaluation and judgment, "selling" the ideas, and implementation? Make a sketch of the process using the four brain quadrants of thinking preferences.

CHAPTER 8 — SUMMARY

Your Role as Producer: To reap the problem-solving payoff, follow a work plan; have courage; do not give up; maintain good communications and an effective support system. Practice good time management. Putting an idea into action is hard work and may take more time than exploration, idea generation, idea engineering, and idea judgment put together. Idea implementation takes careful planning and persistence; this process cannot be rushed.

Selling Your Idea: Analyze the context—develop your selling strategy. Develop a list of benefits; consider your audience. Use effective selling techniques: (1) Don't oversell—be moderate, don't brag. (2) Don't give up too soon; it takes time to build support for your idea. (3) Watch your timing; it is difficult to sell to a person who is annoyed or not feeling well. (4) Be brief and to the point; don't bore people. (5) Plan your presentation carefully and use visual aids for greater impact. (6) Make your ideas easy to accept; emphasize the benefits. (7) Avoid controversy; be courteous; be prepared for compromise. Idea selling requires quadrant C mode.

Reasons for Opposition: People don't understand your idea—they see no direct benefit to themselves. You are trying to do something outside the accepted "rules." People may be prejudiced or have other loyalties and commitments. People may feel that your idea is critical of their way of doing things. People fear the changes and loss of status that your idea may bring.

Working for Acceptance: Be agreeable toward your opposition and try to understand their point of view. Listen carefully and look for ideas that will make your plan even better. Don't take criticism personally. Be calm and friendly; don't attack or ridicule the opposition. Stay casual. Get people to cooperate and work with you since this will be to their advantage. Use humor. Have ideas that are "tryable" and innovation that is reversible or can be done in small steps. Build on what is already there; aim for tangible results. Find out if you could do the project successfully on your own. Motivate yourself—be a champion of your idea.

The Work Plan: Its purpose is to make sure that the idea will be put into action according to schedule and within budget. It specifies who does what, when, and why and can include risk analysis. Quadrant B thinking is used to develop work plans. The 5-W method and the copycat approach are simple work plans. Time/task analysis and flowcharts help identify overlapping activities and visualize the implementation tasks sequentially. PERT charts (often computerized) are used for planning as well as for progress monitoring of large implementation projects.

Implementation Monitoring and Final Evaluation: Implementation monitoring is a check to make sure the solution is working right. The first review may occur within two weeks after implementation, with a follow-up six to twelve months later. Specific testing may be needed to confirm the results. Procedures and documents are updated, and customer feedback is collected. In conclusion, write a brief summary of the final results and your experience with creative problem solving. Review what the process has done for you. What have you learned? Give positive feedback to the team members who worked with you.

REVIEW OF PART 2—APPLICATIONS

The questions and problems below will help you apply and practice what you have learned about creative problem solving and critical thinking. If you are in a formal class, selected problems from this review may be part of your final exam.

PERSONAL APPLICATION

It has been said that people can be divided into four categories: those who make things happen, those who watch things happen, those who wonder what happened, and those who are unaware that anything happened. Your studying this book should prevent you from ever belonging to the third and fourth group—you now have an increased awareness about problems and how they are solved; you have an increased understanding of how the thinking abilities of both hemispheres of your brain can be used more effectively; you will be on the lookout for trends, opportunities, and new and creative ways of doing things. The question before you is: Will you be an observer, or will you be a producer, a person of action, someone who applies and benefits from what has been learned?

1. *Describe three areas where you see a potential for applying creative problem solving in your personal life. If possible, express these as positive goals.*

2. *Indicate three areas where you could possibly apply creative thinking skills in your workplace or in your schoolwork.*

3. *Develop a concise work plan for accomplishing one of the applications from Problems 1 or 2 above. Include when, where, why, how, and what you are planning to do, and who else might be involved in helping you reach the stated goal. Also brainstorm a list of benefits.*

4. *You have had the opportunity to learn some valuable thinking skills. Now think about accountability. You have some choices to make—you can proceed to forget what you have learned as quickly as possible; you can make an effort to apply your knowledge in many different situations, and you can pass on this knowledge and teach some of the skills to other people. You can continue to improve your skills. For example, will you work with a group of other students or teach brainstorming skills to a friend or family member? How will you review your accomplishments and chart your progress over the next year? Make up a monitoring and evaluation plan.*

5. *Brainstorm some positive slogans that will encourage you to use creative thinking and creative problem solving. Draw a circle around the statement you like best. Jot it down on a separate note card. Post it where you will see it often.*

The biggest room in your house is the room for improvement;
the largest window is the window of opportunity.
Roger Milliken, CEO of a Fortune 500 company.

6. *This is a team project. Write down a list of problems that would be fun to solve creatively with your family or a small circle of friends on a weekend. Then pick one of the projects, using the following list as your selection criteria:*

    a. If you are a tense person, find a project that will teach you to be more relaxed. Make an implementation plan; have your family and friends help you carry it out.

    b. If you are inclined to be lazy, find a small project that will require you to be organized and energetic. Make a work plan—include means that will make you accountable for your progress to your family or friends, as well as a reward structure for progress and completion.

    c. If you are in a rut or depressed, try something new and creative, or something in which you will be helping someone else, especially if it involves physical activity. Identify a project that you can do with one other person. Make an implementation plan, tell the idea to that person ,and enlist his or her cooperation.

    Do the project; then write a one-page summary of the results from the point of view of all four quadrants of thinking preferences.

CRITICAL EVALUATION

In Bloom's taxonomy, judgment is the highest level of thinking. Now you can demonstrate what you have learned about critical judgment.

7. *Describe your interest and involvement in the creative problem-solving book.so far. In what way has the material met (or not met) your expectations?*

8. *Evaluate the amount of work required to learn each step. Would you have liked more explanations, activities, examples, questions, group discussions, problems, and applications? Which activities did you like best—and why? Which material did you skip?*

9 . *Describe in a one-paragraph summary the highlights of what you have learned. What type learner/thinker are you? Which activities taught you the most? Think about and discuss how creative problem solving has improved your thinking skills.*

10. *Brainstorm and creatively evaluate a list of criteria that can be used to judge each topic (chapter) presented in this book (or material in any workshop or class that you are taking).*

    You will have opportunities to complete evaluation forms in college classes, at conferences, or at professional workshops. Many forms use some type of scale. For example, you may have a seven-point scale to answer a number of questions about the course material and the speaker or presentation. The scale may be labeled "poor" at one end and "excellent" at the other. Or you may be given a choice of four or five statements to chose from, such as strongly agree, agree, disagree, strongly disagree, not applicable. In our view, these types of evaluation forms have serious shortcomings because the evaluator is not being asked to use explicit criteria or explain the answer, which can be colored by personal bias or misunderstanding. Evaluations are helpful when they show why an item has been rated high or low. Good instructors will be encouraged by knowing what has been most helpful to you and why—and how they can become even better. Those who are not rated very high will learn about possible causes of difficulties—and they will be given positive ideas and concrete suggestions on how these areas could be improved.

11. *Review all the chapters that you have read or studied in this book. For each chapter, select something that you especially liked and something that you would score low— using your list of criteria from Problem 10. Next, look over your selections and choose three high-ranking items and three low-ranking items to be used in the following two problems.*

12. *For each of the low-ranking items, explain why you rated the material or presentation low. Indicate what criteria you used to make your judgment. Then describe how you would improve the situation, taking the context of the problem into account. Also think of some conjectures of what might be possible causes of the deficiency. How do you think your solutions will be received by other students or readers? Again, give some supporting evidence for your opinions. What do you think about this piece of advice: Critics are not allowed to "rewrite the play"?*

13. *For each of the high-ranking items, explain the reasons for your favorable rating. If you like, you may offer suggestions and ideas on how these items could be made even better. Give an assessment of how people with different thinking preferences might judge the same items.*

14. *How successful were the authors—and your instructor if you were in a formal class or workshop—in (a) challenging your thinking; (b) expanding your horizons; (c) getting you to appreciate different viewpoints; (d) making you more comfortable about taking risks with ideas; (e) getting you involved in group projects and group study; (f) getting you to assume responsibility for your own learning; (g) challenging you to "claim your space" and develop thinking skills in all four quadrants; (h) improving your visualization and memory skills; (i) encouraging you to think positively and appreciate your abilities; and (j) teaching you to think critically?*

15. *Should an evaluation be submitted anonymously or with the evaluator's name? Make a case with supporting arguments for both views.*

We believe that thoughtfully working through these problems will give you practice in doing critical evaluations. In life, you will have many occasions to evaluate materials, ideas, people's performance, project proposals, building plans, and your own accomplishments. Evaluation is a serious responsibility: jobs, promotions, project funding, innovation, and even the survival of an organization may depend on the results of your assessment and critical thinking; thus do not take this activity lightly. Use your whole brain—analysis, intuition, values, and imagination—to make good judgments.

Part Three

Applications

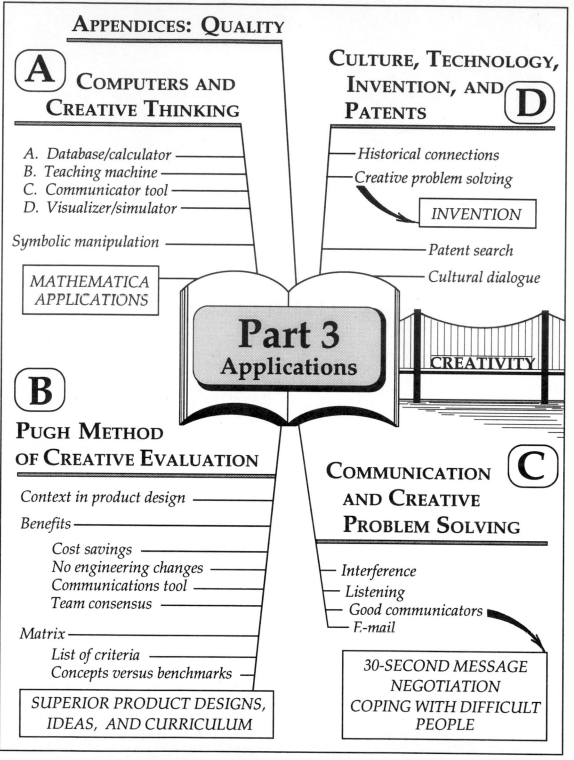

APPENDICES: QUALITY

A COMPUTERS AND CREATIVE THINKING

A. Database/calculator
B. Teaching machine
C. Communicator tool
D. Visualizer/simulator

Symbolic manipulation

MATHEMATICA APPLICATIONS

CULTURE, TECHNOLOGY, INVENTION, AND PATENTS **D**

Historical connections
Creative problem solving

INVENTION

Patent search
Cultural dialogue

Part 3 Applications

CREATIVITY

B

PUGH METHOD OF CREATIVE EVALUATION

Context in product design
Benefits
  Cost savings
  No engineering changes
  Communications tool
  Team consensus
Matrix
  List of criteria
  Concepts versus benchmarks

SUPERIOR PRODUCT DESIGNS, IDEAS, AND CURRICULUM

COMMUNICATION AND CREATIVE PROBLEM SOLVING **C**

Interference
Listening
Good communicators
E-mail

30-SECOND MESSAGE NEGOTIATION COPING WITH DIFFICULT PEOPLE

Mindmap of Part 3 (Chapters 9 through 12). Refer to Chapter 2 (pp. 55-56) to learn more about mindmapping.

9

COMPUTERS AND CREATIVE THINKING

WHAT YOU CAN LEARN FROM THIS CHAPTER:
- *The relationship between computers, communications, and problem solving; computers—a tool for enhanced thinking in all four quadrants.*
- *How are computers used in teaching math?*
- *An introduction to MATHEMATICA®, its uses and its power.*
- *Using MATHEMATICA with creative problem solving—detailed examples.*
- *Further learning: practice problems in the text and bibliography.*

COMPUTERS AS A THINKING TOOL

When we apply what we have learned to new situations, we progress to a higher level of knowledge. Remember the learning cone from Figure 2-2 and the top eleven subjects that professionals in technical fields ought to know, as listed in Table 1-1? We asked ourselves: "Which topics would be best to demonstrate applications of creative problem solving?" We arranged the top eleven subjects into categories and then visualized the relationship between these categories. The resulting diagram looked like this, with the connecting lines indicating paths of communications (verbal, written, visual):

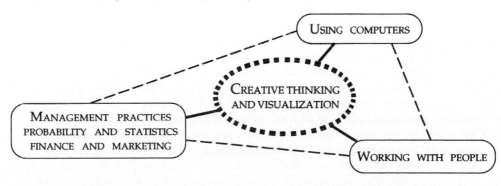

317

This diagram shows that creative thinking can enhance our use of computers and the way we interact with people. It can improve how we integrate various business techniques with computers in the service of the customer and colleagues, where communications is the important connecting link. This made it easy to determine two topics for Part 3: computers and communications—both are enhanced through creative thinking, visualization, and creative problem solving. For a "practice" that would be useful not only to engineers but to anyone working with people in committees, families, friends, or work teams—wherever different options have to be evaluated fairly—we chose the Pugh method. The Pugh method not only improves the design and development of products, it has a much wider application to "engineering" ideas, optimizing solutions, and working in teams.

We have shown how to come up with creative ideas and develop them into good solutions. What if one of these ideas were a patentable invention? What is the wider context—how are culture, invention, and technology connected over the span of history in various parts of our globe? Providing this broad view of invention, together with practical hints about the patenting process, was the rationale for choosing this topic for the last chapter. Each of the topics in Part 3 involves the whole brain but has a primary focus in one of the brain quadrants, as illustrated by the mindmap. Although Chapter 12, with its emphasis on creativity, ties the book together, it is not a close, but a beckoning bridge to encourage further learning and applications. Finally, the Appendix contains material to increase understanding about advanced techniques now coming into use in U.S. industries to improve product quality and thus their ability to compete in the global marketplace.

Pick and choose as you read through Part 3. Although these subjects are connected—as the diagram on the previous page illustrates—they do not need to be studied in any particular sequence. Instead of continuing with this chapter on computers, you may want to start with communications (Chapter 11) or the Pugh Method (Chapter 10), especially if you need these skills for the "producer" activities in Part 2. Would you like to enhance your visualization and mathematical problem-solving skills? If your previous experience with algebra, calculus, or science computation was less than ideal, try to supplement analytical problem solving with an interesting computer tool and a creative approach that combines contextual thinking and visualization—you may be surprised at how enjoyable this learning can become. So without further ado, let's start with the subject of computers as a tool for enhancing our thinking skills.

Computers are the best analogy to the functioning of the human brain. The desire to understand computers better has motivated increased brain research. However, remember that many essential differences exist between the human brain and a computer—a key difference is that the brain can synthesize information and think creatively. Computers constitute an irreversible change in society and culture, because—as stated by Robert O. McClintock of Columbia University in the *Teachers College Record* of Spring 1988—with binary coding we now have a basis for storing and retrieving all the contents of our culture, and it also makes possible the development of intelligence external to the human mind. The question that we—

personally and organizationally—need to think about is: What are we going to do with computer technology? In the context of teaching, learning, and thinking there are at least four distinctly different ways that we can use computers.

USE OF COMPUTERS IN THE LEARNING ENVIRONMENT

Database and Data Processor (Calculator)
Teaching Machine
Communications Tool
Simulator and Visualizer (Graphics)

The computer as a database and data processor (calculator): For analytical problem solving and data processing, many people get their introduction to computer workstations and powerful software (including spreadsheets and business graphics) at their workplace or through career training. Computers are excellent for processing large amounts of data and for "crunching" numbers—they can do complicated calculations much more quickly and accurately than the human brain if they are programmed with the correct procedures. Computers can be used for inquiry—for example, for doing a library search on a specific topic or for accessing national databases via telecommunications networks. Strong A-quadrant thinkers thrive when they can work with computers as an analytical tool; they like the command-line interface and disdain icons as a means of communicating with computers; they are fascinated by the microprocessor technology itself.

The computer as teaching machine: When computers are used with good-quality interactive software, they can make excellent teachers. They never get tired and cross with the student. They can transmit information and concepts; they can be used for practice and testing of mastery (comprehension). In drill and practice, they offer instant feedback, but they are not very suitable for teaching higher-level thinking skills. When teaching routine procedures and problem-solving techniques, they make good one-on-one tutors for B-quadrant, sequential learners but are not as effective with students having right-brain learning styles. Writing computer programs is excellent training for quadrant B thinking. Software tools for evaluation (i.e., spell and grammar checkers) also appeal to quadrant B people.

The computer as a tool for communicating: Computers can be used in many ways to enhance communications. First, as a word processor, they can help with writing skills. Students are more likely to use computers as a resource and writing tool if the software is user-friendly. Text and visual information can be combined for communicating powerful messages. Computer networks are an efficient and amazing way to move information and feedback between people, as discussed in more detail in Chapter 11. Software now available allows people to brainstorm as individuals or in groups. As a communications tool, computers are limited in that they do not transmit nonverbal clues. On the other hand, a common knowledge level about computers can create a bond among people that facilitates communication. People with C-quadrant thinking preferences become "friends" with computers as a helpful tool that lets them interact with other people.

The material being taught is becoming more and more complex,
so instructors need to have more powerful teaching tools to meet instructional needs.
Klaus-Peter Beier, associate director, Laboratory for Scientific Computation, University of Michigan.

The computer as simulator/visualizer: Computers perform a vital teaching function when they are used as simulators. They can realistically represent many situations that would be too dangerous to duplicate in real life, such as in pilot and air traffic controller training. They are being successfully used to demonstrate and model real-life situations in physics, math, chemistry, biology, engineering, and even the social sciences. Through scientific visualization, relationships between parameters are presented visually or dynamically. Computers can integrate many different sources and techniques in truly multimedia presentations in high-tech classrooms. Good software packages with interesting problems give students an active role in problem analysis and problem solving. The underlying assumptions in these programs, however, are not always obvious and have to be carefully explored. Spreadsheets are increasingly employed to ask what-if questions. When used as simulators, computers are a good learning tool for people with D-quadrant (global) learning and thinking preferences. Computer capabilities for playing with what-if questions and presenting data graphically—thus helping to visualize problems and results—offer many valuable ways to enhance quadrant D thinking. Communication as expressed in desktop publishing and as an artistic medium is explored by D-quadrant thinkers; they like to "play" with sketching and painting utilities, 3-D models, and animation; they also like nonlinear, nonviolent computer games with multiple ways to win. We have noted that quadrant D people seem to prefer an icon-driven environment—a more visually-oriented way of communicating with the computer.

Integration—the computer as a whole-brain thinking tool: The word processing, database and spreadsheet capabilities, together with graphics packages—can be used for integrated applications. The HyperCard program is one of these integrated software packages. HyperCard can even be used to control a videodisk player. It thus expands the computer's capability to a multimedia concept. A multimedia presentation can include sound, graphics, animation, dynamic simulations, and video, thus meeting the needs of students with different learning styles. Instructors can create their own materials, or they can use packages prepared by experts for particular topics. The large amount of computer memory required for multimedia is provided by CD-ROM. In the classroom of the future (by the late 1990s) students will use their own personal portable computers to interact with the instructor and classmates, "taking notes" from the lecture and sharing information with the class. With such a system, lectures can be more fully integrated with homework, encouraging students to explore topics in more depth than is possible today when books and computers are not linked. Computers can make colorful transparencies (slides or overhead) and enhance teaching materials in many other ways, letting instructors prepare effective visual aids, handouts, up-to-date workbooks, and text-and-software combinations. For instructors to explore and make use of these presently available or upcoming new technologies will require creative thinking and problem solving; it will ask them to take risks with new paradigms to increase the quality of teaching. Having the tools available is not enough—we must learn the specific skills for using them; we have to understand what they can do and why we want to use them to enhance our creativity and our work. Computers have become an exciting and creative tool for artists as well as scientists; thus they have allowed for synthesis between the two. One example is the area of fractals, chaos, and the Mandelbrot sets. When depicted on a color monitor, these equations produce incredibly beautiful designs. Artists, sculptors, architects, and designers have learned to use computers as a tool for expressing (and changing) their ideas—in addition to using pencil and paper. This allows them to experiment with many more ideas than was possible before. We think that other creative uses will be discovered for computers—we are just beginning to explore the possibilities.

Here is one look, not too far into the future. What does a multimedia classroom lecture look like? The instructor stands behind a console on the right side (as seen by the students), controlling the projection on the wall or screen to left of center in the room—the best location for optimum learning and processing visual information in the right brain. With a click of a mouse, text can be displayed, a video clip can be shown, a computer simulation can be projected, or a live broadcast (via C-Span) or from a classroom across the country can be accessed. Let's say you want to explain the design of a photovoltaic (PV) power system and how it would be installed in a home. You could project the three-dimensional architectural rendering of the house—how it looks from the street. In a different color, the solar cell panels would appear on the roof. Then the image of the house would become transparent, so the wiring to the inverter located in the garage and the connections to the house electrical system, the utility power grid, and the lightning rod could be seen as a flow diagram. A brief video clip could show some close-ups of workers doing an actual installation, followed by examples of houses with PV systems from all over the world. If C-Span happened to carry a live broadcast of a congressional hearing on energy conservation and renewable energy sources, students could listen in for a while, before they are asked to brainstorm the broader context of PV systems—what are the trends, why would people use them, etc.? Next, the theory and mathematical models can be covered with an interactive computer program and graphics. Then, applying creative problem solving, teams of students design and optimize a system for a particular application, using solar insolation data and manufacturer's specs accessed through a computer database.

For efficiency—as for any other tool—computer usage requires instruction and practice. With computers, however, this is not a one-time activity because a learning curve is involved with each new program or system. Some user-friendly software may only need a few minutes of instruction, whereas complicated programs require weeks or months. However, once the initial investment in time for learning has been made, computer proficiency will free up time for more creative activities (in place of drudgery). In practice, though, the time savings usually quickly disappear because people discover many new uses for their new skills. Thus the end result is not necessarily in time saved, but in increased quality and creativity of their work. In what ways are you using computers as a thinking tool? Identify an area where you need to learn more—then make a plan on how to do it.

COMPUTERS AND LEARNING MATHEMATICS

Why is the fear of computer technology keeping many people from advancing in their careers or becoming more effective communicators, teachers, or thinkers? Since many people have acquired a dislike of math, we want to show how computers and creative thinking can enhance the understanding and enjoyment of mathematics and problem solving. First, let's look at barriers and trends—the context. In 1958, Dr. Louis Pipes of Harvard University wrote in his second edition of the book *Applied Mathematics for Engineers and Physicists,*

> In the twelve years that have elapsed since the publication of the first edition of this book, there has been a tremendous development of high-speed computing devices. These modern calculating machines have made possible the solution of problems in engineering and the physical sciences that had been considered too formidable for analytical solutions.

Thirty-three years later, in 1991, Dr. Navaratna S. Rajaram of the University of Houston wrote (in a paper he sent to us on the subject):

> The classroom has so far resisted the computer invasion; science and math are taught no differently today than fifty or even a hundred years ago, except possibly worse. A majority of math teachers, especially at universities, are overtly hostile to the use of computers and even calculators. At my university, students are not permitted to use calculators in their calculus courses, the argument being that their use prevents students from learning the fundamentals.

Although some changes are taking place, we have observed similar attitudes at many other universities. Math professors still avoid computers, and engineering professors use computers for number crunching and not for the improvement of teaching, thinking, and learning. Consider calculus. The knowledge of calculus is essential to an understanding of science and technology in today's world. The subject of calculus already occupies a central position; three-quarters of college-level math is calculus, and it is the last math course taken by a majority of our national leaders (lawyers, doctors, leading educators, and business managers). Yet there is a widespread view that most of what students learn in calculus is irrelevant to the workplace. At any one time, 300,000 high school students are enrolled in calculus (about 60,000 in advanced placement courses for possible college credit); 100,000 students are enrolled in two-year colleges, and 600,000 in four-year colleges or universities—of these, about half are enrolled in "engineering" calculus, the other half in "soft" calculus. Less than 5 percent of these calculus courses require the use of a computer, and about half do not permit calculators on final exams. Only about half the students enrolled in calculus successfully complete the course— all too often, the course is designed to "weed out" students instead of helping them develop their mathematical abilities by teaching to their thinking preferences.

The calculus curriculum has essentially remained stagnant. Because of the many class sections needed, it is often taught by part-time instructors or by the least-qualified, unmotivated teachers. Homework is rarely assigned and even more seldomly collected and graded, thus providing little motivation and feedback to the students. Many changes will be required to improve the situation, ranging from a thorough revision of the entire curriculum to generating new textbooks and software. Until this happens on a broad scale, what are students and people who use math frequently in their work to do? We can take the initiative and learn how to apply computers on our own, not only for eliminating the mechanical (plug and chug) part of problem solving, but to enhance thinking. The time that is no longer needed for doing computations (since computers can do these very quickly) can now be used to focus on the fundamentals, to investigate and define the problem and its broader context, to model the problem, to visualize the problem and results graphically—all of which enhance learning and make it fun. The interactive nature of the software allows us to ask what-if questions, thus encouraging creative thinking which can generate multiple solutions to real-world problems.

Computer Literacy

Let us assume we agree that knowing how to use computers can enhance learning and thinking. But what kind of computer skills does this involve—aren't students already required to take computer courses in high school and college? To answer this question, let's look at the current situation. Many universities teach a course in Fortran, Basic, Pascal, C, or Lisp language to engineering and science students at the freshman or sophomore level, in the belief that this skill will be used later in their college program, yet few students use this knowledge in their later courses. What they have learned is quickly forgotten through disuse; few can write a program to deal with anything but the simplest technical applications. Perhaps only students with career goals in programming need to study these languages. The introduction to computer languages received in high school may be sufficient to let students understand and appreciate the thinking skills needed to write code.

In problem solving, creative idea evaluation requires analytical, mathematical, or physical modeling of the problem before judging and selecting the best solution for implementation. This is the step in which the solution to the problem using a declarative language (such as MACSYMA or MATHEMATICA) rather than a procedural language (such as Fortran) becomes very important. Graphical output (or animation in the case of a dynamic system) is needed to examine the solution. Having to write a computer program at this stage (or having an expert write one) detracts from the focus of finding an optimum solution. Until a few years ago, procedural computer languages were the only tools available to model problems. However, today engineers and scientists can model and graphically display the results with very little knowledge of programming by using **symbolic manipulation programs (SMP).** These symbolic programs are easy to use and are a powerful tool for asking what-if questions. They reduce tedious mathematical manipulation and change the focus of the problem from computation (the "how") to contextual problem solving (the "why"). Instead of looking at the steps of how to get to "the" answer, solutions can be examined in-depth to find the best answer, thus coming to a thorough understanding of the broad problem. SMP programs encourage young (and older) students as well as inexperienced adults to overcome the fear of this technology, as it helps them understand mathematical applications.

The analogy of travel can explain the difference between using symbolic and procedural languages. When pioneers traveled across the continent in wagon trains, the process of getting to the destination (if one made it there at all) was so time-consuming and difficult that people did not make too many plans about the details of what they would do once they reached the destination. Today, with jet travel, the journey is quick and routine, and we spend our time preparing for what we will do after we arrive. The symbolic software that we have available now enables us to "travel quickly" through the drudgery of computation and then explore the wider realms of the problem and solution in many different ways.

Symbolic manipulation programming started at MIT with MACSYMA. Today, additional symbolic programs are available, including DERIVE, MAPLE, THEORIST, and MATHEMATICA. We have chosen MATHEMATICA for this book primarily because of its widespread use, its capability for problem solving, relative ease in terms of purchase, reasonable cost, and our familiarity with the program. A detailed comparison of SMPs is given in the March 14, 1989, issue of *PC Magazine*. Many excellent problem-solving software packages based on declarative language programs are available, such as MATHCAD; however, these solve problems numerically, not symbolically. The problems and examples shown in the remainder of this chapter could similarly be solved with other SMP packages—by using MATHEMATICA as an example, we want to demonstrate the power and usefulness of symbolic language for enhanced thinking and problem solving.

> *Computers will become tools and companions of the everyday educated person.*
> *The computer revolution will change not only the way we live, but ultimately the way we think.*
> *Joseph Deken*, The Electronic Cottage, *1982.*

USING MATHEMATICA

First, we will provide some basic facts about MATHEMATICA, and then we will present an introduction on how to use this software—what you should know to get started. Since the best way to learn is by doing, we want to encourage you to find a computer with MATHEMATICA installed, and then do the example problems and assignments. You will be surprised how easy it is to do math!

MATHEMATICA is available for Apple Macintosh and for 386-based MS-DOS systems, as well as for many different workstations. The program has 1.5 to 3 megabytes of compiled code (depending on the computer system) and requires at least one megabyte of working space. Here is a summary of the things it can do:

Numerical computation: The program handles the following mathematical functions: elementary transcendental, orthogonal polynomials (Legendre, Gegenbauer, Chebyshev, Hermite, Laguerre, Jacobi) gamma, beta, polygamma, Riemann zeta, Lerch, polylogarithm, exponential integral, logarithmic integral, error, Bessel, Airy, Legendre, hypergeometric, elliptic integrals, Jacobi and Weierstrass elliptic functions, where all functions are evaluated to arbitrary precision for any complex defined parameters.

The program does numerical matrix operations (inverse, determinant, null space, eigenvalues, eigenvectors, singular value decomposition, pseudo-inverse) and data analysis with generalized least-squares fit and Fourier transform. It can do numerical operations on fuctions: integration, summation, products, root finding, and minima.

Symbolic computation: The program can do polynomial operations—expansion, factoring, resultant, decomposition. It can do common denominators, partial fractions, and heuristic simplification with rational functions. In calculus, it can do partial and total differentiation, integration, power series, and limits. It can find analytic solutions to linear and polynomial equations and systems of equations, when possible. It can perform symbolic matrix operations: inverse, determinant, null space, eigenvalues, and eigenvectors. It does list operations: subsequence extraction, removal and replacement, union, intersection, complement, sort, flatten, partition, permutations. Also, mathematical funtions can be used on lists. Generalized tensor operations can also be performed.

Graphics: The program can execute two-dimensional, contour, density, and three-dimensional graphics in black and white and color. It can plot single functions, multiple functions, and parametric curves with many options (such as contour, surface, and density plots with special shading and lighting effects). It can do animation with MATHEMATICA-generated plots or imported graphics.

Different versions of MATHEMATICA are available for different types of computers. The most inexpensive package is the student version for personal computers.

Getting Started with MATHEMATICA

There are two ways of learning to use MATHEMATICA—you can do it by trial-and-error, or you can begin by learning some of its rules of operation, or "conventions." Knowing the rules (especially the naming conventions) will save you much time. In this discussion we will show what you type on the keyboard (your input) in boldface Courier font and the computer's response (the output) in regular Times font. MATHEMATICA numbers each set of input and output statements, but to keep the text uncluttered, we have deleted these italicized, sequentially numbered statements.

MATHEMATICA conventions:

1. Built-in commands all begin with a capital letter . If two words are joined to make a command, each is capitalized. These are the only words that are capitalized.

   Examples: **Solve** or **ParametricPlot** .

2. Operation signs are: + (addition), – (subtraction), * (multiplication), / (division), ^ (raising to the power). A space between letters and numerals also denotes multiplication. When variables are multiplied, they must have a space or * (asterisk) between them.

   Examples:
   2x, 2 x, and **2*x** all denote the product of 2 and x.
   a*x, x y t are correct; **ax, xyt** are incorrect and would not be evaluated.

3. Square brackets [ ] are used to enclose the arguments of functions; curly brackets { } denote lists and matrices. Double square brackets are used to extract elements of lists; parentheses ( ) indicate group symbols and order operations.

   Examples:
   Entering   **f[x]**   returns the value of f for the variable x.
   Entering **v = {a,b,c}** defines the list v to consist of symbols a, b, and c.
   Entering **v[[1]]** returns a, **v[[2]]** returns b, and **v[[3]]** returns c if v has been defined as above.

4. MATHEMATICA has a number of ways of specifying equality:
   x = y means substituting x for y immediately, as defined.
   x == y means true equality, as defined in logic. In particular, it is used for equations with the **Solve** command.
   x := y means "delayed" use of the definition. The definition is remembered, however, and when x is called, the definition y is substituted.
   x != y means x is not equal to y, similar to the mathematical inequality ( ).

5. The symbol % stands for the output of the previous result. **N[%]** means "Give the numerical equivalent of the previous expression."

6. Expressions in quotation marks are not evaluated but appear as text. The semicolon (;) is used to suppress the preceding statement; it is also used as an end-of-line marker. The symbol (/.) means "replace all." The double slash symbol (//) means reversing the parts of the preceding statement.

7. Independent variables are indicated with an underline bar immediately following the symbol, i.e., **x_** .

8. **Clear[f]** means that any previous definition for a function f in the computer's memory is deleted; you can now write a new specification for f.

9. MATHEMATICA provides online help; you can ask it to explain the use of a symbol, command, function, etc. To get this information, you enter a question mark followed by the symbol or name of the object.

   Examples:

   ?Solve
   Solve [eqns, vars] attempts to solve an equation or set of equations for the variables vars. Solve[eqns, vars, elims] attempts to solve the equations for vars, eliminating the variables elims.

   ?ListPlot
   ListPlot [{y1, y2, ...}] plots a list of values. The x coordinates for each point are taken to be 1, 2, ...
   ListPlot [{{x1, y1}, {x2, y2}, ...}] plots a list of values with specified x and y coordinates.

Explanations of additional MATHEMATICA symbols and conventions can be found in the glossary included in the MATHEMATICA textbooks listed in the references or in the user's manual furnished with the software package.

Experimenting with MATHEMATICA

Now turn on the computer; find the MATHEMATICA file, and open it. Then double-click on the MATHEMATICA icon (on a Macintosh). For IBM-type computers, follow the appropriate procedure for that system. We have listed a few sample problems with answers. To prepare the screen for each problem, press the space bar once. Then type the problem statement *exactly* as it appears (including the spacing). After typing the command, press and hold down the SHIFT key while pressing the RETURN for Macintosh or the ENTER key for IBM. Verify that you are getting the same answer and output for the computation and graphics problems. If not, double-check your entry. Did you leave out a bracket? Did you use unconventional capitalization? Did you add extra spaces? MATHEMATICA will usually give you some clues on-screen of what you might have mistyped if it is unable to execute the command.

Exercise Problems—Computation

PROBLEM	SOLUTION
1. `(26+33-14)/8`	$\dfrac{45}{8}$
2. `N[%]`	5.625
3. `Expand[(2x+7y)*(5x-6y)]`	$10\,x2 + 23\,x\,y - 42\,y2$
4. `N[Pi,20]`	3.1415926535897932385
5. `1771561^(1/3)`	121
6. `34611 4795`	165659745
7. `Tan[3 Pi/4]`	-1
8. `Factor[12x^2+16x y-3y^2]`	$(6\,x - y)\ (2\,x + 3\,y)$
9. `Solve[{3x+2y==16,4x+5y==33},{x,y}]`	$\{\{x \rightarrow 2\}, \{y \rightarrow 5\}\}$
10. `FindRoot[x^2-6x+5==0, {x,0.5}]`	$\{x \rightarrow 1.\}$
11. `Limit[(3x^3+5x^2-7)/(10x^3-11x+5), x->Infinity]`	$\dfrac{3}{10}$

Exercise Problems—Graphics

1. `Plot[(3x^3+5x^2-7)/(10x^3-11x+5),{x,-2,5}]`

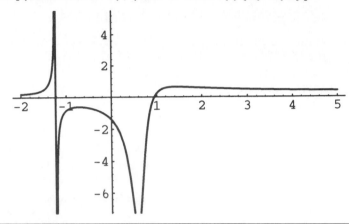

2. `ParametricPlot3D[{Sin[v] Cos[u],Sin[v] Sin[u], Cos[v]},`
   `{u,0,2Pi},{v,0, Pi/2}]`

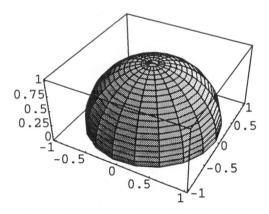

3. `Plot[{x^3+x,2},{x,-3,3}]`

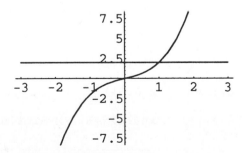

MATHEMATICA AND CREATIVE PROBLEM SOLVING

We will now show examples of the thinking used (including computation and visualization with computers) in the various steps in problem solving.

Example 1— Textbook Problem 1-16a

> *The sum of two whole numbers is 24. One of the numbers is three times larger than the other. What are the numbers?*

Problem definition: As detectives, let's ask ourselves what we know about the problem. We know the numbers are integers—the problem said "whole numbers." Thus, two integers, x and y, are equal to 24, where y is equal to 3x. Since the sum is positive, and $y = 3x$, both x and y must be positive, with y larger than x. We also know that both x and y are numbers that are less than 24.

Idea generation: Take five minutes to quickly jot down ideas for solving this problem. Do not judge your ideas yet. Some ideas are:

1. Make a table in which we list a number in the first column, three times that number in the second column, and their sum in the third column. If we start with one and do this with consecutive integers, eventually we will find the right combination that sums to 24.

2. Use a similar idea, but with weights in one-pound increments. Weigh a one-pound weight with a three-pound weight, a two-pound weight with a six-pound weight, and so on. Do this until the scale reads 24 pounds. Record the weights used to get the required answer.

3. Take 24 jelly beans. Divide them into two piles. One pile is for you and one for your little brother. Every time you put a bean in a pile for him, give yourself three. Do this until the beans are all allocated. Then count the number of beans in your brother's pile (name it x) and in your pile (name it y).

4. Get several cats or dogs in a room together (or do it in your imagination!). Each animal has one left front leg and three other legs. How many animals should there be in the room so that there are twenty-four total legs? The number of left front legs is the number x, the number of all other legs is $24 - x$. Can you "see" the answer: x is equal to the number of animals?

5. Guess the answers.

6. Ask someone else for the answer (or for help).

7. Try to find a book that has this problem worked out; look up this answer.

8. Use algebra to solve the system of equations $x + y = 24$; $y = 3x$. Use a matrix technique or a graphical technique.

9. Use algebra to solve (or plot) the equation $x + 3x = 24$.

Idea evaluation: Some of the above ideas are not practical and may not easily lead to a solution (like 5 or 7). Also, it is no fun or challenge to do 6—although that is a good problem-solving strategy when you are dealing with a difficult problem. It would be fun to do 3 or 4. Since 2 needs a scale and weights, it might be very difficult to do. Since 1, 8, and 9 are mathematical approaches where we can use the computer, we will continue with these ideas.

Idea judgment: We have three options (1, 8, or 9). What are some criteria that we can use to evaluate them? Ease in making the calculations and the accuracy of the solution would be two important items. On the basis of these two criteria, all the ideas will lead to the correct solution, but 9 is easiest, 8 less easy, and 1 involves unnecessary work that does not directly contribute to the solution. But to demonstrate the use of MATHEMATICA, we will show all three solutions.

Implementation of Idea 1: Make a 3-column table of integers: for x, $3x$, and $x+3x$, where $x=1, 2, 3 \ldots$. When the sum of the integer and three times the integer is exactly equal to 24, we will be able to see the answer to the problem. The MATHEMATICA commands to do the table are:

```
TableForm [Table[{x,3x,x+3x},{x,0,8}],
TableHeadings->{None,{"x","3x","x+3x"}}]
```

x	3x	x + 3x
0	0	0
1	3	4
2	6	8
3	9	12
4	12	16
5	15	20
6	18	24
7	21	28
8	24	32

By inspection we can see that the required numbers are 6 and 18. To set up the table, we had to take a guess at the range of values we needed for x. If the problem had been more complicated, we could easily have underestimated or grossly overestimated; in both cases, this would have wasted some time and effort before we got to the right answer.

Implementation of Idea 8: We can solve the system of equations, $x + y = 24$, $y = 3x$, in three ways: (a) by using the command **Solve**, (b) by plotting, or (c) by matrix techniques.

Here is the first way:

```
Solve[{x+y==24, y==3x},{x,y}]
{{x -> 6}, {y -> 18}}
```

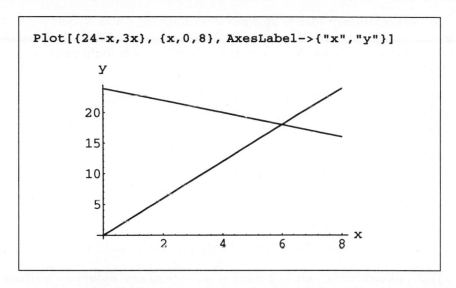

```
Plot[{24-x,3x}, {x,0,8}, AxesLabel->{"x","y"}]
```

Where do the lines cross? On a Macintosh, move the cursor (without depressing the button on the mouse) until it is on top of one of the plot lines. Click the mouse button once. A frame will appear around the graph. Move the mouse until the crosshairs are on top of the intersection point. Push the Apple key, then look at the lower left corner of the screen. It shows the {x,y} coordinates of the point. In the example, you should see {6,18}. Depending on the accuracy with which the computer reads the grid, the numbers may show (6.02, 17.98), for example. However, since you know that the answers must be integers, the closest whole numbers (rounded up or down) are your answers. Did you enjoy the graphical approach?

Finally, let's try a matrix approach. We can set up the system

$$\begin{bmatrix} 1 & 1 \\ 3 & 1 \end{bmatrix} \begin{bmatrix} x \\ y \end{bmatrix} = \begin{bmatrix} 24 \\ 0 \end{bmatrix}$$

to represent the equations $x + y = 24$ and $-3x + y = 0$. The system can be solved in a variety of ways, such as by Cramer's Rule, by matrix inversion, or by Gaussian elimination. MATHEMATICA has commands that make matrix solving very easy. We will use **LinearSolve**, which is a command for solving matrices. We will tell MATHEMATICA to solve the system as follows: the first (2x2) matrix is called matrix1, and the (2x1) matrix on the right is called matrix2. To enter the matrices, we use a set of curly brackets around each row. If the matrix has more than one row, the entire set is enclosed in another pair of brackets.

```
matrix1={{1,1},{-3,1}};
matrix2={24,0};
LinearSolve[matrix1,matrix2]
```

(6, 18)

Implementation of Idea 9: By direct substitution of $y = 3x$ into the equation $x + y = 24$, we get $x + 3x = 24$, or $4x = 24$. This is easy enough to do by hand (the animal problem is a visualization of this), but here is how it could be done on the computer:

```
y=3x;
x+y==24
```

$4x == 24$

```
result=Solve[4x==24, x]
```

$\{\{x \rightarrow 6\}\}$

```
y/. result
```

$\{18\}$

These answers tell us that the first integer is 6, the second 18.

This problem illustrates that there is more than one right way to solve a problem, even a simple math problem. Also, creative thinking and visualization can help us "see" alternatives, and even the computer can be used in different ways to calculate the numerical answer. Perhaps a student who prefers thinking in quadrant B might do this problem using Idea 3; a student with quadrant D dominance might like the graphical approach in 8 (or the animals in a room visualization), and a student with quadrant C preferences might use the jelly beans (sharing the treat with a friend afterwards), while any one of the computational methods will appeal to students who prefer quadrant A thinking. We chose this simple example of mathematical computation for students who have not yet learned much algebra, and for adults who may not remember much of what they learned perhaps a long time ago. The next example demonstrates MATHEMATICA as a database—a resource for information.

Example 2 — Chemical Elements

In Chapter 2, we talked about memorizing chemical elements, their atomic numbers, and their symbols. To do this, it helps to have a chart of elements handy. For your convenience, such a chart is built into MATHEMATICA. To access it, type:

```
Needs["Miscellaneous`ChemicalElements`"]
```

This command loads the program for execution of further commands—it will not yet show anything on the screen. If you want a list of elements, type **List[Elements]**. The resulting output will give the elements in order of ascending atomic number. To see them in alphabetical order, use the command, **Sort[Elements]**. What if you want the symbol for potassium? To do this, type **Abbreviation[Potassium]**. You should get K as the output. Next, try this command,

```
AtomicNumber[Potassium]
```

Did you get 19 as your answer? Now make up an image (and a sketch) to help you link and memorize this information about potassium. Do you remember the peg method? Here, check to see if you can decode this image (invented by Bryan J. Peters, one of our engineering students): "Einstein's equation-solving pup." What is the element, the symbol, and the atomic number?

MATHEMATICA can give you much more detailed information about chemical elements. Here are some examples; try these commands for different elements.

ElectronConfiguration[Potassium]

{{2}, {2,6}, {2,6}, {1}}

Density[Potassium]

862. K/Meter^3

BoilingPoint[Krypton]

1047. Kelvin

Example 3 — Textbook Problem 2-5

Problem definition: To solve this money problem involving five children (Arnold, Becky, Cory, Dotty, and Ernie), let A, B, C, D, and E represent the amount of money each child has, respectively. The given information can then be represented mathematically by the system of equations:

$$B + C = 3A$$
$$D = 2E$$
$$A = 1.5D$$
$$C + D = B + 2E$$
$$A + B + C + D + E = 60$$

Brainstorming, idea evaluation, and judgment: There are several ways in which this problem might be solved. Can you think of some interesting ideas? However, this particular problem is well-suited for an analytical approach and a computer solution. Speed of solution is an important judgment criterion; thus we will use MATHEMATICA to solve the problem. Even here, we have to make choices as to a particular approach. We will choose the **Solve** function, since from experience we judge it to be the most direct and fastest way for this small system of equations. For a larger system, we may prefer another method.

Implementation:

```
Solve[{B+C==3A,D==2E,A==1.5D,C+D==B+2E,
A+B+C+D+E==60},{A,B,C,D,E}]
```

{{A –> 12.}, {B –> 18.}, {C –> 18.}, {D –> 8.}, {E –> 4.}}

Thus we find that Becky and Cory have the most money at $18 each; Arnold has $12, Dotty $8, and Ernie only $4.

Example 4 — The Truck Problem (Example 4 in Chapter 5)

Let's supply some context to this problem. You own a trucking company and are responsible for making deliveries from one location to another. The costs of operating the truck on the road include gasoline, taxes, insurance (on the rig, the driver, and the cargo), maintenance, union dues (which you are paying for the drivers), and road tolls. From experience, you have found that these costs can be lumped together at 50 cents per mile traveled, independent of the speed driven. But since wear and tear goes up at higher speeds, you had to add an adjustment factor that does depend on the speed and the distance driven. You estimate that this factor is equal to the speed times 1/8 cents a mile (per mph). For example, for a 60-mile trip with an average speed of 48 mph, this cost would be $3.60. Your labor costs are inversely related to the speed, since the drivers are paid a flat rate of $10 per hour. The total cost of operating a truck is the sum of all three types of costs discussed:

$$C = 50 + S / 8 + 1000 / S$$

where C is the cost in cents per mile, and S is the speed in miles per hour.

The problem is to find the most economical speed at which to operate the truck, or in mathematical terms, to minimize C (cost) with respect to S (speed). To solve the problem, we need to differentiate C with respect to S and solve for the value of S when the derivative is set equal to zero, or $dC/dS = 0$.

```
Cost[S_]=50+S/8+1000/S;
D[Cost[S]==0,S]
```

$$\left\{ \frac{1}{8} - \frac{1000}{S^2} \right\}$$

```
Solve[N[%]]
```

$$\{\{S \rightarrow 89.4427\}, \{S \rightarrow -89.4427\}\}$$

What does the numerical solution tell you? It says the most economical speed to operate the truck is at nearly 90 mph! Why is this? More importantly, is this a reasonable answer? Does the answer agree with the assumptions made in the model, or are there some factors that are not included in the model? What about the additional risks of speeding tickets and the chance of accidents when you drive above the speed limit? Thus this mathematically correct answer is not a good answer to the original problem—it makes no sense. Let's ask a question—how much money is really being "saved" in going from 65 to 90 mph? To evaluate the function, write:

```
Cost[65]//N
```

73.5096

```
Cost[90]//N
```

72.3611

Thus the cost differential is only a little more than 1 cent per mile, an insignificant amount when weighed against the increased risks of driving over the speed limit.

You can confirm this by asking for the graphical representation of the equation:

```
Plot[Cost[S],{S,10,140},
PlotRange->{70,100},
AxesLabel->{"S,mph","C,¢/mile"}]
```

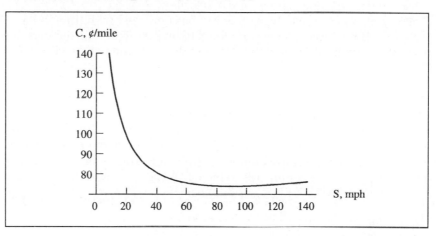

Example 5 — The Handshake Problem

Several people are attending a social gathering. Each person shakes hands once with every other person. Assuming that people do not shake hands with themselves, how many handshakes occur for a total of n people? Specifically, how many handshakes occur among ten people? This typifies a class of problems that require some creative thinking skills for finding a solution. It would be difficult if not impossible to solve this problem by "traditional" means. Rather, a few cases must be tried out and a pattern sought. Once the pattern has been found, the solution of the problem becomes very easy. Let us apply the creative problem-solving mindsets to solve the problem.

First, define the problem: Given n people, how many handshakes H(n) occur among them? Brainstorming the solution may result in the idea of getting ten peple together and asking them to shake hands. An observer could count the number of handshakes. However, this method would not be practical for a larger set of people and would not immediately yield the desired general solution. Nor is it an efficient approach. A more practical way of solving the problem involves making the analogy between people and vertices of polygons, with the number of handshakes being represented by lines connecting the vertices. We will test a few cases and then try to develop a general pattern from our observations.

CASE 1—**Handshakes among n = 2 people**
On a piece of paper, draw two points representing the number of people. How many ways are there to connect these points with straight lines? The answer is one, the same number of handshakes that can occur between two people. Below are the commands if you want to have MATHEMATICA draw the line:

```
pts={Point[{0,0}], Point[{1,1}]}
Show[Graphics[{{PointSize[0.03],pts},
Line[{{0,0},{1,1}}]}]]
```

CASE 2—Handshakes among n = 3 people
Three points not on a straight line can be connected with three lines. Draw them to see it. If the three people were lined up in a straight line, the person in the center would stand in the way of the two outsiders shaking hands with each other.

CASE 3—Handshakes among n = 4 people
The number of possible handshakes is getting more difficult to visualize without a sketch. Draw four dots arranged in a square. How many lines does it take to connect them? The answer is $4 + 2$ (the four exterior sides of the polygon, and two diagonals). The commands to have MATHEMATICA draw this are:

```
pts={Point[{0,0}],Point[{0,1}],Point[{1,1}],Point[{1,0}]}
Show[Graphics[{{PointSize[0.03],pts},
Line[{{0,0},{0,1}}],Line[{{0,1},{1,1}}],Line[{{1,1},{1,0}}],
Line[{{1,0},{0,0}}],Line[{{1,1},{0,0}}],Line[{{0,1},{1,0}}]}]]
```

CASE 4—Handshakes among n = 5 people
The number of possible handshakes here will be—what? You know you will have the five outside lines. If you made a sketch with the five dots evenly spaced around a circle, the interior diagonals will form a five-pointed star made with five lines. So the total number of lines is 10. Draw it to check it out!

CASE 5—Handshakes among n = 6 people
The number of possible handshakes is getting more and more difficult to visualize in the mind. If you draw a free-hand sketch connecting six dots spaced more or less evenly around an imaginary circle, you will have the six outside lines connecting adjacent dots, as well as interior lines forming two overlapping triangles connecting every other dot. Then you have three additional lines connecting opposite dots across the center, for a total of $6 + 3 + 3 + 3 = 15$ lines.

Let's tabulate what we have found so far:

Number of people:	2	3	4	5	6
Number of handshakes	1	3	6	10	15

Sometimes a helpful tool to detect patterns is to look at the differences between successive numbers. What is the difference between the number of handshakes?

Difference	2	3	4	5

What if we calculated the difference again?

Difference of difference:	1	1	1

This format makes the pattern easy to detect. Each time a person is added, the difference increases by one. Thus the next difference would be six. Six when added to 15 would give 21 handshakes for seven people. The number of handshakes for eight people would be equal to the number of handshakes for seven people plus seven, or 28, and so forth. This situation, where results depend directly upon previous results, is called *recursion*. This is a very powerful tool, and many mathematical and engineering functions are defined this way. Now that we have detected the pattern, we can write a general formula,

$$H(n) = H(n-1) + (n-1)$$

where n is the number of people and H(n) is the number of handshakes for n people. The previous number is then expressed as n –1. Let's test this formula to see if it works. First, we will define the function in MATHEMATICA. Be sure to use the := symbol for the definition. Then we have to give MATHEMATICA a starting point: the number of handshakes for one person is zero since people are not allowed to shake hands with themselves. How do you want your data displayed— what do you want to know? Let's ask for a table showing the number of hand- shakes for up to six people.

```
H[n_]:=H[n-1]+n-1
H[1]=0
Table[{n,H[n]},{n,1,6}]//TableForm
```

```
1    0
2    1
3    3
4    6
5    10
6    15
```

To find the number of handshakes for ten people, just ask in this simple way:

```
H[10]
```

```
45
```

There are 45 handshakes occurring among ten people (or, in other words, 45 different combinations of people shaking hands). This is a number that would have been very tedious to compute by hand. What would be the answer for 20 people? And for 30 people? We will leave it to you to find out.

Example 6 —Simple Harmonic Motion

The study of simple harmonic motion is very important in physics and engineering. Simple harmonic motion occurs when an elastic restoring force acting on a system varies proportionally to the displacement about an equilibrium posi- tion. That's not an easy definition to understand at first glance. Let's see if some examples will help us visualize the situation. First, we make a simplifying assump- tion: we are going to ignore the effect of friction.

We can "see" or feel simple harmonic motion when we:
- Pull or push a weight that is hanging quietly from a spring, then let go.
- Push a pendulum of a clock to get it going.
- Push or pull a child on a swing and then let go.
- Blow into a tuba, trombone, saxophone, or flute. The air column inside the instrument begins to vibrate; this vibration makes the musical tone.
- Push a marble up against the side of a round bowl, then let go.
- Move a rocking chair or a cradle.
- Watch a buoy bobbing up and down in the ocean swells.

In general, the displacement of an object about its equilibrium position in simple harmonic motion is

$$x = A \sin(\omega t + \phi)$$

where A = amplitude of the oscillation, ω = circular frequency, ϕ = phase angle, t = time, and x = displacement.

If we want to investigate this motion as it changes with time, calculating and plotting this data for a variety of changing conditions would be very tedious to do by hand. With MATHEMATICA (or a plotting calculator such as HP48SZ) it is easy to visualize what is happening when we change parameters.

In many vibrating systems, the amplitude is small. The frequency of oscillation does not depend on the strength or the velocity with which the vibration is started (a very good thing—imagine what would happen if the sound frequency of a musical instrument depended on how hard it was struck, plucked, or blown). In vibrations, it is important to understand normal modes in harmonic simple motion, because these normal modes describe the possible shapes and amplitudes of vibrating systems. Sometimes it is desirable to have vibrations, as for example when playing a musical instrument. But many times we want to be able to stop the vibration because it is annoying—especially if it causes undesirable noise at high frequencies. Or imagine how unpleasant an airplane trip would be if the plane shook constantly. Harmonic motion in buildings during earthquakes and in bridges during high winds can make the structures collapse. Automobile engineers study vibrations so they can give us quiet cars and smooth rides.

Consider the case of a vibrating string undergoing simple harmonic motion. You can find the first three normal modes with the following MATHEMATICA commands. The first mode is called the fundamental mode ($n = 1$), the higher-order modes are called harmonics or overtones. Musically, they give the octaves. You may want to investigate how the graph of the fundamental mode changes if the value of A is increased from 1 to 4—the command is `Plot[4 Sin[Pi*x],{x,0,1}]`.

```
Plot[Sin[Pi*x],{x,0,1}]

Plot[Sin[2Pi*x],{x,0,1}]

Plot[Sin[3Pi*x],{x,0,1}]
```

For the case of a plate or membrane that is clamped at the edges, the possible shapes of the membrane as it vibrates with simple harmonic motion are given by the equation:

$$\eta \,=\, A\,\Psi_{m,n}\,(x,y)\,\cos\,(2\pi\,v_{m,n}\,t - \phi), \qquad m = 1, 2, 3 \ldots, \; n = 1, 2, 3, \ldots$$

$$\Psi \,=\, \sin\,(\pi\,mx/a)\,\sin\,(\pi\,ny/b)$$

$$v_{m,n} \,=\, 1/2\,(T/\sigma)^{1/2}\,[(m/a)^2 + (n/b)^2]^{1/2}$$

where T = tension, s = mass/unit area, a = width of plate in the x-direction, b = depth of plate in the y-direction.

It is very difficult to visualize the normal modes on a two-dimensional surface. Diaphragms of ordinary telephone transmitters and receivers are examples of this type of structure. Plotting a graph of these vibrations without a calculator or computer would be extremely time-consuming. Yet in MATHEMATICA, the modes of this type of vibration can be visualized with a few simple lines of code. This visualization lets us investigate what-if questions such as changing the weight of the membrane, the tension, or the geometric shape. The changes can be explored very quickly, and the graphical output can give us a better understanding of the problem. We will use m to denote the modes parallel to the x-axis and n for the nodes parallel to the y-axis. In MATHEMATICA, these equations are written as:

```
Clear[psi, nu, eta]
nu := 0.5 Sqrt[ToverSigma ((m/a^2+(n/b)^2))]
psi[x_,y_] := Sin[Pi m x/a] Sin[Pi n y/b]
eta[x_,y_,t_] :=A psi[x,y] Cos[2 Pi nu t-phi]
```

Definition of parameters:
```
ToverSigma = 1;
A = 1;
phi = 0;
t = 1 (*time, s*)
nu = 0;
a = b = 1;

Do[m=n=i;Plot3D[psi[x,y],{x,0,1},{y,0,1},
PlotRange->{-1,1},PlotPoints->15],{i,10}]
```

If you are working in MATHEMATICA version 2.0 or later, this should compute without any problems. If you have a Sun workstation or other machines without the notebook interface (and versions later than 1.1), you can try to generate animations:

```
Needs["Graphics`Animation`"]
Clear[m,n]
m := i; n := i;
Animate[Plot3D[psi[x,y],{x,0,1},{y,0,1},
PlotRange->{-1,1}, PlotPoints->15],{i,1,10}]
```

If you have an earlier version, here is an easier equation to work with. We first substituted the parameters in nu above to come up with a simplified statement. Try plotting this equation for the case of m = 1, n = 1:

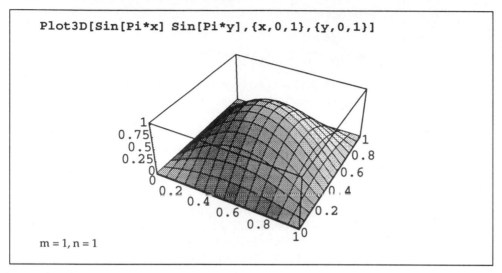

```
Plot3D[Sin[Pi*x] Sin[Pi*y],{x,0,1},{y,0,1}]
```

m = 1, n = 1

Next, plot the following two graphs (m = 2, n = 1 and m = 2, n = 2). Can you take a guess at what they might look like before you get the output?

```
Plot3D[Sin[2 Pi*x] Sin[Pi*y],{x,0,1},{y,0,1}]
```

```
Plot3D[Sin[2 Pi*x] Sin[2 Pi*y],{x,0,1},{y,0,1}]
```

For those readers who do not have a computer handy at the moment, here is what the command and plot for m = 3, n = 2 looks like:

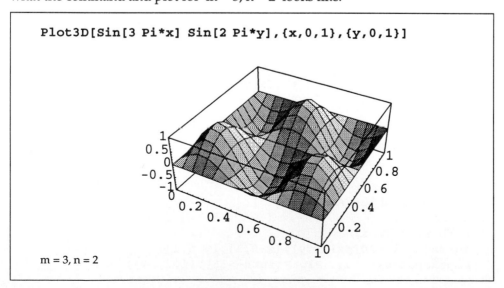

```
Plot3D[Sin[3 Pi*x] Sin[2 Pi*y],{x,0,1},{y,0,1}]
```

m = 3, n = 2

From the pattern that you have seen developing this far when changing m or n, can you predict what the following two graphs will look like?

```
Plot3D[Sin[4 Pi*x] Sin[4 Pi*y],{x,0,1},{y,0,1}]
```

```
Plot3D[Sin[2 Pi*x] Sin[10 Pi*y],{x,0,1},{y,0,1}]
```

Engineers out in the work force know something few students recognize: it is one thing to solve "clean" problems in a mathematics course and another to solve "dirty" real-life problems. Solutions to some of the simpler engineering problems sometimes require some not-so-trivial programming steps. Occasionally, using MATHEMATICA seems to be more complicated than using other methods of computation. However, at no time were we ever tempted to go back and use procedural languages (such as Fortran or Pascal) to solve our problems. When the solution to a complex problem finally appears with a few lines of code, the feeling of pleasure and accomplishment leads to renewed commitment to learn more about programming in MATHEMATICA in order to adapt this powerful tool to specific applications in one's field.

ACTIVITY 9-1: EXERCISE, APPLICATION, AND OUTREACH

You have a double assignment —> learn how to use computers effectively; then help others, especially youngsters.

If you are already using computers at work or for pleasure, learn to use them in a new way. Explore a new math or graphics program. Get involved in your public school systems to see that they make full use of the computers there. Are computer labs accessible to students before and after school, as well as evenings and weekends? Make sure that all students, especially women and minorities, take advantage of opportunities to learn with computers. Get parents involved. Educate the public about the need for schools to have enough basic equipment to give all students a good introduction to and appreciation of this technology. Use creative problem solving to find an area where you can become involved. If you are a college student, explore the computer resources at your institution. If computer-based calculus is not yet offered on campus, teach yourself MATHEMATICA, then apply it in your math, science, engineering, economics, and other suitable courses.

RESOURCES FOR FURTHER LEARNING

Textbooks and Courseware

9-1   Nancy Blachman, *Mathematica: A Practical Approach*, Prentice-Hall, Englewood Cliffs, New Jersey, 1991. This tutorial (with problem sets in each chapter) teaches how to use MATHEMATICA interactively and how to program.

9-2   Donald Brown, Horacio Porta, and Jerry Uhl, *Calculus & Mathematica, Part I*, Addison-Wesley, New York, 1991. Book and Disk. This package covers the first two terms of calculus. Each unit contains basics, tutorials, and "give it a try" homework assignments.

9-3   D. C. M. Burbulla and C. T. J. Dodson, *Self-Tutor for Computer Calculus Using Mathematica*, Prentice-Hall, Englewood Cliffs, New Jersey, 1992. With its self-instructional format, this soft-cover book can serve as an introduction to MATHEMATICA. It encourages active investigation and the use of graphics.

9-4   Roman Maeder, *Programming in Mathematica*, Addison-Wesley, New York, 1991. This book is recommended as a companion to the Wolfram book below for individuals who want to develop their own notebooks in MATHEMATICA for application to specific fields and problems. The second edition incorporates the improvements and new functions included in MATHEMATICA 2.0.

9-5   Stephen Wolfram, *Mathematica: A System for Doing Mathematics by Computer*, second edition, Addison-Wesley, 1991. This is *the* book on MATHEMATICA written by its inventor and is a must for anyone who wants to get into the subject. It is available in hardcover or paperback; it contains an updated reference section as well as many full-color graphics.

Applications

9-6   Martha Abell and James Braselton, *Mathematica by Example*, Academic Press, San Diego, 1992. This paperback book gives careful instructions to beginners for applying MATHEMATICA in calculus, linear algebra, ordinary and partial differential equations, and discrete mathematics. Also by the same authors is the *Mathematica Handbook*, Academic Press, 1992; it provides a convenient reference of all built-in MATHEMATICA 2.0 objects for beginners and advanced users.

9-7   Richard Crandall, *Mathematica for the Sciences*, Addison-Wesley, New York, 1991. This book is suitable for students and researchers in biology, chemistry, engineering, mathematics, and physics; it provides a variety of examples and applications in these fields.

9-8   Theodore Gray and Jerry Glynn, *Exploring Mathematics with Mathematica*, Addison-Wesley, New York, 1991. This package comes with a CD-ROM and a paperback or hardcover book. All MATHEMATICA expressions, graphics, and sounds are given in notebook form; it is a step-by-step illustration of how to apply MATHEMATICA to real-world problems.

9-9   Ilan Vardi, *Computational Recreation in Mathematica*, Addison-Wesley, New York, 1991. This book investigates some common as well as challenging topics in mathematics. Exercises include sequences, the n-queens problem, digital computing, blackjack, etc.

Computers, Learning, and Creativity

9-10   Joseph Deken, *The Electronic Cottage*, Morrow, New York, 1982. This book is by no means outdated: it gives a pleasant introduction to everyday living with a personal computer. It discusses how computers work as well as how to use them for many practical tasks and fun.

9-11   Barry Heermann, *Teaching and Learning with Computers*, Jossey-Bass, San Francisco, 1988. This book examines important educational issues regarding computers in college instruction—how computers can enhance or undermine educational goals.

9-12   Clifford A. Pickover, *Computers and the Imagination: Visual Adventures Beyond the Edge*, St. Martin's Press, New York, 1991. This paperback is a combination of graphics, problems, puzzles, and essays; it encourages the reader to explore the way computer images relate to music, art, math, science, technology, and human creativity.

9-13   Mark von Wodtke, *Mind Over Media: Creative Thinking Skills for Electronic Media*, McGraw-Hill, New York, 1993. This book emphasizes creative thinking and focuses on the interaction between the mind and the electronic media. It includes the integration of creativity with team-work, with computer tools, and with written text, graphics, and other visualizations. Multimedia are described as tools and toys for analysis, synthesis, and evaluation, with applications in art, architecture, planning, and engineering. The book is organized to present visuals on the left-hand and verbal information on the right-hand side of each page.

10

THE PUGH METHOD

What you can learn from this chapter:
- *Creativity in product and process design; steps in the product development process.*
- *Economic benefits of the Pugh method of creative evaluation.*
- *Overview of the evaluation matrix and procedure: list of criteria, datum, evaluation scale, idea synthesis. Example: design of a school locker. Convergence to a "best" solution.*
- *Additional examples of student projects and an application to course design.*
- *Further learning: references and exercise.*

The Pugh method of creative design concept evaluation was developed by Stuart Pugh, a design and project engineer with many years of practice in industry. He later became professor and head of the design division at the University of Strathclyde in Scotland. He came to recognize that designs done purely by analysis were "somewhat less than adequate" because it took a long cycle of modifications to satisfy the customer. He realized that engineers need to see the whole picture in product design and development; they need an integrated approach to be competitive. Although the Pugh method has its most direct application in product design, the procedure and thinking skills used can be applied to many other situations where different ideas and options have to be evaluated to find the optimum solution. If you are not an engineer, you may want to skip the sections on the design and product development contexts and move ahead to the overview and examples.

CREATIVITY IN PRODUCT AND PROCESS DESIGN

Table 10-1 illustrates how the Pugh method relates to the steps of the design process. Rectangles represent a period of activity or phase; the ovals denote the output at the end of the activity. Design generally begins with a need or an opportunity. This need may already be met by existing designs; in this case our problem is to satisfy this need with better value or at lower cost. About 70 percent of the cost of a product is determined during the design concept and development phase—thus creative thinking is essential for finding the best design at lowest cost.

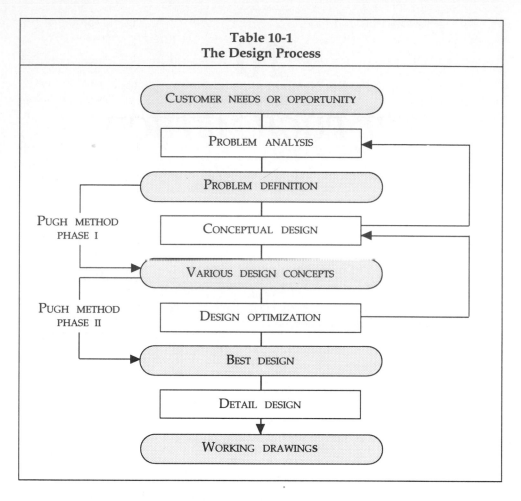

Table 10-1
The Design Process

CUSTOMER NEEDS OR OPPORTUNITY

PROBLEM ANALYSIS

PROBLEM DEFINITION

PUGH METHOD PHASE I — CONCEPTUAL DESIGN

VARIOUS DESIGN CONCEPTS

PUGH METHOD PHASE II — DESIGN OPTIMIZATION

BEST DESIGN

DETAIL DESIGN

WORKING DRAWINGS

A conceptual design is an outline solution to the design problem, in which the rough sizes and structural relationships among the major parts have been determined and it has been decided how each major function will be performed. A conceptual design is worked out in sufficient detail to allow estimates of cost, weight, and overall dimensions to assure at least a reasonable probability that the product is feasible. The design process culminates in a set of working drawings and other instructions that contain all the information needed to make the product.

Various methods can be used for problem analysis, depending on the scope of the project; the purpose is to obtain a clear idea of the objectives the new design is supposed to meet. These objectives are best assembled in the form of a list of criteria. As we shall see later, criteria will undergo revision during the evaluation process as the team members gain a better understanding of the conceptual designs and of the problem they are attempting to solve, so that a strong feedback loop exists between conceptual design and problem analysis (see Table 10-1). Four criteria that should always be included right from the beginning are quality, low cost, manufacturability, and environmental impact (during manufacture, in the product's life cycle, and at final disposal). Which manufacturing process will cause the least amount of pollutants? Can materials be recycled? Does the product's use encourage energy conservation? Through the Taguchi methods, engineers in the United States are learning that lowest cost does not mean a shoddy product when the cost to society caused by low quality is factored into the cost equation. Thus low cost means product excellence, which is designed into the product from the start.

Another purpose of the first feedback loop in Phase I is that it will bring out major points in the product features that need to be decided as well as the rationale for making these decisions. An arbitrary decision, especially early in the design process, means a wasted opportunity to increase quality and decrease cost. When we use an established idea or concept without questioning, we are in essence making an arbitrary decision. During the conceptual design phase, we must question the "accepted" or conventional way of doing things and look for alternatives. During Phase II of the Pugh method, as the conceptual designs are being optimized, a second strong feedback loop exists; it may be repeated several times over weeks or months until it is no longer possible to improve the best design. Finally, during the detailing phase, the creative potential of making decisions about the smaller points in the design should not be overlooked, especially with respect to maintaining the quality of the product. Computers are used in this phase to reduce the plain drudgery of detail design and the chance of making errors.

During the conceptual design phase, broad solutions to the defined problem are developed in the form of design concepts or options. This phase places the greatest demands on the designer in terms of creative thinking. This phase is also the place where the most striking innovations can originate, not only in terms of the product alone but also with regard to process and production, since manufacturability is considered right along with product development. When the designers choose a shape for a part, they must at the same time also think about how this part, this shape, is going to be manufactured. Is there an easier way to achieve the same purpose with a different material, a different shape, a different way of making it?

As the conceptual designs are improved and optimized during Phase II, new ideas and concepts will still be generated and developed. Experience has shown that the final designs usually originate during this synthesis phase, and the best designs may have very little in common with the original conceptual schemes (which, however, serve a very useful function as stepping-stones in the creative thinking process). Thus the second feedback loop in the design process is very important and should not be rushed or skipped.

Improving the practice of engineering design in U.S. firms is essential
to industrial excellence and national competitiveness.
National Research Council, 1991.

The question we need to ask is if we, as a nation, can afford to keep the teaching
of creativity a seldomly used option in our design courses.
Edward Lumsdaine, 1991.

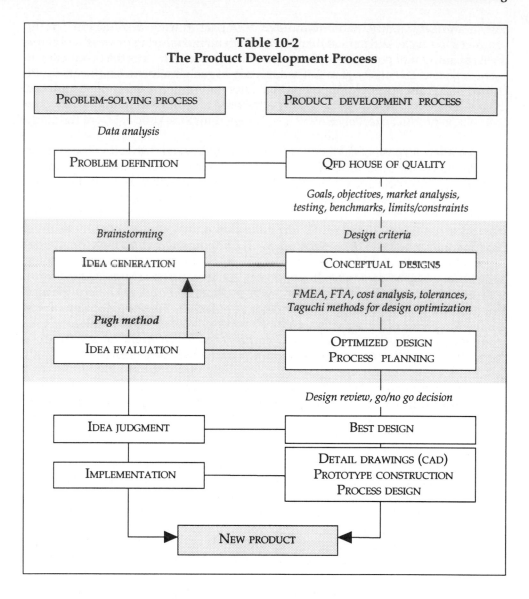

Table 10-2
The Product Development Process

PROBLEM-SOLVING PROCESS	PRODUCT DEVELOPMENT PROCESS
Data analysis	
PROBLEM DEFINITION	QFD HOUSE OF QUALITY
	Goals, objectives, market analysis, testing, benchmarks, limits/constraints
Brainstorming	*Design criteria*
IDEA GENERATION	CONCEPTUAL DESIGNS
	FMEA, FTA, cost analysis, tolerances, Taguchi methods for design optimization
Pugh method	
IDEA EVALUATION	OPTIMIZED DESIGN PROCESS PLANNING
	Design review, go/no go decision
IDEA JUDGMENT	BEST DESIGN
IMPLEMENTATION	DETAIL DRAWINGS (CAD) PROTOTYPE CONSTRUCTION PROCESS DESIGN
	NEW PRODUCT

The Product Development Process

In Table 10-2, product development is compared to the creative problem-solving process. Each step in creative problem solving results in an output, which is indicated as a boxed item on the right-hand side of the table. We can thus visualize how the Pugh method fits within the context of the entire product development process. The goal of problem definition is to come up with a comprehensive list of design criteria. Brainstorming with team members from several departments (including design, manufacturing, sales, and service) is done during this stage. The customer's voice is critical in this data analysis process, and it must be part of the evaluation criteria. Benchmarking is often used; this is the process of comparing your existing product against the best competitor in its class—and then setting goals for doing better for each important product feature involved in satisfying the customer. The QFD House of Quality is an excellent tool for identifying critical product features and their negative interactions. These factors must be addressed and overcome through creative thinking to achieve customer satisfaction.

Basically, the Pugh method is a technique whereby creative ideas are obtained by a team making a conscious effort to overcome the negative features and shortcomings of proposed designs. Because the ideas and concepts are placed together on a matrix, the process encourages force-fitting and synthesis at a conscious as well as at a subconscious level in the team members' minds. The conceptual designs are based on the criteria and generated by using the artist's mindset. Here, the product designers and engineers can really play around with many different ideas and approaches on how to satisfy the criteria. At this stage, the emphasis should be more on exploring alternatives than on finding one adequate solution. Only after a number of very different concepts are on hand should the process move on to evaluation in the engineer's mindset. The ability to obtain insight into the problem, to think of many different approaches, and to use stepping-stones to solve the problem—these skills can be improved with a number of techniques as part of creative problem solving. For example, design engineers can use morphological creativity, function and limits analysis, mathematical modeling, logic chains, experiments with models, and brainstorming with force-fitting techniques to come up with innovative concepts. Dynamic computer models or physical models of the proposed concepts can yield useful information at this stage that could not be obtained with a static model. Most of all, the team members must continue to ask questions with an open mind.

After the first phase of the Pugh method has reduced the number of concepts and has produced superior designs, additional studies can be made, such as cost analysis, FMEA, FTA, and Taguchi methods, with the goal of optimizing the most promising designs while simultaneously initiating process planning (with feedback to product design through revised criteria). This is known as "total design" or "concurrent engineering." At the close of Phase II, one or at the most two superior concepts will emerge. The design and evaluation process is then followed by a design review and a go/no go decision conducted by management. If the design is approved, detail drawings and prototype construction are authorized, and the manufacturing process design is completed. Refer to the Appendix for a description of FMEA, FTA, QFD, Taguchi methods, and total quality management (TQM).

The difficulty in concept evaluation is that we must choose which concepts to spend time developing when we still have very limited data on which to base the selection.
David G. Ullman, The Mechanical Design Process.

These deficiencies [of not considering people's aspirations and needs] have to be recognized, otherwise misdirected engineering rigor will always give rise to bad total design. This implies that design teams should always include non-engineers.
Stuart Pugh, Total Design.

ECONOMIC BENEFITS OF THE PUGH METHOD

Why should the Pugh method be used? If a company wants to produce a product that is best, it has to start by selecting the best design concept. The Pugh method helps the team select the best design without leaving any questions or doubts. It prevents a company from making costly mistakes in the choice of products. If you have to convince managers to use the Pugh method, you can emphasize the cost savings. The material in this section is an excerpt from a lecture that Dr. Don P. Clausing has developed. Formerly vice president at Xerox Corporation, he is now professor at MIT and consultant with the American Supplier Institute in Dearborn, Michigan. He points out ten areas where companies waste money during the product development cycle; using the Pugh method will help avoid many of these pitfalls.

THE TEN CASH DRAINS

Technology push—but where's the pull?
Disregarding the voice of the customer.
The eureka concept.
Pretend designs.
The pampered product.
Hardware swamps.
Here's the product; where's the factory?
We've always made it this way.
The need for inspection.
Give me my targets; let me do my thing.

---

1. Technology push—but where's the pull? It is generally agreed that the United States is very strong in inventing technology. But Americans tend to be infatuated with technology for its own sake, and major resources are spent on developing technology even where no discernible market needs have been identified. This danger exists particularly when we have a very clever or attractive idea. But if it cannot satisfy an important, identified customer requirement, money spent on development is wasted. Conversely, we can have a strong market need without responsive, *ongoing* research and development activities. If a significant new technological concept is being developed concurrently with the product, the result is rarely successful. It should be a strong operating principle that such new, significantly different concepts not be selected during system design unless prior research activities have developed the concept to a sufficient level of maturity.

In addition, even when we have identified a strong market need and have developed new technology through our ongoing R&D activities that may be capable of addressing that need, we can have a problem in that the technology is not being transferred into the system design activities. What we need to remember is that our technological development activities should be guided by long-range planning that will specifically address customer requirements for our products. This is done very effectively with QFD, for instance. And we must allow sufficient funding and time for the new technological concepts to mature before we apply them to our products. When we have a mature concept, we then must make sure that the information will be transferred to our design staff. The Pugh method design review team (which should have people from the technology development area) can guard against the use of improper technology when setting up the list of criteria and during the evaluation process.

2. Disregarding the voice of the customer. The most important concept that American industry has learned from Japanese manufacturers is that the needs of the customer must drive the activities of the entire organization. In the Pugh method, the voice of the customer is expressed in engineering terms in the list of criteria. When we ignore or merely assume we know what the customer wants, we may be in for a rude shock when our product reaches the market.

3. The eureka concept. This is when someone has this "great new idea" that becomes the only concept that is considered seriously. Unfortunately, such concepts are often proven very vulnerable by the time they reach the market. This wastes money in two ways: the expenditures in developing the product are wasted—and the opportunity to develop a better product and retain market share is lost. The Pugh method is specifically aimed at preventing this particular cash drain by objectively assuring that only well thought-out concepts are developed, not ideas that people fall in love with.

4. Pretend designs. Here the emphasis is simply on being new, not necessarily better. As a manager, as an engineer, or simply as a consumer, how often have you encountered products that exhibit this phenomenon? In industry, the focus is on experimental hardware, not on obtaining a production-intent design. When we compare the new design against a competing benchmark, we will be able to avoid developing an inferior product just for the sake of novelty. Many engineers do not like to use someone else's good ideas, so they will be strongly motivated to develop their own better solutions that will beat the competition. The Pugh method, together with the go/no go design review, will help overcome this cash drain.

5. The pampered product. We see an example of this when a product is developed and built to look good on demonstrations to the vice president. We have a similar situation when new cars are given to automaker executives; these cars receive special daily attention and care to keep them in top running condition—and they are frequently replaced by even newer models. Thus, these executives have no idea how their products stand up when used by the average consumer for a longer period of time. We can prevent pampered products when we do optimization studies and when we make the product robust and reliable, for example, by using Taguchi methods (especially through the signal-to-noise ratio analysis). The objective here is to reduce variance. Then, after the prototype is built, the calculated results can be verified experimentally. Thus Taguchi methods serve to challenge a new product instead of pampering it; and they ensure that the design is protected as much as possible from suffering all identified potential critical problems.

6. Hardware swamps. This phenomenon occurs when we have so much prototype iteration that the entire team becomes swamped with the chore of debugging and maintaining the experimental hardware—to the point where no time remains to improve the design. The Pugh method allows us to eliminate many poor concepts before they are built into prototypes. Once the best design has been selected, it will have been developed and improved by the team to the point where at most four iterations of the prototype are needed for successful completion of the optimization and verification process. When used with the Pugh method, prototypes are not needed for experimental optimization; they are only needed for confirmation and fine-tuning. Hardware swamps are usually accompanied by data swamps; a huge quantity of data is generated that just stacks up, and nobody gets around to doing proper evaluation. With the Pugh method, simple products will only need two iterations. Sufficient time becomes available between test runs to properly incorporate the results into the model for the next round of testing.

7. Here's the product; where's the factory? We can avoid this cash drain if we develop production capability simultaneously with product design. Manufacturability is an important criterion that must be introduced very early in the design concept phase. As factories use more and more automation, these resources must be employed to minimize manufacturing costs and increase quality. Environmental considerations must also be taken into account. Can we use a process that creates less toxic wastes? Can we use materials that are completely and easily reclaimed or that are biodegradable?

8. We've always made it this way. This cash drain not only holds for product design, it especially applies to process planning and design when the operating points of the manufacturing processes are being specified. The values of the process parameters are often chosen rather arbitrarily by experience, or they are selected based on scanty development work. To achieve minimum cycle time and maximum quality, these process operating points can be optimized with QFD and Taguchi methods. In product design, the group discussions and the creative thinking involved in the Pugh method create an environment in which "the old way of doing things" will be questioned and improved if possible.

9. The need for inspection. The need for extensive inspection is minimized through the use of on-line quality control. The control chart (one of the tools in SPC) has been used until quite recently to good benefit. However, Dr. Taguchi's method of optimal checking and adjusting can minimize the direct and indirect costs of inspection even further. When production approaches zero defects, frequent inspection obviously will become superfluous. Manufacturers can also insist that their suppliers in turn manufacture their parts with a quality level approaching zero defects.

10. Give me my targets; let me do my thing. When people in organizations work in isolation—when they work without looking at the context of the entire process or product—the result is a subsystem that cannot be integrated or a product design that cannot be produced. Successful manufacturing requires teamwork and cooperation between horizontal and vertical levels in the organization. This cash drain can be overcome with teamwork and with competitive benchmarking: how can we do better than our competition? Again, this cash drain is avoided when the Pugh method is used with a team that includes members from manufacturing and sales in addition to product design and development. No one involved with the product works in isolation. However, a balance is needed; too much supervision and paperwork requirements imposed by management can impede the quality of the design process and the efficiency of the product development team.

A major benefit of the Pugh method is that it eliminates engineering changes late in the product development process or, what is even more costly, after the start of production. Through intensive discussion and analysis, no flaws are overlooked. The Pugh method is an effective communications tool; through the team discussions about the criteria and the concepts, the team members gain a common understanding of the problem and the different options and solutions. When the best solution emerges after several rounds, every person on the team understands why this solution is best; each person is ready to champion this concept.

OVERVIEW OF PUGH MATRIX AND PROCEDURE

Table 10-3 outlines the key steps in the Pugh method of creative design concept (or idea) evaluation. The matrix is basically an advantage/disadvantage evaluation scheme similar to that illustrated in Figure 7-2, except that a three-way evaluation scale is used and each concept is evaluated against a datum (most commonly an existing idea, product, or process). During the early parts of Phase I, the number of concepts under consideration increases because each alteration to a concept is considered to be a new concept. In later rounds, especially in Phase II, the number of concepts carried forward to another round of improvements decreases, since ideas are merged and synthesized. Weaker concepts drop out as the quality of the remaining concepts increases. Developing and understanding the list of criteria is a primary task for the evaluation team.

The List of Criteria

The criteria used in the evaluation matrix are comprehensive, relevant, and explicit; they must incorporate the objectives of the planned product, its purpose, and its targeted market. Performance specifications have been established through testing or a benchmark analysis of the competition. Constraints (such as cost ceilings and government regulations) are identified. In the case of a component, specifications or tolerance limits are set to make the part fit into the context of the entire product. When evaluating nondesign ideas, the criteria may not be as technical but they still must include all important aspects of what the solution needs to accomplish to solve the problem. When selecting the datum against which the new concepts will be evaluated, choose the best existing solution, not the "problem" item, if a better product or idea is already available.

**Table 10-3
Pugh Method Overview**

PHASE I

1. The list of evaluation criteria is developed through team discussion. A benchmark or datum is selected, usually the "best" existing product.

2. Original design concepts are generated by individuals or small teams.

3. Each design concept is discussed and evaluated against the datum. The top-ranking concept is selected as the datum for the next round.

4. Through the discussion, new concepts emerge; they are added to the matrix and evaluated. Also, the list of criteria is being refined and better understood.

5. During an incubation period, the teams improve the original design concepts by borrowing ideas and components from each other, as well as through additional creative thinking. Then Steps 3 and 4 are repeated with these designs.

Phase II

6. The weakest designs are dropped; the improvement process is continued for additional rounds with fewer but increasingly better concepts. During the process, the strong, surviving concepts are engineered to more detail; the criteria are expanded and further refined. The weak points of the concepts are being eliminated. The team gains insight into the entire problem and solution.

7. The process converges to a strong consensus concept that cannot be overturned by a "better idea." The team is committed to this superior design and wants to see it succeed.

In the early stages of idea generation and development, it is better not to have too many constraining or detailed criteria. Did you know that the first successful modern airliner was built within two years from a one-page list of performance specifications? This was the DC-3, perhaps the most successful airplane ever. In comparison, the specs for a new aircraft today might fill several trucks. Different organizations, especially large companies such as Ford or General Motors, usually have established procedures to come up with the design criteria. These design criteria traditionally have reflected the "voice of the boss" or the "voice of the engineers." The Japanese have perfected the art of collecting the voice of the customer through such methods as quality function deployment (QFD). This technique efficiently culminates in a list of criteria which assures that the designed product is responsive to the market needs and customer wants, not engineering or technology requirements. For example, in the United States we have spent much effort and money on R&D in ceramics technology to make grain size smaller and more uniform, since grain size seems to be related to the ability to withstand higher temperatures. The Japanese on the other hand concentrated on the requirements of the customer—where will the ceramics actually be applied? They investigated means to fulfill those requirements in a number of much less expensive and more appropriate ways. This is why Americans have the technology—and the Japanese have the market. QFD assures that customer needs are not lost somewhere between the design shop and the factory floor. Each worker understands how his or her job contributes to meeting the customer needs.

The Evaluation Process

After the design criteria have been handed out to the designers (which can be individuals or, more commonly, teams of engineers), conceptual ideas are worked out over a period of weeks. Then the conceptual designs are submitted for evaluation. A group meeting is called for all those involved in product development. The meeting is held in an ample conference room with a large board covering an entire long wall. An evaluation matrix is set up on the board, with the design criteria listed in the left-hand column of the matrix. The large drawings of the design concepts are posted across the top of the evaluation matrix. Depending on the number of design concepts submitted, the matrix may take up the entire board or wall of the conference room. Each concept is presented to the whole evaluation team to explain the main features; immediately after the presentation, each concept is evaluated against the datum using a three-way rating scale:

PUGH METHOD EVALUATION SCALE

+   *means substantially better.*
−   *means clearly worse (or this item needs attention).*
S   *means more or less the same.*

If the design concepts are improvements over an existing product, the existing product is entered as the first concept on the matrix, and the new concepts are judged against this datum. A "plus" mark is given when the new concept is markedly better, a "minus" mark when it is definitely worse, or an "S" when it is about the same as the datum. If the concepts are for a new product, any one of the concepts can be chosen as the datum. The three-way evaluation scale may appear rather primitive, but it is easy to do with a team, and the results are effective, because the objective is not quantitative, precise information but a movement toward increasing quality and superior satisfaction of all criteria. Inexperienced group members may be very defensive and protective of their design and will argue about every minus mark. However, remember that this marking serves to point out weaknesses or potential problems in the design that must be overcome for a product to be competitive. The judgment only determines (as objectively as possible) if the concept is better or worse than the datum or benchmark. But because the negative sign has an emotional impact on people (since it is seen as criticism), **we propose using a different symbol for items that fall short and thus need attention—the delta (Δ) which gives a more positive message —> change!**

So that the evaluation team can judge each concept carefully, the concept is drawn large enough to be visible to everyone in the room. At this early stage, the conceptual designs are outline solutions only. To allow for a fair comparison among the concepts, the drawings should be carried out to a similar level of detail and follow the same format. It is difficult to compare concepts if one resembles a sketch on the back of an envelope and another is a beautiful artist's perspective or a detailed CAD drawing. Questions and disagreements about the criteria often occur during this comparison process, due to differing interpretation. The open discussion during the creative evaluation helps clarify and resolve these ambiguities in the criteria, and the criteria become increasingly better defined and useful.

When the first-round matrix has been completed, the results are critically examined. Is there a criterion that received no plus signs all across the board? This indicates that none of the new concepts addressed an important customer concern. The teams must do additional creative thinking to improve on the datum in this area. Criteria that have been satisfied uniformly by all the improved concepts or that were found to be irrelevant can be dropped in future rounds to simplify the evaluation. What new criteria were brought out in the discussions? If they are important or clarify an ambiguity, they must be added to the list. Typically, the customers become more precisely identified. For example, if we are working on ideas to improve a curriculum, the customers are not only the students, but the parents, the teachers, the profession (business, law, engineering, medicine, etc.), the future employers of the students, the taxpayers, and perhaps even society as a whole, with each customer group having different requirements.

The scores in each column are now added separately for the positives and the negatives (or deltas). Positives never cancel out shortcomings—thus do not add them together mathematically to obtain an overall ranking of the concepts. The ultimate goal of the process is to obtain concepts all of whose shortcomings have been eliminated with improved ideas. With the results of the matrix, the design teams can now take their concepts back to the drawing board and target their improvements to the identified weaknesses in preparation for the next round of evaluation. The concept that had the highest number of positives becomes the datum for the next round. Its creators in the meantime will try to improve this design even further by eliminating the identified shortcomings.

What if a negative or delta cannot be eliminated? This could be the case for high manufacturing costs. If this is for an item where a competitive price is very important, other concepts that do not have this "flaw" will have to be pursued. Or this could be an area where the development of new technology is needed or a new paradigm must be discovered. Teams must not discard low-scoring concepts too quickly—they may contain valuable stepping-stone ideas that can be merged with some of the other concepts. The evaluation process ends in Phase II when it is no longer possible to improve the best concept by a superior idea. This means that the best solution has been found. This also means that the team knows why this is the best solution. The final concept will have no negatives that may have been overlooked during a more traditional, more cursory design review.

To show how the Pugh method actually works, we will now present an example taken through two rounds of Phase I: the design of an improved school locker. Then we will add some comments on Phase II. Further examples are given from projects done by secondary school students and college freshmen; an application to course syllabus design is also shown. As an engineer, you may be interested in a more advanced illustration. The classic "teaching" example used by Professor Pugh is the design of an automobile horn; it can be found in his book (Ref. 10-3). But to really experience the process and learn how powerful this technique is, you must make up a team and go through two or three rounds in your own evaluation exercise. This may take half a day or more, depending on how much time is spent in thinking up and sketching improved concepts, as well as on the total number of concepts that are being evaluated and the number of criteria used. A good starting point is having four different concepts with ten important criteria.

Example: Design of a Better School Locker

We did the following exercise with secondary school students in the Level 2 pilot of our math/science Saturday academy during the fall of 1991. This outreach program focused on creative problem solving as a thinking tool for everyone. The parents who sat in the back of the room as observers were at first rather skeptical when the students selected the locker topic for the exercise—they thought it would be too difficult since these students did not know anything about design. The amazing outcome proved that creative thinking and teamwork are more important for developing conceptual designs than technical drawing skills.

Problem briefing: The most important problems with lockers were determined through a survey and Pareto analysis. The students who came from city, suburban, and rural schools interviewed their classmates with a questionnaire to find the real problems with lockers. The problems were ranked as follows: (1) open door blocks access to neighboring locker, (2) not enough space, (3) not enough shelves; (4) noise; (5) ugly; (6) trouble with lock operation; (7) damage to books, clothes, or students' skin. The students brainstormed a list of design criteria, developed their design concepts, and then presented these designs to the entire class for evaluation.

Design criteria: (a) Efficient arrangement of groups of lockers. (b) Shape and size of individual locker for retrofit if possible. (c) Interior space divisions. (d) Door redesign. (e) Improved lock. (f) Recycled or recyclable materials; easy to manufacture. (g) Acceptable because of good looks, low noise, and easy use.

Advantages of Round 1 locker design concepts:
1a Shelves and drawers; foam edge; I.D. card or fingerprint lock.
1b Desks with individual lockers; separate small locker for coats; card lock.
2 Double width; many drawers + shelves, rolltop door; button combination lock.
3 Door opens 90°, then slides straight back; extra shelves; ABC lock.
4 Retrofitable to present lockers; stopper at 90° for door; extra shelves.
5 Foam rubber door gasket; floor drain; "laser beam" door.

Pugh method Phase I—Round 1:

Evaluation Criteria	Design: Old	1a	1b	2	3	4	5
1. Easy to use (door)	|	S	S	S	+	S	+
2. Nice looking	|	+	+	+	S	S	S
3. Reasonable cost	D	–	–	–	S	+	S
4. Acceptable to administrators	A	+	–	–	S	+	–
5. Better storage (students)	T	+	+	+	+	S	S
6. Improved materials	U	S	S	+	+	S	S
7. Locker arrangement in hall	M	S	+	–	S	S	S
8. Lock design	|	+	+	+	+	S	+
9. Convenience	|	S	–	+	S	S	+
TOTAL +		4	4	5	4	2	3
TOTAL –		1	3	3	0	0	1

This was a nice selection of different concepts, from retrofit improvement of the present "old" locker to futuristic innovation. At this point, no single concept was "the winner"—the design teams had to return to the drawing board to see how they could improve their designs by borrowing ideas from others and adding new ideas. We emphasized that the + and – must not be added up; the goal was to have as many + as possible, with all – eliminated. Concept 3 was designated as the datum for the second round, since it had the most improvements. Concept 1b was set aside since it involved redesign of desks (a different problem). Discussion during Round 1 brought out some changes for the criteria as well. Lock design was now eliminated as a criterion since it was taken for granted that all concepts would incorporate an advanced lock, either integral or detached. Designs for Round 2 were engineered to more detail (for example, rough dimensions were required.)

Pugh method Phase I—Round 2:

Evaluation Criteria	Design: Datum	1	2	3	4	5	6
1. Easy to use (door)		S	S	S	S	S/–	–
2. Nice looking		S	+	S	+	S	S
3. Reasonable cost		S	–	S	S	+	–
4. Acceptable to administrators		S	–	S	S	+	–
5. Better storage (students)		S	S	S	S	S	S
6. Improved materials		S	S	S	S	S/–	S
7. Locker arrangement in hall		S	S	S	–	–/S	S
8. Cleaning		+	S	S	+	+	–
9. Convenience		S	S	S	+	S	–
10. Safety		S	–	S	+	S	–
11. Long-term development/innovation		S	S	S	+	–	S
12. Short-term improvement/retrofit		–	–	–	–	+	–
13. Space required (floor area)		S	–	–	–	S	S
TOTAL +		1	1	0	5	4	0
TOTAL –		1	5	2	3	4	7

Advantages of Round 2 improved locker designs:

1. *"The Slider"*: Door opens 90°, slides back; lunch compartment; bottom grid shelf for boots; umbrella hook; 2 top shelves; pencil/pen holder; recycled plastic in various colors; mirror-reverse units of two; standard size, floor mat for muddy boots.

2. *Roll-Away Locker:* Rolltop door; recycled plastic; lunch shelf at top, book drawer with lower front edge (for visibility) next, pencil drawer in middle, vents at bottom section; each compartment pulls out separately with release button to give access to space in rear (for coat and umbrella), thus locker is deeper than standard size.

3. *"The LETLOCK":* Design like datum, except available in many color combinations; door opens 90°, then slides back; shelf and drawer at top, shelf and drawer at bottom; vent at top; standard-size interior (door "pocket" requires 1" extra for storage), thus locker requires more space for installation; with "alphabet" lock.

4. *The Wider Locker:* Standard width extended by 8"; increased coat hook size; foam door gasket, extra shelf and drawer at bottom; floor mat for muddy boots; door hinges let door open to only 90°; assorted colors in recycled material; integral lock sunk into door—no protruding parts; door will spring open when lock is released.

5. *"The Stopper Locker":* Foam gasket around door; top compartments for lunch and miscellaneous, followed by book shelf, 3 hooks, pocket for pens and pencils, bottom shelf above removable perforated boot shelf; recycled plastic; all doors open to right to 90° only; width 20" (or 12" for retrofit); recycled plastic replacement door available.

6. *"The Convenience Master":* Standard size for retrofit; top shelf adjustable 6" up or down, with hook on shelf; 8" shelf and drawer below, with 12" book drawer, and drain at the bottom; recycled plastic in standard blue or choice of colors; "garage-type" push-up door with vents.

Concept 6 received very favorable comments from the audience of students and parents, even though it did not score well on the evaluation. One reason for this is that it was a new concept that could benefit from a round of improvement. Also remember that it was not compared to the existing locker but against the top-ranking concept from the first round. The comparatively large number of negative marks does not mean it is an inferior design; it just means that the design needs more work or that the criteria may have to be reviewed to make them more accepting of new, creative concepts.

Traditionally, we hesitate to change criteria midstream in an evaluation to keep the judgment fair for everyone. When the results of an evaluation do not agree with our intuitive feelings about what the outcome should be, this usually indicates that we need to change the criteria to bring them into line with our improved understanding. This is an important benefit of the Pugh method in that it encorages flexibility as well as understanding and consensus about the criteria.

If the locker design project had been continued beyond this point, the students would have had to go back to their teams to further improve their designs and expand and refine the criteria. Drawings would have become more detailed, with exact dimensions and specifications. The design teams would still have tried to borrow ideas from each other; they would also have involved sales people, perhaps a school administrator and some students as well as a recycled-plastics expert. Concept 4 would have become the new datum; all improved concepts would then have been evaluated against this "standard." The process could have been repeated for several additional rounds until one design emerged that could not be improved any further, and the team members would have agreed and understood why this was the best possible design. This would most likely have been an integrated design incorporating several ideas that first appeared in some of the early concepts as well as creative ideas engineered later through this process.

We did not have time in this class to take the locker designs through Phase II; we just wanted to give the students an introduction to the Pugh method for evaluating ideas. The process was somewhat slow because the class had 30 students, and it was important for the development of their communication skills and self-confidence that we allowed them enough time to explain their designs, to critique each other's ideas, and to defend the different concepts. The final class assignment was for each design team to prepare a group presentation that involved "selling" the improved locker to school administrators or taxpayers. The Pugh method helped them identify potential problems with implementation and acceptance; it also gave them a clear understanding of the benefits. It is very interesting to us to note that students will almost always make further improvements to their designs for the final team presentation, in essence taking on aspects of Phase II on their own.

Phase II: Convergence to a Superior Concept

We want to close the discussion of the Pugh method with a few general comments about Phase II. (You may want to refer back to Tables 10-2 And 10-3.) Now imagine that several weeks have gone by since the first phase of the Pugh method has been conducted in two or more rounds. The design engineers have had time to incubate the revised list of criteria and come up with further improvements to the designs. The purpose of Phase II is to work with these improved designs and make them even better by synthesizing the most promising concepts.

In Phase II, the highest-scoring design concept out of Phase I is taken as the datum. The competition sharpens since weaker concepts are now dropped and only the strongest are carried forward for further development. As before, the teams can devise modifications to the designs and submit these as additional, new concepts. Phase II is continued through several iterations and incubation periods over the span of weeks or months, depending on the complexity of the product that is being developed. The list of criteria undergoes continuous refinement. After each evaluation session, the weakest designs are eliminated; the remaining designs are improved through further creative thinking and engineered in more detail. It is through this process of discussion, review, and evaluation for the purpose of improvement that the team members grow to understand why the solution that finally emerges is best: all good points of the design have been defended and all the negative points eliminated. Only at the very end may a weighting system be used on the criteria for confirmation. With the absence of negative points in the top designs, weighting factors will usually not be able to add additional insight to the evaluation, except perhaps in the case when two very different designs are emerging that appear to be almost equally strong.

Finally, Phase II ends when convergence occurs to a strong consensus concept that cannot be overturned by a superior concept; the team has a strong commitment to this design and is confident that this concept will succeed. This matter of commitment is very important, and this is one reason why the presence of managers from different departments (manufacturing and sales) is crucial during this product development activity. After all, the whole purpose of conceptual design is to find a product that can be manufactured competitively and that will sell so that the company can profitably stay in business.

Before the first prototype can be constructed, the superior concept is thoroughly analyzed with one or more of the following tools: FMEA, FTA, value engineering (see Appendix E for a summary of VA/VE), engineering analysis, and cost analysis. The drawings are now in the piece-part design phase, where suppliers are consulted. And finally, a go/no go design review is conducted to answer three vital questions: (1) Is the design inherently superior to competitive benchmarks? (2) Does the design meet consumer requirements? (3) Will the new product be timely, or is it already outdated? Only if all three questions above are answered in the affirmative are detail drawings and prototype construction authorized. These drawings are production-intent and precise since all foreseeable problems have been solved, thus eliminating the need for engineering changes after the start of production. The purpose of the prototype is to confirm the design, not to identify and correct problems.

Let's return for a moment to a question that was asked earlier—why should the Pugh method be used? In answer, we stated if a company wants to produce a product that is best, it has to start by selecting the best design concept. This statement assumes that companies want to produce a "best" product. However, as pointed out by Jim Hibbits, former president of Monarch Analytical Laboratories in Toledo, Ohio, such an assumption may not be valid. From his experience, he has found that engineers generally are not searching for the "best"—which is relatively easy. They are encouraged by their management to find something that is better than the competition, that can be made cheaper, and that will yield a profit and result in greater market share. There is rarely a reward for manufacturing the "best." Yet it is our conviction that U.S. manufacturing can regain a leading position if we widely adopt the goal of meeting the customer's needs with "best" products, "best" service, and extra value. When successful Japanese products are critically examined, it can be seen that they are by no means the best possible designs—they are just better than what we are accustomed to making. In the service sector, many businesses could improve with staff providing "best" service.

What makes the Japanese difficult to surpass in excellence is that they have introduced the concept of quality as an expandable commodity, not a fixed standard. When we talk about a "best" product, design or service, this means only with presently available, cost-effective technology and methods. A technological breakthrough or paradigm shift can quickly increase customer expectations and expand the requirements that constitute acceptable quality. Therefore, getting the Pugh method accepted routinely into organizational procedures will not only involve training for the employees, it will require support and commitment from management and an attitude that keeps working for continuous improvement.

EXAMPLES OF OTHER PUGH METHOD EXERCISES AND APPLICATIONS

The first example illustrates the Pugh method with a problem that did not involve engineering design; it was conducted during the spring of 1992 in our math/science Saturday academy. The second example summarizes a project done by talented and gifted high school students in a one-week summer institute. The third example shows an improved product—a luggage carrier—from a Pugh method exercise with engineering students, and the fourth example illustrates how the Pugh method can be used to improve a course syllabus or an entire curriculum.

Example 1—Solving Parent/Kids Conflict (Control of Car Radio)

Problem briefing:
Parents and kids have conflict when in a car together, because of different tastes in music. How would you redesign the car radio (or solve the problem in another way)?

Evaluation matrix, Round 1:

Criteria: Concepts:	Datum 1	2	3	4	5	6	7	8
1. Easy to use	S	S	Δ	Δ	Δ	+	+	S
2. Reduces conflict	+	S	S	Δ/+	Δ	+	+	+
3. Delivers music/news	S	S	S	S	+	+	S	S
4. Safe for driver	S	S	S	Δ/+	S	+	S	S
5. Safe for passengers	Δ	S	+	S	S	Δ	S	S
6. Creative	S	+	+	+	+	S	+	+
7. Retrofit	Δ	+	Δ	+	Δ	+	Δ	Δ
8. Cost	Δ	+	Δ	+	Δ	+	Δ	S
9. Attractive to buyer	Δ	+/Δ	Δ	+	Δ	+	Δ	Δ
10. Attractive to seller	Δ	+	Δ	Δ/+	+	+	Δ/+	S
11. Robust quality	S	S	S	S	S	Δ	S	+
TOTAL (+)	1	5	2	7	3	8	4	3
TOTAL (Δ)	5	1	5	4	5	2	4	2

Concepts for Round 1:

1. *Major Threat:* Two radios—one has a headphone jack for kids, one is cheap (for parents). The kids' radio is in the front seat (passenger side).
2. *Mix Master:* Audiotape has a combination of parents' and kids' music.
3. *Master Blaster:* Has intercom system controlled by parents. One radio is in the backseat, the other in the dash (able to monitor the sound level of the other).
4. *Cooperation:* Have parents and children agree to specific rules for radio use.
5. *Car-Gear Deluxe Set:* A game system that plugs into the lighter socket, with TV screen built into the back of driver and passenger seats.
6. *Walkman:* Buy a *Walkman* set for each child.
7. *Individual Channel Jacks:* A system of four radios that allows each person to plug into the channel he or she wants (like people on airplanes).
8. *Portable Unit:* A second radio (with jacks, like a *Walkman*) that can fit into a slot in the backseat area of the car but connects to the car antenna.

Concepts for Round 2:
Concept 6 was the new datum; criteria were discussed and expanded to include long-term benefits. The six improved designs are:

A. *SEGA* (2 video game machines): One machine is built into the back of each front seat. This is marketed as "free with $50 donation to C.A.R.E.S."
B. *Master Blaster II:* Two different radios. Parents can control the kids' radio in the back via their radio—the kids' radio has 3 headphone jacks. Cost is $125.
C. *The Innovator:* Quadruple radio with headphones; fits into existing radio socket. Each passenger can thus select his or her own station. Cost is $600.
D. *Mix Master Tapes:* Tape alternates parents' and kids' music selections and builds appreciation for different styles; it encourages mutual understanding.
E. *Parent/Kid Radio Unit:* Double radio with headphone jacks. Cost is $10.
F. *Walkman Choices:* Several models, i.e., AM/FM, AM/FM + tapedeck, AM/FM + tapedeck + CD player, all very affordable, so each family can buy one for each child. These can be used away from the car and without headphones.

Evaluation Matrix, Round 2:

Criteria:	Concepts: Datum	A	B	C	D	E	F
1. Easy to use/install		S/Δ	S/Δ	Δ/Δ	+/+	S/S	S/S
2. Reduces conflict		S	S	S	+	S	S
3. Delivers music/news, etc.		+	S	S	S	S	S
4. Safe for driver		S	S	S	S	S	S
5. Safe for passengers		Δ	S	Δ	S	S	S
6. Creative		+	S	S	+	S	S
7. Robust quality/simplicity		S/Δ	S/S	S/Δ	S/+	S/Δ	+/S
8. Reasonable cost: buy/operate		Δ/Δ	Δ/+	Δ/Δ	+/+	Δ/S	S/S
9. Attractive to parents		Δ	Δ	Δ	+	S	S
10. Marketable/easy to manufacture		Δ/Δ	S/S	Δ/Δ	+/+	S/S	S/S
11. Long-term benefits to society		S	S	S	+	S	S
TOTAL (+)		2	1	0	11	0	1
TOTAL (−)		8	3	9	0	2	0

Round 2 more clearly identified the *Mix Master* as a superior solution (over the *Walkman*). If costs can be brought down radically through a breakthrough innovation, some of the other concepts might have a great potential market. This exercise brought out very good discussions and thinking by designers and evaluators.

Example 2—Design of an Improved Lamp

This team of high school students selected the problem of inconvenient lamp switches as their design project. During problem definition, they realized that the problem had to be expanded to include the design of the whole lamp. The students conducted a customer survey, did a Pareto analysis, brainstormed ideas for solving the identified problems, and thus developed a list of design criteria. The students then formed small teams to come up with different design concepts for the Pugh method exercise. Basically, three categories of designs appeared: (1) improvements over existing table lamps; (2) innovative—completely new—concepts; and (3) novelty lamps. The novelty lamps did not score well in the first-round evaluation since they cost more than most people would be willing to pay; also, they did not address some other important customer needs. For the second round, a traditional lamp with many improvements (such as a built-in timer switch in the base, flexible shade, retractable cord, and fluorescent bulb) was the datum. An innovative design scored very high: it could be used as a table lamp or a pole lamp, and it had an interesting shade that could be fixed for up-down indirect or task lighting or expanded to expose a lighted column for room lighting. The novelty lamp in the shape of a tree with jade leaves and little lighted apples was evaluated with almost all negative marks, yet this team remained steadfast in not wanting to change its concept: it had "fallen in love" with its design. This is one of the fatal dangers in product design and thus demonstrated the utility of the Pugh method in identifying this flawed design. Lack of time and resources prevented the high-scoring team from pursuing its innovative design. Several people who saw the team's presentation on this design felt that it had potential to be patented—it was a real invention.

Example 3—Design of a Luggage Carrier

Introduction: When asked to suggest a team exercise for a creative problem solving class at Michigan Technological University, we had just experienced trouble with a product: our luggage carrier broke during the second week of use (even though we had purchased it because of its sturdy appearance). The spot-weld that gave way was impossible to repair; the carrier had become useless. So we turned this problem into an exercise. A team was given a design assignment and a briefing. The students inspected the broken carrier as well as two other models that had different shortcomings and brainstormed some ideas about other potential users. From this input, we, as the clients, made up the briefing document.

Briefing document:
1. Description of problems with existing carriers: Current models of luggage carriers cannot cope with heavy luggage: the handles bend so they cannot be retracted, or the base collapses. Wheels are too small to negotiate obstacles (stairs and the

threshold of most airplanes); suitcases drag on the floor. The loaded carrier cannot be lifted easily; also, it is prone to tipping over. Bungee cords are a problem (they snap back if not hooked on right; they do not fit; they become weak too quickly, and they easily get tangled up in the folded carrier.

1. Description of customer needs: Flight connections are made either through Detroit or Minneapolis—both airports require a long walk from the commuter terminal to the gates of the major airlines, as well as negotiating a flight of stairs, escalators, and moving walkways. The luggage carrier must be lifted onto a bus without unloading the luggage—present models do not have a convenient handle to do this. Quick loading and unloading capability is important at security checkpoints. The travelers are middle-aged and have back problems. Older travelers are primary customers; parents with small children are another target.

2. Description of luggage: (a) Bulky, soft-sided garment bag. (b) Pilot's case (18" wide, 9" deep, and 13" high) loaded with books and workshop supplies weighing 40 lbs or more. (c) One or two regular briefcases. (d) Padded computer bag with a computer and printer. (e) Large gym bag packed with clothing.

3. Design specifications:
• Collapsible (or foldable), to fit in the overhead bin of an aircraft.
• Can be set up or unloaded quickly.
• Can carry 120 lbs easily and securely.
• Reasonably light weight when empty.
• Maneuverable when fully loaded; rolls easily; can negotiate small obstacles.
• Can be lifted by two people when fully loaded.
• Luggage should be kept off floor and securely fastened; carrier must be stable.
• Comfortable handle; also, most of the weight is supported by the wheels.
• Purchase cost: $50 to $80 maximum (with 2-year guarantee).
• Can carry a child with or without a car safety seat.
• Comfortable for people from 5' to 6'6" tall.

4. Problem definition statement: Design a heavy-duty, improved luggage carrier.

Creative problem solving: The students brainstormed 70 ideas, which were sorted into categories; the categories were assigned to teams of five students each: Team 1—structural (folding model); Team 2—structural (telescoping type); Team 3—wheels and security; and Team 4—handle, strapping, and frills. This was an unusual outcome of idea sorting in that the different categories did not each result in a complete solution to the original problem; the teams were focusing on components. Each team synthesized several practical ideas within their category and then incorporated the highest-ranking concept or ideas into a conceptual design for the Pugh method evaluation. The teams were encouraged to consult each other.

Features of Round 1 designs:
Concept 1—"The Milk Crate": Molded fiberglass/graphite frame; golf-cart type third wheel and telescoping side supports in the base; the base folds to fit inside the back. Back/handle combination has adjustable height, with a fold-out shelf or child's seat. Reversible handle.

Concept 2—The "Fold-and-Tote": Collapsible shelves; easy fold-up for storage; adjustable base width, telescoping handle; fold-out second set of wheels from back for horizontal loading or negotiating steep stairs.

Concept 3—"B.F. Goodcart": I.D. tag with engraved name. Large, locking wheels with good bearings; three-wheel design (one in front, two in back), swiveling front wheel. Air-filled tires. Telescoping handle bent backward.

Concept 4—'The Easy Cart": Soft, easy-grip, curved handle with good weight distribution over wheels; adjustable for length; adjustable for width (for two people); 2 grips at front of base for lifting the loaded cart. Cargo net or three-point snap-in harness straps for luggage. Webbing between the frame to hold luggage. Small storage in handle. Small fold-out seat for tired travelers.

Round 1 evaluation matrix:

Criteria Concepts:	Datum	1	2	3	4
1. Carry 120 lbs securely		+	+	S	+
2. Quick load/unload/stow		S	S	+	S
3. Stable/safe		S	Δ	S	+
4. Maneuverable when loaded		Δ	+	S	S
5. Negotiates stairs		+	+	+	+
6. User-friendly handle		+	S	+	+
7. Manufacturability		Δ	+	Δ	Δ
8. Cost		S	S	Δ	Δ
9. Innovative		+	+	+	+
10. Value added		+	S	S	+
TOTAL (+)		5	5	4	6
TOTAL (Δ)		2	1	2	2

Summary of design process and results: At about the same time the teams carried out the first Phase I evaluation, one student team made up of a representative from each design team was invited to learn about patent searching on the Mead Data Central's LEXIS automated information retrieval service located in business administration and management. The students were amazed to find almost 200 patents listed under the hand truck category. Some of the recent patents on luggage carriers are held by Lucy Esposito who kept making improvements to her patents.

The Round 1 evaluations were somewhat difficult to do, since the four teams had concentrated on different parts of the cart. The teams were asked to improve their designs for Round 2. The evaluation criteria for Round 2 were: (a) Carry 120 lbs. (b) Quickly folds to regulation size. (c) Can be loaded or unloaded in less than 30 seconds. (d) Stable when loaded. (e) Safe under all conditions. (e) User-friendly (handle and maneuvering). (f) Attractive appearance. (g) Recycled materials (or recyclable materials). (h) Easy to manufacture. (i) Retail costs $50 to $80, with good profit margin for retailer. (j) Innovative. (k) Foolproof strapping system. Unfortunately, the term ended before the Round 2 evaluation could be completed.

What we want to show you in Figure 10-1 is the final design of one of the teams, since they continued working on the design as a topic for their team presentation to a large audience. The streamlined, space-age design no longer bears much resemblance to a milk crate. It is to be made of moulded, recycled plastic. Small hooks on the sides allow better ways of fastening bungee cords or a stretchable harness (unless the "quick-strap" system of Design 4 is used). A small bag is provided near the handle for stowing the cords or net. The students demonstrated the strength of moulded plastic by having a heavy student jump on a flat crate made of moulded plastic. Although this final design exhibited the most innovative thinking, it adopted many ideas from the other teams, such as the large air wheels, the soft, reversible handle, the hand grips in the wheel base, and the cord storage pocket. The team paid particular attention to having the center of gravity centered over the wheels in the loaded tilt position. This final design is thus a good illustration of a compound solution, the ultimate outcome of the Pugh method.

Figure 10-1
Improved luggage carrier developed by a whole-brain team of Michigan Tech students in a creative problem-solving class. Design by Levant Engin, Nils Johnson, Richard Lewnau, Chris Rust, and Anthony Zink; drawing by Nils Johnson.

One of the aspects of doing a Pugh method exercise with a group of students is the amount of discussion that it generates. It helps to divide the class into teams (with one design per team) to limit the number of design concepts that will be submitted. Even then, some teams come up with multiple ideas which they want to have evaluated. Students get upset if the evaluation process is hurried: the facilitator must remain strictly neutral and not make unilateral decisions about criteria and the evaluation marks. Ask for a vote on a particular evaluation when there is no consensus. Another difficulty is keeping track of the data. Designate an assistant to take notes when the evaluation is conducted on the blackboard, or use large sheets of paper as a permanent record of the evaluation. It is also a good idea to ask each team to submit a one-paragraph description of their design for each round (this will make it easier to write up a project summary later).

Most of our student teams have come up with very interesting concepts to an assigned or self-selected problem. They are disappointed that this introductory course does not provide for building a prototype or for implementing the superior solution (although most engineering students can look forward to a complete project in their senior capstone course). Since students of all ages usually want to keep their conceptual sketches, it is a good idea to ask them to use CAD for the drawings so that they can print out a copy for the instructors. We have found in our classes and workshops that teams will typically keep working with their designs to have a strong product to "sell" in their team presentation as part of their final exam—they gain experience with an attitude of continuous improvement.

Example 4—Application to Course Design and Curriculum Restructuring

With teamwork skills being demanded by industry, and with team exercises and projects taking up extra class time, how can instructors make room in an already overburdened class syllabus for new activities? It is clearly impossible to cover everything—thus the important question is how to determine what to cover and what to leave out. Students must still be taught the fundamentals. But students do not thrive when we "load dump"—they want to learn the important things well; they want to reflect and make connections to what they are learning in other courses. Which topics should be covered in class, and which topics can be assigned for self-study or computer-based instruction? The Pugh method is an excellent tool for identifying topics that can be dropped from a syllabus and for ranking those that need to be emphasized. An example is given in Table 10-4; it was first published by Edward Lumsdaine and Jennifer Voitle in a paper entitled "Contextual problem solving in heat transfer and fluid mechanics," AIChE Symposium Series, Vol. 89, No. 295, 1993, pp. 540-548.

The Pugh evaluation matrix of Table 10-4 is an excerpt from a much more extensive list since a valid course evaluation must include all the course topics. Instructors can apply their own criteria and priorities. The two heat transfer instructors (who team-teach the course) were surprised at some of the results of their analysis. For example, they had always included the product solution method; yet undergraduate engineering students have little opportunity to apply the method and gain proficiency. The method solves only a very limited class of problems (encountered only in textbooks, not out in the field). Being able to use the method does not increase understanding of heat transfer principles or phenomena. But because of its mathematical elegance (and because the instructors understood the method so well), "it was one of our favorite topics to teach, even though it left many students frustrated and confused." Also, a survey of the entire curriculum showed that only two elective undergraduate courses taught the method, and previous knowlege was not needed. Cutting this topic from the syllabus freed up two weeks.

The Pugh method can be used in conjunction with QFD and Taguchi methods to evaluate an entire curriculum. Each topic in a course can be judged against a list of criteria based on requirements from different customer sectors—students, alumni,

---

Table 10-4
Pugh Method Evaluation for Heat Transfer Course Syllabus (Excerpt)

Criteria Topic:	a	b	c	d	e	f	g	h
Relevance to subject	+	S	+	+	+	+	+	+
Usefulness	+	−	+	+	+	S	+	+
Teachability	+	−	+	+	+	S	+	+
Duplication	S	S	S	S	S	S	+	+
Fit with context	S	S	S	S	+	S	+	+
Need in subsequent courses	+	−	+	+	+	−	S	S
Need in industry	+	−	+	+	+	−	+	+
Need in design	+	−	+	S	+	S	+	+
Integration with other courses	+	−	S	+	S	S	S	S
Integration with computers	S	−	−	S	+	−	+	S
TOTAL	7+	0+	5+	6+	8+	1+	8+	7+
	0−	7−	1−	0−	0−	3−	0−	0−

Topics in course syllabus:

a	=	energy balance	e	=	finite difference methods
b	=	product solution, $T(x,y) = X''Y''$	f	=	boundary layer theory
c	=	use of temperature charts	g	=	heat exchangers
d	=	dimensionless parameters	h	=	convection correlations

Evaluation Key: S = satisfactory   + = advantage   − = disadvantage

---

professors teaching follow-on or prerequisite courses, accrediting agencies, and employers in industry. This process is changing the focus of these courses since it critically examines *what* we teach. When this approach is combined with an analysis of each topic against the Herrmann four-quadrant brain dominance model, we address the *how* and the *why*. Student learning improves considerably as the courses are streamlined, fundamentals are emphasized, many different hands-on activities are added, and realistic team projects are required for applied creative problem solving and computer use.

FURTHER LEARNING

References

10-1 Michael J. French, *Conceptual Design for Engineers*, Springer-Verlag, New York, 1985. This book, by a British engineering educator, links the creative design function with analytical engineering by emphasizing synthesis; it is illustrated with a wide variety of examples and problems. It is written for engineering sophomores; the vocabulary is interesting but differs from American usage. Also, *Invention and Evolution: Design in Nature and Engineering*, Cambridge University Press, New York, 1988, by the same author, is recommended for engineering and biology students.

10-2   Sidney F. Love, *Planning and Creating Successful Engineered Designs: Managing the Design Process,* revised edition, Advanced Professional Development, Los Angeles, 1986. This easy-to-read book focuses on the iterative principle of design. It makes both the systematic procedure and the creativity of the engineering design process understandable to managers and engineers as well as the management of engineering design . It includes a chapter on computer-aided design.

10-3   Stuart Pugh, *Total Design: Integrated Methods for Successful Product Engineering,* Addison-Wesley, New York, 1991. This book provides the framework of a disciplined design and evaluation method for creating products that satisfy the needs of the customer. It includes many examples from a variety of fields and a wide selection of design exercises.

10-4   David G. Ullman, *The Mechanical Design Process,* McGraw-Hill, New York, 1992. This book includes a discussion of the human interface with mechanical products. It includes the application of quality function deployment, concurrent design, robust design (Taguchi methods), function mapping, and the Pugh method.

THE PUGH METHOD EXERCISE

Homework assignment: (1) Select a topic from one of the assignments below. (2) In teams of three to five people, define the problem, collect data, prepare a briefing, and brainstorm design criteria. (3) Brainstorm conceptual ideas based on the criteria. (4) Evaluate these ideas using a creative idea evaluation technique; then develop the most meritorious ideas into a conceptual design. Try to have several very different ideas for the first round (one or more per team, depending on how many teams you have). Remember to use a dark marker; the drawing must be recognizable from a good distance away (about the size of half a flip-chart page). Specify only nontoxic materials; give rough dimensions. (5) Brainstorm a list of evaluation criteria.

Preparation: Select a large room with a long blackboard or blank wall. Have flip-chart sheets, tape, and markers available. Write the initial list of evaluation criteria on the board (or sheets) to the left. Then post the conceptual designs (including one for the datum) above the evaluation matrix. Invite nonengineers to participate as evaluators. Have an assistant to help keep track of the results if possible. Have a table with extra sheets of paper and markers ready so new design ideas can be sketched and added to the matrix immediately.

Evaluation: (1) Briefly review the features and problems of the datum. (2) A representative of the team explains the advantages and features of the first new design concept. (3) Then the concept is evaluated—through group consensus—against the datum. Lively discussion usually arises; through this process, the criteria become better defined. Team members must not become defensive of their concepts; look at the negative marks (or deltas) as opportunities for improvement; examine the other concepts for ideas to borrow. (4) Repeat the explanation and evaluation for all the concepts posted. As the concepts are discussed, the team members will think of new ideas. These should be sketched, added to the matrix, and evaluated. (5) Use positive thinking—put-downs and sarcasm are out. Look at all ideas as stepping-stones for further creative thinking.

Results and further activities: (1) As an engineer, do not feel proprietary about your design; the objective is to find the best solution for the entire team. If you are not an engineer, you can make especially valuable contributions. You may have a broader viewpoint; you haven't yet learned what doesn't work. At the least, you are a potential customer. (2) Add up the totals for the positives and negatives (or deltas). (3) The concept with the highest number of positives (and, in case of a tie, the least number of negatives) becomes the datum for the next round. (4) The teams now have a new assignment: to improve the designs in view of the evaluation results and ideas seen. (5) Repeat the process for one or two more rounds. Then have the teams prepare a final evaluation with the superior concepts.

Comments and hints: (1) Leave plenty of time for the teams to thoroughly discuss the concepts and criteria. For optimum learning, this activity cannot be rushed. (2) Remember that this exercise teaches the process—the end product and its quality are incidental the first time around. In an invention project or upper-division design course where the focus is on producing a truly superior solution, a quality outcome becomes important. (3) In subsequent rounds, add more detail to the drawings. Also, obtain cost estimates; do a risk assessment; do other design analyses as required to evaluate and improve the design. (4) Finally, write a brief summary about your experience with the Pugh method—what have you learned?

Preliminary Thinking Assignments

10-1 PROBLEM DEFINITION AND BRIEFING DOCUMENT
Identify a simple product in your daily life that could benefit from creative thinking and improved design. Prepare a briefing document and a preliminary list of design criteria for this product.

10-2 IDENTIFY PRODUCTS THAT NEED IMPROVEMENT
Observe your daily activities and note how often you wish some item were designed differently or could perform better. Assemble a list of at least five items.

10-3 THE MISSING PRODUCT
Observe your daily activities and note occasions when you wish out loud for something that could make your task easier. Identify the problem and brainstorm some possible solutions. Could any of these possibilities be turned into a useful product? Discuss ideas in a team, then collectively vote to select one of the ideas as your design project.

Design Topics

10-4 TRAVELER'S SPILL-PROOF CUP
Examine a plastic traveler's coffee cup (typically given away by donut shops). Fill it with warm water and experiment with the cup. Can you identify some problems? Does it spill easily? Does it splatter when being filled? Is it comfortable to hold and to drink from? Does it keep its contents warm? Is it spill-proof? Can you improve on the design? Look at the following list of criteria and see if any of these criteria need to be improved.

CRITERIA: (1) Spill-proof during fill. (2) Spill-proof during use. (3) Easy to open and close (if applicable). (4) Equally suitable for left- and right-handers. (5) Stable in moving vehicle. (6) Easy to clean. (7) Won't burn fingers. (8) Comfortable. (9) Nice appearance. (10) Easy to manufacture. (11) Competitive cost. (12) Special selling points (such as space for advertisement or logo). (13) Recyclable material.

10-5 REUSABLE CEREAL BOX
The objective of this design exercise is to reduce trash. Design a cereal box that has other uses so that the packaging will not be thrown away. As a datum, take a box of any of the well-known ready-to-eat breakfast cereals like Cheerios, Corn Flakes, *etc. At this point you are not yet given a long list of criteria. The assignment is thus very open and will allow you to think of some very innovative approaches to the problem.*

10-6 TOY FOR A SMALL CHILD
Make a conceptual design for a child age 3-5, following the list of criteria. The conceptual design does not have to be a completely novel idea; it may be an improvement of an existing product. The datum is a nice, cuddly teddy bear.

CRITERIA: *(1) The toy encourages creative play in teams. (2) The toy encourages each child's imagination. (3) Selling cost is less than $20. (4) The toy must be robust. (5) The toy must be cleanable. (6) The toy must be compact for storage. (7) The toy must be easy to manufacture.*

10-7 EDUCATIONAL TOY FOR GRADE-SCHOOL KIDS
Brainstorm a list of criteria and develop conceptual designs for an educational toy suitable for children from kindergarten through grade six. Important criteria are that the toy encourage imaginative play and cooperation among students.

10-8 MOSQUITO TRAP
Design a quiet suction device that will trap and destroy mosquitoes and similar insects.

10-9 TOBOGGANS AND SLEDS WITH BRAKES
Each winter season, people are killed in toboggan or sled accidents. Design a braking system for sleds and toboggans. Also incorporate a steering mechanism if possible.

10-10 VISUAL AID FOR AN ENGINEERING OR SCIENCE CONCEPT
A science principle can be selected (by vote). Develop a list of criteria, then discuss and modify the list before doing conceptual designs for a visual aid to help teach the principle.

10-11 REDESIGN OF HOUR GLASS
Design a simple hour glass that makes a sound when time is up. Alternately, design a simple hour glass (to be used when playing Pictionary®, for example) that can be "dumped" instantly, so people do not have to wait for the glass to empty itself before starting a new play.

10-12 IMPROVE A BUSINESS PROCEDURE OR A WAY OF DOING SOMETHING
Think of a business procedure, paperwork, or way of doing something that could be simplified, improved, and made more efficient through creative thinking. Brainstorm the criteria and use the Pugh method to identify the best concept.

10-13 THE PROBLEM WITH CRUTCHES
People on crutches cannot carry a beverage or a plate with food. That's just one of the problems. Do a customer survey; then, using the results, come up with ideas and design concepts that might solve the problem.

10-14 CHILD-PROOF CONTAINER CAP
Medicine bottles and other containers with child-proof caps are difficult to open for some people, especially those with arthritis. Investigate the problem—then design a storage system that is accessible to adults but not to children, or design an improved bottle cap or a bottle-cap opening tool.

10-15 CREATIVE IDEA EVALUATION
Use the Pugh method to "engineer" nondesign brainstorming ideas in creative problem solving, in place of some of the other techniques discussed in Chapter 7.

11

COMMUNICATIONS AND CREATIVE PROBLEM SOLVING

WHAT YOU CAN LEARN FROM THIS CHAPTER:
- *Good communications as the key to teamwork. What makes a good communicator? The importance of developing listening skills.*
- *How to get your point across in 30 seconds.*
- *Negotiating an outcome in which everybody wins.*
- *Coping with difficult people.*
- *High-tech communicating in real-time: cyberspace, Internet, E-mail.*
- *Further learning: reading references and other resources; exercises.*

Volumes have been written about communication and getting along with people. The single most frequent reason by far why people are fired from their jobs is because they do not get along with their colleagues or their bosses. Having good communication skills—knowing how to interact with people positively—is very important to productive teamwork. With creative thinking we can improve communication. Negotiation and difficult people give us many opportunities for applying our creative problem-solving skills. Do not just read about these techniques, but put them into practice! Make an implementation plan to acquire one new skill at a time; then give yourself at least three weeks of practice before going on to another item. Continue to hone your skills by delving deeper into the subject—study one of the books listed in the references at the end of this chapter, or research a particular aspect by scanning through books on communications in a library.

GOOD COMMUNICATIONS

In creative problem solving, we focused on a problem and then developed the best solutions that could be found for it. Basically, communicators also start with a problem—then they use verbal or written communication skills to ask for some change. To be good communicators, we must make sure the ideas or messages we are transmitting are received, understood, and acted upon. When we communicate, we do not merely pass on facts and know-how—the package includes feelings, attitudes, values, hopes, and dreams. Communication is easier when people have a common language, thinking preference, culture, and memories. However, during times of change, we cannot assume that we have many common bonds or that they operate as reliably as during "business as usual." Yet good communication is especially critical during times of change, when our success depends on our ability to "sell" our ideas and solutions.

Communication involves a sender, a receiver, and a message, as diagrammed in Figure 11-1. As the sender, we must encode our message to attract maximum attention and generate the desired motivation. The receiver must have enough time and information to decode the message properly. Then the receiver in turn becomes the sender and transmits feedback so that both parties can verify that the message has been properly understood and will result in the desired change. Both the sender and the receiver must be aware that the message is affected by two sets of screens or filters as well as by direct interference. Many different filters can be involved: language, culture, values, bias, memory, previous experience, emotions, paradigms, time pressures, the lack of speaking and listening skills, motivation, attitude, physical well-being, etc.

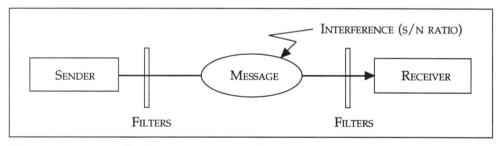

Figure 11-1   Factors affecting the transmission of a message.

The interference affecting a message directly is often defined by a technical term—the signal-to-noise ratio. It indicates how clearly a signal is coming through an environment filled with competing signals, which are called "noise." Thus the clarity of a message can be affected by the background noise in a room or other distractions and messages arriving simultaneously (for example, a ringing telephone, a secretary entering the room, another person talking simultaneously, a passing siren, a thunderclap, music, or a blaring television set). Filters are internal signals that can distort the message; the S/N ratio has to do with interference to the message through influences in the physical environment surrounding the speaker and listener. The S/N ratio analogy has been borrowed from the field of radio. When electrical appliances and high-intensity lamps generate electrical fields that interfere with signal transmission, when aluminum siding (or a mountain, in the case of an FM-signal) blocks clear reception, or when people communicating by cordless telephone or cellular car phone hear messages from another phone using the same band width—these are examples of low signal-to-noise ratios.

One-way communication from a speaker to a listener or audience appears to be simple and easy. At first glance, taking the time for giving and listening to feedback seems to complicate matters. Why can't a boss simply tell the employees what to do, either verbally or by memo? For routine tasks, this approach may be adequate, but in new situations, increased variability occurs in the way the message is understood as well as in the values and priorities of the different people involved. For this reason, two-way communication (even though it is slower and often messy) becomes especially important in times of major change. Teamwork is built on good communication, and for long-term benefits, these skills must be carefully developed and nurtured.

We have learned earlier that in most circumstances—barring a very unpleasant or boring original situation—people do not like change. If priorities and values must be changed, intense communication is necessary to achieve these objectives, especially when people have to be convinced that they must change their roles. But in any organization, communication across disciplines is unusually difficult because people's minds are not particularly eager to learn new jargon and techniques. People may not know or want to admit that learning and growth are needed. They do not like to be perceived as ignorant and will be reluctant to ask questions. But if we want to innovate, we simply must take the time to communicate well.

What is good communication? When a group of 14-year-old students from Detroit inner-city schools brainstormed this question, they came up with the list of characteristics shown in Table 11-1.

Table 11-1
To Be a Good Communicator

- Get to know people before judging.
- Respect others; appreciate them as they are.
- Spend time together.
- Take time to listen to each other's point of view.
- Don't try to be the leader (or the person in control) all the time.
- Have communications umpires.

- Be yourself.
- Be comfortable.
- Be open, kind, and caring.
- Watch body language.
- Be understanding.
- Talk one-on-one.
- Be positive and supportive.
- Keep a sense of humor.
- Learn how to have a "fair" fight.

The two items on the bottom line are intriguing, don't you think? They are one reason why we are including the topic of communication in this book. These students felt that their school environment (and their lives) would be much improved if they—and their teachers—were taught communication skills. The students came up with many of the same ideas that Dale Carnegie promotes in his approach in *How to Win Friends and Influence People*. Dale Carnegie courses are still in demand all over the world, and his books have been reprinted scores of times. Table 11-2 gives a summary of some of his ideas.

Table 11-2
Communication Ideas from Dale Carnegie

- Remember people's names.
- Become interested in other people.
- Talk in terms of the other person's interests.
- Respect other people's opinions.
- Ask questions; don't give direct orders.
- Call attention to your own mistakes before criticizing others.
- Admit when you are wrong.
- Praise improvement and give people a reputation to live up to.

- If you want to gather honey, don't kick over the beehive.
- Smile.
- Make the other person feel important, and do it sincerely.
- Give heartfelt praise and honest appreciation.
- Be courteous.
- Encourage people to talk about themselves.
- Let others save face.
- Give encouragement.

Since thinking about interpersonal relationships is a right-brain (quadrant C) ability, it is not surprising that this subject is neglected in our left-brain-biased education systems. Being accepted and liked by people is important for self-confidence. Research done at Arizona State University showed that good communication lowered stress. We have learned earlier that our minds work better and can think more creatively when we are not under stress. We can control how we react to the environment around us by maintaining a positive, caring attitude toward others. We can learn and practice good communication skills, such as listening and giving thoughtful feedback—these skills involve creative and critical thinking.

> *When I listen, I have the power. When I speak, I give it away.*
> *Voltaire, French philosopher.*

Developing Listening Skills

Ron Meiss is a communications expert and consultant with a background in sales, broadcasting, teaching, psychology, and human resource development and training. During the Fourth Annual Teaching Excellence Seminar in the College of Engineering at the University of Toledo, he spoke on the topic of communication skills for enhanced learning. His talk focused on the listener. Taking this point of view has at least two immediate benefits: First, we can learn to become better listeners; second, as speakers, we need to make it easy to listen to our message—we must learn *how* to say what we want to say. We will discuss the preparation of concise messages in the next section.

If we want the listener to hear us, we not only must speak loud enough to be heard, we must use the correct language; we must attract the listener's attention. If we want the message to be understood, we first must know something about the listener's thinking preferences, as well as the level of previous knowledge and cultural experiences. We must clearly speak in terms of the listener's interests. To get a response, we must speak in specifics and invite a response.

A Harvard study in the 1970s found that 9 percent of communication time is used for writing, 16 percent for reading, 30 percent for speaking, but 45 percent for listening. Another interesting statistic is that we can speak at a speed of about 120 to 140 words a minute, yet we can hear as much as 600 words a minute (if we concentrate). A study at the University of California at Los Angeles found that 7 percent of a message comes from words, 38 percent of a message comes from the tone, pitch, inflection, rate, and emphasis, and 45 percent of the message comes from body language. Thus listening involves more than just paying attention to the words of a message. Also, we do not learn very well when we are merely told something; we learn when we ask questions and discover the answers for ourselves. As a speaker, how can we get our listeners to ask questions? Let's ask some questions about communications.

> **TEAM ACTIVITY 11-1: QUESTIONS ABOUT COMMUNICATIONS**
> *In a small group, discuss the following questions:*
>
> • *Who do you think has the responsibility for good communications? Is it shared fifty–fifty between the speaker and listener? Or do you think each is 100 percent responsible to make sure mutual understanding has occurred?*
>
> • *When do we communicate? Does communication occur mostly verbally? Can we not communicate?*
>
> • *How do these six "people" increase the complexity of good communications?*
>   1. *The way I see myself.*   2. *The way I see you.*
>   3. *The way you react to me.*   4. *The way you see yourself.*
>   5. *The way you see me.*   6. *The way you see me reacting back to you.*

> *What you are, stands over you the while, and thunders so*
> *that I cannot hear what you say.*
> *Ralph Waldo Emerson.*

> *Perhaps the biggest barrier to communication is the assumption*
> *that it has taken place.*
> *Ron Meiss.*

What makes a good listener? Some ideas are summarized in Table 11-3. These hints concentrate mostly on the things the listener can do to improve communication—getting the message and giving feedback.

> **TEAM ACTIVITY 11-2: BODY LANGUAGE**
> *Investigate what body language is used in different cultures to indicate that the person is listening. How is nonverbal feedback given for agreement or disagreement? Try to interview people from at least three different cultural backgrounds: Native American, Asian (Chinese, Japanese, Malayan, Indian, Pakistani, etc.), Latin American, African, Middle Eastern, or Eastern European.*

Table 11-3
Characteristics of Good Listeners

- They want to hear what others have to say.
- They want to help with the problem.
- They accept the feelings of others as genuine (although they understand that feelings are transitory).
- They trust the other person to think and solve their own problems.
- They listen to understand and do not judge negatively.
- They do not "correct" the message or change the subject.
- They focus on the goal, not minor issues.
- They know that first impressions or appearance can be deceiving.
- They do not finish the speaker's sentences!
- They do not jump to conclusions.
- They do not prejudge: "I've heard this before; it's boring; it's too hard."
- They pay attention; they maintain eye contact (in Western culture only—elsewhere it may be considered rude); they smile if appropriate.
- They give feedback to show they are listening with appropriate body language.
- They ask questions.
- They can summarize the facts and meaning of what was said.
- They can pick up on the nonverbal message.
- They are not distracted by unconventional behavior or anger.
- They are in control of their own behavior and focus on solutions.
- They can give supportive feedback (as well as constructive criticism if asked).

[Employees] are expected to have greater ability to communicate and "sell" their own ideas—orally, electronically, and on paper—not only among fellow employees, but to suppliers and customers.
James Braham, senior editor, Machine Design.

HOW TO GET YOUR POINT ACROSS IN THIRTY SECONDS

Now we want to discuss another aspect of communication that is especially relevant in our busy times, and that is preparing an effective message that will not take more than 30 seconds at most to hear or read. People are busy. Time is at a premium. We have so many labor-saving devices nowadays—yet people seem to have less and less time to spend on communication. We are living in the information age, and thus we are being bombarded with messages from everywhere. Think back—by how much would you say your junk mail has increased over the last two or three years? If you are on a computer network and can't read your E-mail for a week, how many messages would have piled up? Just to cope and preserve our sanity, we are learning to "tune out." In this kind of environment, where our messages have to compete with a lot of information "noise," how can we make sure that we are being heard? How can we become efficient communicators without wasting our efforts?

Milo O. Frank, a business communications consultant, has written a poignant book on *How to Get Your Point Across in 30 Seconds—or Less.* We have found this approach very useful and would like to share some of the important concepts of this technique. Among the benefits of this technique are that you will be able to:

- Focus your thinking, writing, and speaking.
- Be logical and concise; have better meetings and interviews.
- Improve listening and keep conversations on track.
- Make better presentations; be more successful in "selling" ideas.
- Use questions and answers to make a point more effectively.
- Have increased self-confidence and achieve your objectives.

This approach is especially useful when you want a specific response from people— when you are asking them to do something for you, or when you want them to react and get involved on some issues. We will discuss these nine subtopics:

HOW TO GET YOUR POINT ACROSS IN 30 SECONDS

Preparation: Objective, Audience, and Approach.
Message: Hook, Subject, Close.
Presentation: Style, Acting, Speaking.

Why 30 seconds or less? Why not one minute or two minutes, or even five minutes? We would like to submit the following reasons for your consideration:

✦ Memos and letters of request are too long—just check over your junk mail.
✦ The attention span of the average person is 30 seconds.
✦ Doctors listen to their patients for an average of only 19 seconds before they start making a diagnosis, use instruments, and proceed with the physical exam.
✦ Your are allowed to add an explanation of 100 words to your credit report.
✦ E-mail messages are more effective when they are sized to fit on a computer screen without scrolling.
✦ TV commercials do a good job of getting their message across in 30 seconds. It cost $850,000 to air a 30-second commercial during the 1993 Super Bowl; thus many companies presented their messages in 15-second "half spots."
✦ TV news "sound bites" are 30 seconds long or they do not get air time. Reporters spend about 30 seconds introducing the subject. Then the topic or sound bite is shown, followed by a summary not exceeding 30 seconds.
✦ If you can't say it in 30 seconds, you probably are not thinking about your message clearly. You may need more time to present supplementary information (if asked), but the main thrust of your message should be very concise. President Abraham Lincoln's Gettysburg address and President George Washington's inaugural speech are brief but extremely effective messages.

The following discussion will give you the steps for preparing a 30-second message. When we teach this subject, people at first think that it will take them only 30 seconds to prepare the message. This is a misconception—preparing an effective message takes much thinking and creative problem solving and could easily take an hour or more, especially at the beginning, when this is a new activity for you. Thirty-second messages can be verbal or written—they can be telephone requests and messages left with answering machines or secretaries; memos, letters, fax messages, and thank-you notes; abstracts for scholarly papers and work proposals; formal presentations at meetings; interviews; a request or sale; social situations with superiors, chance meetings, and giving toasts. The 30 seconds in an elevator may be all the time you have to present a creative idea to your company's president. And it is a sad commentary on our times that you may only have the 30 seconds of a commercial break on television to present an urgent request to a family member.

Preparation

As you prepare your talk, phone message, or letter, you must determine your objective, your audience, and your strategy.

Objective: What do you want to achieve? Why? Ask yourself these types of questions: Why do I want to have this conversation? Why do I want to write this memo? Why do I want to meet with this person? Why do I want that interview? Why do I want to leave this note? Why do I want to speak to this group?

> *Your aim is to have a single, clear-cut, specific objective.*

Audience: You must address your request to the right person. Who is the target of your message? Get to the person or group who can get you what you want—otherwise you are just wasting time. Learn something about that person or group; address their needs. For instance, when you have a complaint, ask to see the store manager; do not waste your time talking to a salesperson or clerk.

> *Know what your audience is going to want from you.*

Approach: Think about the approach or strategy that you want to use. This is the sales pitch. Brainstorm different ideas, then select the one that meets the objective best. Here you select the form as well as the content. Should you make a phone call, meet in person, send a telegram, a letter, or a memo? Or would a formal presentation, a newspaper ad, or TV commercial serve your purpose best? Ask yourself: What am I talking about? What's the basis of my game plan? What is the heart of my 30-second message? What is the single best statement that will lead to what I want? Can I comfortably build a case around this statement? How will this statement relate to the needs and interests of the audience?

> *How will you get what you want—what's the best approach?*

Play around with different ideas and scenarios, then select the approach that fits your objective and audience best. Here is an illustration:

First Approach:

Blondie comic strip of September 14, 1991 reprinted by special permission of King Features Syndicate.

Second Approach:

Blondie comic strip of September 15, 1991 reprinted by special permission of King Features Syndicate.

What is the difference between the first and second approach? Why is the second approach successful? Do you suppose the list of benefits directed at the "audience" has something to do with it?

Message

After you have settled on the approach you want to use, you need to work on the actual 30-second message. This will have three parts: the hook, the subject, and the close. You could also think of it as the "hook, line, and sinker."

Hook: To get attention, you must use a hook. If possible, state the hook in form of a question. You can also use humor (at your own expense only). Or come up with a visual aid. Keep a notebook of good hook ideas, quotes, stories, and jokes.

Make sure the hook that you choose relates to your objective, listeners, and approach—it is a bridge connecting the audience's interests to what you want. If you have a very brief message, your entire message can be a hook.

> *Connect the hook to your objective, audience, and approach.*

Subject: Focus on the subject by answering who, what, where, when, why, and how as they relate directly to your objective. Ask yourself: What am I talking about? Who is involved? Where is it? When is it? Why is this being done? How do I do it? How should the audience get involved? Then check your answers against these questions: Am I reinforcing my objective (the objective may be mentioned explicitly or hidden)? Does the message correspond with my approach?

> *Make the subject relevant to your audience.*

Close: This is the bottom line. Be forceful or subtle, depending on your audience and how well you know them. Demand a specific action within a stated time frame, or ask for a reaction through the power of suggestion. If you do not ask for what you want, you have wasted an opportunity. The first three paragraphs in this book's preface are a 30-second message with an indirect close—we are asking the reader to buy and study this book to learn to be effective problem solvers.

> *Directly or indirectly, ask for what you want.*

Message checklist: Use your creative, critical, and interpersonal thinking skills to make your message more effective and memorable. Use this checklist to flesh out and improve your message. In a widely competitive environment, quality (even in communication) has to meet ever-expanding standards and expectations.

— *CLARITY? Do you have concise facts for quadrant A thinkers? Pay attention to semantics—will the audience have the same understanding of the words you are using as you do? Do not make any errors in logic if you are presenting arguments in support of a position.*

— *ACTION PLAN? Does your request ask for implementation that is well organized for quadrant B thinkers?*

— *IMAGERY? Are you using quadrant D thinking in painting a creative word picture to make your message easy to remember? Or use a visual aid. We still remember "Pickle," the green parakeet brought to class by one of our students to emphasize his sales pitch.*

— *PERSONAL STORIES AND EMOTIONAL APPEAL? Are you building relationships and reaching the heart of the listener by sharing emotions in quadrant C mode? You can establish your credibility by making yourself human with talk about your own failures.*

— *Finally, know when to stop.*

> *Create a message that will be remembered.*

We tend to draw conclusions, make assumptions, evaluate and judge rather than
observe behavior and report what we see, hear and feel.
Lt. Col. H. A. Staley, The Tongue and Quill, *Maxwell Air Force Base, 1977.*

Presentation

Neatness, good grammar and spelling, and an appropriate format (conservative or creative) count. In a verbal presentation, style and appearance, acting, and mode of speaking are important since they transmit part of the 30-second message.

Style: Give some thought to your personal style and image. What nonverbal message are you conveying? Monitor your body language. Practice delivering your message in front of friends who can critique you in a supportive way. Better yet, have someone videotape your presentation, then use critical thinking to evaluate your performance. Examine your facial expressions, eye contact, posture, gestures, movement, and tone of voice. Check your appearance. Do you know what kind of clothes make you look your best? Please yourself, but realize that in some situations it does matter what others think. Being considerate of others has preference over your own tastes. Wear clothing that will draw the audience's attention to what you are saying, not to itself. Be creative; subtly, and as much as possible, be yourself. Courtesy and good taste in clothes show that you care about other people and about yourself. Be clean and well-groomed. Whether you are making a verbal or written presentation, be sure to follow the rules of etiquette. These have not been devised to make your life difficult; social interaction is made more comfortable when everyone knows what is acceptable behavior. Are you familiar with some of the recent changes for the 1990s?

Acting: Are you conveying a positive attitude? If you "act" friendly, this will make you feel friendly. Smile. Use eye contact; focus on different people in the audience while you speak. Do not read off a script or memorize your speech. You may feel that you are being asked to pretend, to do play-acting. To some degree, that's what good communicators do. A prime example is former President Ronald Reagan. Have a friend videotape you in a situation where you are interacting with people—then evaluate your acting skills. You may want to take some private lessons (or a class in theater) if you find that you need considerable help

Speaking: It helps to show surprise, puzzlement, or concern in your facial expression and voice as you speak. Do not use distracting body language (like pulling on your fingers, tapping your knuckles, jingling coins in your pocket, or turning your wedding band). If you usually speak in a monotone, learn to modulate your voice. Use strategic pauses. Practice breathing and relaxation techniques prior to the start to reduce your stress level and thus have your voice sound more natural. Start on time. Respond directly to questions from the audience, but don't be carried away. Finishing within the allocated time is usually much appreciated.

> *Your appearance and style speak louder than words.*

> ⏱ TEAM ACTIVITY 11-3: THIRTY-SECOND MESSAGE
>
> *This activity requires three or more people. Each person prepares a 30-second message on a common topic (or alternatively, on a topic of choice). A good subject could be the topic of the thinking report or the "selling" of the best solution of the class exercise.*
>
> *Then each person presents the 30-second message to the others. The audience has to give positive feedback on what they think worked especially well in the message.*

> ⏱ TEAM ACTIVITY 11-4: QUESTIONS OF ETIQUETTE
>
> *In a small group, discuss the following scenarios and the proper way to respond:*
>
> a. *You are in the hall talking to a colleague, when a visitor—a good acquaintance of yours—walks by and greets you. This person does not know your colleague.*
>
> b. *You are conducting a business meeting. Some people from the outside have been invited to attend, and they are about to enter the room.*
>
> c. *You have an appointment with someone. You realize that you will be delayed.*
>
> d. *You have dialed a wrong number.*
>
> e. *You have someone in your office who made an appointment to see you. Suddenly, the secretary interrupts you to tell you that you have an important telephone call.*
>
> f. *You are speaking or writing to a person who has a professional degree or an affiliation after the surname. Make up some specific examples. How would you address these people?*
>
> g. *Julia Montez Smith is married to Sidney W. Smith. How should she be addressed in her personal life and in her workplace? What if she were to be divorced or widowed?*
>
> h. *Imagine that you are the chief executive officer in a company. Come up with the ten most important "rules" or guidelines for projecting a well-polished image to the community and its customers.*

NEGOTIATION

So far we have discussed communication from the angle of sending and receiving a message and then responding to that message appropriately and concisely. Many times, our interaction with people is more complicated, especially when a conflict is involved that needs to be resolved. In such a situation, we have to negotiate to come up with a resolution. Roger Fisher and William Ury of the Harvard Negotiation Project, in their book *Getting to Yes—Negotiating Agreement without Giving In*, demonstrate that negotiating does not have to be an adversarial battle but can be a productive problem-solving process. A summary of their technique is given in this section.

Let us never negotiate out of fear, but let us never fear to negotiate.
John F. Kennedy, Inaugural Address, 1961.

Negotiating is done all the time. According to the Ury-Fisher model, there are basically three approaches to negotiation: soft negotiation, hard negotiation, and principled negotiation. Here is a brief description of the three cases:

SOFT NEGOTIATION. One person wants to avoid personal conflict and makes concessions quickly to reach an agreement, but as a result, this person may eventually feel exploited and become bitter. The participants are friends or family members, or they may have an employer-employee relationship. The balance of power is unequal; one of the parties has a much larger investment or deeper commitment to maintaining the relationship than the other. This person is trusting and flexible; he or she will make offers and change positions to resolve the conflict; he or she will yield to pressure. This person will reveal the bottom line and will accept losses as the price of peace and agreement. The individual's behavior may be guided by cultural pressures or significant personal values. This person is most likely a strong quadrant C thinker who wants to"win" the negotiation through accommodation.

HARD NEGOTIATION. The situation is perceived as a contest of wills. Both parties want to win, at almost any cost. This process is exhausting and can cause serious harm to personal relationships. The participants are adversaries on an equal footing; they demand concessions as the price of maintaining their relationship. They are inflexible; they distrust each other; they make threats, apply pressure, and mislead as to the bottom line. They demand one-sided gains as the price of agreement—compromise is out of the question. They dig into their positions and thus find it very difficult to yield and thereby "lose face" or status. This mindset, too, may be strongly shaped by cultural influences; these people most likely are quadrant B thinkers who try to "win" the negotiation through intimidation.

PRINCIPLED NEGOTIATION. Issues are decided on merit. Both parties work toward an outcome that will benefit everyone concerned. If conflicting interests persist, the solution worked out is based on fair standards. These participants are problem solvers; they seek an outcome that will be mutually agreeable, efficient, and amicable. They are able to separate the issues from personal feelings. They focus on common interests, and they avoid having a bottom line. Through creative thinking, they invent options for mutual gain; the final decisions made are based on agreed-upon objective criteria. They are open to reason and will yield to principles. The participants are whole-brain thinkers who consider values, relationships, the context, the long view, the facts and reasons, as well as the mutual benefits and risks in arriving at the best solution through cooperation as equals. Developing win-win outcomes requires flexibility and a positive attitude. Benefits are maximized for both parties; self-respect is maintained and relationships are strengthened.

A successful outcome in negotiation depends on good communication. However, we can often observe that serious errors in communicating are committed by negotiators, be it in labor-and-management relations or even in the international arena. Can you think of examples of the following situations?

1. Negotiators are not talking to each other or are not understood. Instead, they are playing to the gallery or constituents.
2. Negotiators are not listening; they are not paying attention to what is being said because they are thinking of what to say next.
3. Negotiators are speaking different languages—a situation that lends itself to misunderstanding and misinterpretation.

When a translator is involved in negotiations, special care must be paid to the different cultural meanings that can be attached to words after they are translated. It always amazes us how even common words can have a considerable difference in meaning in another language. Also, people can speak different "languages" even if they use the same tongue—if their cultural background and experiences are different, or if they use different thinking preferences or speaking modes (a subject we will discuss further in a moment). Table 11-4 gives some guidelines for successful communication and negotiation. Ultimate success is not defined in terms of getting your way but in terms of building partnerships.

Table 11-4
Guidelines for Negotiation and Communication

1. *Listen actively and acknowledge what is being said. Provide feedback from the point of view of the other person or group by stating their position in positive terms.*

2. *Speak to be understood. Look at the others as partners for solving a joint problem. The more important the decision, the fewer people should be involved. Two is best for a "summit" meeting.*

3. *Don't condemn. Describe the problem in terms of personal impact. "We feel discriminated against" is better than "You are a racist or oppressor." Try not to provoke a defensive reaction or anger; instead, stick to the objectives.*

4. *Take the long-term view and build relationships. It is possible to win the battle and lose the war.*

5. *Follow creative problem solving: do not judge too soon, look for options and alternatives, do not assume a fixed pie (either/or) concept or act in pure self-interest. Brainstorm—alone, with the other party, or with other interested people—then do a creative evaluation to find the best options. Develop a list of objective criteria.*

6. *If you are negotiating from a weak position, have a Plan B. This way, you will not be tempted or forced into accepting a plan that will put you too much at a disadvantage.*

7. *What if the other party won't play and follow the rules of principled negotiation? In this case, do not attack the opposing position—look behind it. Do not defend your ideas or take the attack personally. Instead, invite criticism and advice. Listen and agree as much as possible. Restate an attack on you as an attack on the problem. Reframe the opposing position by using what-if questions. Build on ideas; make it easy for the other party to gain honor or a good way out. Discuss the cost of drawn-out disagreement. Most of all, treat everyone with respect.*

⌚ *TEAM ACTIVITY 11-5: NEGOTIATION*
 In a group of three, analyze a current negotiation, for example, in labor and management, or on the international scene. Identify the type of negotiation being used. Cite supporting evidence. Discuss how creative thinking could be introduced or strengthened in the situation. How could this affect the outcome? Or describe a case in which you were able to mediate a dispute. What strategies did you use?

COPING WITH DIFFICULT PEOPLE

People are not always reasonable: they have different personalities, different values, different attitudes, habits, manners, expectations, and thinking styles—all of which can make communication and getting along a very complicated process. It is extremely difficult or virtually impossible to change people unless they have a strong motivation to change. The situation, however, can be improved if we can learn to understand and change our own reaction to difficult people—in essence, if we can learn to cope instead of feeling angry, helpless, or defensive. Coping minimizes the impact of the undesirable behavior and lets people get on with their business. It reduces stress and thus helps us deal with the situation more creatively.

Robert M. Bramson in *Coping with Difficult People* outlines the following problem-solving approach:

1. Analyze the situation. Look at the context. Why is the person difficult? What circumstances make the behavior better or worse?
2. Stop wishing the difficult person were different. Stop wasting energy in trying to change the person. Don't take the situation personally.
3. Put some distance between you and the difficult behavior. Try to develop an understanding attitude and an outside perspective.
4. Formulate a coping plan. It is possible to learn to interrupt negative reinforcement. You can have control over the situation. Be positive.
5. Implement your plan. Determine the best timing. Prepare the appropriate approach for the personality trait you are facing. Visualize a good encounter. As the interaction proceeds, be flexible to make changes if necessary.
6. Monitor the effectiveness of your coping strategy. Use creative thinking to find ways of distancing yourself, if nothing else helps.

How do you develop a "difficult person" coping plan? Start by describing in detail the behavior you find so annoying. Include an analysis of what you think are differences in your thinking style and that person's approach—then write down your understanding of that behavior. Resist the temptation to attach a derogatory label to the person; instead, concentrate on the behavioral patterns. Review your own behavior in your interactions with the person in the past. What factors have made the situation better or worse? Creatively think of some coping behaviors that may help you deal with the problem—then analyze your ideas in the light of your experience. Could defensive reactions be involved? As far as you know, have any of the proposed approaches worked in similar situations? What do you need to change in your own reaction to best carry out the coping behavior? What values and strengths in your personality will help you? Which coping behavior do you need to practice? Make an action plan—write down specifically what you will do and when. Mark your calendar! If you are dealing with a very stressful situation which has existed for a long time, you may need to enlist some help through a support group or a counselor, especially if you are dealing with addictive or co-dependent behavior patterns. You may also want to spend some time thinking about the way words are being used in your interaction with the difficult person.

The Power of Words

One problem that can occur in communications is that people are not aware of the power of words; they come across as rude, thoughtless, manipulative, or even hostile because of the presuppositions behind the language they are using. Confrontations and misunderstandings occur when people use language in verbal attacks—purposely or inadvertently. Presuppositions are part of our language and grammar. For example, the sentence, "Even Pat could pass this course" implies that the class is second-rate and so is Pat, although the sentence itself used only a string of six innocent words. Verbal attacks use a lot of hidden presuppositions. How do you cope with such a verbal attack? Suzette Haden Elgin in *Success with the Gentle Art of Verbal Self-Defense* recommends a number of strategies.

Identify the filter: When people have difficulty communicating, it frequently is because they view the world through different filters. Therefore, the first thing that you must do in order to understand what another person is saying is to try to identify the filter being used. You do this by assuming that the statement that was made is true—then try to imagine what situation it could be true of. Once you have identified this context, you will be able to communicate with a common vocabulary and common basis of understanding. This is important in the business world where communication problems between men and women can create real difficulties. Is the person seeing himself as a father talking to a child, as a teacher lecturing to a class of immature students, as a quarterback or a coach leading a team, as a scorekeeper or judge keeping detailed tabs on performance? Is she seeing herself as a caring mother or nurse, a secretary in charge of an office, a scientist investigating and proving a hypothesis, a banker safekeeping an investment? Keep in mind that people change their roles or their "game" in response to different situations and different people.

Identify the sensory mode: Become aware of the kind of sensory mode that is being expressed in people's language. Do they use words that relate to seeing, hearing, touch, smell, or taste? When you match people's mode in reply, they feel understood and are thus more likely to trust you. Also understand that under stress, people become less flexible. For effective communication, match the sensory mode coming at you. If you are unable to do this, use a neutral style without any sensory content at all in your interaction.

 Team Activity 11-6: Sensory Modes

In groups of four, sit around a small table. Write a small script of a complaint and matching reply in three different sensory modes. Then pass your script to the person sitting on your left. Make editorial comments and improvements to the script that you have received. Pass it on in turn, and continue this activity two more times.

Share several scripts with a larger group (if you are doing this activity in a class). Discuss why you think this communication technique "works," citing evidence from the examples or from real life.

Identify the Satir mode: Virginia Satir, a well-known family therapist, discovered that during a confrontation or other important verbal interaction, people's language and nonverbal behavior tended to fall into one of the following five distinct modes or communications preferences (which may become habitual):

1. BLAMING. These people are openly hostile; they make categorical accusations; they give direct orders; they hiss, they speak loudly. They use threatening body language, shake their fists, scowl and loom over people. Consider that they may be trying to cover up insecurity.

2. PLACATING. These people appear very anxious not to offend. They plead; they use excessive praise; they exaggerate and stress words. They cling and fidget; they lean on people. They may pretend not to care, even when they do so desperately.

3. COMPUTING. These people avoid using personal pronouns and speak in abstractions instead. They keep communications as neutral as possible, including a flat tone of voice and minimal (stiff) body language. They are afraid to let emotions show.

4. DISTRACTING. People in this mode come across as unorganized and distracted. They cycle rapidly back and forth through several modes in an attempt to hide their feelings of panic.

5. LEVELING. People in this mode are coping well. They know their own feelings; they use words and body language that are in sync, not in conflict. What they say is simply the truth about the situation and their own feelings.

As you observe a person's communication behavior or mode, you have choices on how you want to interact. If you want escalation, match the person's mode. When you match a person's pattern, you will feed it. Blaming a blamer escalates the violence and causes fights. Placating a placater causes undignified delay; computing causes dignified delay (which can be a useful strategy at times). Distracting a distractor feeds the panic loop. And leveling at a leveler means exchange of truth both ways. If you don't know how to respond in a particular situation, use computer mode. This will gain you time.

Identify stress patterns: Verbal attack occurs when statements and sentences have abnormal stress patterns. You can ignore the bait instead of responding in kind and escalating the mode. You can learn to recognize when you are under attack, what kind of attack you are facing, how to make your defense fit the attack, and how to follow through. You can respond directly to the attack hidden in the presuppositions. Your objective is to say something that will transmit the message: "Don't try that with me. I won't play that game. I refuse to be the victim." When someone makes a stressed statement like: "If you REALLY cared about the sucCESS of our project, YOU'd be doing YOUR share of the WORK around here," you can counter this by addressing the presupposition directly and asking: "When did you begin to think that I don't care about the success of our project?" while carefully using leveler mode without undue stress on words. Alternately, if you prefer to cut the discussion short, you can just reply with an affirmative statement (in leveler mode) that begins with "Of course I care about our project. As a matter of fact, just this morning I got another idea on how we can expand our market." Then you can continue discussing the subject under consideration on your terms, propose new ideas—whatever. The attacker will be caught by surprise by your unexpected response and refusal to take the bait. This stategy can be used for all verbal attacks. In short, when you are having communications problems because of covert or open hostility, try one or more of these techniques: (1) analyze and respond with the appropriate sensory mode; (2) analyze and respond in the appropriate Satir mode, or (3) identify the verbal attacks and defuse them.

Here is a 30-second message: We would like to share with you one final, important thought about communication. What do we do when we are operating a piece of equipment and suddenly find that we are in trouble because something is not working right? The first thing we should do is go and read the manufacturer's instructions. The best teaching on communication and relationships is given by Jesus Christ in the Sermon on the Mount (as recorded in the Gospel of Matthew, Chapters 5 through 7). It all comes together in the Golden Rule:

> *TREAT OTHERS AS YOU WOULD HAVE THEM TREAT YOU.*

TEAM ACTIVITY 11-7: DEFUSING CONFLICT

In a small group, brainstorm some situations in the past that have led to conflict. Concentrate on the verbal expressions used, such as, "You never . . ." or, "You always . . ." Try to make up a skit to demonstrate how the conflict situation could be defused when the "you" statement is replaced by an "I" statement, such as "I feel hurt, when I'm not given a chance to . . ."

COMMUNICATING IN CYBERSPACE

In the last few years, time has become compressed. Instant food, instant news, instant entertainment are all in demand. We can remember in the late 1950s when communication between us and our parents overseas took a month or more—we wrote a letter, sent it airmail, and then awaited the reply. In an emergency or to share news such as the birth of a child, we sent a telegram. Now, we pick up the telephone every few weeks to call family and friends all around the globe. If we do write a letter, it gets sent by fax to those who own a fax machine. Now people carry beepers so they can be reached more quickly, and they own car phones. Others stay in touch with yet another marvel of technology: E-mail.

Developments in technology have led to fundamental changes in communications. Communications satellites have made it possible to reach anyone practically anywhere on our globe instantly and clearly. In the days when telephone signals went by transatlantic cable, appointments (days in advance) were necessary when calling from some countries; then you waited until a line was free—which perhaps would be at 3 a.m. Costs of calls and of equipment (including fax machines) have come down significantly. Fax, phone, and computer messages can be handled by the same telephone line with the right switches or software. You do not even need to have a separate fax machine: computers and printers with the appropriate software can take over this function. These advances all make it possible to have "real-time" communications—messages are sent and received with little time lag, and the rapidly expanding use of fiber-optic cables and networks increases the speed and quantity of messages that can be transmitted at any one time.

In Chapter 9, we showed how computers can help us think through and solve problems—computers are a thinking tool. Computers are also a communications tool—nowhere more so than when used for electronic mail, or E-mail. Quadrant C people especially enjoy connecting with others who share common interests through electronic bulletin boards. First, we will give a brief overview of the three main ways that computers can be connected to allow them to communicate with each other; then we will go exploring in "cyberspace" with Carol Reed, who is a science subject specialist in the William S. Carlson Library of the University of Toledo.

Within organizations, computers can be connected in a **local area network (LAN).** No longer need memos and other pieces of information and communication be duplicated and stuffed into each employee's box; this communication can now be done on-line. In the academic environment, the administration can communicate with the faculty, and even more importantly, the students can link up with the faculty. This is particularly beneficial when courses are taught by extension—thus students can E-mail their questions about homework to the instructor, who can then answer the questions individually or have the questions and answers available for access by the entire class. This type of fast feedback helps the learning process. The system works well when the people involved check their "mailboxes" several times a day. E-mail has several advantages over the telephone, in that replies can be given at the sender's convenience, not interrupting ongoing work, and the communication can be printed out in case a paper copy for traditional record keeping is needed.

But LANs are not the only way computers can communicate with each other. LANs at numerous institutions can be connected in a **wide-area network (WAN)** covering a large geographic region. Two such WANs are BITNET and Internet. BITNET is administered by the Corporation for Research and Educational Networking (CREN), and Internet by the National Science Foundation (NSF). These networks allow U.S. researchers to collaborate with colleagues in Japan, Europe, or Canada. Among the services offered are file transfer, E-mail, and mailing lists on specific subjects. Internet users (who include government agencies and commercial entities as well as academic institutions) can log on to a remote supercomputer. Access to the WAN is through a LAN in your organization. ARPANET is another well-known WAN; it is run by the U.S. Department of Defense. A diagram of the communication links in a WAN is shown in Figure 11-2. Carol Reed will discuss many fascinating features of Internet in some detail a bit later in this section.

You do not need to be a member of an educational institution or a LAN to communicate with a computer miles or a continent away. If you have a modem (a piece of hardware that can modulate/demodulate computer signals into a signal that can travel along standard phone lines), you can access other networks. This dial-in communications is referred to as **telecommunications.** It is accessed through a commercial on-line service such as CompuServe, Prodigy (which began as a joint venture of CBS, Sears, and IBM), AT&T Mail, MCI Mail, Bibliographic Retrieval Services (BRS), or other specialized information services—access from most locations is simply through a local telephone call. CompuServe is more technically oriented, offering information on news, weather, sports, travel, and finance, as well as the forums (a two-way public message system for a broad range of topics), E-mail, and large databases in business, finance, and technical material for research. Prodigy is a more family-oriented reference offering news and information, financial services, on-line shopping, and communication services. These networks and services connect to thousands of databases and bulletin boards. AT&T Mail and MCI Mail allow communications by E-mail, fax, telex, or conventional paper mail.

Computer bulletin boards (BBS) allow people with a modem to call into the system and leave messages for other callers and to read their own messages, just like a bulletin board in an office or a school. They have been expanded to allow file sharing. Most of these bulletin boards are operated by private individuals; they are frequently devoted to a single topic. Both CompuServe and Prodigy carry indexes of active bulletin boards. Some BBSs are free; others charge a subscription fee to help defray the costs of additional phone lines and equipment. BBSs are commonly reached by long-distance call via modem, not through a LAN. Special software is available if you want to run a bulletin board yourself.

> _In the future, when everyone knows and converses with several hundred citizens on a regular basis, computer bulletin boards may be seen as the most significant result of the personal computer revolution. At the very least, it means that no one need ever be alone again._
> Alfred Glossbrenner, The Complete Handbook of Personal Computer Communications, 1983.

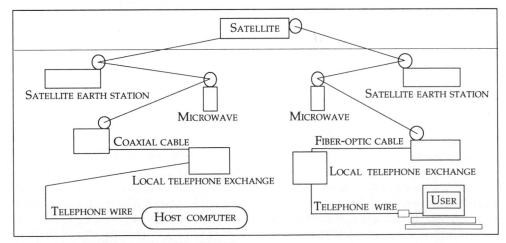

Figure 11-2   _Wide-area networking (WAN)._

Now here is our guest writer, Carol Reed:

Welcome to a new reality: **cyberspace**! You are already familiar with the physical environment for learning: the sounds and sights of classrooms, libraries, and friends. We want to introduce you to another world, a world of interconnecting webs and zipping electrons, and yet (in a figurative sense) still a world of places to learn (like a classroom), of information resources (like a library), and of people to communicate with (like your friends). This world surrounds you now, though rarely in a definite form. You reach it through a computer keyboard and screen.

Anyone with access to a computer has a small segment of cyberspace at his or her fingertips. This small playground can be enough for many people, and often they never move beyond this arena. But there is a much larger world available to those who can connect their computer to other computers. Through these computer networks, computer users can locate the vast amounts of information available on-line, can meet up with other computer users who share their interests, and can pull all these disparate bits of information and many minds together to create new knowledge and synthesize creative ideas. We will look at these two functions—computer networks as a resource and computer networks as a communications tool—in some detail, but first I want to give you a few hints on where to go for assistance and information to help you navigate in cyberspace.

What I am presenting here is just a glimpse of some of the myriad resources available in cyberspace to those who go looking for them. I will focus on the exploration of one arena, the Internet. Once you have become adept at maneuvering around the Internet, you will encounter other areas, and you will feel comfortable enough to be able to find your way to them on your own. One problem with trying to create an introductory guide to cyberspace such as this is that cyberspace is not a stable landscape with definite landmarks that can be pointed out as guideposts for the journey. Things change all the time in this world: new computers are added, while old computers crash and may disappear from the landscape temporarily or permanently. New files and resources are being added continuously, while older resources may be moved to other locations. Thus we have tried to select resources and locations for discussion here that have been in place for a while and should still be available when you go looking for them.

Printed resources become quickly outdated on many of the smaller details of cyberspace because of its fluid, ever-changing nature. This is also true of the Internet in particular. Thus we cannot hope to cover all the possible circumstances you might find yourself in. If you have questions about this new land, contact your local experts for up-to-date information. Usually these people are found in your computer services department—they can show you how to make the local connection between your computer and the larger network and will authorize your access to this larger world. Internet access is not automatic in most places; you will have to set up an account and be given an address before you can venture out into this realm. These people are also available to answer the 1001 questions you are sure to have as you begin your journey. Thus do not hesitate to use this resource!

Another place to look for help is among other users, especially those with some experience. Check in your own organization—you are most likely not the first person to have run into a particular problem, so one of your colleagues may already know an easy solution. Another option (and often an easier way) is to ask questions about the network of people on the network. Although we will introduce you to some of the static resources—databases—available through the Internet (as well as to some searching tools that will make your journey to these resources easier), it is even more important to your mastering the Internet if you become familiar with the dynamic resources surrounding you in this new world—the other Internet users. Get to know them: listen to them, and learn from their experiences. Ask your first questions politely and explain your unfamiliarity, and you will receive considerate and helpful answers in return.

A Resource for Information

What is the Internet? It is an interconnected web of hundreds of computer networks, both in the United States and around the world, composed of millions of different computers of different sizes and running all sorts of different operating systems. There is no one Internet; in fact, even the definition of the Internet has changed within the past few years. Previously, the Internet could be defined as those computers running a protocol known as **TCP/IP** (Transmission Control Protocol/Internet Protocol), so that they could all interact with each other in real time. Now, major segments of the larger Internet do not run this protocol. Instead, they are connected to the Internet protocol sections of the network through gateways (if they only want to transfer electronic mail between them) or full service translators (if they wish to interact in real time).

How do you find your way around on the Internet? How do you know where you are? Every computer connected to the Internet has a unique **address**, and each user has a different user identification. So moving around the Internet merely involves connecting from one computer address to another, and finding a particular user means locating that user's identification name (or number) on his or her home system. This is much like traveling to another location, making a telephone call, or sending a letter by mail or fax to communicate with a friend in another location—you need to have an address or contact number.

When you get a computer account that allows you Internet access, you will be given an Internet address where you can be reached. If you aren't told, request it (and an explanation if needed)! E-mail is getting to be so common that most people will include their electronic address on their business cards or stationery—ask them for it. Messages posted on listservs (yes, that's how it's spelled) or newsgroups (more on both of these below) will normally contain the sender's address somewhere in the header material, even if the sender forgot to put it in the body of the message. But it is a good habit to put your address in the body of the message—like giving your telephone number when leaving a message on an answering machine (since you cannot assume that the person will remember your number or will be within reach of a directory).

For a more general explanation of computer names and addresses, as well as a more complete description of any of the online resources listed below, look in the relevant sections of References 11-9 and 11-10 at the end of this chapter. On-line, you can also find electronic documents that explain network addresses (in addition to many other aspects of the Internet). Look for "The Hitchhiker's Guide to the Internet" – RFC 1118 (written by Ed Krol), or "FYI on Questions and Answers— Answers to Commonly Asked 'New Internet User' Questions" – RFC 1325. The RFC documents are a large series of electronic publications on various aspects of the Internet. They are available via anonymous FTP (more on that below) from the Internet computer known as **nic.ddn.mil** (one of the main coordination points for the Internet).

What are some of the specific resources you can reach once you are connected to Internet? Here is a brief summary of some, just to get you started.

Remote computing through telnet: Since the main purpose of connecting various types of computers together originally was to allow researchers access to computing power without having to physically travel to the location of the more powerful computer, let's look at that function first. Remote computing is accomplished through an application called **telnet**. When you connect with the remote computer via telnet, it is as though your keyboard were directly connected to that other computer, and you will be able to do anything on that computer that any direct connection would allow you to do. This might include logging onto an account you might have on that computer, or using any of the special services that computer may have available, such as accessing a library catalog or a public database. Many of the resources described below can be accessed by telnetting to remote computers if they are not directly accessible on your local computer.

File transfer through anonymous FTP: FTP is an acronym that you will encounter frequently. It stands for "file transfer protocol" and is the application used to move files from one computer to another via the Internet. It does not matter where the computers are located or what kind of computers they are; if they can use the FTP protocol and have access to the Internet, then files can be passed between them. Each system, of course, may have a slightly different setup, so if you have any problems, always check with your local experts. Most often when you are dealing with FTP, you will be logging onto a remote computer and getting files transferred to your local computer. FTP also allows you to put files onto another computer, so it can be very useful for researchers who are collaborating on projects.

For students, perhaps the most useful feature involving FTP are the **anonymous FTP** archives available on many computers around the world. *Any type of file you can possibly imagine is available via anonymous FTP from somewhere:* software for all types of computers, computer games, text documents explaining almost any topic (including the RFC documents mentioned above), song lyrics, recipe collections, etc.). Usually, anonymous FTP is restricted to getting files only; rarely would you be able to input a file via anonymous FTP, mainly because of concerns about system integrity. There is no such thing as a comprehensive list of anonymous FTP archives—the archives and the documents contained therein are changing all the time. The **nic.ddn.mil** site mentioned above is a valuable source for many different types of documents about the Internet. Three other popular sites are **gatekeeper.dec.com** (Digital Equipment Corporation's corporate gateway to the Internet), **wuarchive.wustl.edu** (at Washington University in St. Louis), and **archive.cis.ohio-state.edu** (at Ohio State University).

Finding a particular file—Archie: Trying to find a particular file in all these sites can be nearly impossible if you don't know where to look. **Archie** was created to help alleviate this problem. Archie is a system that searches indexes of files that are available on public computer sites on the Internet. You can ask it for the names of files which have particular character strings within them, or for files whose descriptions contain a certain word. Once it completes its search, it will bring back a list of actual file names, along with their location on the Internet. You can select the files that most closely fit your needs; then retrieve them from their location via anonymous FTP. Some Archie setups even go and get the file for you—all you do is specify the destination file name on your home computer. Be forewarned: if you ask for an Archie search that is fairly complex or that has many common words in it, the search may take a while to complete; thus be patient!

To use Archie, you must choose an Archie server. There are many different Archie servers available, and all of them contain the same information. Your main concern should be to try to use the network resources as conservatively as possible, that is, to try to pick a server close to you geographically, so as to avoid unnecessarily bogging down the network with network traffic. This is just considered good "netiquette" = polite behavior. Think of access to the Internet as a privilege, not a right, and try to make your presence on the net as pleasant for others as possible. Three common servers often used in the United States and the areas they normally serve are **archie.rutgers.edu** (which covers the northeastern United States), **archie.sura.net** (which provides access for the southeastern United States), and **archie.unl.edu** (which covers the western states).

Resource access via Gopher and Veronica: While Archie helps make locating files much easier, there are still many other resources available on the Internet that are very difficult to locate if you do not already know where they are. At the University of Minnesota, the home of the "Golden Gophers," the Microcomputer-Workstation- Networks Center noticed this problem, and in April 1991 created a computerized campus information service for their university community based on easily negotiated menu layers. Thus was the **Gopher** system born. Gopher uses many of the applications available on the Internet—telnetting, FTP, and others—to retrieve all sorts of different resources for you, and it does so automatically, without you having to know where the resources are located or how to retrieve them. There are now hundreds of systems, called Gopher servers, available throughout the world, and you can access them by way of Gopher clients. A Gopher client can be installed on any computer that is connected to the Internet by TCP/IP protocol, and Gopher clients are available for a wide variety of computer operating systems (DOS, Macintosh, UNIX, VAX/VMS, VM/CMS, etc.).

Many universities have Gopher systems installed on their campus mainframes, and more are adding them all the time. You may want to try connecting with your local Internet computer system and see if they have a Gopher client available yet. (Try typing "gopher" once you are connected, and see if anything happens.) If your local system does not yet have a Gopher client and you have the ability to telnet, you can try out a public Gopher client by telnetting to **consultant.micro.umn.edu** (at the University of Minnesota), **ux1.cso.uiuc.edu** or **gopher.uiuc.edu** (at the University of Illinois at Urbana-Champaign), or **gopher.msu.edu** (at Michigan State University). Remember to try to use the Gopher client closest to your geographic location.

While the Gopher system is quite easy to use, it has grown quickly into a very large worldwide network; thus it can still be difficult to locate a particular item if you do not know its Gopher home. So once again a resource has been created: this time to help locate Gopher menu items. Much as Archie is used for locating files for anonymous FTP, **Veronica** allows keyword searching of most Gopher-server menu titles throughout the worldwide Gopher net. A Veronica search is done through a Gopher client and produces a customized menu of Gopher items, each of which points directly to a Gopher data source. As with Archie, if you are asking for a complicated Veronica search, or a search with common words in it, expect the search to take a while to complete—after all, it is searching the entire world for you. By the way, Veronica stands for: Very Easy Rodent-Oriented Netwide Index to Computerized Archives—a nice touch of humor, don't you think?

On-line libraries and reference information: Many libraries, especially academic research libraries, have made their on-line catalogs accessible via the Internet. Most of these will be available through a local Gopher server or through an anonymous log-in once you telnet to the library's computer. Lists of these on-line libraries are available on-line, as well. Two of the better-known ones are compiled by Art St. George and Ron Larsen, and Billy Barron. Look for these using Archie— try the keywords **St-George** or **Billy-Barron** to find them. Two catalogs in particular that you may want to explore on-line are the CARL (Colorado Association of Research Libraries) catalog, which you can reach by telnetting to **pac.carl.org,** or the Library of Congress, which is at **locis.loc.gov.**

A Tool for Communication

We are a society of communicators and use many different methods for getting our message out and interacting with one another. The Internet has given us a new way to communicate. It is not necessarily better or worse than any other way, just different. It is faster (often much faster) than correspondence using the postal system, as has already been discussed earlier with the example of staying in touch with family overseas. It is still slower than just calling someone on the telephone (if the person is there to answer when you call). It produces a written record of all that transpired between the two parties, since it is a visual communication system, not an audio system, but it is not necessarily a permanent record since electronic messages are surprisingly easy to "lose" and delete. This type of communication also lacks the normal visual cues we use when communicating face-to-face (remember, the largest part of the message is conveyed nonverbally). It is not a secure system for communication—the usual caution given is to make sure that you do not use any form of electronic communication for any material that you would not want broadcast from the most exposed, crowded street corner.

E-mail: Originally, the Internet was intended for sharing files and computing resources. Electronic mail was added almost as an afterthought and was expected to be only a small part of normal Internet activity. It quickly swamped the network and now easily constitutes the majority of network traffic. One of the most useful aspects of electronic mail (and also sometimes the most frustrating) is that the sending and receiving machines for the message do not have to be able to communicate directly with each other. The messages are gathered up at one point (much as normal mail is), then forwarded to another place (presumably closer to their destination), until ultimately they reach their recipients. They may have to pass from one network to another through "gateways" set up between the different networks that act as translators for the different formats each network requires. Every electronic mail system has its own particular quirks and features, but all systems do ask for the electronic address of the receiver of the message, and most ask for a subject header (a short title that will describe the contents of the message). Some systems do limit the allowable length, so it is considered polite to check for maximum length before sending very long messages.

Two main problems are involved in getting E-mail started in an organization or for an individual: the proper technical equipment must be selected and bought, and the people have to learn how to use it. Since the available technology is changing rapidly, and the choice depends on many parameters, we recommend that anyone considering getting into E-mail consult experts (or at least other users in a similar situation) about the peripherals and hardware, software, and networking choices. Important criteria for most users are fast speed, easy to learn and use, and low cost. Today, anyone who wants to be an effective communicator must know how to communicate by E-mail. The best way to learn how is to get started: have an enthusiast teach you. If your organization already has the equipment, your next step is to start using it! For information about your local system, check with your computer support people for any documentation they may have already created that would be of help to you. In larger organizations, regular training sessions to instruct new users on the basics of the system may be held periodically.

One major benefit of E-mail is that it has not yet developed traditional formats and clichés; thus more real communication occurs. People tend to come right to the point of what they want. E-mail communication encourages instant feedback; this is much easier than writing a conventional letter, and such "conversations" can go back and forth for many rounds. In a way, E-mail combines aspects of informal conversation and more formal, written communication. But because the communication lacks nonverbal cues, take extra care in the way you present your message. Your idea of a short, funny comment may be read by someone else as a sarcastic slap at a most cherished belief. Read everything over twice before you send it out, especially if you are responding with some heat to another's posting. Make sure that what you are saying is really what you mean (and that it cannot be misunderstood). It is very difficult to express emotions electronically. If you want to be doubly sure about what you are saying, try letting a trusted friend read your message before you send it to your "subject of wrath." Accept counsel—you may be using more force than is really necessary—there is another human being at the other end of the wires, after all.

Mailing lists and BITNET listservs: One feature of many electronic mail systems is the ability to create lists of electronic addresses, so that you can broadcast one message out to many people at once. This ability can be useful if you find yourself discussing the same topic with several different people, since you no longer have to repeat yourself to each person. For example, this would be a very useful feature for a professor to have when discussing homework problems by E-mail with one student—the information could be sent to all students in the class, to give them an equal advantage. With this system, the message is sent to the mailing list server which in turn sends it on to all the members on the list. Hundreds of mailing lists are floating around on the Internet, covering thousands of topics. Often, a list is begun by a group of people who all want to join together for a discussion on a specific topic. Some of these mailing lists can be quite small (with perhaps two dozen members); others can have thousands of members. BITNET listservs are probably the best known of these mailing lists and are controlled by a particular computer program, rather than by an overworked human being. A list of all the available BITNET lists can be obtained by sending a message to **listserv@bitnic.bitnet** with the body of the message having the command: **list global**. Check around your local site before you send this messge, though, as the list you receive will be VERY long and may well be already available either on your local computer or in printed form.

USENET news: The mailing lists and BITNET listservs discussed above are one step away from private electronic mail communication. They assume that only the people on the mailing lists will be receiving the messages, although anyone is usually free to subscribe to or be canceled from the lists at any time. The lists are, however, a public forum and should be treated as such, since methods are available that can find out who is subscribed to any list at any time. The mailing lists have the handicap that everyone receives every message sent to the list, and on some high-volume lists, the electronic mail load can become overwhelming at times. Quite often, also, many of the messages on these lists will not be of interest to everyone, so the ability to pick and choose which messages you actually want to read, without having to delete all the messages you are not interested in, can be quite handy. This ability to pick and choose might be likened to reading messages that have all been tacked up on a bulletin board, an even more public forum. This is the niche filled by the **USENET** news service.

USENET itself has been around since before the Internet actually existed, when the connections between the machines were made by dial-up lines and modems. These machines exchange articles tagged with one or more universally recognized labels and organized into specific areas of concentration. USENET newsgroups are the Internet equivalent of an open discussion group or the private dial-up bulletin board systems (BBS) many new computer users start out on. Several thousand different newsgroups—some moderated, many totally unmoderated—exist with potentially millions of readers and posters worldwide. Every computer site that chooses to become a USENET news site appoints an administrator to select the particular newsgroups to be received by that site and to handle the many questions and problems that the readers at that site will have. Check with your local service for more information about access at your site. Or, if your site does not get the newsgroups, you can reach them at another site using the Gopher system, as many Gophers have the newsgroups available on their systems.

Electronic journals and books: The communication systems discussed above are all rather informal methods of human interaction. The rapid growth of the Internet has seen strong growth in the past few years of more formal methods of interaction on-line. Several on-line journals have begun "publication," and more will no doubt appear shortly. More and more books are also available on-line, such as newly published books with searching capabilities, making the on-line versions in many cases more popular than the print equivalent (especially in the case of encyclopedias). Older classics are being made available on-line by volunteers who type them in and proofread them in the interests of greater access for all. The most ambitious of these efforts is Project Gutenberg, which seeks to give away a trillion electronic texts by December 31, 2001. The works they have already typed in are available via anonymous FTP from many sites, and many have been added to the Gopher systems around the country. Look for them!

Synergy in cyberspace: Now go out and explore this new world of cyberspace! The best way to learn is to do it—so poke around; look in the various nooks and crannies that we have just pointed out to you. Use the guidebooks. Remember the Golden Rule as you interact with those you meet on your journey. Internet is a powerful and ever-growing tool for creativity. As you browse through the material available on-line, you will find that new ideas will spring up from many expected (and unexpected) places. The synergy made possible by the convergence of so many interested people can be absolutely amazing. Enjoy!

Final comment: We appreciate Carol Reed's contribution to this text in the area of electronic communication and resources, an area in which we do not yet have much expertise. We want to add one thought—a brief look at the concept of "synergy." What is synergy? In the lead article, "Whole-Brain Creativity: Eight Key Conclusions," published in the *International Brain Dominance Review* (the networking journal on brain dominance technology), Spring/Summer 1993, pp. 4-5, Ned Herrmann writes:

> Synergy is a key ingredient in the creative mental process. By synergy, I mean the mental result of interaction between different specialized parts of the interconnected brain—the creative ideas that can result from the interaction between the differing modes of analysis and synthesis, between rational processing and intuitive processing, between facts and feelings, between linear processing modes and global thinking.

When people are able to break down the walls within their own minds between the different modes of mental processes beyond acquiring tools, techniques, and skills, the result is true applied creativity. Good communication helps us interact and interconnect with the minds of others, which in turn can enhance our own creative thinking and problem-solving outcomes. This synergy is simply impossible to achieve when people work in isolation.

> *Vision without action is merely a dream. Action without vision just passes the time. Vision with action can change the world.*
> *Joel Barker*, The Power of Vision.

FURTHER LEARNING

Books, Software, Videos, and Organizations

11-1   Joel Barker, *Discovering the Future: The Power of Vision* (see Ref. 1-4). This videotape focuses on vision as the motivation for change. For change to happen, the organizational vision initiated by a leader must be supported and shared by the community; it must be comprehensive as well as detailed, positive, inspiring, challenging, and worth the effort.

11-2   Kenneth Blanchard and Spencer Johnson, *The One-Minute Manager*, Morrow, New York, 1982. This book teaches goal-setting, praising, and reprimanding as one-minute communication; it makes an interesting companion piece to Reference 11-6.

11-3   Robert M. Bramson, *Coping with Difficult People*, Doubleday, Garden City, New York, 1981. The major part of this book deals with identifying different types of behaviors of difficult people and presents techniques for coping with these behaviors.

11-4   Suzette Haden Elgin, *Success with the Gentle Art of Verbal Self-Defense*, Prentice-Hall, Englewood Cliffs, New Jersey, 1989. This book, by a noted communications consultant, focuses on identifying and eliminating patterns of verbal abuse; thus manipulative behavior is replaced with clear, effective, courteous communication. This book has interesting exercises as well as an extended bibliography with many recommended books.

11-5   Roger Fisher and William Ury, *Getting to Yes—Negotiating Agreement without Giving In*, Houghton Mifflin, Boston, 1981. This book presents a concise, proven, commonsense method of negotiation that will help you get along with people while pursuing your goals.

11-6   Milo O. Frank, *How to Get Your Point Across in 30 Seconds—or Less*, Simon & Schuster, New York, 1986. The author presents his discovery of the 30-second message that is at the heart of effective business communication.

11-7   Les Freed and Frank J. Derfler, Jr., *PC Magazine Guide to Modem Communications*, Ziff-Davis Press, Emeryville, California, 1992. This softcover book comes with a disk (MS-DOS) with several utilities, including a program for access to MCI Mail. It also provides detailed information on communications software, hardware, peripherals, and connecting diagrams.

11-8   Robert Hemfelt, Frank Minirth, Paul Meier, *Love Is a Choice: Recovery for Codependent Relationships*, Thomas Nelson, Nashville, 1989. This paperback book is a guide on how to establish healthy interpersonal relationships and an independent identity.

11-9   Brendan P. Kehoe, *Zen and the Art of the Internet: A Beginner's Guide*, Prentice-Hall, Englewood Cliffs, New Jersey, 1993. This neat little book provides a plethora of basic, practical, and detailed information (including computer names and addresses) that will help in accessing and exploring Internet resources.

11-10 Ed Krol, *The Whole Internet User's Guide & Catalog*, O'Reilly & Associates, Sebastopol, California, 1992. The "catalog" portion lists many valuable sources of information on Internet and FTP archives. Although some of the information may be a bit dated, the *Guide* is still very useful for exploring network addresses.

11-11 Judith Martin, *Miss Manners: Guide for the Turn-of-the-Millennium*, Simon & Schuster, New York, 1989. This large softcover "Definitive Reference for Civilized Behavior" gives explicit, practical, and entertaining advice on social, business, and personal etiquette.

11-12 *New Testament* (any easy-to-read version). The Gospel of Matthew (the tax collector) and the Gospel of Luke (the physician and scientist of his day) are good places to start your discovery if you have never read the Bible before. The teachings of Jesus Christ not only have a lot to say about the relationship between God and human beings but also between people themselves.

11-13 Mark K. Schoenfield and Rick M. Schoenfield, *The McGraw-Hill 36-Hour Negotiating Course*, McGraw-Hill, New York, 1992. This is a 320-page self-study program that includes using (or countering) 52 separate negotiating tactics.

11-14 William Strunk, Jr., and E. B. White, *The Elements of Style*, third edition, Macmillan, New York, 1979. Eighty-five pages of examples are given for improving written expression for clarity and brevity—a very useful and enjoyable "little" book.

11-15 Dudley Weeks, *The Eight Essential Steps to Conflict Resolution: Preserving Relationships at Work, at Home, and in the Community*, Jeremy Tarcher, Los Angeles, 1992. This book presents a step-by-step approach to resolving conflict that emphasizes collaboration, mutual benefit, and the strengthening of relationships; it includes many case studies.

Dale Carnegie's *How to Win Friends and Influence People*, (Ref. 8-2) has been mentioned earlier as an excellent resource for help with communication skills. By 1966, the paperback version was already in its 79th printing. The text is somewhat dated now, but the advice is still valid. Two executives of Dale Carnegie & Associates, Stuart R. Levine and Michael A. Crom, have recently published a version that addresses current issues, *The Leader in You: How to Win Friends, Influence People, and Succeed in a Changing World*. Two additional references mentioned earlier are also relevant to this chapter. *Writing in the Computer Age: Word Processing Skills and Style for Every Writer* by Andrew Fluegelman and Jeremy Joan Hewes (Ref. 5-3) shows how word processors can be used as tools for creating and refining written works (it covers equipment and the mechanics, including maintaining files, as well as strategies for writing and editing.) *The New Oxford Guide to Writing* by Thomas S. Kane (Ref. 5-6) demonstrates how to write more creatively, develop an idea from scratch, and express it clearly and elegantly, using the rules of good grammar—in essence, it teaches writing for business and fun.

Toastmasters — An excellent resource for learning speaking and communication skills is *Toastmasters International*, an organization dedicated to helping its members improve their ability to express themselves clearly and concisely, develop and strengthen their leadership and executive potential and achieve whatever self-development goals they may have set for themselves. For information on local chapters, contact *Toastmasters* at 23182 Arroyo Vista, Rancho Santa Margarita, California 92688.

Technical Writing and Word Processing—If you anticipate that you may have to do technical writing in your field, take an appropriate course while still in college (or later through continuing education). Continuously update your computer writing and visualization skills. Explore other ways of communicating creatively through the written word: poetry, letters, essays, novels, and multimedia; explore how these can be enhanced through various types of graphics.

Telecommunications Networks

Your computer vendor can advise you on the peripherals (modem, etc.) that will be compatible with your computer, as well as the software and commercial services that are available. Here are some addresses and phone numbers that might be useful:

1. BITNET, 1112 12th Street NW, Suite 600, Washington, D.C. 20036. (202) 872-4200.

2. CompuServe, 5000 Arlington Center Boulevard, P.O. Box 20212, Columbus, Ohio 43220. 1-800-848-8199.

3. PRODIGY Services Company, P.O. Box 791, White Plains, New York 10601. 1-800-284-5933.

4. BRS Information Technologies, A Division of Maxwell Online, 8000 Westpark Drive, McLean, Virginia 22102. 1-800-289-4277.

EXERCISES FOR COMMUNICATORS

11-1 **DEFINITIONS**
a. *The concept of democracy is frequently in the news lately. Do some research—how is this concept understood in three or four different parts of the world, such as Central or South America, Eastern Europe, Western Europe, China, Haiti, India, Sri Lanka?*
b. *Make up your own definition of power and compare it with those of your friends and family members.*
c. *Make up your own definition of negotiation. Then ask a male and a female friend each to define the word also. Note the similarities and differences among the three definitions.*

11-2 **SATIR MODES**
Over the span of two or three weeks, identify occurrences in which you had difficulties with communication. Analyze the situations using the scheme of Satir modes.

11-3 **BODY LANGUAGE**
Do some research and find ten expressions in body language that have very different meanings in different cultures. Visualize some misunderstandings in communications that could arise if people are not aware of these different meanings.

The book Success with the Gentle Art of Verbal Self-Defense *has many interesting assignments. Here is an example on the power of body language: Get a videotape of the congressional hearing on Lt. Col. Oliver North in July 1987. First read the transcript, then listen to the audio portion, and finally view the video. How did he manage to convince the public that he now spoke the truth—even though he had to admit over and over again that he had lied?*

11-4 *TRANSLATION*
Interview a person who has only recently learned how to speak English. Explore this person's understanding of some idioms. When idioms are translated literally, misunderstandings occur very easily. For example, "The spirit is willing, but the flesh is weak" could be understood as: "The wine is strong, but the meat is rotten." Does "Don't worry" have the same meaning as: "Let not your heart be troubled"?

---

*Asterisks denote advanced exercises that require extra time but provide more in-depth learning.

11-5   DON'T SAY "NO"
The Japanese use a strategy in their conversation that avoids a "no" response from people. Try this approach for a day. Was it difficult to make this adjustment? How did people react?

11-6   GRAMMAR
First-generation Chinese Americans frequently will say "he" when they mean "she" and vice versa. This has nothing to do with their level of education. Investigate this curious phenomenon—what could be the reasons for their difficulties with English grammar?

11-7   DISAGREEMENT
Next time you have a serious argument or disagreement with a person close to you, approach the situation differently. Take a time-out to identify at least ten factors and goals involved in the situation on which you are in agreement. Return to the problem at hand—do you find it easier now to focus on a cooperative solution?

11-8   SENSORY-MODE AD
Find an advertisement that is a good example of writing in a particular sensory mode. Then rewrite the ad in another mode.

11-9   THINKING PREFERENCES
Write a 30-second message about something that is important to you. Write it in four different ways—to reach quadrant A, B, C, and D thinkers in turn.

11-10   NONVERBAL COMMUNICATION
Scenario: You have a throat infection that has affected your vocal cords. The doctor has absolutely forbidden you to talk or use the telephone. How would you communicate the following messages to (1) someone in the same room, (2) to someone in the same city, (3) to someone abroad, and (4) to someone not speaking your language? Using the creative problem solving process (including visualization), mix and match the listener and the situation; make up at least six different messages:

a. The birth of a daughter.
b. Your car is in the shop, and you need a ride to go and pick it up.
c. Your wallet was stolen; you need to borrow some emergency funds.
d. You just baked a very "interesting" batch of cookies and want this person to try them.

11-11   *FAMILY COMMUNICATION*
Analyze how your family members communicate with one another. How do you resolve conflict? What type of negotiation are you using? Make two lists—approaches, techniques, and things that you have found work well, and areas where you have problems. Select one problem area, then alone or with one or more family members, use creative problem solving to identify a better way. Make a work plan, "sell" the idea to the others, then make a commitment to try the new approach for three weeks. Follow up with an evaluation—how did the project turn out? In what way could it be improved? If it worked well, do you want to try solving another communication problem? Are you listening to everyone's suggestions?

11-12   PROVERBS AND COMMUNICATION
In a group of three or four, do some research into proverbs that have to do with communication. Discuss under what circumstances they may be true, and when they are a gross simplification. Example: "Sticks and stones may break my bones but words will never hurt me." What is required to make this a true statement? Under what circumstances is this false?

How would an A-quadrant thinker define communication? What about a B-quadrant thinker? Can some of the proverbs be identified with particular brain quadrants? Make up four different definitions for communication, one for each quadrant of thinking.

12

CULTURE, TECHNOLOGY, INVENTIONS, AND PATENTS

WHAT YOU CAN LEARN FROM THIS CHAPTER:
- *Thoughts about the relationship between culture, technology, and invention in the context of history.*
- *Creative problem solving in the invention process.*
- *How to do a patent search (and other information about patents).*
- *Inventors and inventions—case study assignment.*
- *Building bridges and authors' personal postscript; reading list and other resources.*

Inventions and innovation do not happen in isolation; technology is not something that is developed apart from its cultural setting. And culture does not stand alone, unaffected by technological innovation. What are these relationships? Is there common ground between the artist and the engineer, the scientist and the poet, the farmer and the manufacturer? Why do people invent? How do they use creativity? Where do their ideas originate? Can ideas be protected? These are some of the questions we want to explore in this chapter.

CULTURE, TECHNOLOGY, AND INVENTION IN THE HISTORICAL CONTEXT

Let us begin with some definitions. **Culture** as defined by Webster has quite a breadth of meaning, stemming from the Latin agricultural origin of tilling and cultivating (soils, plants, animals), then extending to the development, improvement, or refinement of the mind, manners, taste, and body through special training or care, culminating in the ideas, customs, skills, and arts of a given people in a given period. In the broadest view, culture is civilization. **Technology** is a word with interesting roots in the Greek language: art combined with systematic treatment—it is the science or study of the practical or industrial arts, the applied sciences, or a method, process, etc., for handling a specific technical problem. **Art**

403

(from Latin) is defined as human creativity, skill or its application, any craft or profession, or its principles; the making or doing of things that have form and beauty; creative work or any branch of creative work; products of creative work, or a branch of learning—the liberal arts as distinguished from the sciences. Last, it is also defined as cunning, trick, and wile. **Invention** (also of Latin origin) is defined as the power of inventing—thinking up or devising in the mind; producing a new device; devising something for the first time—it is ingenuity. The term is also used in music for a short composition developing a motif in counterpoint. **Innovation** is an invention that has been applied and accepted for some human purpose—it is an integration of what is technically possible and what people think and feel. Since engineering and technology are concepts that are frequently interchanged, let's also consider the definition for **engineer**. The word comes from the latin *ingeniatorem*, which means "being ingenious at devising." Why then do many people think of engineers as "train drivers," including the engineering students in the exercise described in Chapter 3? This is a peculiarity of the English language—in other languages, engineers are distinguished linguistically from the operators of their machines and inventions. In English, the word *engine*, meaning "ingenious device," comes from the same Latin root as the word for engineer. When the suffix -eer, indicating operator or driver, is attached to engine (similar to musketeer or puppeteer) it hides the original connection to inventive thought or creator.

Think about these definitions. Try making a sketch of the relationshipsl What do culture and technology have in common? In what way are they different? We will discuss the subject of invention in more detail later, but looking at the definitions, it seems that we can have inventors both on the culture side as well as on the technology side. The one major difference seems to be that technology places a stronger emphasis on **science**—which is the systematized knowledge derived from observation, study, and experimentation. Invention is both a process and a product or artifact; the process can be structured or unstructured, planned or serendipitous.

> *If technology may be defined as the systematic study of techniques for making and doing things, then the history of technology is, in a sense, the history of man.*
> Encyclopaedia Britannica, *18:24, 1978.*

The history of technology and the development of culture make fascinating reading. However, the filter or bias of the authors must be taken into account. Most of these histories are written from the Western European point of view and ignore or gloss over the remarkable achievements of cultures in the Far East, the Americas, and Africa. Also, early history depends on archaeological evidence, finding and dating artifacts, and then drawing conjectures about these cultures and these people's thinking and beliefs based on our own worldview and conditioning. Dates cited by different authors disagree, at times considerably. Thus, what we expected to be an easy writing task turned into a rather complex and at times frustrating undertaking with no easy answers and data available at one's fingertips. History is commonly taught as an isolated subject, with a magnifying glass focused on a particular geographic area, nation, political system, or battle. This makes it difficult to gain an understanding of the broad connections and important relationships among economic conditions, social structures, religious beliefs, technology, and the arts, including architecture, literature, music, sculpture, pottery, weaving, and painting.

One problem that we, as products of the twentieth century and its teaching of evolution have, is an outlook that sees our achievements as advanced. In comparison, anything extending back prior to the Industrial Revolution is perceived as primitive and somehow attributed to inferior intelligence. What are the subtle implications in the way Disney's EPCOT Center presents the development of technology? To truly understand either technology or culture, and especially their interdependence, we need to think about and question some of our assumptions. For example, do we see the civilizations arising on different continents isolated by forbidding oceans? Or should we—as eloquently argued by Thor Heyerdahl—perceive oceans as wide-open highways connecting diverse cultures? How we answer such basic questions colors how other data is interpreted. What is the popular image of neolithic people? Do we credit them as endowed with the capability of complex thought about the cosmos and astrology expressed through inventive myths we are no longer able to decode in our time? This is a fascinating subject worthy of exploration; the reading list at the end of this chapter is just a starting point for expanding your horizons.

It is quite possible for a student in engineering today to complete the requirements for a doctorate believing that most inventions and technology have originated in the West (in Europe and later in the United States). Yet the modern world is a unique synthesis of Eastern and Western ideas and inventions. This truth was discovered by the distinguished British scholar and biochemist Dr. Joseph Needham. When in the 1930s he built a close relationship with Chinese scientists studying at Cambridge, he began hearing about some of the fascinating early discoveries in Chinese science. In the following years he learned to speak and read Chinese. He then spent several years in China researching the subject and assembled what is now the finest library outside China on the history of Chinese science, technology, and medicine at the Needham Research Institute in Cambridge. He and his fellow researchers have been at work on a monumental project of writing a 25-volume set on *Science and Civilization in China,* with some 15 volumes published so far. Robert Temple worked with Dr. Needham on the condensed version, *The Genius of China,* for the general public since the scholarly set is beyond the economic means of most libraries. Important information, as yet unpublished by the Needham group, is incorporated into this informative reference book, the source of much of our information on Chinese technology cited in the following discussion.

Technology in Ancient Times

We will very briefly summarize some of the highlights of cultural and technological achievements, with approximate millennia based on archaeological evidence and reasonable conjectures. The African origin of the bow and arrow many millennia preceding the neolithic cultures is well documented. Oil lamps from an even earlier period have been found in Mesopotamia and Europe. To examine the development of technology, we can begin with what is termed the *neolithic revolution*—when people changed from hunting, fishing, and gathering to farming and living in towns, with a corresponding flourishing of invention, especially in the construction of buildings, and innovation in agricultural implements and techniques. Technology can be seen as humankind's creative attempts to modify the environment to increase survival and improve life-style. Pottery was one of the earliest inventions present in each neolithic civilization; the earliest dated pottery of circa 9000 B.C. has been found in Japan. A boat paddle found in England has been dated to 7500 B.C. Copper was used for knives in Turkey by 6000 B.C. Textiles, spindles, and simple looms can be traced as far back as 5000 B.C. Wheels and sailboats were in use by 4000 B.C. in Egypt and Mesopotamia, and Sumerian writing for tax records and commerce by 3500 B.C. In *Turing's Man*, J. David Bolter uses the term *defining technology* for an invention or artifact that is not only in common use during an age, but is also taken as a metaphor to describe the worldview of the culture at that time. For the ancient world, the defining technology was the drop spindle and loom. Spindles of many different designs and shapes but with a common purpose have been found in all neolithic cultures.

Around 3000 B.C., people began to live in cities. This social change has been named the urban revolution, and the period to about 500 B.C. is denoted as the Bronze Age after its most important material. The major civilizations established at the beginning of the Bronze Age were located in Egypt, Crete, Bahrain, Mesopotamia, India, Sri Lanka, and China—in favorable climates on islands or along great rivers. These conditions made it possible to accumulate the resources for the building of cities and a corresponding increase in technological innovation. Since metals were scarce in the great river valleys and islands, elaborate trade networks had to be established to maintain supplies, contributing to a spread and exchange of ideas and technology. Several important developments arose around 3000 B.C.: the potter's wheel, mining and metallurgy, and standard weights and measures. The pyramids in Egypt date from that time. Developments in astronomy and time measurements aided in the construction of intricate irrigation systems in Mesopotamia, where agriculture benefited from the invention of the plow and architecture from the invention of the true arch. Paved roads and sewage systems existed in Crete by 2000 B.C. Iron weapons, pontoon bridges, armored wagons, and battering rams were the military hardware developed by the Assyrians in the last millennium before the birth of Christ.

Most of these innovations are quite well known, and we have taken it for granted that they provided the basis for the Iron Age—the flourishing of the Greek and Roman civilizations from around 500 B.C. to roughly A.D. 500. Three engineers stand out during this time period: The Greek mathematician Archimedes with his invention of the screw for irrigation, the discovery of the buoyancy principle, and work with lever and pulley; Hero, an engineer from Alexandria who invented pumps, organs, and compressed-air engines; and Vitruvius, a Roman engineer who wrote the first engineering textbook. The beauty of Greek temple architecture has been widely admired over the centuries. However, these edifices do not represent innovation since the building technology employed was copied from Egypt. The Romans were mostly interested in practical applications of their discoveries of arches, vaults, domes, and underwater cement in civil engineering works such as watermills, aqueducts, bridges, and tunnels. They were the builders of an amazing road system for military purposes extending over a 50,000-mile network to the farthest reaches of the empire and still in use today in some places.

But now let's take a look at the East—China in particular. Table 12-1 is a list of Chinese inventions and discoveries from 1500 B.C. to A.D. 500. The advances in agriculture and the invention of the magnetic compass and ship's rudder were to have a profound influence on technological developments and exploration in the West many hundreds of years later. The invention of the stirrup and the crossbow gave a powerful advantage to invading armies and changed the course of history.

Table 12-1
Early Inventions in China

AGRICULTURE:
Row cultivation of crops and intensive hoeing. The iron plow. The trace and collar horse harnesses. The rotary winnowing fan. The "modern" seed drill.

ENGINEERING:
Cast iron and steel manufacture. Manned flight with kites. Double-acting piston bellows. The first transport canal. Belt drive. Parachute and hot-air balloon. Deep drilling for natural gas. The rudder. Water power and chain pump. Suspension bridge. The first cybernetic machine. Helicopter rotor and propeller. Paddle-wheel boat. Essentials of the steam engine.

INDUSTRY AND TECHNOLOGY:
Lacquer (the first use of plastic). The magnetic compass. Use of natural gas and oil as fuel. Chemical warfare. Crossbow. Paper. Wheelbarrow. Fishing reel. Stirrup. Porcelain. Dial-and-pointer devices. Seismograph. Biological pest control.

MATHEMATICS:
The decimal system. A place for zero. Negative numbers. Extraction of higher roots. Decimal fractions. Using algebra in geometry. A refined value for pi.

Technology in the Middle Ages

The history of technology in the next thousand years—the Middle Ages—shows a preservation, recovery, and modification of the earlier achievements, as the barbarian invaders were assimilated into the local populations of Western Europe. Because written records are scarce, it is difficult to document how particular innovations were introduced. What is believed is that some inventions during this period developed in widely separated locations independently and spontaneously. Other inventions were improved and transferred through the strong Islamic civilization arising in the eighth century; this civilization in the Persian Gulf formed a bridge between China, India, and Europe since it had trade routes by land as well as by sea with both East and West. This culture was a key in shipbuilding and navigational technology transfer, which then made possible the great voyages of discovery at the end of the age. Some technology transfer also occurred via trade routes from Central Asia through Byzantium.

Most advances in Europe, however, were made in response to local needs and conditions, one of which was a labor shortage. Horse power—once the stirrup and rigid collar were available—became important, especially for drawing heavy plows. Wind and water power were harnessed, and the coal industry developed in the north. Soap was invented for scouring textiles; it was applied for personal hygiene only much later. The first universities were founded in the twelfth century. The notable architectural achievement was the flying buttress in the construction of Gothic cathedrals. In the fifteenth century, the development of the blast furnace made it possible to manufacture cannons. Work on the defining technology of this age—the mechanical, weight-driven clock—began in various places during the fourteenth century. Another significant development was the movable-type printing press by Johannes Gutenberg, an invention made possible by experience with block printing and agricultural presses and probably also by vague knowledge of Chinese technology. Papermaking had spread to Europe from China by skilled workers via the Persian Gulf. These advances in technology were supported by strong growth in population and in industrial, commercial, and cultural activities.

Let us now compare the technological development in Europe during the Middle Ages with what was happening in China during the corresponding period, as summarized in Table 12-2.

Table 12-2
Developments of Technology in China (A.D. 500–1500)

MATHEMATICS AND CARTOGRAPHY:
"Mercator" map projection. "Pascal's" triangle of binomial coefficients by Liu Ju-Hsieh, before A.D. 1100. Magnetic declination.

ENGINEERING:
The segmented arch bridge. The chain drive. Underwater salvage operations. Canal pound locks.

INDUSTRY AND TECHNOLOGY:
The umbrella. Matches. The mechanical clock. Movable printing (400 years before Gutenberg). Paper money. Permanent lamps. Spinning wheels. Phosphorescent paint. Gunpowder, flame throwers, flares, and fireworks. Bombs and grenades, land and sea mines, multistage rockets, guns and cannons (450 years ahead of Europe).

It is now accepted by most historians that iron casting, the wheelbarrow, the sternpost rudder, the crossbow, gunpowder, paper, silk-working machinery, and printing with wood and metal blocks all arrived in Europe, having originated in China during the thirteenth and fourteenth century, although the route of transmission is not always clear. The years around the turn of the millennium were especially productive in China. The population balance had shifted to the south, and the introduction of a new variety of rice from Vietnam made it possible to cultivate two crops per year. Many agricultural implements were adapted to rice culture and processing. Iron production and commerce increased, as did canal construction. Technological innovation (especially relating to clocks and military hardware) originated within the bureaucracy, with other innovations coming from the commercial sector and the Buddhist monasteries. Arnold Pacey, in *Technology in World Civilization: A Thousand-Year History*, states:

> In 1100, China was undoubtedly the most technically "advanced" region in the world, particularly with regard to the use of coke in iron smelting, canal transport and farm implements. Bridge design and textile machinery had also been developing rapidly. In all these fields, there were techniques in use in eleventh-century China which had no parallel in Europe until around 1700.

China had flourishing trade contacts with the Persian Gulf region and with Southeast Asia. Traveling Buddhist monks provided additional links with India. China had learned to cultivate and process cotton from India around A.D. 600. There was an active "invention exchange" between the two countries in crop plants, textiles, and mineral processing. The monks introduced the idea of printing books on paper, and books were soon used to spread technical information within China. In the area of hydraulic engineering and machines, a major exhange of ideas occurred between China and Iran. For example, the idea of the Iranian windmill became known in China, where it led to the development of a different, but related, windmill. The irrigation and hydraulics technologies—canals in China and dams in Iraq—developed because of need and the presence of appropriate institutions and social structure. Because of the advanced state of hydraulic technology in the countries bordering the Indian Ocean, these civilizations are identified as hydraulic cultures. When the social structures suffered upheaval in China during the fourteenth century, technological development experienced loss and setbacks. It is interesting to note that the achievement of two women has come down from this period, as told by Arnold Pacey. In 1315, a road-building project through the mountains of Fujian province was directed by a woman engineer. In 1337, a monument was erected to honor Huang Dao-po, a teacher who traveled widely, instructing people on how to use improved spinning and weaving technology, possibly including the cotton gin.

What is the evidence that technological artifacts, or at least tales and descriptions of technology and inventions, could have reached Europe from China during the Middle Ages? Here are just two brief but interesting glimpses of how this could have happened. First, consider the Maldive Islands in the Indian Ocean. They were also known as the money islands, because they had a monopoly on the cultivation of cowry shells. These shells were used as money in trade going as far back as the Bronze Age. Now, potsherds of Chinese origin have been found on the Maldives, and what is very surprising, cowry shells have been found in Finland (which had a trade route via Bulgaria to the Black Sea and Bizantium, and on to the Persian Gulf) as well as in Sweden and Norway as far north as the Arctic Circle. Second, consider the technology transfer involved in the invention of the clock.

Case Study: The Invention of the Clock

In China, society revolved around the emperor, the Son of Heaven, who had many wives and concubines. It was the astrologers' responsibility to designate a successor from among the emperor's many sons. For making this determination, the exact time of conception had to be known, so this stimulated the invention of an accurate clock. But even though the Chinese were an ingenious people, they were not the first to invent time-measuring devices. Shadow clocks consisting of a vertical post date back to 3500 B.C. Water clocks existed in Egypt by 1400 B.C. and sundials a few hundred years later. Ctesibius of Alexandria is reported as having developed rack and pinion gears for a water clock. The Babylonians used hemispherical sundials and water clocks, and the Chinese most likely imported this technology (as they did many astronomical ideas) from this Middle Eastern region. The Chinese improved the water clock and even used mercury in a portable model. The idea for the clock dial that they used originated in Greece or Rome.

The problem with inventing a mechanical clock was that it required an escapement— a way to turn a wheel in regular increments in exact synchronization with the daily turning of the Earth. The escapement is the key component in a mechanical clock. It consists of a toothed escape wheel combined with an anchor or peg which periodically stops the escape wheel from rotating. The jerky action of the escapement is still noticeable in modern mechanical watches where the second hand advances in five tiny jerks each second. The first person on record as having invented an escapement is the Buddhist monk and Chinese mathematician I-Hsing in A.D. 725 when he completed an astronomical instrument which had a drum beat for the quarter hours, a bell striking the hour, and other mechanisms showing the phases of the moon and the orbit of the sun over the year. Unfortunately, the bronze and iron mechanism rusted within a few years of outdoor exhibition, and the instrument was placed in a museum from where it eventually disappeared. The escapement for the clock consisted of a vertical waterwheel with cups that were filled by water dripping from a water clock. A big problem with this system was winter, when torches had to be kept burning to prevent the water from freezing. I-Hsing's death soon after the clock was completed prevented immediate improvements. It is interesting to speculate on the reason why water was chosen to power the clock. However, if we remember that this was a hydraulic culture, this is understandable; its cultural philosophy saw the flow and fall of water as a metaphor for the perpetual turning of the heavens.

An improved clock model was built by Chang Hsu-Hsün in A.D. 976. This clock, which was over 30 feet high, used mercury as the fluid. The clock, in addition to chiming the quarter hours and making a complete revolution each 24 hours, also showed the movement of sun, moon, and five planets. The greatest of all Chinese medieval clocks was constructed in 1092 by Su Sung, based on Chang's earlier work. Su's book, *New Design for a Mechanized Armillary Sphere and Celestial Globe*, has fortunately been preserved and describes the design and construction of the great clock in full detail. The clock itself only survived for about forty years, when it was destroyed by an opposing political group. Joseph Needham has translated and published the original text in the book *Heavenly Clockwork.* Robert Temple, in *The Genius of China,* writes, "The stimulus for [achieving the invention of the clock by 1310 in Europe] seems to have been some garbled accounts of Chinese mechanical clocks which came to the West by way of traders."

A very early device of a clock is described in a book compiled by scholars in Spain late in the thirteenth century. This clock did not have an escapement but combined the idea of a falling weight with a counterbalance of trickling mercury. Within 100 years, European clocks were built that could strike the hours; these clocks incorporated a verge escapement with falling weights. Figure 12-1 compares the geometries of three different escapements as clocks evolved to modern times.

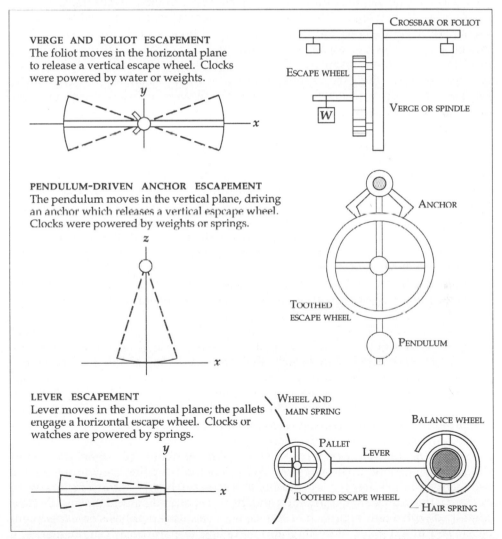

Figure 12-1   Clock escapement geometries.

The famous clock at Rouen, France, was the first in Europe to use quarter-hour strikes. Built in 1389, it still exists today. The first spring-driven clock was invented about the time Columbus set off on his voyages. These early clocks were not very accurate, especially not for astronomers like Tycho Brahe and Johann Kepler. Jobst Burgi, a Swiss instrument maker working with these two scientists, devised a number of improvements, such as combining weights with a spring and inventing the cross-beat escapement. These were soon outdone, however, by a new paradigm—the pendulum. Galileo invented the pinwheel escapement with a pendulum of small amplitude. He was blind by this time, and his son Vincenzio died before he was able to complete the clock. It took more than a century before this invention was widely applied. The Dutchman Christiaan Huygens used the verge escapement with his pendulum; he published two books on horology and the discoveries that he made. By 1700, the anchor escapement invented by William Clement (see Figure 12-1) rapidly displaced the Huygens mechanism. The escapement used in most quality watches until the invention of the quartz movement was the lever escapement invented by Thomas Mudge in 1755, although it did not become common until the early nineteenth century, when it was adopted by English and Swiss watchmakers. These later improvements were all driven by the need for increased accuracy in timekeeping.

Technological Developments in Modern Times

Technological development and its interaction with the cultural, social, economic, and historical events in Western Europe is quite well known; thus we will only mention a few items in passing and leave it to you to explore areas that you are not yet familiar with. The Renaissance saw a tremendous advance in scientific knowledge. The defining technology for the Industrial Revolution was the steam engine or heat engine; the first commercially successful model was built in England by Thomas Newcomen. No longer was nature seen as an ordered mechanism like the clock; the cosmos now was viewed as "running down"—entropy or the second law of thermodynamics had entered the human consciousness.

Finally, with the advent of computers, we have the Information Age. The computer is no longer an invention made by a single individual; it is a team effort. It is not a fixed artifact in that it can be easily changed and altered. It is versatile—like the human mind. Computers as very powerful metaphors have affected our culture, and nature is now seen as information to be processed or decoded. Computers are very much a part of "high technology" and have become embedded in education, in industry and commerce, and in everyday living.

The question many people ask when learning about the advanced state of Chinese science and technology is, "Why did the Scientific Revolution of the eighteenth century happen in Europe, and not in China?" Many factors were involved: the Protestant Reformation and the rise of capitalism and the middle class combined with a new approach to science with its strong emphasis on experimentation and a systematic decoding and modeling of nature with mathematics, a universal language. But Joseph Needham (in the foreword to Ref. 12-24) cautions,

The sciences of China and of Islam never dreamed of divorcing science from ethics, but when at the Scientific Revolution ... ethics [was] chased out of science, things became very different, and more menacing. This was good in so far as it clarified and discriminated between the great forms of human experience, but very bad and dangerous when it opened the way for evil men to use the great discoveries of modern science and activities disastrous for humanity. Science needs to be lived alongside religion, philosophy, history and aesthetic experience; alone it can lead to great harm.

Because of computers and other high-tech achievements such as communications satellites, MRI scans, and Tomahawk missiles, we may feel superior to other cultures. Is such an attitude justified? Why have the United States and the Soviet Union not been very successful in transferring technology to developing countries? We tend to view a country that lacks widespread use of machinery as "backward." Francesca Bray, who has studied agricultural methods in the wheat-growing north and the rice-growing south of China, found that each region developed its own technology complex in adaptation to climatic and social conditions for survival and producing life's necessities. These conditions and available resources determined both the direction and the scale of the engineering works that were developed, for example, in the "high-tech" Islamic and Northern Chinese cultures, compared to the "low-tech" Indian and South Asian cultures. When technology is imported to a new area, it is rarely adopted in its original form—either it will be abandoned when its teachers leave, or it will undergo radical adaptation to local conditions. In turn, it can change the local culture. Willingness to take risks with a new technology and make innovations depends on how closely the economic survival of the people is linked to the success of their current technology and the resources available to bridge the learning curve or paradigm shift. Imported technology must have a good fit with existing survival technology to be acceptable and understood locally. Only then will technology transfer have a chance to succeed.

Intercultural Technology Transfer

Does technology transfer occur without the presence of workers and artifacts? Does similarity in invention in widely separate locations imply that some cultural contact has occurred? These very interesting questions have by no means received a complete answer. Historical records document that the migration of skilled workers was instrumental in the transfer of papermaking technology from East to West. Another route is through books and their translators, as happened when a large store of Islamic technical books became accessible to Christians in Spain at the fall of Toledo in A.D. 1085. But what about the similarity of terrace farming in Peru to that practiced in China? At this time, no proof of direct voyages has been found, although some technology from Asia could have reached the Americas through diffusion via Polynesia. The engineering and agricultural achievements of the Aztec, Inca, and Mayan civilizations are astonishing in the absence of iron and the wheel. This brings up another interesting question—if contact existed between the early American civilizations with civilizations either across the Atlantic or across the Pacific, why was technology (or idea) transfer so selective and did not include the wheel? Was this paradigm shift too "foreign" to the mindset of the time, or was the circle a concept too sacred for mundane applications in those cultures?

Yet connections—or dialogue—between cultures seem to have occurred more frequently than most people realize. And some ideas and products dispersed and were adopted very rapidly. For example, the sweet potato brought back to Europe by Columbus himself had within a hundred years spread as far as the Philippines. During a famine in Fujian province of China the governor sent out a delegation to search for new crop plants. As a result, the sweet potato was introduced to China in 1594 and grown so successfully that it soon reached Taiwan and then Japan. Europe, the Americas, and the Far East had an interesting relationship in the sixteenth and seventeenth centuries. Do you know what was one of the driving forces that led to the exploration and colonization of the American continent? If you say that it was the search for riches—for gold and silver—you would be right, but this is not the complete answer. Why was there such a need for gold and silver—what was the economic reason? The answer sounds very much like a modern problem—it was the huge trade imbalance that existed between Europe and Asia because few European products could be sold in Asia, which had a much superior manufacturing technology. Thus the goods bought in India and China for European consumption had to be paid for in gold and silver bullion or coin. This need drove the development of gold and silver mining and the corresponding transfer of mining technology, especially to Mexico and Peru.

The high quality of cotton textiles produced and dyed in India during the eighteenth century were an incentive to manufacturers in England and elsewhere on the European continent for continuous innovation and improvements lasting over a century. In turn, shipbuilding techniques from Europe led to flourishing developments of the shipbuilding industry in India, since it was assimilated into a tradition of innovation and skills already present. The dialogue between East and West has continued, albeit with major interruptions due to wars and politics. The Chinese have been quite adept at integrating or adapting Western technology to smaller-scale applications requiring a lower level of skills (for example, in hydroelectric installations). Another example is the way Japan has been able to take technical innovations and develop them into many practical, high-quality products, another kind of creativity that involves processes and organization coupled with a very different allocation of engineering and research to consumer products, in contrast to America, where military research has long been a priority. At this time, the West can still learn much about appropriate technology and integrated approaches—for example, agriculture is not separate from forestry or fishery in many developing countries. These "primitive" areas in the world, especially in arid Africa, are far advanced of our wasteful ways in what is termed environmental technology, and we must learn from them. In Arnold Pacey's words,

> There is also a dialogue within each society between people with different kinds of knowledge and experience, women as well as men, artisans (and farmers) as well as scientists. Related to this is a dialectic within science and technology between the universal and the particular; between survival technology and spectacular, symbol-creating developments; and between dreams and practical needs. In the past, much innovation has been stimulated by interactions of these kinds between and within cultures. The dialogue continues. It is of particular relevance for environmental and human aspects of technology, but is a source of creativity in high technology also.

Technology is neither good nor bad; nor is it neutral.
It has short-range and long-range impacts. Impacts may differ according to the scale
at which a technology is applied. Technology always entails trade-offs.
In short, technology has different results in different contexts.
Melvin Kranzberg, founding editor of Technology and Culture.

CREATIVE PROBLEM SOLVING AND INVENTING

From our brief overview and discussion of some of the connections between culture, inventions, and technology, let us now focus on the invention process and the individual inventor. When does invention happen in a society? As we have seen, the process can be driven by social need, including survival (food and shelter technologies), commerce, and warfare, supported by resources and an openness to new ideas and change. But why does an individual or a small group of people invent—what gets the process started? Some of the possible reasons are listed in Table 12-3—see if you can identify additional resulting products.

Table 12-3
Reasons and Motivation for Inventing New Products

- As a response to threat —> radar, weapons.
- As a response to existing need —> can opener.
- As a response to a future need —> high-temperature ceramics.
- For the fun of it, as an expression of creativity.
- To satisfy intellectual curiosity.
- As a response to an emergency —> Band-Aid.
- To increase one's chance of survival and security.
- To increase comfort and luxury in life-style.
- Through better problem solving —> hydraulic propulsion system.
- Through turning a failure into a success —> Post-it notes.
- To overcome flaws —> Magic tape.
- Accidentally, on the way to researching something else —> polyethylene.
- As a deliberate synthesis —> carbon brakes for aircraft.
- Through brainstorming with experts or outsiders —> courseware.
- From studying trends, demographic data, and customer surveys.
- Through cost reduction and quality improvement efforts —> float glass.
- Through finding new uses for waste products —> aluminum flakes in roofing.
- Through continuous improvement of work done by others.
- Through having new process technology —> proteins from hydrocarbons.
- By finding new applications for existing technology.
- Through having new materials available.
- To win a prize or recognition —> human-powered aircraft.
- Meeting tougher legal and legislated requirements —> catalytic converter.
- By having research funds available to solve a specific problem —> Kevlar.
- By being a dissatisfied user of a product.
- Responding to a challenge or assignment.
- Because "it's my job—I do it for a living."
- To improve the organization's competitive position.
- To get around someone else's patent.

As we have already seen throughout this book, having an idea is just the first step. How do we translate an idea into an actual invention? First of all, having a supportive environment helps. This can include committed management, an interested supervisor, encouraging parents, imaginative technical staff, good non-technical communications, free time, and champions. Second, we can learn the actual steps, starting with creative problem solving—we set a goal, answer the question "why," visualize an outcome, imagine a new use; we develop specific technical targets or determine the customer's needs. Then we use our flexibility and entrepreneural skills, together with techniques such as the Pugh method and market analysis, to prevent errors in judgment. Finally, we need to be familiar with the process that is available to protect an invention.

> _Invention, its development, testing, and implementation, always involved adventures._
> _To have a victory over a technical problem takes flexibility of the brain and bravery._
> _If you are looking today for adventures that are useful for the human race, invent!_
> _[For technical creativity] . . . you have to start to prepare yourself from an early age._
> _Henry Altov (Altshuller), president of the Inventor's Association of the former Soviet Union._

To be an inventor requires quadrant D thinking skills—we first have to start with a creative idea. To convert this idea into a marketable product and then sell it requires additional thinking skills—quadrant B organizational ability, quadrant C people and communications skills, and quadrant A analysis and judgment. If you as an inventor do not have these skills, you must find someone who does to support your efforts. The difference between an "inventor" and an "entrepreneur" seems to be at the business end: both excel at quadrant D thinking, but the entrepreneur has much stronger skills in the producer's mindset and is thus able to carry an idea through to fabrication and marketing. If you as the inventor do not want to be bothered with doing the marketing research or the patent search, you must be prepared to hire someone to do these tasks for you—at a price, of course. Perhaps you have a sponsor, for example, your employer, who will support you with the necessary resources. However, inventors strongly advise from experience that you always pay anyone who is assisting you with implementing an invention—even relatives—and get a receipt; this is important to protect your patent rights. Only about one out of five inventions today are submitted by an independent inventor; the patent rights of the others are owned by the companies who have supported them during the invention process. People with quadrant A thinking preferences can also be inventors. These thinkers usually work on improving an existing product through analysis.

If you are interested in getting a patent for your invention and in making a commercial success of it, you can save yourself countless hours of time and money if you first read a book or two written by inventors who are able to share much valuable advice. This will also put the undertaking into perspective. For example, Gordon D. Griffin, the inventor of the automatic seat belt and author of _How to Be a Successful Inventor: Turn Your Ideas into Profit_, supposes that "only about one out of a thousand original ideas can be developed into a viable, financially successful invention." The problem that most inventors have is that they have too many ideas. About 1 out of 100 ideas gets developed; of those, about 1 in 10 will eventually be a financial success. Actually, when you think about it, these are pretty good odds—much better than winning a lottery, and with a payoff probably equal to that of a lottery. With creative problem solving, critical thinking, and persistence, we believe these odds can be further improved in your favor.

Here is another very important quadrant B tip listed as Item 4 in Table 12-4. This is emphasized in *The Inventive Thinking Curriculum Project,* an outreach program of the United States Patent and Trademark Office which shows teachers how to encourage students of all ages to invent. This program is divided into a sequence of activities: (a) an introduction to inventive thinking, (b) brainstorming, (c) elaboration, which is part of the engineer's mindset in creative problem solving and part of Osborn's nine thought-starter questions, (d) going through a practice problem with the class, and (e) developing an invention idea with the steps given in Table 12-4. Creative problem solving can enhance this process and outcome.

Many resources and networks are available to help parents and teachers encourage students to become inventors. These are listed in *The Inventive Thinking Resource Directory* under *Project XL—A Quest for Excellence* sponsored by the U.S. Patent and Trademark Office and involving government, professional societies, business, parents, students, and educators. If you are involved in such a program, it is important to have a means of recognizing the students' inventive thinking—have a fair or "Inventor's Day" and issue certificates of achievement (available with the U.S. Patent and Trademark Office logo). Provide motivation; share stories with students of young inventors, women inventors, minority inventors, persistent inventors. Encourage students to read about inventors. Show a video or film about inventing. Exhibit the invention models. Model building has two purposes: it is for sharing ideas and explaining the invention, but even more importantly, it is for proving and demonstrating that the invention actually works.

The steps listed in Table 12-4 can be enhanced with the techniques learned in creative problem solving, such as using the Pugh method of idea evaluation, doing a customer survey to collect data about the problem, etc. The process of inventing is a wonderful learning tool: it encourages a flexible mindset for coping with rapid changes in our technological world; it motivates students to learn; and it builds communication. Students read to gain background information; they study science to understand the problem and the best solutions; they apply math to figure out the dimensions of their design and then build the model; and they sketch and write to communicate their invention to others. They may explore economics, history, the arts, as well as music and verbal presentation skills in the broader context of developing the invention, and they will learn how to interact with other people in the process. Students see—perhaps for the first time—the connection between what they are learning in school to how it can be applied in real life.

How do we find out if an inventive concept really has merit?
A truly significant inventive concept will use its new combination of scientific principles
to relieve or avoid major constraints inherent in the previous art.
George R. White and Margaret B. W. Graham, "How to Spot a Technological Winner,"
Harvard Business Review, *March/April 1978, p. 147.*

Table 12-4
Developing an Invention Idea

1. Conduct a survey in the search for a problem that needs solving. Interview a wide variety of people at work, at school, at home, at play.

2. Pool these problem ideas.

3. Use judgment techniques (such as the advantage/disadvantage listing) on each idea, to find those that represent the best options for an inventive solution within the scope of the people involved.

4. Begin an **Inventor's Log** or Journal. This has two purposes—to help think through and develop ideas, and then to protect the completed invention.

    Rules:
    *   *Using a bound notebook, make notes each day about the things you do and learn while working on your invention.*
    *   *Record your idea and how you got it.*
    *   *Write about problems you have and how you solve them.*
    *   *Write in ink and do not erase.*
    *   *Add sketches and drawings to make things clear.*
    *   *List all parts, sources, and costs of materials.*
    *   *Sign and date all entries at the time they are made and have them witnessed.*

5. Analyze the most promising problems; then select one to work on. Brainstorm creative solutions to the chosen problem. Then select one or more of these possible solutions to work on; refine and improve your ideas.

6. Apply critical thinking to narrow down the possible solutions. Answer these questions: Is my idea practical? Can it be made easily? Is it as simple as possible? Is it safe? Will it cost too much to make or use? Is my idea really new? Will it withstand use, or will it break easily? Is my idea similar to something else? Will people really use my invention (do a survey)?

7. Complete the invention—make and follow a work plan which should include these steps:
    *   Identify the problem and possible solution. Name your invention.
    *   List the materials needed to make your drawings as well as a model of your invention.
    *   List, in order, the steps needed to carry out your invention.
    *   Think of possible problems that might happen and how to prevent or solve them.
    *   Carry out the steps. Ask for help with the model from your parents, friends, teachers, and other resource persons in your community.

8. Optionally, create a logo, an ad, or a jingle to sell your invention.

PATENTS

Sooner or later, an inventor will want to think about getting his or her ideas protected with a patent, especially if it looks like the invention may have potential economic success. The following summary is excerpted primarily from information supplied by the Toledo Public Library under the patent depository library program. U.S. law recognizes three types of property: real property (land, buildings), personal property (cars, furniture, jewelry, equipment, books, etc.), and intellectual property (inventions, designs, and writing).

Intellectual property is protected in one of four ways:
1. Patents—for an invention or discovery.
2. Copyright—for an artistic or literary work (text and drawings).
3. Trademarks—for signs or symbols on products or associated with names.
4. Trade secrets—for processes that a company or individual does not want to disclose.

There are three types of patents:
1. Utility patents cover useful processes, machines, manufactured devices, or the composition of matter. They focus on the function of the item and grant exclusive rights to the inventor (or inventors) for 17 years.
2. Design patents cover new, original, and ornamental designs; their purpose is to cover only the outward appearance of an item. Terms are 3, 7, or 14 years.
3. Plant patents cover asexually reproduced distinct and new varieties of plants.

So, what is a patent?

DEFINITION

A patent is a grant of a property right
by the government to the inventor
"to exclude others from making, using, or selling the invention."

To be patentable, the invention has to be **new** and **original**. It has to be **non-obvious** to experts in the field and the patent examiner. It also has to make a **full disclosure** of patent and legal claims for what is to be protected. Today, inventors no longer must demonstrate that the invention is **useful** (since uses for products are often discovered years after the product itself was invented). The Patent Office has very detailed requirements for the format to be used in writing and making the drawings for the patent application. If you do not think you can or want to do this work yourself, specialists can be hired. Some successful inventors do not want anyone's help; others equally successful cannot be bothered with the details and gladly hire the needed assistance so the patent document will closely follow the prescribed format. The main parts are the drawings, an explanation of the drawings, a description of the advantages over prior art (with information from a very thorough patent search), the specifications, and the claims. The patent attorney is a professional who provides services in searching as well as in preparing the patent application. Fees vary widely but are highest in metropolitan areas. An attorney is most valuable for wording the claim of the invention, especially for covering as broad an area as possible, since individual inventors tend to make narrow, specific claims. Claims that are too restricted invite others to come up with a new improvement under a separate patent. Patent attorneys are also experienced in overcoming objections made by the patent examiner. It is common for a patent application to be rejected in its initial round. Examiners will cite a plethora of references including newly issued patents, foreign patents, and other obscure sources; this information is very helpful to the inventor and attorney in improving the claims statement, which should eventually result in a much stronger patent.

If you are thinking of inventing something in a specific problem area, there are a number of reasons listed in Table 12-5 why we are urging you to do a patent search, even if you don't think you would ever want to file for a patent.

Table 12-5
Reasons for Doing a Patent Search

---

1. The U.S. patent file of around 5 million patents is the world's largest storehouse of technological information. Four million of these patents are not described anywhere else in any literature. Some people think that Albert Einstein's years of working in the Swiss Patent Office were a waste of his time. We maintain that this occupation provided excellent training for innovative thinking.

2. You can save yourself a lot of trouble and time before you begin your invention if you first make sure that your idea has not already been invented by someone else. This is called a "prior art search." At the least, it will give you information on the state of the art today—your starting point. If you are preparing a patent application, having this information is vital to support your claim of being unique and different.

3. Examining patents is a thought-starter tool. Perhaps nine out of ten patents are not completely new ideas but substantial improvements over existing patents on products or processes. You might get a great new idea by seeing what other people have invented.

4. This investigation provides the historical background. How did earlier people solve the particular problem? You can also discover other inventions that are related to the problem.

5. Patents can be a fascinating learning tool into "how things work" and how technical problems have been solved by inventors. You may also be surprised to discover things are invented and patented that require very little technical knowledge.

6. A patent search can give you information on your competition. It can also give you clues to innovative companies you may want to work for.

Doing a patent search takes patience and perseverance. Do not restrict yourself too narrowly to patents whose accompanying drawings resemble your invention. You must pay particular attention to the paragraphs toward the end that describe the inventor's claims. Investigate patents that only seem vaguely connected to your idea to make sure that none of these incorporate essential parts of your idea. Buy copies of those patents that relate closely to your idea, so you can study them in more depth at home. Patent searches can be done in the search room at the Patent Office located in the U.S. Department of Commerce Building on 14th and E Streets in Washington, D.C. It is open to the public into the evening during the workweek and a few hours on the weekend. This is the only location where patents are bundled on the shelves by classification. The patent depository libraries located in each state have bound volumes of patents in numerical order. This means they are arranged by date of issue, not by subject. Not all libraries have complete collections. However, it is still possible to do a patent search in a repository library, if you follow a systematic procedure. The key is to find the classification number of your area of interest, and for this you need to use several **publications issued by the U.S. Patent Office**.

Your search must begin with the *Index to the U.S. Patent Classification*. This index is in general, everyday terms and can guide the investigator toward the correct classification numbers. When searching through the index, also explore the synonyms in your subject area. The index provides class and subclass numbers. Next, you need to move on to the *Manual of Classification,* which provides a more detailed definition on the classes and their subclasses. To obtain a proper field of search, you must check the classification definitions that clarify the brief phrases given in the manual. This study provides important clues to the location (by number) of related subjects. Use your imagination to come up with subclass designations that might relate to your topic of invention. Finally, check the *Classification Orders* issued periodically to give updates on classification changes made by the Patent and Trademark Office. If you want to get a feel for the types of new inventions that are coming out, read through the *Official Gazette* being issued every Tuesday and made available at the repository libraries. It contains an abstract of each patent granted, with representative claims and a selected drawing. The gazette also includes general information and notices; it is fun to browse through an issue in the explorer's mindset. Even young students find this interesting—as a teacher, you may want to check out and prescreen some of the issues first before handing them to the students.

The classification system for patents is based on "claimed disclosure" and focuses on proximate function. The classes are mutually exclusive. The subclasses are not mutually exclusive; this means that the same designation for subclasses can be found under different classes. There are over 400 classes and over 110,000 subclasses. Once you have determined the class and subclass(es) of interest to your invention topic, you can use the **on-line computer at the depository** to get a printout of all patent numbers relating to the subclass. This list enables you to go to the bound volumes and begin exploring the individual patents. Keep an organized worksheet to record your progress.

Using a systematic approach and keeping records will save you time and avoid duplication or omissions. You can design your own worksheet format, or you can use prepared sheets such as those given in *Patent It Yourself* (Ref. 12-18). Begin by listing your name(s), and a description of your invention using key words and synonyms. These words will direct you to classifications that might be applicable. The next task is to make a list of the classes and subclasses with their numbers that you have identified from the *Index* (see Table 12-6 for an example of subclasses). You may want to leave some space next to each entry for some comments after you have checked out these classes and subclasses in the *Manual*. Which of these classes or subclasses seem to be most closely related to your invention? After you have a comprehensive list of subclasses, you can move on to the on-line computer at the depository library (or through a commercial system—but check out the costs first) to get the list of all patent numbers in your listed subclasses. With this information, you can now begin investigating individual patents. For this phase, again keep good notes; for each patent that you are checking, include the patent number, the name, date of issue, class/subclass, and your comments. As should be a good habit for any researcher, jot down your name and date as you complete the record.

Table 12-6
Classification Example

CLASS 36   BOOTS, SHOES AND LEGGINGS   (DECEMBER 1990)

83	BOOTS AND SHOES
84	. Made of material other than leather
85	. . Rigid material (e.g., metal, wood, etc.)
86	. . . Wood
87	. . Plastic
88	. Foot-supporting or foot-conforming feature

:
:

110	. Shoe for cast on foot
111	. Animal shoe
113	. Shoe for children or dolls
113	. Occupational (e.g., Roof Climbing, Gardening, etc.) or Athletic Shoe
114	. . Athletic shoe or attachment therefor
115	. . . For sports (e.g., skating, etc.) featuring relative movement between shoe and ground
116	. . . . For walking in shifting media
117	. . . . . Ski boot
118	. . . . . . With leg extension
119	. . . . . . With interior foot-retaining means
120	. . . . . . With separate pivotal portion
121	. . . . Cuff-restraining means
122	. . . . . Snow shoe (e.g., having binding, harness, etc.)

continued

From the *Manual of Classification*, U.S. Department of Commerce, Patent and Trademark Office, p. 36-1.

The form of made things is always subject to change in response to their real or perceived shortcomings, their failure to function properly. This principle governs invention, innovation, and ingenuity; it is what drives all inventors, innovators, and engineers. And there follows a corollary: Since nothing is perfect, and, indeed, since even our ideas of perfection are not static, everything is subject to change over time.
Henry Petroski, The Evolution of Useful Things.

The greatest invention of the nineteenth century was the invention of the method of invention.
Alfred North Whitehead, Science and the Modern World.

Human subtlety . . . will never devise an invention more beautiful, more simple or more direct than does nature, because in her inventions nothing is lacking, and nothing is superfluous.
Leonardo da Vinci, The Notebooks, Vol. 1, Ch. 1.

How do you read the *Manual of Classification?* Table 12-6 gives an illustration for a ski boot. Start with the furthest indentation, then search upward for the next outer category. Thus, beginning with Subclass 121, we read: Cuff-restraining means (move up to 120) with separate pivotal portion (move up to 117) ski boot (move up to 116) for walking in shifting media (move up to 115) for sports featuring relative movement between shoe and ground (move up to 114) athletic shoe or attachment (move up to 113) occupational or athletic shoe (move up to main subclass 83 BOOTS AND SHOES of Class 36, BOOTS, SHOES AND LEGGINGS. Don't forget to check related subclasses, such as 84, 88, 110, etc.

If you are doing a patent search because you are working on an invention you are intending to patent yourself, you should use a library with a complete file of patents. However, since there are only about 60 patent depository libraries in the United States, what do you do if you merely want to check out some representative samples of patents in a particular classification, or if you want to do a preliminary search to identify the applicable subclasses for your invention? Many university libraries have a government documents section. Although they will not have complete files, they may have the *Official Gazette.* You can search past issues of the gazette for the most recent patents issued in the class and subclasses you are interested in. To supplement your manual search, a computerized database is available from the Patent and Trademark Office, the Classification and Search Support Information System (CASSIS). Although it is accessible on-line from the patent depositories only, it is available as a CD-ROM product by subscription for personal computers. This database makes it possible to search for patents by words contained in the abstract (for patents issued since 1986) and title (for patents since 1969). However, consider the database as a useful time-saver, not a substitute for a thorough manual search.

◔ *TEAM ACTIVITY 12-1: PATENT SEARCH*

Check out the resources available in your school, organization, or community for doing a patent search. Even if your nearest library is not a depository, it may still have a section on government documents and patents. Pick an interesting product related to your creative problem-solving project and examine at least three patents in the subclass of the product.

INVENTORS AND INVENTIONS—
CASE STUDY ASSIGNMENT

In the days before we had books and schools, how did people learn? Children worked alongside their parents, apprentices and students worked alongside their masters and teachers observing and copying everything that was done. Could this approach work if you wanted to learn more about being an inventor? How could you observe a master inventor? Since we now have books—and there seems to be a renewed interest lately in writing about inventions and inventors—we highly recommend that you explore some of these very interesting, entertaining, and challenging texts. Look for imagination. Observe the flow of ideas—where might they have come from? How did they get synthesized into the product? What are the connections to other ideas, to the culture, to the contextual situation? This study can have two very different goals—you can focus on the inventor, or you can focus on the invention. You can study the life and thinking of the inventor, or you can investigate the history of a product which may involve a long time span and a whole series of inventors, as we have seen with the example of the clock.

Another way you can learn from inventors about the process of inventing is by talking to them. Ask questions. Find out how and why they invented. What happened as they went through the process? What thinking skills did they use or recommend that you use? It would probably help your case in getting the interview and holding the inventor's interest if you prepared ahead of time by researching one or more of his or her inventions so you can ask more specific questions. Where can you find inventors? Probably one of the best places is to start with the Yellow Pages and look for patent attorneys; they will be able to refer you to inventors right in your community. The people on their list of inventors may surprise you—it will likely include students, housewives, farmers, mechanics—a lot of "regular folk" in addition to engineers and other "learned" people. Patents have been issued to children ten years old or even younger.

For starters (and perhaps for giving you a topic for the thinking report mentioned at the very end of Part 1 in this book) you may want to do a library search. You can scan through Tables 12-7 and 12-8 to see if one or more topics sound interesting to you. If not, start in the library's catalog (cards or computer) under "inventions" or "inventors" or even more broadly, under "history of technology." Look for the location numbers of some of these books, then go to the stacks where you will find many others covering the same or related topics. Now you can browse through these, then select two or three that catch your fancy. Since learning will be easier and exploration more interesting if you prime your brain, write down a list of questions that you would like to have answered about the topic that you selected. For example, if you are investigating one or two particular inventors, you may want to know such things as: "Why did they become inventors? When did they find out they were creative? Where did their ideas for the invention come from? How did they sell their ideas? What steps did they follow to get their original ideas through the patenting process? How did they come out financially?

Table 12-7
A Sample List of Inventors

---

1. Thomas Edison, holder of 1093 U.S. patents.

2. Alexander Graham Bell (who had his first invention at the age of 14).

3. Igor Sikorsky, the inventor of the modern helicopter.

4. Robert Fulton—how many different things did he do with boats?

5. Harold Edgerton—inventor of the strobe light.

6. Catherine Greene—how did she and Eli Whitney invent the cotton gin?

7. Margaret Knight, holder of 26 patents in diverse fields.

8. Sarah Breedlove Walker, daughter of former slaves and the first American woman self-made millionaire.

9. Bette Graham, a housewife and typist, invented "liquid paper"—her company in 1980 sold for over $47 million.

10. Ann Moore, a Peace Corps volunteer, adapted an idea from African women to invent a baby carrier, as well as various means to carry oxygen cylinders.

11. Mary Anderson received a patent for windshield wipers in 1903.

12. Josie Stuart, also in 1903, patented a dandruff shampoo.

13. Josephine Cochrane patented a dishwasher in 1914.

14. Marion Donovan received a patent for the first disposable diaper in 1951.

15. Rose Totino patented a dough for frozen pizza in 1979.

16. Elijah McCoy had over 50 patents for his inventions; he was the "real McCoy."

17. Granville Woods had more than 60 patents; he invented airbrakes, improved Bell's telegraph, and invented an electric motor suitable for subway trains.

18. Garrett Morgan invented the traffic signal. From Chapter 2, do you remember when and where it was invented?

19. George Washington Carver was a scientist and inventor. Although mostly known for his work with peanuts, he also created many uses for sweet potatoes and in general improved agriculture in the southern United States.

20. Keith D. Elwick of Vinton, Iowa, invented a manure spreader that became so popular in England that he received a Silver Medal and Trophy from Queen Elizabeth for his creativity.

21. Robert G. LeTourneau of Longview, Texas, received 187 patents and founded a college, even though he only had a seventh-grade education. He also wrote an autobiography, *Mover of Men and Mountains*.

22. Guglielmo Marconi is often called the inventor of radio. Why does he have a monument dedicated to him on a mountain peak in Rio de Janeiro?

23. Michael Faraday has been called the greatest experimenter in electromagneticism.

24. James Clark Maxwell, Heinrich Hertz, John Fleming, Lee De Forest, Irving Langmuir—what were their contributions to the field of radio and electronics?

Table 12-8
A Sample List of Inventions

1. Earmuffs were invented in 1873 by 13-year-old Chester Greenwood.

2. Band-Aids were invented by an employee of Johnson & Johnson.

3. LifeSavers® were invented in 1913 by Clarence Crane.

4. Stephanie Kwolek, a chemist with the DuPont company, invented the miracle fiber, Kevlar. What is the connection to Paul MacCready's Gossamer Albatross?

5. The *Bobcat* loader was invented by Louis Keller, a farmer from North Dakota who had dropped out of school after the eighth grade.

6. The jet engine was invented in two places: in England by Frank Whittle, a RAF pilot, and in Germany by Hans von Ohain, an engineer with the Heinkel airplane manufacturing company.

7. Orville and Wilbur Wright invented the first airplane, but they built on knowledge and experiences of others. Investigate the early history of manned flight.

8. Henry Ford's assembly line for automobile manufacturing was not his invention. Where did the idea originate?

9. The pop-up toaster was patented by Charles Strite in 1918.

10. The ballpoint pen was patented in 1938 by two Hungarian chemists, George and Ladislao Biro.

11. It took Guideon Sundback 30 years to perfect the slide fastener, now known as the zipper and patented in 1913.

12. The lead pencil is about 450 years old.

13. The mousetrap was invented by Charles F. Nelson of Galesbury, Illinois (Patent Number 661068—you might want to look it up to see the original model).

14. Eating utensils developed very differently in the East and West. Compare the invention of the chopstick with the invention of the fork.

15. How was the microwave oven invented?

16. How was bar coding and scanning invented?

17. How was the Velcro fastener invented?

18. How was magnetic resonance imaging (MRI) invented, and what is it used for?

19. How was the camcorder invented?

20. How was the home smoke alarm invented?

21. What are some new developments and applications for lasers?

22. Philo T. Farnsworth, Paul Nipkow, Vladimir Kosma Zworykin, and John Logie Baird—what are their contributions to the invention of television?

23. Charles Babbage and Ada, Lady Lovelace (Byron's daughter) were the first to understand and invent machine computation—the first computer. However, their designs could not be built with the technology available in 1833.

24. What are some important inventions made in Japan in the last ten years?

Who helped them?" If you are exploring the development of a particular product, include questions that will help you look for connections and context. In your library search, you will notice that quite a bit of material will be available on inventors from Europe and the United States, whereas information on inventors from other parts of the world will be almost nonexistent. Dr. Yoshiro NakaMats holds more than 2300 patents (more than twice as many as Thomas Edison), and he is still continuing to invent. He invented the floppy disk (which he has licensed to IBM), yet it is difficult to find an encyclopedia that mentions his name as a separate entry or in connection with the floppy disk. Check it out!

Another topic that you can explore is the time lag between related inventions, as well as the process of continuous improvement. For example, the tin can for conserving food was in use for almost half a century before the first true can opener was invented. In 1809, Nicolas Appert in Paris, France, perfected the process of conserving cooked food in glass bottles. But since glass bottles have the disadvantage of being breakable, Peter Durand, in London, invented the tin canister. Instructions for opening usually advised to "use a chisel and hammer." William Lyman from Connecticut patented the first can opener with a wheel in 1870, with the now familiar, improved style of opener receiving a patent in 1925. What other ways of opening cans have been devised and improved in the last 30 years?

🔑 ⌚ *ASSIGNMENT 12-2: INVENTION, INVENTORS, INVENT!*

After your preliminary survey of this chapter, select a specific invention as your topic. Do an in-depth study of the subject from library materials as well as by interviewing people inventing, developing, manufacturing, or using the invention or product.

Alternatively, select an inventor as the subject of your investigation. You can select a special focus: thinking style, environment, motivation, as well as the process, effort, and final outcome. What were the costs (and benefits) to the inventor?

Alone or in a small team, create an invention! Brainstorm ideas, then focus on one area. Use the creative problem-solving process to develop your ideas. Then build a prototype to demonstrate that the invention actually works.

Case Study: Invention and Development of Window Glass

Sir Alistair Pilkington, the inventor of the process used to produce float glass, was at the Eitel Institute for Silicate Research at the University of Toledo in the fall of 1990. He gave a speech entitled "Glass and Windows" which we would like to summarize here, since it illustrates many different aspects of creative thinking, invention, and technology.

Glass is an amazing product; it can be high-tech or common place; it is 5000 years old, yet at the forefront of developments in energy conservation, communications, and space exploration. The heat-resistant tiles protecting the space shuttle are made of glass ceramics. The solar cells used in communications satellites are protected with a very thin layer of glass. Glass fibers increase the speed and quantity of signal transmission by cable. Glass has become a versatile building material and, in the form of fiberglass, an insulating material. As Pilkington describes it, "Most glass begins with sand, soda and lime which are fused together under very high temperatures. It is a super-cooled inorganic liquid which can be pressed like butter, molded like jelly, rolled and cut like pastry, drawn like hot toffee, spun like candy floss and blown like a bubble."

Glassmaking was known in the Middle East by 3000 B.C. The Egyptians made glass jars for cosmetics. The Romans were able to blow, mold, cut, and decorate glass. It is amazing to see the art and versatility achieved by these early glass-makers. The British Museum has a great collection of early glass. Another nice collection is in the Toledo Museum of Art (endowed by the glass-manufacturing Libbey family). One of the most useful applications of glass is as a material for windows—it provides both a link and a barrier between people and nature since it keeps out precipitation, wind, heat, and cold, yet admits sunlight and a view. Glass windows were in use by the Romans. The technique known as the crown process appeared in the Near East in the fourth century. In this process, a hand-blown bubble of glass was spun into a disc with a diameter of about 8 inches. By the seventeenth century, crown panes for windows measured 10 inches by 13 inches; they were cut from discs that commonly had a diameter of about 54 inches. Larger window sizes became possible when the blown-cylinder process was introduced in the middle of the nineteenth century. Glass cylinders up to 13 feet long with a 2-foot diameter could be blown. After the glass cooled, the cylinder was slit, and the resulting glass sheet was flattened. This yielded panes that were 3 feet wide and considerably longer, but the process required great skill and hard labor.

By the early 1900s, manufacturers began to think about mechanizing the process. First, blown cylinders were drawn with a metal rim from large pools of molten glass. This allowed glass of even larger size, but it was not a continuous process, and the flattening harmed the brilliance of the surface. Thus it was a logical step to draw a continuous flat sheet from the molten pool of glass in the furnace. This process was invented by Fourcault in Belgium by drawing a sheet of glass through a slot up to an annealing tower, as well as by Libbey Owens in America, where the sheet was drawn vertically but was then passed over rollers into a horizontal position for annealing. One disadvantage of this procedure was that contact with the rollers destroyed the brilliance in the glass. The Pittsburgh Plate Glass Company patented a process of drawing the sheet directly from the surface of the molten glass in the furnace in 1926, thus maintaining the natural brilliant fire finish. This process was inexpensive—why then make more improvements?

Although sheet glass was very acceptable for windows, it was not sufficiently distortion-free for applications such as mirrors and large shop windows (and later car windows). Late in the seventeenth century, the French had begun making glass with parallel flat surfaces, thus eliminating distortion. In this process, a molten mass of glass was poured onto metal tables and then rolled into a plate. Because the glass was in contact with metal while being rolled, it became marked. Thus it had to be polished on both sides to restore transparency. This was a time-consuming, multistage process. In 1919, Ford Motor Company began to mechanize the process of rolling glass continuously. Alistair Pilkington successfully combined a continuous melting surface with continuous rolling of glass, and this technology was then sold back to America. In the early 1930s, Pilkington added a grinding machine to the continuous glass production line, and this innovation became licensed throughout the world. By grinding off about 20 percent of the surfaces, beautiful, high-quality glass was produced, yet more advanced technology was developed. Why?

This illustrates one of the driving features of innovation—the continuous correction of "failure" to achieve a perfect product, while the idea of quality and perfection keeps changing. What if the best features of both processes could be combined—the fire polish of the sheet with the distortion-free plate in a waste-free process? This was the dream of many glassmakers. Alistair Pilkington gives us some insight into the actual circumstances present when the idea of float glass came to him. This was in June 1952 when he had some "thinking" time in a new position. The actual idea occurred while he was helping his wife wash dishes (though—as he says—the dishwashing in itself had nothing in common with the conscious idea). We think it is possible that the "dish bath" and perhaps the presence of tin in the kitchen might have triggered the idea for the float process in his subconscious mind. But it took a lot of hard work by teams of people before the conceptual idea became a workable process resulting in an improved consumer product. The float glass process has now replaced all previous processes throughout the world. "It makes glass better than the best plate and cheaper than the cheapest sheet glass."

In the 1990 Philips Lecture published in Great Britain in *Science and Public Affairs,* (Vol. 6, No. 1, p. 58), Alistair Pilkington describes the float process:

> A continuous ribbon of molten glass moves out of the melting furnace and floats along the surface of an enclosed bath of molten tin. The beauty of the float process is that it works with nature, and the forces of both gravity and surface tension flatten the glass. The ribbon is held in a chemically controlled atmosphere at a high enough temperature for a long enough time for the irregularities to melt out and for the surfaces to become flat and parallel. Because the surface of the molten tin is dead flat the glass also becomes flat. Not, incidentally, totally flat, the glass actually follows the curvature of the Earth. ... Tin is an ideal substance for making glass. In fact I often think tin was created by the Almighty especially for glass making! ... The ribbon of glass is then cooled down while still advancing across the molten tin until the surfaces are hard enough for it to be taken out of the bath without the rollers marking the bottom surface; so a glass is produced with uniform thickness and bright fire-polished surfaces without any need for grinding and polishing.

This sounds easy, but in actuality it took seven years, teamwork by six people working in secret, and a huge investment to develop the process into a technical and commercial success. One of the most difficult problems to be solved was the exclusion of oxygen from the atmosphere of the bath. Another key ingredient to success was the optimism that was maintained by the team while struggling to overcome problem after problem after problem. Today, the process is completely computerized. Over 115 float lines are operating around the world today, producing a ribbon that could circle the Earth thirty-four times each year.

Now the focus of further developments in glassmaking moves back to the United States. What technique would you use if you wanted curved, not perfectly flat, glass, as for example in airplane or automobile windshields? While working at Libbey-Owens-Ford, Harold McMaster, an inventor living in Toledo, Ohio, led a team of researchers in the development of "gravity sag" technology used on furnaces to bend glass for windows on bombers. He and his designers hold 350 patents in the area of glass manufacturing. Later, he founded his own company, Glasstech, Inc., which is now conducting research and development in the area of manufacturing solar cells. His advice to young people about creative thinking is quoted in Chapter 4.

Why do you think windows for homes and businesses are still being improved? Since clear window glass allows about 75 percent of the sun's heat to enter buildings, coatings and additives have been developed to reduce the transmission of solar radiation. Double or triple glass filled with an inert gas such as argon is used to produce a low-emissivity pane, which means that heat from inside the building is not radiated to the outside, thus helping to conserve energy without the need for tight-fitting window curtains or shutters. The glassmaker, the architect, the engineer, the heating and lighting expert, and the designer all work together to integrate a building technology that creates a comfortable interior environment while taking local climatic conditions, people's needs, space usage, and energy conservation into account.

Case Study: Design of the Centripetal Rider

The following is a description of a report submitted by a group of six junior high and high school students (Rob Bennett, Brian Herr, Matt Huss, Kara Iskyan, Heidi Sungurlu, and Sylvia Wu) working with two teachers (Pat McNichols and Dave Simmons) and an engineering faculty (John Rich) on an invention project as part of the Level 3 math/science academy at the University of Toledo during one summer week in 1992. The project report was accompanied by an operations manual covering the materials list; costs; directions for construction of the base frame, wood ball bearing, platform, and motor support; construction drawings, and a list of teaching ideas. A photo of the completed device being demonstrated by some of the students is shown in Figure 12-2.

Physical science has been difficult to experience in a classroom setting. Since it is challenging to understand some concepts, we have decided to design an instrument to assist in teaching centripetal forces. The summer academy class brainstormed over 50 ideas for devices that would help teachers teach math and science. The top three choices for our team were: merry-go-round (centripetal force), wind tunnel, and simple transparent machines for demonstration with overhead projectors. After using the Pugh method, we agreed on the centripetal force problem. We began our quest by brainstorming the problem. First, the team immediately wanted to build a merry-go-round, but Professor Rich reminded us to focus on the science problem to be solved rather than on a specific device to be built. As we brainstormed the problem, we arrived at this problem statement:

> Teachers need an inexpensive device (presently unavailable)
> to use in the classroom to enable students to experience
> and experiment with rotational motion and the forces involved.

From there, we thought of thirteen solutions to the problem statement. There were many choices ranging from going to Cedar Point amusement park to swinging a student on skates to riding ponies in a circle. We narrowed down our ideas to making a rotating platform on which students could ride. Overall, our final design is a device that will be used for experiencing rotational motion. This will help students feel the apparent centrifugal force when moving in a circle. The design criteria asked that the device be inexpensive (less than $100), reusable, storable, versatile, safe, user-friendly, and reproducible.

Figure 12-2
The Centripetal
Rider.

First, Professor John Rich made a giant ball bearing. Using particle board as a bottom platform and a slightly smaller board as the cover, he made a matching circular groove in each board in which he placed sixty marbles. [Marbles were inexpensive and easily replaceable if necessary.] The two halves of the wood ball bearing were connected with a 3/4-inch bolt. The bottom of the bolt passed through a wheel bearing to allow it to rotate. Next, the team placed a 4-foot circular piece of wood with a groove around the circumference (the rider platform) on top of the ball bearing. In order to make the device rotate, we installed a 1/4-horsepower motor. The A-C motor, which was salvaged from a clothes dryer, operates at 115 volts and rotates at 1725 rpm. The major problem we ran into was how to get the motor to turn the platform. The options ranged from a bicycle chain to a V-belt to a nylon rope. We decided to drive the platform with the motor and a rubber belt. We found that the cost of the V-belt was quite high, but Professor Rich found a company willing to donate a rubber belt to the project. As a team, we developed finishing touches to the design and decided on materials. During the project, we used calculations done with MathCAD on a computer, as well as results from the student team working on the friction machine, to calculate the radii of the platform and belt drive so that students would not slip off the platform when riding on it. Finally, we had to add diagonal blocking to the entire assembly to raise it to the correct height needed, so the belt drive from the motor would be at the same height as the groove in the platform.

Pat McNichols was our secretary and did many mathematical computations. Dave Simmons came up with the idea of a rotational force device and assisted in combining different designs. Rob Bennett drew design concepts, designed the motor encasement, and helped build the prototype model. Brian Herr sketched designs and was our personal thesaurus. Matt Huss assisted with electrical matters, helped build the prototype, and helped with specific data. Kara Iskyan assisted in writing this project report and the operations manual for the Centripetal Rider. Heidi Sungurlu labeled drawings and assisted in writing the report. Sylvia Wu did the preliminary drawings and assisted in writing both papers. Everyone contributed many ideas to this team project. John Rich, assistant professor of engineering technology, was the project adviser. Funding for this project was provided by the University of Toledo College of Engineering and the Ohio Board of Regents under the Dwight D. Eisenhower program for math and science education.

> *Instant communication does not in and of itself create understanding. Thinking about the whole explosion of information … one might visualize a pyramid: At the bottom of the pyramid are data; the next layer up is information culled from all the data; the next layer is our experience. … It is information filtered through that experience that in the best of circumstances creates wisdom at the top of the pyramid. … Advanced technology does not produce wisdom; it does not change human nature; it does not make our problems go away. But it does and will speed us on our journey toward more human freedom.*
> Walter B. Wriston, *former chairman of Citicorp, in* The Twilight of Sovereignty.

EPILOGUE

We opened the preface of our book with the analogy of using different materials in constructing a building. Now we want to conclude with another metaphorical image—a bridge. Bridges are more than just an invitation to cross over to the other side. Bridges make connections, not only between two sides of a chasm or body of water, but between people. Bridges connect to the past—some bridges are centuries old and still in use; many ancient bridge-building techniques continue to be used. Some bridges form a connection to nature, such as those hewn out of local rock, fitted from logs, or woven together with sisal or grass fibers. Some bridges are rather utilitarian; they carry streams of automotive traffic, pedestrians, railways, or pipelines. And then we have the bridges that soar, that have become objects of beauty, not just monuments of technological achievement. Examples are the cable-stayed Sunshine Skyway Bridge in Tampa Bay; the Ganter Bridge, a combined cantilever-cable-stay design in concrete traversing a Swiss valley near Simplon Pass; and perhaps the most photographed bridge in the world, the Golden Gate, where the stunning setting of hills and ocean enhances the aesthetic appeal of the red-painted lofty towers and steel cables supporting a thin deck. Finally, we have the "bridge to the stars"—our explorer spacecraft where the bridge is for information and dreams, and not yet (or maybe never) for people transport.

Bridges have strategic value; in war they are the first targets to be destroyed. They have symbolic value; they are the first infrastructure to be rebuilt, witness Baghdad and Saddam Hussein in 1991. Creative problem solving is about building bridges—from a problem to a solution. Language and communication are bridges between different people. Knowledge and learning are also bridges—to the past and to the future. In his book *Disappearing Through the Skylight*, O. B. Hardison, Jr., a professor at Georgetown University in Washington, D.C., argues that our connections to nature are disappearing in our culture. In some respects he is right, and he presents impressive evidence. We have found evidence to the contrary—yet we should take his viewpoint as a warning. If nature has disappeared from our vision, we are not seeing the degradation in our environment.

This was brought home to us vividly this past summer when we retraced our 1959 honeymoon visit to the Golden Gate. We have photographs from several trips during those years to this part of San Francisco, and the bridge is etched crisply in each one of them (as well as in our memory). This time, from the north end outlook, it was barely possible to see either San Francisco or the south tower, not because of the fog which can romantically shroud the bridge, but because of murky air pollution. People have been complacent; it is high time that collectively we begin to solve the pollution problems of our living space. How encouraging it is to hear about the recently completed I-70 Glenwood Canyon highway project. The Colorado legislature was adamant; they specified that the expanded highway be "so designed that ... the wonder of human engineering will be tastefully blended with the wonders of nature." Thirty-nine bridges with a total length of over six miles were needed for the new roadway which blends into the landscape. With the new leadership in the White House and looking at present trends in the world scene, the picture for cooperative projects in support of nature globally is beginning to improve.

Is nature disappearing from the music of our culture through synthesizers and other technological gimmickry? But blue grass, country music, and soul music are becoming more and more popular. If history is disappearing, we have no one to blame but ourselves, because we have treated it as a boring subject void of relevant connections to our students' lives. Ditto for language, which is the bridge connecting us to the global community as well as to our neighbors. We can use computers to help with grammar and spelling—the mechanics—thus we can turn our attention to the beauty and subtlety of language and to learning different languages and cultures, many closer to nature than we are.

A powerful source of development is the exchange
of all things original that are created independently by each nation.
Mikhail Gorbachev, speech to the United Nations, December 8, 1988.

What about art, painting, or sculpture? Here again, the electronic media are tools that do influence artistic expression but are not the message. Art (with or without the computer) still continues to express visually and intuitively what we feel and "see" as truth. What strikes us as beautiful in computer-generated pictures is only what resonates with something within us, with our spirit and experiences—as happens with any other artistic expression. Artists now have more freedom to create a synthesis using many different media and materials. An example is the moving sculpture of "Noah's Raven" about humans and the natural world created by Mary Lucier and shown at the Toledo Museum of Art in early 1993.

Professor Hardison says that science has lost its connection with reality; it has become playful and more like a game. What do you think of naming an invisible particle "quark"? But can we not consider chaos theory as a turn toward a different, "fractal" understanding of nature, where the neatly described one-, two-, or three-dimensional world is an artificial construct (although quite useful in the past)? More frightening is how video games and television entertainment have created an artificial world free from the natural law of consequences. Bullets miss the hero or heroine—or if not, the actors appear alive again later in another show. Gravity-free skiing in virtual reality may be an interesting experience, but it is a poor substitute for gliding down a mountain through glistening snow, with the sun tracing blue shadows and the biting wind swishing nylon-layered clothing. Is bungee-jumping a protest against the artificially risk-free world of electronics wizardry?

Yet nature does not let us be "safe" for long—floods and volcanoes, fires and hurricanes, droughts and tornadoes, as well as the consequences of thoughtless behavior remind us that living involves risks, change, and responsibility. To us personally, nature did not "disappear through the skylight"—skylights brought us closer to nature because our passive solar home in New Mexico had three enormous skylights. We became aware of changes in the weather, of the contributions of the sun to lighting and heating and thus of our comfort. We heard the drumming of the rain on the glass; we saw clouds, the phases of the moon, lightning, and especially the seasonally changing solar angle and its effect on us and our atrium plants.

Our involvement with creative problem solving has certainly changed us. First of all, we deeply regret that we did not know—that we were never taught—this thinking skill. Both of us, in various ways, were driven through ridicule to hide and lock away our creativity by the time we were 15, until we no longer even remembered that we once had this ability. As an engineer, Ed used his analytical skills to solve problems and come up with some improvements to products. But when he had a creative idea for an invention, it was set aside, because he did not know how to appreciate it, much less how to develop it into a usable product. Even raising children would have been easier had we known then what we know now.

We believe that the last few years have been the most productive and the most rewarding in our lives, because now we are applying creative problem solving. We have created this book through a multistage process of continuous improvement— yet also by incorporating new insight and synthesis. Just compare the first edition with this version to see some of this developing process. Ed has several projects going in developing integrated courseware for engineering math, heat transfer, and vibrations, with additional projects waiting until he has more free time. We are involved in teaching others this thinking process through workshops and courses, from sixth-grade students and their parents, teachers, engineering students, faculty and staff to people in industry. Ed has worked with a colleague on an invention, and they now have a patent pending. Monika has designed and made a new type of wall hanging that combines knitting with quilting; she has also developed a creative problem-solving seminar based on the Bible, and she found that she can write poetry. By teaching, we are learning more ourselves; we continue to explore new subjects. We are glad there is no limit to what we may yet discover. We are seeing that our example is helping make a change in how students are taught in schools and universities—this is our greatest reward.

> *It is no accident, then, that computer control and simulation figure so largely*
> *in attempts by government and industry to manage shrinking resources:*
> *the computer was made for such work, just as the steam engine was perfectly*
> *suited to the expanding national and imperial economies of the last century.*
> J. David Bolter, Turing's Man.

> *It is better to light one candle than curse the darkness.*
> *Adlai E. Stevenson, in an eulogy to Eleanor Roosevelt, 1962.*

The dreams of space travel notwithstanding, we only have one world, and we must take care of it. We must learn to get along with each other. No one group or nation is superior—we all have things to share; we all can learn from each other; we all have unique abilities and achievements that can be appreciated and further developed. Travel and spend some time in another part of the world. Live in another culture for a while—learn the language. Find out how different peoples solve problems; find out what values you have in common. Our world requires the service of all our energies, our talents, our minds as well as our hearts—then we will have the power to make changes for the better. Learn to make wise choices—spend your time and your money on what has lasting value. Be a unique individual as well as a team member and a citizen of the global community. Be involved in teaching, because it is your responsibility to pass on the values of your culture past and present in a framework understandable to the young. You must build the bridges and make the connections—computers and television will not do this for you. Model whole-brain critical thinking by the choices you make in your life and the responsibilities you assume for yourself and others.

ASSIGNMENT 12-3: LIFELONG APPLIED CREATIVITY

We have given you a match, now light a candle. Determine to expand and nurture your thinking skills and creativity. Apply whole-brain creative and critical thinking to solve problems, to help others, and to preserve the environment for a future generation. Be a bridge builder. Strengthen the values that have undergirded your culture. The state of the world at the beginning of 1994 could give cause for pessimism; yet we believe that with positive thinking and creative problem solving we can make a difference right where we are. So can you!

FURTHER LEARNING AND RESOURCES

Reading List

12-1 James L. Adams, *Flying Buttresses, Entropy, and O-Rings: The World of an Engineer*, Harvard University Press, Cambridge, Massachusetts, 1991. This contextual overview presents a brief history of technology, diversity engineering, who determines the direction of problem solving, and the design and invention process. Further topics are: mathematics; science and research; development, test, and failure; manufacturing and assembly; money and business.

12-2 Henry G. Altov (Altshuller), *And Suddenly the Inventor Appeared: How to Invent and How to Solve Technical Problems*, 1984; translated and adapted from the Russian by Lev Shulyak, Technical Innovation Center, Worcester, Massachusetts, 1992. This presents the basic concepts of the theory of solving inventive problems (TRIZ is the Russian abbreviation), an approach taught in the former USSR and other European countries from fifth grade on up. It is based on the experiences of people who have solved real technical problems and requires some knowledge of physics.

12-3 J. David Bolter, *Turing's Man: Western Culture in the Computer Age*, University of North Carolina Press, Chapel Hill, 1984. The author, with a background in the classics as well as computer science, examines the relationship between cultures and technology. He identifies a defining technology to characterize specific ages.

12-4 McKinley Burt, Jr., *Black Inventors of America*, National Book Company, Portland, Oregon, 1989. This book contains copies of successful patent applications by many African-American inventors. Minority students will appreciate learning about these role models.

12-5 David Gelernter, *The Muse in the Machine: Computerizing the Poetry of Human Thought*, McMillan, New York , 1994. This expert in parallel computing believes that not logic, but emotions are most important for doing creative activities—metaphor adds richness and value. These thoughts are connected to how computers learn and may develop in the future.

12-6 Giorgio de Santillana and Hertha von Dechend, *Hamlet's Mill: An Essay on Myth and the Frame of Time*, Gambit, Boston, 1969. The authors present an exploration of the origins of human knowledge in the preliterate world. They see myth as a language for the perpetuation of a vast and complex body of astrological knowledge.

12-7 Mike Gray, *Angle of Attack: Harrison Storms and the Race to the Moon*, Norton, New York, 1992. This book about Project Apollo centers on North American Rockwell engineer Harrison Storms, Jr., and how he inspired his team to surmount every obstacle toward winning the NASA contract. Written beautifully in nontechnical language, it gives a vivid picture of 1960s America.

12-8 Gordon D. Griffin, *How to Be a Successful Inventor: Turn Your Ideas into Profit*, Wiley, New York, 1991. This book is filled with good advice on how to recognize the economic potential of ideas, how to develop a mind for the business side of an invention, how to write licensing agreements, how to patent and market an invention, etc. The author holds a variety of patents.

12-9 O. B. Hardison, Jr., *Disappearing Through the Skylight: Culture and Technology in the Twentieth Century*, Viking Penguin, New York, 1989. This thought-provoking book—which we highly recommend although we do not agree with some of its conclusions—draws interesting connections between the development of technology, our view of nature, and its expression in culture (history, art, language, architecture, and music).

12-10 Thor Heyerdahl, *The Maldive Mystery*, Adler and Adler, Bethesda, Maryland, 1986. The famous explorer of *Kon-Tiki* and *Ra* fame details his discoveries of early Maldive history and artifacts which prove these islands to have been a crucial crossroads between the civilizations of Sumer, Mesopotamia, Bahrain, Sri Lanka, the Indus Valley, and beyond. We also recommend his books describing his voyages with *Kon-Tiki* and *Ra*, and *Early Man and the Ocean: A Search for the Beginnings of Navigation and Seaborne Civilizations*, Doubleday, Garden City, New York, 1979.

12-11 *How in the World—A Fascinating Journey through the World of Human Ingenuity,* Reader's Digest, 1990. This reference book, in the format of a newsmagazine with many color photographs, answers many "how" questions about the world of made things: everyday products, organizational achievements, ingenious solutions, artful ideas, scientific and medical "miracles," feats of engineering, and mysteries of the past.

12-12 Tracy Kidder, *The Soul of a New Machine,* Avon Books, New York, 1981. James L. Adams calls this story of the development of a new computer by Data General Corporation a must-read.

12-13 David Macaulay, *The Way Things Work : From Levers to Lasers, Cars to Computers—A Visual Guide to the World of Machines,* Houghton Mifflin, Boston, 1988. This is a fun book for nontechnical readers of all ages, giving an overview of technology and how things work. Explanations of physical principles are enlivened by means of a "wooly mammoth," and the colorful sketches and drawings are marvelously inventive.

12-14 Lewis Mumford, *Technics and Civilizations,* Harcourt, New York, 1934. This classic work draws strong connections between technology and culture from the Middle Ages to the beginning of the twentieth century.

12-15 Elizabeth L. Newhouse, ed., *The Builders: Marvels of Engineering,* National Geographic Society, 1992. This book is a visual feast of human ingenuity as expressed in large structures: roads, canals, bridges, railroads, pipelines; towers, tunnels, skyscrapers, arenas, exposition halls; walls, dams, and fortresses; pyramids, temples, domes, and cathedrals; wind, solar, and hydroelectric installations.

12-16 Elizabeth L. Newhouse, ed., *Inventors and Discoverers: Changing Our World,* National Geographic Society, 1988. This book contains individual essays (with many illustrations) on these topics: discoverers and inventors; the power of steam; the age of electricity; on wheels and wings; a world of new materials; capturing the image; messages by wireless; power particles; computers and chips; engineering life; and technology of tomorrow.

12-17 Henry Petroski, *The Evolution of Useful Things,* Knopf, New York, 1992. Products get changed and improved because existing devices fail to live up to their promise—this is the premise that is illustrated with case studies on the invention and development of everyday things from fork to paperclip and zipper. Another recommended book by the same author and publisher is *The Pencil: A History of Design and Circumstance,* 1990.

12-18 David Pressman, *Patent It Yourself,* third edition, Nolo Press, Berkeley, California, 1991. This book contains useful hints and forms for those who want to apply for their own patents.

12-19 Struan Reid, *Handbook of Invention and Discovery,* Usborne, London, 1986. This softcover reference book with colorful illustrations presents condensed information about the most important inventions and discoveries, beginning with the Stone Age. The format used clearly brings out the development over time of a particular invention and related products and processes.

12-20 Moshe F. Rubinstein, *Patterns of Problem Solving,* Prentice-Hall, Englewood Cliffs, New Jersey, 1975. This textbook opens with an interesting discussion of the relationship of culture, values, and problem solving. Its last chapter discusses values and models of behavior; it includes examples and problems. It also shows an application of the Delphi method in the development of an interdisciplinary course.

12-21 Charles Singer and Trevor I. Wiliams, eds., *A History of Technology,* Oxford University Press, New York,1954-1984. The first five volumes describe in painstaking detail the development of technology and its artifacts from ancient history to the end of the nineteenth century; they include many photographs, drawings, and illustrations and were published in the 1950s. Three additional volumes covering the twentieth century were published more recently.

12-22 Arnold Skromme, *Memorization Is Not Enough: The 7-Ability Plan,* Self-Confidence Press, Moline, Illinois, 1989. This monograph begins with interviews with inventors of farm machinery.

The author was astounded to discover the great creativity of the people who often did not have schooling beyond the eighth grade. He describes how to help schools develop the creativity, dexterity, empathy, judgment, motivation, and personality of students, not just academic abilities.

12-23 Autumn Stanley, *Mothers and Daughters of Invention*, Scarecrow Press, Metuchen, New Jersey, 1992. The upcoming publication of this book was announced at the end of an article by the same author about "The Champion of Women Inventors—Charlotte Smith" in *Invention & Technology*, Summer 1992. It brings out the difficulties most women have had in getting their talents and inventions recognized. This is recommended for those who want to encourage women of all ages to develop their talents and anyone interested in exploring the influence of social constraints on invention.

12-24 Robert Temple, *The Genius of China: 3000 Years of Science, Discovery and Invention*, Simon & Schuster, New York, 1986. This beautifully illustrated volume summarizes the discoveries of Joseph Needham about the technological achievements of China between 1300 B.C. and A.D. 1500. It is organized by inventions within broad categories.

12-25 Charles Thompson, *What a Great Idea! The Key Steps Creative People Take*, Harper Perennial, New York, 1992. Exercises in this workbook help overcome barriers to creativity. It includes an interview with Yoshira NakaMats, inventor of the floppy disk, digital watch, and compact disc player and holder of thousands of patents.

12-26 Donald J. Treffinger, Patricia McEwen, and Carol Wittig, *Using Creative Problem Solving in Inventing*, Center for Creative Learning, Honeoye, New York 14471, December 1989. This monograph is an aid for making connections between creative problem solving and inventing.

12-27 W. Vincenti, *What Engineers Know and How They Know It*, Johns Hopkins University Press, Baltimore, 1990. Several examples illustrate the engineering approach to problem solving.

12-28 *The Way Things Work: An Illustrated Encyclopedia of Technology*, 2 volumes, Simon & Schuster, New York. Translated from the original German edition, *Wie funktioniert das?* (1963). With diagrams on one page and a description on the facing page, this book was "designed to give the layperson an understanding of how things work, from the simplest mechanical functions of modern life to the most basic scientific principles and complex industrial processes."

12-29 Robert J. Weber, *Forks, Phonographs, and Hot-Air Balloons: A Field Guide to Inventive Thinking*, Oxford University Press, New York, 1993. The human mind's problem-solving abilities are examined across the span of two million years of inventions. The focus is on the ideas that shape inventions, and it should stimulate the reader's own creativity.

We are also indebted to Arnold Pacey's *Technology in World Civilization: A Thousand-Year History* (Ref. 1-14). This book, more than any other, gave us a better understanding of the connections between cultures, innovation, and technology in the historical context. It introduced us to the concept of survival technology. Also, *American Heritage of Invention and Technology* magazine (Ref. 1-3) has fascinating articles about inventors and the history of technological achievements (and failures). The focus is on the process of invention, not necessarily on the end product.

Other Resources

12-30 *From Dreams to Reality: A Tribute to Minority Inventors* is a video available at no charge from the Patent and Trademark Office or from your nearest Patent Depository Library.

12-31 The National Innovation Workshop is a cooperative effort of federal and regional programs to provide practical guidance and information to innovative businesses, entrepreneurs, and inventors. The workshop is sponsored by the Department of Energy and held periodically at different locations around the country.

12-32 The Minnesota Historical Society has an exhibit, "Her Works Praise Her, Inventions by Women." For further information on women inventors, contact the Society's Traveling Exhibition Department at (612) 297-4497.

12-33 Ramon D. Foltz and Thomas A. Penn, *Understanding Patents and Other Protection for Intellectual Property*, 1990, Penn Institute, Cleveland, Ohio 44141, (800) 426-7495. This summary text is suitable for college students. Even though it provides the legal knowledge they will need as engineers or scientists in industry, the text is written in nonlegal language and is easy to understand. Instructors receive a copy of the industrial version, *Protecting Engineering Ideas and Inventions*, as well as a complimentary set of overhead transparencies.

The U.S. Department of Commerce, Patent and Trademark Office has published *The Inventive Thinking Curriculum Project* and *The Inventive Thinking Resource Directory* under *Project XL—A Quest for Excellence* (Ref. 4-17). The two volumes are excellent resources and guides for teachers who want to encourage their students to learn by inventing. Check with your local school systems to find out whether they are using any of the creative thinking and invention programs listed in the *Directory*, or if they have developed their own invention projects, courses, or extracurricular activities. An additional source of information about programs in area communities is the person in charge of the patent section at your nearest Patent Depository Library.

> *Consider travel as a tool for awareness on how other cultures go about the physical and mental tasks*
> *of solving problems. This type of awareness increases a person's decision making ability*
> *by accounting for the many ways that culture influences different perceptions of the same situation.*
> *Mauricio Gonzalez, Assistant Vice President,*
> *Office of Multicultural Student Development, The University of Toledo.*

Do not overlook people as a resource for information on inventions (or for a possible field trip), especially if you are teaching young or older students. Patent attorneys can speak about inventions and inventors. For example, in Toledo we have met a patent attorney who is not only working with an inventions program in area schools—he also has a daughter who received her first patent at the age of 14 for inventing a notepad that glows in the dark. Your local newspaper may have information about inventors in your community. The engineering and technology faculty in your community colleges and universities will most likely number several inventors and patent holders among its faculty. And last but not least, you may want to contact innovative companies in your community; they may not only have inventors willing to share their stories about patents, they may also have managers eager to share their ideas on how they are encouraging all their employees to think creatively about everything they do. After all, it is within people's minds that all invention ultimately originates.

⌚ *TEAM ACTIVITY 12-4: CULTURE, TECHNOLOGY, AND CREATIVITY*

With a team, brainstorm and then discuss the values in your culture that support creativity and inventiveness. How can these values be strengthened? How do they support the integration of technology with the culture? Focus your discussion on one or more of the following: your family culture, your organizational culture, the culture of your local community (political unit, educational system, etc.), or the culture of your ethnic community.

Appendices

QUALITY FUNCTION DEPLOYMENT (QFD)

WHAT YOU CAN LEARN FROM APPENDIX A:
- *The scope of QFD: customer-driven quality control.*
- *The QFD House of Quality product planning chart.*
- *Further QFD activities: component, process, and production planning.*
- *Two important principles and resource materials for further learning.*

We have seen in Chapter 10 how the Pugh method of creative design concept evaluation can help us develop a superior product. Creative thinking and problem solving are also needed at various stages throughout a company's manufacturing, sales, and service activities. To achieve companywide or total quality control, organizations have devised various procedures. Among these is quality function deployment (QFD). QFD is a very structured team approach to quality control that offers many opportunities for creative thinking and brainstorming.

SCOPE OF QFD

Basically, QFD is a mechanism to ensure that customer needs drive the entire product design and production process in a company, including:

> **Marketing Strategies and Planning**
> **Product Design and Engineering**
> **Process Development and Prototype Evaluation**
> **Production, Sales, and Service**

QFD is an extremely complicated procedure that originated in Japan, where it has been very successful in companies such as Toyota and many of its suppliers. So far, no American company has incorporated the entire process, although some are learning to use parts of it, notably the House of Quality. Also efforts are under way to adapt portions and features of the technique to the American organizational culture and American procedures and approaches to quality control. Workshops of various lengths on QFD are available through the American Supplier Institute in

Dearborn, Michigan; they provide a thorough introduction as well as additional training in this technique of total quality control. Why are Japanese industries sharing this technique with their competitors? One reason is that Japanese companies are increasingly becoming purchasers of U.S.-made components for their American manufacturing plants and products; another reason is that the Japanese have progressed far beyond QFD in their quality control efforts.

QFD is based on the Japanese philosophy of quality, which considers the loss or cost to society if something does not have perfect quality. Efforts that will reduce these costs result in an increase in quality. In engineering terms, hitting the target is quality in Japan, whereas in the United States, we still think of quality as falling within a range or band of tolerances around a specified value. Many of us operate on the principle that a certain number of defects is acceptable. In Japan, no defect is acceptable. How does QFD differ from quality control in U.S. companies? The major difference lies in what drives the company's quality control efforts. Traditionally in the United States, this has been the "voice of the company" (as shown in Table A-1) and involves mostly manufacturing process quality control. In QFD, it is the voice of the customer; it is centered on quality in product development. Quality is designed into the product; it is not something that enters the picture only during the manufacturing phase.

Quality function deployment has four distinct phases: Phase 1—product planning—translates the customer's wants into design requirements through an analysis matrix called the House of Quality. Phase 2—part-deployment or component planning—takes the critically important design requirements down to the level of part characteristics in a scaled-down House of Quality. Phase 3—process planning—identifies key process operations that are related to the important part

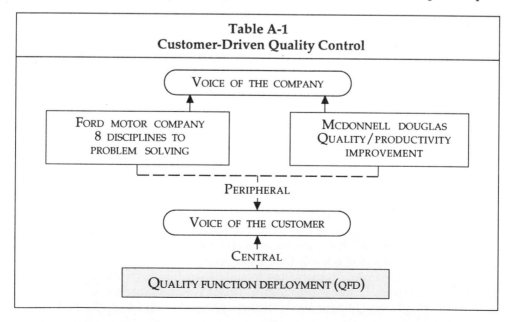

characteristics. Phase 4—production planning—relates the key process operations to production requirements (through the operating and control charts) and results in prototype construction and production start-up. Operational factors such as process monitoring functions and worker training requirements are also identified.

THE FOUR PHASES OF QFD

Phase 1—Product Planning
Phase 2—Part Deployment (Component Planning)
Phase 3—Process Planning
Phase 4—Production Planning

A primary benefit of QFD is its function as a comprehensive database; all important information about a product is collected and stored in one place. This enables new staff to be informed quickly about all aspects of a product, beginning with objectives and moving all the way through production, sales, and service. Restructuring the handling of a product or the entire workings of a company according to QFD requires a tremendous initial effort; however, once this database and the procedure have been established, further development and changes are very easy to make. QFD can identify targets for improvement and areas in which technological development is required; these again are directly linked to consumer needs. In other words, the effort will immediately result in quality improvements that relate directly to the product and the market. QFD disperses all essential information about the product and product quality horizontally as well as vertically throughout the company—there is a common understanding of the quality issues. QFD information flow is shown schematically in Table A-2.

THE QFD HOUSE OF QUALITY PRODUCT PLANNING CHART

Remember the evaluation matrix that was used in the Pugh method exercise? The House of Quality is developed with a similar matrix. A number of steps are involved in building the House of Quality planning chart (and the subsequent QFD charts). Three major activities are associated with each step and identified with the following activities (and thinking mode) code in Tables A-3, A-5, A-6, and A-7:

ACTIVITIES CODE

A = Data Analysis
C = Creative Thinking and Brainstorming
D = Data Collection

We will briefly survey items in the QFD charts that are identified with creative thinking. In Step 1 of the House of Quality (Table A-3), customer data are collected through brainstorming with the sales department and through interviews, questionnaires, and other surveys. Problem definition considers such questions as:

Table A-2 QFD Information Flow (Schematic)

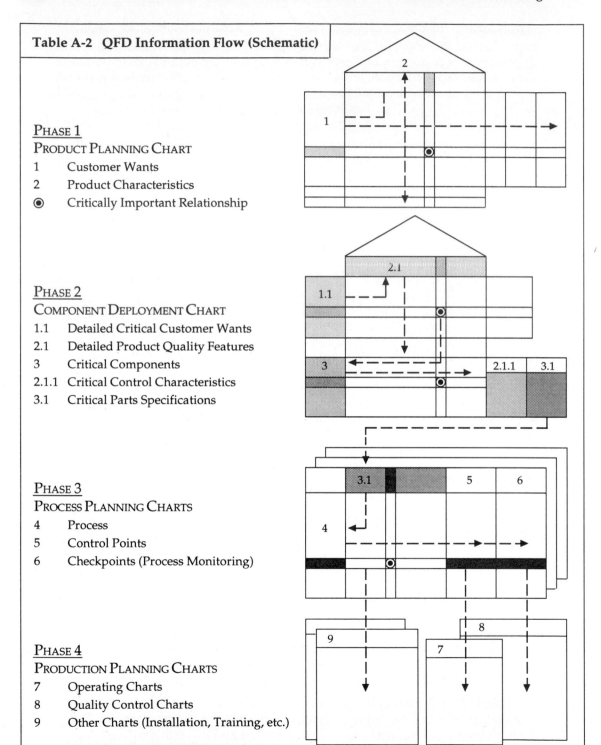

PHASE 1

PRODUCT PLANNING CHART

1   Customer Wants

2   Product Characteristics

◉   Critically Important Relationship

PHASE 2

COMPONENT DEPLOYMENT CHART

1.1   Detailed Critical Customer Wants

2.1   Detailed Product Quality Features

3   Critical Components

2.1.1   Critical Control Characteristics

3.1   Critical Parts Specifications

PHASE 3

PROCESS PLANNING CHARTS

4   Process

5   Control Points

6   Checkpoints (Process Monitoring)

PHASE 4

PRODUCTION PLANNING CHARTS

7   Operating Charts

8   Quality Control Charts

9   Other Charts (Installation, Training, etc.)

Table A-3
Initial Steps in the House of Quality Planning Chart

C,A,D   1. Data on the "voice of the customer" are collected, sorted into tiers of related categories and entered as customer wants in the left-hand column. EXAMPLE: 4—long life, 4.2—strong car, 4.2.1—many years durability.

C,D   2. The customer needs are translated into product characteristics, features, or requirements and entered as headings across the matrix. EXAMPLE: 7—rust prevention, 7.1—initial-stage rust.

A,C   3. The relationship matrix between customer wants and product features is completed. EXAMPLE: There is a strong, direct relationship between vehicle durability and rust prevention.

A,C   4. The crucially important interaction matrix between the product features is completed. EXAMPLE: Preventing dust and rust during manufacturing will result in increased painting surface quality —> positive interaction. Washing the car (to reduce dust) during manufacturing can result in initial-stage rust —> negative interaction.

- Who will be the users of the planned product?
- Who will be the purchasers?
- What is the product expected to accomplish? How should it perform?
- Are there any warranty claims against a similar, "old" product?
- Why do users use this product?
- Why are others not using it?
- What features would turn the nonusers into users?
- How does the product compare to the competition?

The customer's "wants" data are sorted into related categories in a process similar to the one we used during the idea evaluation phase in creative problem solving. The Japanese call this process the affinity diagram or the KJ method. An example for a car door is shown in Table A-4. A complicated product such as a car may have as many as nine levels of categories. What is important here, though, is that none of the customer requirements are either left out inadvertently or ignored on purpose. These groupings, as well as all customer wants, are listed in the left-hand column of the evaluation matrix. In the House of Quality, the first three tiers of customer wants are listed; in the subsequent component characteristics deployment matrix, the relevant customer wants are listed from the third tier on down to the last item in each subcategory to ensure that not a single identified customer want that is related to a critical component is left out.

In Step 2 of Table A-3, the customer wants are translated into product features or characteristics by a staff of mostly engineers, also by using brainstorming. These product characteristics should be expressed in measurable engineering terms. This

Table A-4
Affinity Diagram (Car Door Example)

Doesn't leak when washing car	Easy to service	
Window easy to operate	Quiet when driving	No noise at city driving speed
		Easy to maintain
Provides safety in a collision		Won't rust
Doesn't rattle	Doesn't leak in heavy rain	
Quiet when engine is idling		
Long-lasting operating mechanism	Doesn't drip water or snow when opened	
Difficult to dent	No noise at highway driving speed	etc.

Individual Customer Wants Sorted into Related Categories

Tertiary Grouping	Secondary		Primary
Doesn't leak in heavy rain Doesn't leak when washing car Doesn't drip water when opened	Water leaks		Sealing
No noise at city driving speed No noise at highway driving speed	Wind noise	Noise	
Does not rattle Quiet when driving Quiet when engine is idling	Road noise		
Will not rust Long-lasting operating mechanism Easy to maintain Easy to service	Easy maintenance		Durability
Provides safety in collision Difficult to dent	Strength		
Window easy to operate …			

activity can be quite time-consuming; however, it does get easier with practice. The difficulty here is in identifying which characteristics are the important ones that actually will cover the customer wants or needs. These product features are listed across the top of the House of Quality evaluation matrix.

In Step 3, the matrix is evaluated by the team by indicating whether there is a strong relationship, a weak relationship, or no relationship between each product characteristic and each customer need. If only a few strong relationships are identified, this means that the product is not doing a good job of meeting the customer's needs. Another important task is to identify conflicting requirements. This is done in Step 4 with a second matrix above the product characteristics, making up the "roof" of the House of Quality. Here, each product feature is compared with all the others to see if there is a positive (reinforcing) relationship, no relationship, or a negative (conflicting) relationship. If a conflicting relationship is identified, measures will have to be taken to resolve this conflict. This is an area for creative problem solving because the design must be optimized to meet target values without traditional compromise and trade-off between cost and quality.

When it is impossible to find a satisfactory balance between conflicting requirements for an important product feature, the development or application of new technology is indicated. With creative problem solving, promising ideas for investigation and development can be generated. For example, increasing the thickness of a metal panel could increase the strength of a weak component, but this would cause the hemming process to be more difficult—a negative interaction with present technology. Are there other ways of increasing the strength of the component—other materials or other metal-working processes that could increase panel strength or would make hemming unnecessary in this part? Is this part necessary—could its function be taken over by some other component? Thus these identified negative relationships are starting points for creative thinking that can lead to innovation in a company's efforts to achieve highest quality at lowest cost.

The House of Quality is continued with a number of additional analyses (see Table A-5). In Step 5 (or the "back porch"), the marketing analysis is listed, including a customer importance rating and an analysis of competing products with reference to customer wants. Strong selling points are noted in a separate column (Step 10). In the "basement," competing products are evaluated against the product characteristics. The targets or goals that the planned product has to satisfy are established, and the constraints and specifications that must be met are also indicated. These targets can form the basis of an excellent list of criteria for the Pugh method (Steps 7, 8, 9, 11). The House of Quality analysis is important because it determines the product characteristics that are essential or critical for meeting the customer wants, those that are strong selling points, and those that have difficult targets (Step 12). Only these critically important product characteristics are carried forward for further deployment at the component level. Each model year, four or five items are typically selected for further deployment. But when this procedure is repeated year after year, it becomes a process of continuous improvement that results in quality (and innovation) that is very difficult to beat.

Table A-5
Additional Steps in the House of Quality

D,A,C   5.   The market analysis against customer wants is completed (including warranty data, customer importance rating, and competitive evaluation).
EXAMPLE: Customers have lodged many complaints about rusting, and three competing vehicles have outperformed the company's best product.

D   6.   Engineering (technical) standards in quantitative terms are listed as horizontal "basement" headings.
EXAMPLE: Initial-stage rust prevention, minimum 30 cycles.

D,A   7.   Competitive benchmarks (test data) are listed and evaluated against product features and technical standards (this listing is also known as the product evaluation chart).
EXAMPLE: The company's best product has the worst showing for rust.

D   8.   Regulatory, warranty, liability, and other constraints and control items are listed in the lower left-hand column.
EXAMPLE: Important warranty component: rust-resistant paint job.

A   9.   The relationship (technical) matrix between the constraints and technical standards is completed.
EXAMPLE: There is a strong relationship between initial-stage rust prevention and a rust-resistant paint job.

C,A   10. The key selling points are developed (from a comparison study of market analysis, the product evaluation chart, and the technical matrix) and listed to the right of the "back porch."
EXAMPLE: In this particular model, no selling points related to rust were listed; however, new features were a sunroof and a streamlined back-door damper.

C,A   11. Targets are developed from the market analysis, the product evaluation chart, the technical matrix, and the key selling points; these targets are entered on the product evaluation chart.
EXAMPLE: The rust performance target is set to substantially exceed that of the best competitor.

A,C   12. The critical product features are selected from the importance indications in the relationship matrix, in the market analysis, and in the targets; these critical targets are "flagged" in the bottom line of the House of Quality.
EXAMPLE: Rust prevention is selected as one of the critical consumer wants because of high warranty claims and low performance when compared to competitive benchmarks.

FURTHER QFD ACTIVITIES

The House of Quality is only the initial document in QFD. Additional documents are required to assure that the voice of the customer via the identified critical product characteristics is carried through every step of product and process development and continued into production, marketing, and sales. We will briefly look at a summary and survey of each of these steps.

ADDITIONAL QFD ACTIVITIES

PHASE 2
The component deployment chart is completed.

PHASE 3
For each component,
process planning and quality control charts are developed.

PHASE 4
For each process, control and operating charts, equipment installation
and other instructions, and worker training plans are developed.

QFD Phase 2—Part Deployment or the Component Planning Chart

What parts or components are instrumental for achieving the selected critical product characteristics? Creative design concepts must be developed and evaluated in those important areas identified in the House of Quality. If the first idea that appears to work is taken without developing more (and most likely better options), this will usually result in a rush to experimental hardware and wasted, unfocused effort. As we have seen, the Pugh method is a very useful tool that can take the critical product characteristics developed in the House of Quality and thus creatively generate and evaluate a variety of design concepts until a superior design emerges. This design is then processed further and culminates in the prototype.

QFD Phase 2 resembles a magnifying glass in that the QFD analysis zeroes in on detailed customer needs and the related detailed quality characteristics of the critical targets and product features identified in Phase 1. Another House of Quality analysis is conducted for the selected items, resulting in further "distillation" or clarification of critically important quality characteristics. The subsystems, components, and parts involved in these characteristics are then identified and the final, critical control characteristics are selected and parts specifications developed. Particular attention is focused on those parts that are critical to safety and to the quality performance or functioning of the product. Testing needs for the selected critical areas are established (if necessary), and process capability and R&D requirements are assessed. These steps are listed in more detail in Table A-6. It is interesting to note that the Japanese do not use the Pugh method—thus this is an area where creative thinking can give a competitive advantage.

Table A-6
QFD PHASE 2—Steps in the Component Deployment Chart

D   1. Consumer wants are listed in detail for the critical items selected in the last step of the main House of Quality in the left-hand column of the component's House of Quality.
EXAMPLE: For rust prevention, 53 original customer wants are listed.

D,C   2. More detailed product quality requirements for the critical items above are listed as matrix headings; sometimes, weighting factors are assigned to these.
EXAMPLE: The initial three rust characteristics are expanded to six; resistance against spot rusting is targeted as most important.

A,C   3. The relationship matrix and the interaction matrix are completed.

D,C   4. The market quality evaluation (related to the customer wants) is graphically noted on the "back porch," together with the target values.

A,D,C   5. The competitive evaluation from test data and targets are entered.

A,C   6. The technical difficulty in achieving the important quality requirements is assessed and noted at the bottom of the chart. Also, areas where R&D is needed are pinpointed.

A,C   7. Testing specifications for problem clarification are developed (if applicable) and entered below the technical difficulty assessment.

A,D,C   8. The critical components, subassemblies, and parts related to the important quality characteristics are identified and entered in the left-hand column in the bottom section of the chart.

A   9. The relationship matrix between the critical components and the important quality characteristics is completed. Only those components or parts that show critical relationships are deployed further. In particular, critical safety parts and critical function parts are identified. Value engineering analyses, FMEAs, and cost analyses are performed as appropriate on parts that have high warranty claims, high potential for cost reduction, and high variance.

A,D,C   10. Detailed final quality (or control) characteristics are entered for the critical parts.

C,A,D   11. Parts specifications are entered for each critical part (including numerical values and sketches), addressing the important quality requirements identified in the matrix. This activity especially demands creative problem-solving skills from the group because here the group specifies "how" the problem is to be solved in order to meet the customer needs and quality requirements.

QFD Phase 3—Process Planning and QFD Phase 4—Production Planning

Process planning and production planning activities occur concurrently with product planning and part deployment (product design). The process planning and quality control charts that are developed in Phase 3 identify critical product and process parameters for each critical part, together with their control points and checkpoints. This information is used to set up and prove out the production lines. Optimum processes and sequencing can be developed with brainstorming. Process planning is usually done concurrently with product evaluation, and creative ideas need to be exchanged between the teams working on these two activities. The steps needed to develop the process planning charts are described in Table A-7.

Table A-7
Steps for the Production and Process Planning Charts

PROCESS PLANNING CHARTS (QFD PHASE 3)

D   1. The final critical quality (or control) characteristics for each critical part are entered from the component deployment chart as headings in a new matrix. Each part has a separate chart.

A,C,D   2. The processes needed to accomplish the control characteristics in the matrix heading are listed in the left-hand column.

A   3. The relationship matrix is completed.

A,C,D   4. The control points are identified for each process and control characteristic.

A,D,C   5. The process checkpoints are identified from the results of the matrix.

A,D,C   6. Specifications for process monitoring are established for the process checkpoints (method, schedule, etc.).

QUALITY CONTROL CHARTS (QFD PHASE 4)

D   7. The control points are listed (in the second column) from the process planning chart. For each process, a separate quality control chart is developed.

A,C,D   8. The process flow schematic diagram is developed and sketched in the left-hand column of the quality control chart.

A,D   9. The control method, sample size/frequency, and check method are determined and indicated for each control point.

ADDITIONAL CHARTS

A,D,C   10. Operating charts, equipment installation charts, personnel training charts, etc., are developed as needed with information from the quality control charts, the process planning charts, and the component deployment chart to complete the QFD procedure.

In QFD Phase 4, the control and operating charts for each process identified in Phase 3 are developed; these charts specify operations and checks to be done by plant personnel to assure that all important customer needs are achieved, including SPC activities to monitor and solve problems of product variance. The necessary plans for equipment installation and personnel training are also developed. These large QFD charts, including the House of Quality, are posted in the factories where the shop floor workers study and understand them. They can see the direct relationship between their job and achieving the quality that the customer wants and buys. Production planning—although still conducted as a team activity—is primarily an analytical process supplemented with creative thinking to avoid arbitrary specifications. Table A-7 summarizes the steps involved in the Phase 4 charts.

SUMMARY AND RESOURCES

From this brief overview of quality function deployment, remember these:

TWO KEY PRINCIPLES

1. *The context is never irrelevant. When we develop a product or a service, the customer's needs and wishes contribute as much to the context as do cost, production or implementation, innovation, and environmental effects. To do the job well, we cannot work in isolation when we develop our plans.*

2. *The design or planning and development phase of a product or service is extremely important to quality and offers many opportunities for the productive use of team creative problem solving, even if we are not using QFD.*

The following resource material on QFD is available from the American Supplier Institute, Dearborn, Michigan, (313) 336-8877:

A-1   William E. Eureka and Nancy E. Ryan, *The Customer-Driven Company: Managerial Perspectives on QFD*, second edition, ASI Press, 1994.

A-2   *QFD DESIGNER*, Interactive Software by Qualisoft, Version 2.1 for windows—IBM-PC and compatibles (an Apple version is under development)—requires DOS 3.0 or higher. A demonstration package can be ordered from ASI.

A-3   Nancy E. Ryan, ed., *Taguchi Methods and QFD: Hows and Whys for Management*, ASI Press, 1988.

The Proceedings of the Sixth Symposium on Quality Function Deployment are a good source of information on QFD applications in various manufacturing and service industries. Selected volumes of earlier symposia are still available. Also, ASI periodically conducts QFD workshops, and the course materials are available from the ASI Press, Dearborn, Michigan.

FAILURE MODE AND EFFECTS ANALYSIS (FMEA)

WHAT YOU CAN LEARN FROM APPENDIX B:
- *Definition and objectives of the FMEA.*
- *Description of steps; occurrence, severity, and detection ranking criteria.*
- *Example.*

This material has been summarized from two Ford Motor Company Documents: (1) Reliability Methods: Failure Mode and Effects Analysis—Module XIV (January 1972) and (2) Potential Failure Mode and Effects Analysis for Manufacturing and Assembly Processes (Process FMEA) Instruction Manual (December 1983).

DEFINITION AND OBJECTIVES

A failure mode and effects analysis (FMEA) is a systematic, analytical technique performed by an experienced design, manufacturing, or quality control engineer on a planned product, manufacturing process, or quality control system to assure that the product characteristics will meet customer needs. The FMEA allows the engineer to assess the probability as well as the effect of a failure; it identifies significant process variables to be controlled. The FMEA should be performed as early as possible in the engineering program and be updated as later information becomes available. By identifying potential problem areas, an early FMEA will aid engineers in directing timely design actions to prevent defects and in planning appropriate test programs. The FMEA also documents the rationale for the manufacturing or assembly process being developed and thus represents a valuable database to track future actions and improvements and to train new personnel.

The main objectives of an FMEA are:
- Identification of potential and known failure modes.
- Identification of the causes and effects of each failure mode.
- Ranking in priority of the identified failure modes according to frequency of occurrence, severity, and detection.
- Specific directions for problem follow-up and corrective action.

DESCRIPTION OF STEPS

Each organization can devise its own FMEA worksheet format. In general, the following sequence of steps is followed in recording the data of the analysis.

a. General identification: These items are usually entered in the space above the worksheet headings and include:

1. Subsystem or process name and numerical designation.
2. Model year, lines that will utilize the product or process.
3. Division or office that has the design responsibility.
4. Other departments involved, including manufacturing and design.
5. Outside suppliers of a major component within the subsystem.
6. Engineers: name and phone number of (a) subsystem, design, or process engineer responsible for the FMEA, (b) system engineering supervisor.
7. Scheduled date for release of complete subsystem.
8. Date of first FMEA and dates of later revisions.

The following items are entered as separate columns in the worksheet:

b. Part or process identification: Specify the assembly, process, or component being analyzed exactly as shown on the design drawings and indicate design level suffixes and change letters or any other identifying information.

c. Part or process function: Concisely describe the function of the part or process being analyzed, to help identify the severity of the consequence of a failure.

d. Failure mode: Describe each possible failure mode by considering the question, "What could possibly go wrong with this part, system, or process?" No judgment is to be made on whether or not it *will* fail, only on how it *could* fail. As a starting point, review past design FMEAs or quality, warranty, durability, and reliability problems on comparable components. Typical failure modes are:

Bent	Cracked	Leaking	Open circuited
Blistered	Damaged	Loose	Porous
Bound	Deformed	Melted	Rough
Brittle	Discolored	Misaligned	Shorted
Broken	Distorted	Misassembled	Tight
Corroded	Grounded	Omitted	Wrinkled

e. Effect of failure: Describe, "What does the customer experience as a result of the failure mode just listed?" Examples are:

Designs: Excessive operating effort. Engine will not start. Fuel fumes. Insufficient A/C cooling. Luggage compartment water leaks. Oil leakage.
Processes: Stops the line. Generates loud noise. Damages parts. Impairs safety.

In the case of a process FMEA, control items are flagged with a special symbol, such as ◆.

f. Causes of failure: Analyze what conditions can bring about the failure mode; list all potential causes for each failure mode. For example, would poor wire insulation cause a short? Would a sharp sheet metal edge cut through the insulation and cause the short? Other examples of causes might be:

Broken wire	Inadequate venting
Damaged part	Incorrect speeds, feeds
Handling damage	Material failure
Heat treat shrinkage	Missing operation
Improper surface preparation	No lubrication
Inaccurate gauging	Out-of-tolerance
Inadequate clamping	Packaging damage
Inadequate control system	Worn tooling

g. Current controls: For a process FMEA only, list all current controls of process variables that are intended to prevent the causes of failure from occurring or to detect the causes of failure or the resultant failure mode.

h. Estimate of occurrence frequency: Estimate the probability that the given failure mode will occur, using an evaluation scale of 1 to 10, with the low number indicating a low probability of occurrence and a 10 indicating near certainty of occurrence as shown in Table B-1. Probability in Table B-1 means the statistical proportion outside the specification limits. Each organization can develop its own ranking tables; the only requirement is that the tables will be used consistently throughout the company's FMEAs.

i. Severity: Evaluate the severity or estimated consequence of the failure on a scale of 1 to 10, with the low number indicating a minor nuisance and a 10 indicating a severe, total failure (as shown in Table B-2).

j. Failure detection: Estimate the probability that the problem will be detected before it reaches the customer. A low number indicates high probability that the failure would be detected; a 10 means that it would be very difficult to detect the failure before the product is shipped or sold (see Table B-3). For example, an electrical connection left open and thus preventing engine start might be assigned a detection of 1; a loose connection causing intermittent no-start might be a 6, whereas a connection that corrodes causing no-start after a period of time might be assigned a 10.

k. Calculation of risk priority number (RPN): When the numbers of (h), (i), and (j) are multiplied, the RPN index is obtained, providing an indication of the relative priority of the failure mode. Regardless of the RPN value, components or systems receiving a high occurrence ranking should be given special attention, and corrective action should be taken to reduce the ranking.

l. Recommended corrective action: The follow-up aspect is critical to the success of the FMEA. Responsible parties and timing for completion should be designated for all corrective actions. Give a brief description of the corrective action or actions recommended to prevent the failure mode, including the person responsible for implementing the solution and the status of the corrective action (transmittal numbers, promise date, etc.).

Based on the analysis, actions can be specified for the following:

- To reduce the probability of **occurrence**, process or design revisions are required. An action-oriented study of a process via SPC should be implemented with an ongoing feedback of information for never-ending improvement and defect prevention.

- To reduce the **severity** of product failure modes, part design actions are required. As a means of highlighting significant process effects, the severity ranking may be increased, depending on the effect of the failure on subsequent process operations in the plant.

- To increase the probability of **detection**, process revisions are required. Generally, improving detection controls is costly and ineffective for quality improvement since inspection is not a positive corrective action. In some cases, a design change to a specific part may be required to assist in detection. Changes to the current control system may be implemented to increase detection probability. However, emphasis must be placed on preventing defects (reducing the occurrence) rather than on detecting them.

Table B-1
Occurrence Ranking Criteria

CRITERIA	RANKING	PROBABILITY
Remote probability of occurrence. Process capability shows at least $x \pm 4\sigma$ within specs.	1	1/10,000
Low probability of occurrence, with process in statistical control. Capability shows at least $x \pm 3\sigma$ within specifications.	2 3 4 5	1/5,000 1/2,000 1/1,000 1/500
Moderate probability of occurrence, for processes experiencing occasional failures. Process is in statistical control, with $x \pm 2.5\sigma$ within specs.	6	1/200
High probability of occurrence, with frequent failures. Process is in statistical control, but capability shows $x \pm 2.5\sigma$ or less within specs.	7 8	1/100 1/50
Very high probability of occurrence. Failure is almost certain to occur sooner or later.	9 10	1/20 1–1/10

Table B-2
Severity Ranking Criteria

CRITERIA	RANKING
Minor nature of failure, no noticeable effect on performance, undetectable by customer.	1
Low severity, causing only slight customer annoyance due to very minor subsystem performance degradation.	2 – 3
Moderate failure causing some customer discomfort, dissatisfaction, and annoyance due to subsystem or total performance degradation.	4 – 6
High degree of customer dissatisfaction due to nature of the failure (inoperable subsystem or total system).	7 – 8
Very high severity ranking for failure mode involving potential safety problems and/or conformance to federal regulations. Nonregulated components with a 9 or 10 severity ranking and occurrence rankings > 1 should be designated as control items (♦).	9 – 10

Table B-3
Detection Ranking Criteria

CRITERIA	RANKING	PROBABILITY
Remote likelihood that product would be shipped containing such an obvious defect, since it is detected by subsequent factory operations.	1	1/10,000
Low likelihood for shipment with defect which is visually obvious or has 100% automatic checking.	2	1/5,000
	3	1/2,000
	4	1/1,000
	5	1/500
Moderate likelihood for shipment with defect, since the defect is easily identifiable through automatic inspection or functional checking.	6	1/200
	7	1/100
	8	1/50
High likelihood of shipping with subtle defect.	9	1/20
Very high likelihood that defect will not be detected prior to shipping or sale (checks are impossible or defect is latent).	10	1 – 1/10

 m. Actions taken: The need for the best possible FMEA has been reinforced by studies of safety and emission recalls. These studies found that in a number of cases a fully implemented FMEA program with follow-up on critical concerns would have made the recall unnecessary. An effective FMEA is a living document and should always reflect the latest process actions and product designs. Therefore, a section on actions taken is added to the worksheet periodically, together with a new listing of occurrence, severity, and detection and the revised PRN calculation, and with identification of the individual responsible for the action.

EXAMPLE

Part/Process Name and Number:
      Manual lever assembly, #xxx-xx.

Part/Process Function:
      Transmit manual selector motion from external linkage to manual valve
      and park linkage.

Failure Mode 1: Plastic lever breaks.
Potential Effect of Failure:
      No drive. Locked in park.

Possible Cause(s) of Failure:
      Overload on lever when disengaging park on grade. Inferior plastic
      material. Material brittle when cold. Material damaged in handling.

Existing Current Controls:
      None.

Rankings:
      Occurrence Ranking = 3.
      Severity Ranking = 10.
      Detection Ranking = 9.
      Risk Priority Number = 270.

Recommended Corrective Action:
      Redesign lever with thicker material and strengthen ribs to carry 100%
      overload.

Actions Taken (specify activity, personnel responsible, and date of completion):
      In progress.

Failure Mode 2: Wear at hole in lever.
… … …
Failure Mode 3: Loose fit at shaft & lever serration.
… … …
etc.

FAULT TREE ANALYSIS (FTA)

WHAT YOU CAN LEARN FROM APPENDIX C:
- *Definition and objectives of FTA; logic gate symbols; events symbols.*
- *Description of steps to construct a fault tree.*
- *Example.*

This material has been condensed from lecture notes prepared by Dr. K. C. Kapur of the American Supplier Institute, Dearborn, Michigan.

DEFINITION AND OBJECTIVES

Fault tree analysis (FTA) is a method of system reliability/safety analysis. The fault tree analysis provides an objective basis for analyzing system design, justifying system changes, performing trade-off studies, analyzing common failure modes, and demonstrating compliance with safety requirements. The concept of fault tree analysis was originated in 1961 by H. A. Watson of Bell Telephone Laboratories to evaluate the safety of the Minuteman launch control system. Fault tree analysis is a deductive analysis that requires considerable information about the system; it is a graphical representation of Boolean logic associated with the development of a particular system failure.

Fault tree analysis accomplishes the following:

- It provides options for qualitative and quantitative reliability analysis.
- It helps analysts to understand system failures deductively.
- It points out the aspects of a system that are important with respect to the failure of interest.
- It provides insight into system behavior.
- It is restricted to the consideration of one undesirable event.
- It graphically communicates the sequence of causes leading to the undesirable event.
- It directs the analysis toward elimination of the undesirable event.
- It is concerned with ensuring that all critical aspects of a system are identified and controlled.

Definitions:

• Fault tree—model that graphically and logically represents combinations of possible normal and fault events leading to the top event.
• Top event—system failure under investigation.
• Primary events—basic failures (causes) leading to the top event.
• Fault event—abnormal system state.
• Normal event—event that is expected to occur.
• Event—dynamic change of state occurring in a system element.
• System element—hardware, software, human, and environmental factors.
• Logic gate and event symbols—defined in Tables C-1 and C-2, respectively.

Table C-1		
Definition of Logic Gate Symbols		
SYMBOL	GATE NAME	CAUSAL RELATIONSHIP
	"And" gate	Output event occurs if all input events occur simultaneously.
	"Or" gate	Output event occurs if any one of the input events occur.
	Inhibit gate	Input produces output only when a conditional event occurs.
	Priority gate	Output event occurs if input events occur in order from left to right.
	Exclusive gate	Output event occurs if one (but not both) of the input events occurs.
	Sample gate	Output event occurs if m out of n input events occur.
• Gate symbols connect events according to their causal relations.		
• A gate may have one or more input events but only one output event.		

Table C-2 **Definition of Event Symbols**	

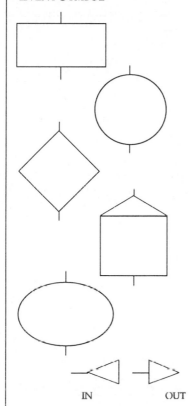

EVENT SYMBOL

MEANING

The rectangle defines an event that is the output of a logic gate; it is dependent on the type of gate and the inputs to the gate.

The circle defines a basic inherent failure of a system when operated under specified conditions —> primary failure. Sufficient data.

The diamond represents a failure other than a primary failure that is purposely not developed further. Insufficient data.

The switch or house represents an event that is expected to occur ("on") or to never occur ("off") because of design and normal operating conditions. It is thus used to examine special cases by forcing some events to occur and other events not to occur.

The oval represents a conditional event used with the inhibit gate.

Triangles (arrowheads) are transfer symbols to and from the fault tree to simplify the representation.

IN   OUT

- The fault tree sequence leads down from the system failure via a sequence of events to basic causes.
- Elimination of the causes eliminates the top event.

DESCRIPTION OF STEPS

Whereas FMEAs consider all possible failure modes for a product or process, fault tree analysis is restricted to the identification of the system elements and events that lead to a single, particular system failure. The steps for fault tree analysis are: (1) Define the top event. (2) Establish boundaries. (3) Understand the system. (4) Construct the fault tree. (5) Analyze the fault tree. (6) Recommend and take corrective action.

Fault Tree Construction: The fault tree is structured in such a way that the sequence of events that leads to the failure is shown below the top event; these events are logically related to the failure by the logic gates. The input events to each logic gate are shown as rectangles; they are usually the output of lower-level logic gates. These events are developed to lower levels still until the sequence of events leads to basic causes of interest (represented in circles). Diamonds at the lower edge of the fault tree indicate the limit of resolution of the analysis.

Three types of causes can contribute to a failure: (1) A primary failure can be due to the internal characteristics of the system under consideration. (2) A secondary failure can be due to excessive environmental or operational stresses. (3) A command fault can result from the inadvertent operation or nonoperation of a system element due to failure of elements that can control or limit the flow of energy to respond to system conditions.

EXAMPLE OF FAULT TREE ANALYSIS

Table C-3 shows a diagram for the operation of an electric motor. The system boundary conditions are:

Top event = Motor overheats.
Not-allowed events = Failures due to effects external to the system.
Existing events = Switch closed.

Table C-4 shows by inductive reasoning that the motor overheats if an electrical overload is supplied to the motor or if a primary failure within the motor causes the overheating (through bearings losing their lubrication or through a short within the motor). From a knowledge of the components, the fault tree construction is continued to two other causes, which in turn are caused by primary failures (in circles).

Table C-5 shows a fault tree for the same system given in Table C-3, but for different boundary conditions:

Top event = Motor does not operate.
Initial condition = Switch closed.
Not-allowed events = Failures due to effects external to the system.
Existing events = None.

Here the diamond symbol has been chosen to show that the "open switch" is a failure external to the system boundaries and that insufficient data are available to develop the event further. Secondary fuse failure can occur if an overload in the circuit occurs, because an overload can cause the fuse to open. However, the fuse does not open every time an overload is present in the circuit because not all conditions of an overload result in sufficient overcurrent to open the fuse. Thus the oval off the diamond inhibit gate gives the condition that would cause the secondary fuse failure. Even though the development and analysis of the fault tree are nominally separate tasks, there is much interaction between the two activities. During the course of analysis, additions and changes are made to the tree as new insight is gained into the failure paths. The analysis must be followed by recommendations for action that will lead to the elimination of the main event failure.

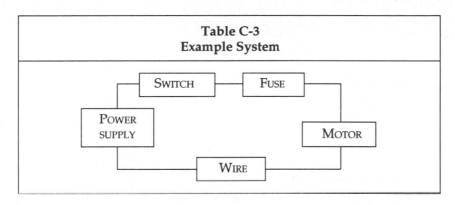

Table C-3
Example System

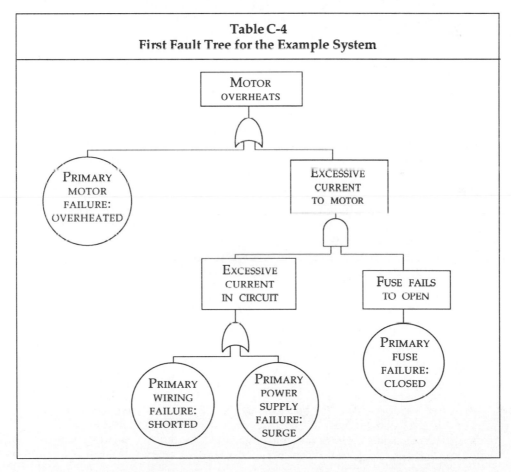

Table C-4
First Fault Tree for the Example System

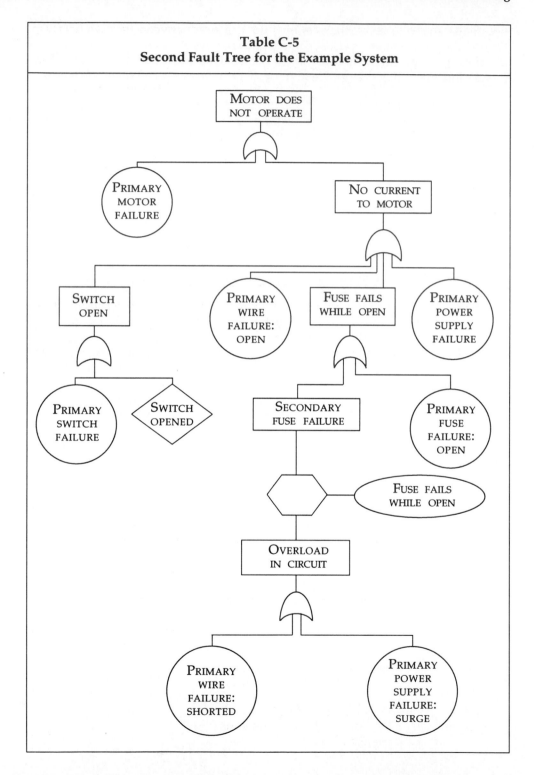

Table C-5
Second Fault Tree for the Example System

Appendix D

TAGUCHI METHODS

WHAT YOU CAN LEARN FROM APPENDIX D:
- What is quality and companywide quality control?
- Overview of the Taguchi loss function.
- Introduction to Taguchi design of experiments; references.

Quality and Companywide Quality Control

Quality, until recently, was perceived very differently by Japanese and U.S. manufacturers. In Japan, producing a quality product has been the concern of all employees from the boardroom to the factory floor; in the United States, quality was supposed to be the concern of the "quality control department." In Japan, the emphasis has been on improving what the customers like; in the United States, quality efforts were aimed at fixing what people didn't like. In Japan, quality is minimizing the cost to society; in the United States it has been mostly conformance to specifications. Table D-1 compares the two approaches; these are essentially successive stages of implementing quality into an entire organization, from the bottom (Stage 1) up. But even as we are writing this book, things are changing, and American manufacturing companies are progressing to the higher levels.

Table D-1 Total Quality Control		
J A P A N	Stage 7:	QFD defines the "voice of the customer" in production terms —> consumer-oriented.
	Stage 6:	The Taguchi loss function links quality to cost —> cost-oriented.
	Stage 5:	Optimized product and process designs result in robust products —> society-oriented.
	Stage 4:	All employees are continuously educated and trained —> people-oriented.
U. S.	Stage 3:	Quality assurance involves all departments —> systems-oriented.
	Stage 2:	Quality assurance concentrates on production —> process-oriented.
	Stage 1:	Inspection and problem analysis are used —> product-oriented.

Companywide quality control is seen as encompassing the quality of management, the quality of the work performed, the quality of the employees, and the quality of the work environment above and beyond the quality of the product or service. In the past, American-style total quality control was limited to the product or service. Japanese managers recognized that it was necessary to change the system if quality was to be improved; defect-correction in itself would not lead to better quality. Japanese companies have been doing their major problem solving earlier in the product development process than many U.S. companies, to achieve a robust product and eliminate defects. Defects are costly—Japanese manufacturers have estimated that the cost of customer dissatisfaction and loss of market share can be as high as four times the cost of warranty claims and repairs. (Warranty costs for the three big U.S. automakers approximated a staggering 8 billion dollars in 1993.) Japanese companies also consider continuous education of their workforce essential because education changes people's thinking. In the late 1950s, Dr. Genichi Taguchi quantified a philosophy and a method for companywide quality control. The loss function used in Stage 6 (see Table D-1) will be described first, followed by the Taguchi method of design of experiments employed in Stage 5.

The Taguchi Loss Function

Quality loss is defined as the financial loss to society after a product has been manufactured; it is related to quantifiable product characteristics and expressed in monetary units. Minimizing the loss to society will almost always significantly cut manufacturing costs. The loss function redefines how we think about quality; it allows us to justify essential quality improvements that do not meet the traditional return on investments and paybacks demanded by finance-oriented management.

The Taguchi loss function is $\quad L(y) = K(y - m^2)$

where L is the loss to society caused by a deviation from a nominal target value m of a product quality characteristic y, and K is a constant established by each company. The value of K is a composite cost made up of all internal costs as well as warranty and field costs; the cost to customers and the cost to society are added as the company gains experience with this method. It is interesting to note that U.S. manufacturers often recognize and thus exclude "the cost to society" in their warranties—they may repair or replace the defective product, but they disclaim any responsibility for the damage that the product caused because of the defect.

The loss function enables us to quantify the benefits of reducing variability in a product. Keeping variability to a minimum by manufacturing products as close to set target values as possible is central to the Japanese concept of quality. This is in contrast to manufacturing parts within specification ranges. These specification limits are irrelevant because the cost to society increases with increasing deviation from the target value, whether the deviation is within specification or not. Also, the loss function can identify priority items for quality improvement—those areas causing the greatest loss need attention ahead of those merely registering large variations (demonstrated by a low process capability factor). Process variation around a target is measured using the C_{pk} index. A C_{pk} of 1.00 means that the process average $\pm 3\,\sigma$ falls just within the specification limits. In Japan, minimum acceptable performance has a C_{pk} of 1.33 (8 sigma); for important quality characteristics, this increases to a C_{pk} of 1.66 (10 sigma). But frequently, values exceeding 3.00, 5.00, and even 8.00 are encountered. Traditional managers are satisfied when they find parts meeting specifications with a $C_{pk} = 1.00$ and thus would not spend money to improve a process that exceeds this performance, not realizing that inspection, scrap, and rework will be eliminated for a process that shows a high process capability index. The index is used for monitoring continuous improvement over periods of time—from month to month and year to year; it thus provides a means to compare performance between groups of workers and different plants.

The determination of the target specifications for the product characteristics is called *tolerancing*. It must be done during product development for all critical points in the production process. It may also be required for improving an existing product, and suppliers need to be educated in this technique. Tolerancing is based on the loss function equation; the driving force in setting each specification is cost. The procedure is widely used in Japan.

Taguchi Design of Experiments

Factors that cause variability in product functions are called *noise factors* (or error factors). External noise factors are variables in the environment, such as humidity, temperature, dust, and differences in people. For example, a product should perform just as well for a left-handed person as for a right-handed person; it should function the same for a child as for a strong, hefty adult. Internal noise can be the deterioration a product suffers over time or the variation from one unit to the next due to uncontrollable variations in the raw materials. Signal factors (or control factors) are parameters that can be arbitrarily changed. Those signal factors that have large linear effects should be chosen for investigation in the design of experiments. The primary objective of the Taguchi methods is to reduce variability around a target value to achieve a robust product—one that performs well under many different conditions. In tolerance design, decisions are made about the amount of variability that is acceptable (through loss function analysis); the variability caused by noise is given by the standard deviation. When more than two factors are involved, orthogonal arrays are used for tolerance design. An analysis of variance (Anova table) is performed with the results of the orthogonal array.

Companies use off-line quality control during the development of products as well as on-line quality control during production to reduce variability. On-line methods are those used during the manufacturing process and include statistical process control (SPC) to maintain the target values. Off-line approaches include: (a) system design and process design focused on technology; (b) parameter design for optimization—by reducing variance as well as cost with design of experiments for the product and the processes, and (c) tolerance design for control of noise factors and determination of precise production processes—also with design of experiments. Keeping tolerances within narrow limits is usually expensive; thus it is important to first minimize the influence of noise factors through parameter design. Process design follows product design, with all specifications developed during product design. The sequence of system design, parameter design, and tolerance design for product and processes results in a robust product. Product design can solve both design problems and production quality problems.

The Taguchi method of design of experiments is a tool for investigating the effect of a large number of variables, interactions, and levels of variables in the industrial environment to determine optimum specifications and tolerances. In the United States, when engineers diagnose an unwanted, negative effect, they search for a specific cause and then try to eliminate it. Frequently, however, effects are not simply the result of one specific cause; they are caused by the interaction of many factors, thus making identification and cure extremely difficult. Taguchi's philosophy is not to try to remove the cause, but to eliminate or dampen the effect. In practical terms, this means that products will be robust against variation for key parameters. The influence of key parameters can be determined statistically; it is not necessary to build and test many prototypes. Data from a few crucial experiments and simulations will be sufficient to determine the optimum levels of the control factors. Brainstorming is used, especially for identifying possible causes of variation; it also helps a team identify and select the most important noise factors and control parameters for the experimental investigation and design of experiments analysis. The Taguchi method of design of experiments improves the quality of the product, significantly cuts down the number of experiments required, reduces development and manufacturing costs, and reduces the time to market.

In the Taguchi design of experiments, several statistical tools are used. Orthogonal arrays allow many design parameters and noise factors to be evaluated simultaneously. In traditional experiments, all factors except one would be strictly controlled (which of course is an unrealistic real-life situation). With the orthogonal array, interaction effects can be taken into account. Two or three different values or levels are assigned to these parameters. With the resulting data, an analysis of variance on the signal-to-noise ratios is conducted. Linear graphs are plotted that identify the most sensitive parameters and the most appropriate settings or specifications to achieve optimum performance. The signal-to-noise ratio measures the sensitivity of an effect to changes in specific control factors. This sensitivity is expressed as a change in the mean as well as a change in the standard deviation (or variability) of a characteristic. Both of these results are important for achieving optimum performance. The objective of these studies is to select levels for the parameters that will minimize variability—the signal-to-noise ratio will be high; sensitivity to variations in the noise factors will be low. Detailed examples are given in these references published by ASI Press, Dearborn, Michigan:

D-1   Lance A. Ealey, *Quality by Design—Taguchi Methods and U.S. Industry,* 1988.

D-2   Genichi Taguchi, *Introduction to Quality Engineering—Designing Quality into Products and Processes,* Asian Productivity Organization, 1989.

D-3   Genichi Taguchi and Yuin Wu, eds., *Taguchi Methods™ (Case Studies from the U.S. and Europe).* The American Supplier Institute (ASI) in Dearborn, Michigan, conducts workshops on Taguchi methods and holds a yearly symposium with proceedings containing many case studies.

Appendix E

VALUE ENGINEERING (VA/VE)

> WHAT YOU CAN LEARN FROM APPENDIX E:
> • Objectives of value engineering.
> • Information phase: data collection, function identification, cost analysis, and user survey.
> • Additional phases: speculation, evaluation/analysis, and implementation.

Objectives of Value Engineering

Like quality function deployment, value engineering (or value analysis)—also known by the initials VA/VE—is a special application of creative problem solving. The basic steps of creative problem solving are incorporated into a specific format and procedure. The purpose of this technique is to create functionally equivalent or improved product designs or services at reduced cost, resulting in a more acceptable, useful product or a more efficient plan or policy when applied to social issue problems. Value engineering employs one or more teams with five to seven people each. Mixed teams of people from different departments or backgrounds have been found to be most successful, after the initial communications barrier has been overcome. A trained leader (if possible a certified value specialist) is essential.

Lawrence D. Miles developed value engineering in 1949 at General Electric. The main focus is to identify the functions of the targeted product or service. Alternative ways of fulfilling each function are brainstormed, and cost reduction is achieved through critical evaluation to find the best (and least expensive) way of meeting those functions. Because the goal is to provide the desired functions at low cost while maintaining or increasing user acceptance, the form of the target is not taken as fixed, whereas in traditional cost reduction programs, the aim is merely on using less expensive parts without changing the basic configuration of the product.

> *Value = (Function + High User Acceptance) / Cost*

A product or service gains value when it performs its basic and supporting functions better than competing products or services. Poor value is designed into a product when these functions are not analyzed, either by following tradition or wrong assumptions or through lack of information or creative thinking. Value engineering has been successful in architectural, chemical, product, and civil engineering, as well as in organizations providing social services. Value engineering is a strategy—a structured plan or sequential process—that consists of four phases: information, speculation, evaluation and analysis, and implementation.

471

The Information Phase

In creative problem solving, the emphasis has been on the creative thinking activities—idea generation and creative idea evaluation. In value engineering, the most crucial step and the hardest work lies at the beginning, in problem definition, or as it is called, in the information phase. The information phase has four distinct tasks which require different thinking skills: data collection, function identification, cost analysis, and user reaction surveys.

Task 1—Data collection: This task focuses on finding accurate user data about the product or service. Here market surveys on consumer satisfaction and attitudes, user expectations and experiences, customer wants and needs, as well as competitive benchmarking are very important for obtaining complete information. Three distinct groups of users must be consulted: (a) consumers, customers, or clients who have experience with the product, (b) management or administrators who are involved in supervising the production or provision of the product, and (c) people who are directly involved in the manufacture or execution of the product. In the case of a kitchen appliance, these people are: (a) cooks, (b) management in the factory (including sales), and (c) the supervisors and workers on the factory floor. In the case of a school system, the customers are: (a) the taxpayers, the citizens of the community, and the future employers, in addition to the students, (b) the school administrators, and (c) the teachers. These groups typically provide very differing perspectives on needs for acceptance of a product.

Task 2—Function identification: Function identification is a difficult and time-consuming task. The key operative is:

> *All product functions are expressed in two-word (noun-verb) phrases.*

Everything the product does or is expected to do has to be expressed in these noun-verb phrases. First, the primary task must be identified; it is the reason for all other basic and supporting functions. For example, the primary task of a school system is to *educate children;* the primary task of a lawn mower is to *cut grass;* the primary task of an engine is to *supply power.* Next, basic and supporting functions must be identified. This activity is surprisingly difficult and takes a lot of brainstorming by the teams. Basic functions are critical to the performance of the primary task. Supporting functions are subordinate and supplemental to the basic functions. In many products, these are "perks" added to increase buyer appeal, but they can add as much as 50 percent to the cost of the product. Basic and supportive functions are then further broken down into primary, secondary, and tertiary groupings. This process is the reverse of the idea grouping activity done for engineering ideas after the first round of brainstorming in creative problem solving.

A large chart is prepared that shows how the different functions are connected. In a vertical arrangement, this is called a function tree. In a horizontal arrangement, it is a FAST (function analysis system technique) diagram. Each phrase moving out toward the branches of the function tree will answer "how" the function directly below it is to be performed; each phrase moving down toward the trunk answers "why." This chart is worked out in a team activity, with each phrase written on a small card; these cards are moved around and new phrases are added until the team is satisfied that all the functions of the product have been identified and organized in their correct relationship on the chart. For example, a basic function in a school system was identified as *develop individuals*. Seven secondary functions were identified on how this could be accomplished; one of these was *encourage individuality*. In turn, the tertiary functions here were *develop uniqueness, stimulate curiosity, foster creativity*, and *develop values*.

Task 3—Cost analysis: Because this task is very detailed and analytical (requiring quadrant A and B thinking), a creative group may find it tedious and boring. Various strategies and costing procedures are available to give a valid allocation of costs to each item in the function tree. If the cost analysis is based on incomplete information, the results will not be reliable and the outcome of the value analysis will be less meaningful. Material, labor, design, and overhead costs must be included. Other factors may be involved, and VA/VE specialists employ cost allocation equations.

Task 4—User reaction surveys: When the function tree has been completed, the team must determine which of the functions are most important to product performance, consumer acceptance, etc. Diagnostic tools here are consumer surveys or interviews. Feedback should be obtained from all three user groups. Functions that are relatively high in cost but are judged unimportant by users are prime candidates for cost reduction. This identification of high, low, and indifferent acceptance of function categories is essential to the next phase.

Additional Phases

Phase 2—Speculation or search: Three additional phases are required to complete value engineering. Creative thinking and brainstorming are used to generate ideas during the second phase, as these questions are answered for each function: What else could do the job? In what other ways can each function be performed? Deferred judgment is important; the group members are given bells or whistles to be used when a group member communicates an evaluation or criticism. All ideas are recorded; individuals are encouraged to jot down their ideas before verbal brainstorming begins. Then these ideas are shared, additional ideas are generated, and all ideas are written on large charts and posted for all to see.

Phase 3—Evaluation and analysis: Here, the brainstorming ideas are evaluated in terms of cost, feasibility, and other relevant criteria (depending on the specific product). The guiding principle for this phase is to make ideas better. Ideas are only eliminated if a better solution is developed. Although critical judgment is used, people are encouraged to maintain a positive attitude. The cost figures and the importance rating developed earlier are now very useful for focusing this effort—especially on the functions that have high acceptance or high costs. This closely resembles the engineer's mindset combined with critical thinking.

Phase 4— Implementation: A primary function here is to sell the results of the value analysis—the identified improvements for the product. For this purpose, a proposal is prepared. This proposal must demonstrate that the proposed changes are feasible from a technical as well as a managerial standpoint, in addition to leading to substantial cost savings. Time schedules and budget schedules for implementation need to be prepared, as well as an assessment of the new skills and employee training that will be required. Both long-range and short-range feedback and monitoring plans need to be established to evaluate the results of the effort. The proposal presentation is then made, using good presentation techniques and colorful visual aids.

The "sales" presentation to management should be brief and concentrate on the highlights and benefits:
- The final product, in terms of its function and manufacturing costs, delivers "better for less." Service also delivers more value per unit cost.
- The high-cost components targeted for redesign or restructuring are identified.
- User acceptance data (the customer's voice) is guiding the process and can form the list of criteria in a Pugh analysis for optimizing the high-acceptance functions.
- Cost savings come from eliminating the areas of low and indifferent customer acceptance by concentrating on those areas that are important.

Through creative thinking, the team is able to devise alternate ways of achieving the important functions, yet the structured approach keeps the team focused on the functions and problem solving. Successful implementation has demonstrated large return on investment. However, trained specialists are required as leaders, a fact that differentiates this process from creative problem solving. An additional advantage of value engineering is that it identifies several customer sectors, not just the end user or purchaser. Implementation is quite difficult when an organization does not yet have a culture supportive of teamwork among departments, because value engineering is truly a multidisciplinary approach. It also demands commitment to a specific problem-solving procedure and allocation of the necessary resources. A variety of tools (forms, data collection and analysis procedures, and computer programs) can be developed within the organization to make the process more efficient. Value engineering is used in QFD for parts that have high costs and the potential for significant cost reduction. QFD and VA/VE are structured techniques that require special training and procedures for implementation. One characteristic that they have in common is that creative thinking and teamwork are an integral requirement. With the global, competitive marketplace, many companies are now becoming interested in using value engineering. For more information consult the following references:

E-1   James Brown, *Value Engineering: A Blueprint,* Industrial Press, New York, 1992.

E-2   Theodore C. Fowler, *Value Analysis in Design,* Van Nostrand Reinhold, New York, 1990.

E-3   Lawrence D. Miles, *Techniques of Value Analysis and Engineering,* McGraw-Hill, New York, 1961. This book is by the inventor of value analysis.

E-4   Larry M. Shillito, *Value: Its Measurement, Design, and Management,* Wiley, New York, 1992.

TOTAL QUALITY MANAGEMENT (TQM)

WHAT YOU CAN LEARN FROM APPENDIX F:
- *Overview of total quality management and its benefits.*
- *The steps needed for implementing TQM.*
- *Reference book list.*

What Is Total Quality Management (TQM)?

Basically, total quality management (TQM) is an attitude and a way of life in an organization, not merely a technology or procedure. With TQM, quality is designed into the entire organization and its products—it is not achieved by inspection. It involves learning and continuing education by everyone. It is different for each organization because it must be custom-fitted to each organization's purposes and culture, to its social and economic context. Creative problem solving can be used for planning and implementing TQM in an organization. Although quality improvements are measured and monitored in a variety of ways, the ultimate success is "voted on" with the customer's pocketbook or market share. Organizations must know who their customers are and be in constant contact with them because customer needs and expectations undergo frequent change. Appendix A on quality function deployment and Appendix D on Taguchi methods include discussions of quality, especially as it relates to the voice of the customer when manufacturing a product or providing a service. The voice of the customer is central to total quality management.

Why should TQM be used? After all, isn't it a lot of trouble for organizations to change the way they are used to doing things? If the world were stable, with conditions much as they were ten, twenty, or more years ago, then doing business the usual way might be adequate. But we live in a rapidly changing world, a world with a global marketplace and increased competition from industrialized as well as developing countries. Sophisticated communications and data-processing technology are increasingly being used as tools for identifying and meeting customer needs. Consumers everywhere now demand quality—it is no longer a luxury reserved for the privileged few. TQM has become essential to the long-term survival of organizations. With TQM, they can learn how to manage change, anticipate future markets, and develop possibilities into innovative products and services. If your organization does not provide an improved product, someone else will. In TQM, employees are seen not as subordinates but as associates who are given decision-making responsibilities and power.

DEFINITION

TQM is a never-ending quest to satisfy the needs of the customer better than anyone else through extra value, no variation in performance, and no defects, and through change and continuous improvement in response to changing conditions and expectations— through all the people in an organization working together.

Every product or service produced is a creation of the whole organization.

Who are the customers? This is not always obvious at first glance. For example, who are the customers of the activities of a dean of engineering? This could be the faculty or the staff in the college; it could be the students or the parents who are footing the bill. It could be the engineering profession or industry who are hiring the graduates. It could be the technician working in one of the laboratories. It could be a professor elsewhere who is using the dean's software or textbook. Or it could be the board of trustees who hired the dean or the state that is paying the salary. Anyone who is a user or who is impacted by an organization's activities is a customer. If your manufacturing activities impact the environment, your community's citizens are your customers, not just the people buying your product. If you are a secretary, your superior as well as other people who interact with you may be your customers—staff members, the visitors to your office, the people you talk to on the telephone, or the departments that deal with your paperwork.

Three people have been crucial in the quality movement: W. Edwards Deming, Joseph Juran, and Philip Crosby. Although they differ somewhat in the specifics on how to achieve quality, it is important to know the main principles that each man has developed to give you a better understanding of TQM.

W. Edwards Deming's fourteen points of quality management are probably best known; they are listed in Table F-1. Deming developed his philosophy while working with teams on the improvement of U.S. military materials during World War II. His ideas were received with open arms by the Japanese who were trying to reconstruct their economy after the war, and his ideas of teamwork were well-suited to their culture. He did not become well-known in the United States until the early 1980s.

If better is possible, then good is not enough.
Ron Meiss.

Quality is the product of everything we do. It is what sets us apart in the way our products perform, in the way our people respond. It is a philosophy and business practice that keeps our customers satisfied and willing to do business with us again and again.
Bill Pittman, CEO, Xerox.

Table F-1
Deming's Fourteen Points of Total Quality Management

1. Create a common goal and constant purpose throughout your organization, beginning with management.
2. Adopt the TQM philosophy.
3. Stop depending on mass inspection to achieve quality.
4. Do not award business on price tag alone; minimize total cost.
5. Constantly improve production processes and service systems.
6. Institute on-the-job training.
7. Institute leadership at all levels.
8. Drive out fear (of change and of risk taking).
9. Break down barriers between departments.
10. Eliminate slogans, exhortations, and numerical targets.
11. Eliminate work standards (quotas) and management by objective.
12. Encourage people to take pride in their workmanship.
13. Institute a vigorous program of education and self-improvement.
14. Everyone in the organization works to accomplish the transformation.

Joseph Juran was a quality expert who worked with Deming and Kaoru Ishikawa in Japan as part of General MacArthur's program to rebuild Japan's industry; together they developed the concepts of employee involvement, just-in-time delivery, and SPC. Juran's emphasis is on building awareness of the need and opportunity for improvement, setting goals, organizing and training teams for problem solving, putting the teams to work on projects, reporting and recognizing progress and results, and making annual improvement a part of all processes and systems of an organization.

Philip Crosby, formerly with ITT, developed his zero-defect movement while at Martin Marietta. He expanded and modified Deming's fourteen points. Many companies have adopted or built on his ideas. The 8-D approach developed at Ford Motor Company depends heavily on his approach, which features committed management; quality improvement teams; measurements that allow objective evaluation; a systematic method for identifying problems and permanently resolving them through corrective action; employee education, especially in defining the type of training needed to carry out the quality improvement process; improved communications; recognition of team results, and continuous improvement. (Note that education and training are emphasized by all three quality leaders.)

Benefits: What benefits can a company expect from its efforts to institute TQM? First, we must ask—what are the costs of *not* using TQM? It ultimately comes down to economic survival, because in today's and tomorrow's business climate, no organization can afford to stand still and rest on the accomplishments of the past. U.S. companies, even such giants as General Motors and Xerox, have been slow to recognize this.

Here are some examples of success stories. Through TQM, Corning doubled its profits in seven years. Ford, Xerox, Motorola, Procter and Gamble, and many others have significantly increased market share. Xerox established a 10-year program to achieve 100 percent customer satisfaction. It spent $125 million on employee education and trained 480 suppliers in quality as well. In 5 years, it achieved a 78 percent decrease in defects of machinery, 40 percent in emergency maintenance, and 27 percent in the response time of its service calls. In 1989, it won the coveted Malcolm Baldrige National Quality Award.

In 1981, someone at Motorola had the courage to tell the board of directors the shocking news: "The quality of your products stinks." In response, the company set a 5-year goal to improve quality by a factor of 10. This was achieved. But when in 1986 its executives went to Japan, they discovered that companies there were 1500 times better in lowering the number of defects per unit of work than Motorola (as reported in the cover story of *Template: The Magazine of Engineering Systems and Solutions*, Volume 5, published by Xerox). With this benchmark, Motorola then set a goal to improve its performance by 68 percent every year to achieve six sigma quality (this means only 3.4 defects per million). As a result of this effort, the company reduced manufacturing costs by $1.5 billion over 4 years! Although some investment may be needed in the beginning to initiate quality improvements—perhaps in the order of 1 percent of revenues—most companies find that this investment in quality will quickly pay back since—as shown by Genichi Taguchi—an increase in quality leads to cost reduction.

Ten Steps for Implementing TQM

Ron Meiss, in his workbook *Total Quality in the Real World (from Ideas to Action)*, discusses ten steps that must be taken by an organization that wants to adopt TQM. Each of the steps is outlined below. The general requirements are that everyone is involved at all levels of the organization. It is a continuous process, with ever-expanding goals as the demand for quality by internal and external customers expands.

Ten Steps for Implementing TQM	
1. Awareness	6. Components
2. Commitment	7. Measurement
3. Involvement	8. Leadership
4. Change	9. Training
5. Improvement	10. Implementation

1. Awareness: As a first step, the organization's awareness about the need for quality improvement must be increased through effective communications. A supportive climate for creativity, risk taking, teamwork, and lifelong education must be nurtured. Paradigm shifters throughout the organization must be recognized and rewarded.

2. Commitment: Fostering commitment to TQM by everyone in the organization is crucial. All must participate and contribute to decision making. This can be accomplished through creative problem-solving teams and will lead to a basic change in the organization's culture. Communication is the key, and change must be championed. This is managed through a whole-brain approach, as shown in Figure F-1 within the Herrmann brain model schematic.

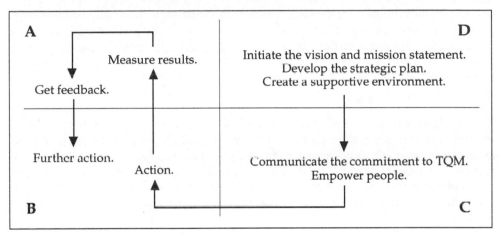

Figure F-1 Implementing TQM in an organization.

3. Involvement: Learning to use and implement TQM requires a change in the mindset of management from control to facilitation. Teamwork, flexibility, and creative problem solving are encouraged, not blind adherence to bureaucracy and pedantic procedures.

4. Change: Change is seen as positive, with concrete benefits to the survival of the organization. A climate for acceptance of new ideas and innovation is fostered. The organization encourages creative problem solving as it implements the philosophy of quality improvement through concrete team projects and processes. People do not fear change; they learn to act as change masters.

5. Improvement: Improvement is achieved in many small steps. Failure is acceptable if it serves as a learning experience and stepping-stone to improvement. People learn from each other. Creative thinking is needed to identify areas and ideas for improvement. Success will usually come easily at first, since obvious quality problems are tackled first. Persistence and hard work are needed to keep the process going and effect permanent change in the organization.

6. Components: Many tools, techniques, and procedures are part of TQM, as it affects all areas of an organization. The overriding maxim is: "The customer is always right!" This attitude involves all areas of communication: the information handled within the organization as well as to the outside; the financial resources and billing practices; supplier partnerships; manufacturing and inventory procedures; the delivery of the product; service, warranty claims, and response to mistakes through apology, restitution, and adding extra value; and the prevention of problems. Each person is responsible—there is no "passing the buck." Each person is qualified to come up with creative ideas on how to improve his or her own job. Even small ideas are important and are recognized and rewarded.

7. Measurement: Success is gauged through quantitative measurements, such as the seven tools of SPC. Reports are made on the actions taken and the results: How many errors have been detected? What is the percentage decrease in errors? How much time has been saved? What were the savings in expenditures? What were the increases in sales or decreases in complaints, etc.? Long-term trends are documented also.

8. Leadership: The essence of TQM leadership is to get people to do what is needed because they want to, not because they have to. Managers should help their employees set goals and succeed in reaching the goals. This takes a whole-brain approach of maintaining a balance between caring and expecting top work. A unique feature is to have a mindset that catches people doing something *right!* Superior performance is easily recognized and rewarded when benchmarks exist with well-publicized standards and criteria.

9. Training: Training in the following TQM skills is especially needed:
✦ Interpersonal—knowing how to work with others.
✦ Working as part of creative problem-solving teams.
✦ Identifying the customers and being sensitive to their needs.
✦ Doing quality work with SPC techniques and other specified procedures.
Each person receives cross-training and continuing education to expand job skills and an understanding of the context of the job within the organization as related to total quality.

10. Implementation: Everyone becomes involved in being producers—in implementing TQM. This is an ongoing process; each person is 100 percent responsible for quality—including setting goals and achieving the customer requirements, as well as maintaining a support network. It is difficult to change an attitude of fault-finding to a mindset that finds solutions when requirements are not met, but with the commitment of the entire team, it can be done. Continuous innovation can keep people excited about maintaining their efforts in meeting the customer requirements.

What *are* the customer requirements? On what criteria do customers judge a product? The absence of actual defects is just the starting point. Performance, extra features, reliability, durability, conformance, serviceability, aesthetics, and reputation are all contributing factors. For service, receiving extra value for the cost is especially important. Communication with the customer is vital because the requirements can change quickly. Tools—such as the QFD House of Quality, the Taguchi loss function analysis, and the Taguchi robust design with its emphasis on uniformity around a target—can help determine quality goals. In summary, quality means anticipating what customers might want in the future, then translating this vision into practical, innovative, dependable products in a system that can produce these at a low cost while generating profits or benefits. To implement this requires creative problem solving and the best thinking skills of everyone!

> *Any institution has to be organized so as to bring out the talent and capabilities within the organization; to encourage people to take initiative, give them a chance to show what they can do, and a scope within which to grow.*
> Peter Drucker, 1946.

> *Challenge your peple to think about the future, to pay attention to marketplace and technological developments and how they will impact the organization. Seeing the challenges and the threats mobilizes action.*
> James A. Belasco, Teaching the Elephant to Dance.

> *Quality isn't an event, a decision, or a program with starting and finishing dates. It is ongoing performance, integrating excellence into marketing, manufacturing, planning, research and development, in our interrelationships with our customers and ourselves.*
> Edgar S. Wollard, Jr., Chairman, DuPont.

With TQM, people take pride in their work—they love their jobs and enjoy their working environment. They have a clear purpose and a commitment to that purpose—and everyone's mission is the same. People take the responsibilty for shifting from outdated problem-solving paradigms to innovation as they continuously try to perfect what they do. They are supportive, they listen, they communicate positive feedback. Where frequent communication between different departments is important, the key people's offices are located in the same area (not more than 75 feet apart). Negative attitudes and criticism are recognized as detrimental to a quality environment; instead, positive reinforcement and rewards are used. Teamwork and cooperation are the key—mutual respect and self-worth result when people can see the value of their contributions to the whole organization. Barriers to creative thinking are removed so people can flourish individually and as teams actively involved in the quality effort.

Anyone who contributes to the creation of a product or service and does his or her best is a producer in the highest sense of the word. To do the best requires creative problem solving by the individual and the teams within the TQM framework of the organization. This is a never-ending activity because every process can be improved; cost can be lowered and productivity increased. This is also a never-ending process of formal education and informal continuous learning which is very much a part of the thinking processes that are taught in this book.

Recommended References

F-1   James A. Belasco, *Teaching the Elephant to Dance: Empowering Change in Your Organization,* Crown, New York, 1990.

F-2   Philip Crosby, *Quality Is Free: The Art of Making Quality Certain,* McGraw-Hill, New York, 1979.

F-3   Michael L. Dertouzos, Richard K. Lester, Robert M. Solow, *Made in America: Regaining the Productive Edge*, MIT Press, Cambridge, Masschusetts, 1989.

F-4   Peter F. Drucker, *Innovation and Entrepreneurship: Practice and Principles*, Harper and Row, New York, 1985.

F-5`   Andrea Gabor, *The Man Who Discovered Quality: How W. Edwards Deming Brought the Quality Revolution to America—The Story of Ford, Xerox, and GM*, Random House, New York, 1990.

F-6   David Haberstam, *The Reckoning*, Morrow, New York, 1986.

F-7   Masaaki Imai, *Kaizen: The Key to Japan's Competitive Success*, Random House, New York, 1986.

F-8   Kaoru Ishikawa (translated by David J. Lu), *What Is Total Quality Control? (The Japanese Way)*, Prentice-Hall, Englewood Cliffs, New Jersey, 1985.

F-9   J.M. Juran and Frank M. Gryna, *Quality Planning and Analysis: From Product Development through Use*, third edition, McGraw-Hill, New York, 1993.

F-10   Ron Meiss, *Total Quality in the Real World: From Ideas to Action*, Ron Meiss and Associates, Kansas City, Missouri, 1991, 64114; (913) 897-6754.

F-11   Rosabeth Moss Kanter, *Change Masters: Innovation for Productivity in the American Corporation*, Simon & Schuster, New York, 1985.

F-12   Tom Peters and Nancy Austin, *A Passion for Excellence*, Warner Books, New York, 1985.

F-13   Richard J. Schonberger, *Japanese Manufacturing Techniques: Nine Hidden Lessons in Simplicity*, Free Press, New York, 1982.

F-14   Peter M. Senge, *The Fifth Discipline: The Art and Practice of the Learning Organization*, Doubleday, Garden City, New York, 1990.

F-15   Robert L. Shook, *Turnaround: The New Ford Motor Company*, Prentice-Hall, Englewood Cliffs, New Jersey, 1990.

F-16   Genichi Taguchi, *Taguchi on Robust Technology Development: Bringing Quality Engineering Upstream*, ASME Press, Fairfield, New Jersey, 1993.

F-17   Robert H. Waterman, Jr., *The Renewal Factor: How the Best Get and Keep the Competitive Edge*, Bantam Books, New York, 1987.

F-18   James P. Womack, David T. Jones, and Daniel Roos, *The Machine That Changed the World (The Story of Lean Production)*, Harper, New York, 1991.

Some of the references listed in Chapter 5 on the topics of quality, innovation, and manufacturing are pertinent to the discussion of TQM presented here, particularly References 5-3, 5-4, and 5-15. *Thriving on Chaos: Handbook for a Management Revolution* by Tom Peters (Ref. 6-7) addresses quality issues as related to management. Be on the lookout for new books and journal articles about TQM implementation case studies—on successes and on failures. This approach to companywide quality is by no means universally accepted in the United States, and even companies that made easy progress at first now find that they are not reaping the expected profits, cooperation, and acceptance. We believe that the implementation of fundamental changes in an organization requires careful preparation—most of all in the area of teaching everyone creative thinking and problem-solving skills.

INDEX

Note: A separate listing of personal names follows this topical index.

List of Personal Names